Sustainable Use of Soils and Water

Sustainable Use of Soils and Water

The Role of Environmental Land Use Conflicts

Special Issue Editor

Fernando António Leal Pacheco

MDPI • Basel • Beijing • Wuhan • Barcelona • Belgrade • Manchester • Tokyo • Cluj • Tianjin

Special Issue Editor
Fernando António Leal Pacheco
University of Trás-os-Montes e Alto Douro
Portugal

Editorial Office
MDPI
St. Alban-Anlage 66
4052 Basel, Switzerland

This is a reprint of articles from the Special Issue published online in the open access journal *Sustainability* (ISSN 2071-1050) (available at: https://www.mdpi.com/journal/sustainability/special_issues/Sustainable_Soils_Water_Environmental_Land_Use).

For citation purposes, cite each article independently as indicated on the article page online and as indicated below:

LastName, A.A.; LastName, B.B.; LastName, C.C. Article Title. *Journal Name* **Year**, *Article Number*, Page Range.

ISBN 978-3-03928-644-7 (Pbk)
ISBN 978-3-03928-645-4 (PDF)

Contents

About the Special Issue Editor

Fernando A.L. Pacheco, Prof., was born in 1967 in Mozambique. He holds a Ph.D. degree in Hydrogeology (2001) from the Trás-os-Montes and Alto Douro University (UTAD, Portugal) and an Aggregation title in Environmental Geochemistry (2011) from the UTAD. He has served as Head of the Geology Department of UTAD since his appointment in 2013. He joined the Vila Real Chemistry Research Centre of UTAD in 2005, where he still conducts most of his research. He has published over 90 research papers in international journals, co-authored by over 90 scientists from various nations (Portugal, The Netherlands, Brazil, Hungary, Germany, Italy, and Spain). Prof. Pacheco is a top reviewer, as certified by Clarivate Analytics Publons. In 2016, he received the "Sentinel of Science Award" as the Top Overall Contributor to the peer review of the field of Earth and Planetary Sciences. In 2017, he was recognized as Top Overall Contributor to peer review of the journal *Science of Total Environment*. He is also Editor of various scientific journals (e.g., *Science of the Total Environment*, *Water*, *Sustainability*, *International Journal of Environmental Research and Public Health*, and *Arabian Journal of Geosciences*). Prof. Pacheco is actively involved in scientific cooperation on land management and water security projects with several Brazilian institutions: Federal Institute of Triângulo Mineiro (IFTM), Regional Coordination of Environmental Justice Prosecutors in the Paranaíba and Baixo Rio Grande River Basins (MPMG), Minas Gerais Institute for Water Management (IGAM), and the San Paulo State University (UNESP).

The research interests and expertise of Prof. Pacheco cover the topics of hydrologic models coupled with weathering algorithms, especially in areas with significant anthropogenic pressure; multivariate statistical and environmental analyses of surface and groundwater databases, with focus on the prevention of surface and groundwater contamination; land degradation and management, as well as the negative impacts of inadequate land uses on soil erosion, surface and groundwater quality; and water security issues, such as conjunctive use of surface and groundwater sources in public water supply systems or the attenuation of hydrologic extremes (floods, droughts) through implementation of detention basins and decentralized rainwater harvesting systems in catchments.

 *sustainability*

Editorial

Sustainable Use of Soils and Water: The Role of Environmental Land Use Conflicts

Fernando A. L. Pacheco

CQVR – Chemistry Research Centre, University of Trás-os-Montes and Alto Douro, Quinta de Prados Ap. 1013, Vila Real 5001-801, Portugal; fpacheco@utad.pt

Received: 3 February 2020; Accepted: 4 February 2020; Published: 6 February 2020

Abstract: Sustainability is a utopia of societies, that could be achieved by a harmonious balance between socio-economic development and environmental protection, including the sustainable exploitation of natural resources. The present Special Issue addresses a multiplicity of realities that confirm a deviation from this utopia in the real world, as well as the concerns of researchers. These scholars point to measures that could help lead the damaged environment to a better status. The studies were focused on sustainable use of soils and water, as well as on land use or occupation changes that can negatively affect the quality of those resources. Some other studies attempt to assess (un)sustainability in specific regions through holistic approaches, like the land carrying capacity, the green gross domestic product or the eco-security models. Overall, the special issue provides a panoramic view of competing interests for land and the consequences for the environment derived therefrom.

Keywords: water resources; soil; land use change; conflicts; environmental degradation; sustainability

---

Competition for land is a worldwide problem affecting developed as well as developing countries, because the economic growth of activity sectors often requires the expansion of occupied land, sometimes to places that overlap different sectors. Besides the social tension and conflicts eventually caused by the competing interests for land, the environmental problems they can trigger and sustain cannot be overlooked. In rural catchments, where land uses and occupations are dominated by agriculture, livestock production and forests, a specific conflict is frequently observed, called "environmental land use conflict". This is known to be the deviation between actual land uses and natural land uses set up on the basis of land capability, and has been amply studied for the severe environmental consequences that these conflicts are responsible for. An immediate consequence is amplified soil erosion, which has been documented in catchments with craggy topography that, in spite of being prone to erosion, are being occupied by vineyards because of their favorable geology (metassediments) and sun exposure [1,2]. This amplification of soil erosion could be related to precedent organic matter declines also caused by the conflicts [3], and to a succeeding cascade of other environmental impacts including the amplification of surface and groundwater quality deterioration [4,5], biodiversity declines [6], and extreme events such as floods [7]. Overall, environmental land use conflicts have been claimed to be a major cause of land degradation [8], and have even become a concern for judicial organizations [9].

The role of land competition for specific uses, as well as the environmental consequences of land use changes and conflicts, can, however, be analyzed from other standpoints, besides the aforementioned environmental land use conflicts. The diversity of those perspectives was, in fact, the purpose of this Special Issue "Sustainable Use of Soils and Water: The Role of Environmental Land Use Conflicts". In this context, the Special Issue tackled the problem of the sustainable use of water resources coupled with the strategies of land planning that can ensure recharge and suitable groundwater resources for multiple uses (drinking water supply, irrigation); water footprints for agriculture and livestock products; land uses, their changes, and the associated consequences for catchment hydrology, as well as for soil and water quality, not ruling out the roles of management practices, climate change and natural hazards; and, finally, balances between resources, the environment and the socio-economic development, based on the concepts of land carrying capacity, green gross domestic product, eco-security, and the overlapping of geosystems. During our working period, we received many submissions, which had significant contributions for the main topics of interest of our special issue. However, only 17 high-quality papers were accepted after several rounds of strict and rigorous review. These 17 contributions are summarized in the forthcoming paragraphs, being integrated in a coherent narrative.

The conjunctive use of surface and groundwater is among the recent strategies to ensure sustainable and secure water supply systems of drinking water. Conjunctive water use may be the way to mitigate overexploitation of groundwater resources, especially in areas where headwater forested catchments surround the urban areas, and therefore can be efficiently used to collect, store and deliver good quality surface water to the population. In Contribution 1, Hugo Henrique Cardoso de Salis, Adriana Monteiro da Costa, Annika Künne, Luís Filipe Sanches Fernandes and Fernando António Leal Pacheco have studied the case of Sete Lagoas town (state of Minas Gerais, Brazil) with ca. 220,000 inhabitants in 2008, which exploits 15.5 hm^3 year^{-1} of karst groundwater for public drinking water supply, while the renewable resources estimated within the enclosing Jequitiba watershed with an area of 571.5 km^2 are solely 6.3 hm^3 year^{-1}. This large difference between annual abstraction and aquifer recharge have caused impressive water table declines and sinkhole development in recent decades in Sete Lagoas. Contribution 1 proposed the storage of quality water in a small dam lake in the Marinheiro catchment, a forested tributary of the Jequitiba basin located near the town. According to the authors, this lake would supply the municipality with a sustainable 4.73 hm^3 year^{-1} of quality surface water, and thus would help to slow down the depletion of groundwater resources in the karst aquifer.

The efficacy of conjunctive water use can be enhanced if the watersheds selected to store the surface water are located where aquifer recharge is favorable. In this case, besides the positive effect of reducing the groundwater abstractions, conjunctive water use would also contribute to increasing the renewable resources through the induced recharge of dam lake water. This was the purpose of Contribution 2, written by Adriana Monteiro da Costa, Hugo Henrique Cardoso de Salis, João Hebert Moreira Viana and Fernando António Leal Pacheco. In that study, a physically based, spatially distributed method was used to evaluate groundwater recharge potential at the aforementioned Jequitiba River basin. The regions where recharge was classified as more favorable comprised flat areas, porous aquifers and forested regions, including the Marinheiro catchment.

The use of induced recharge as tool for water resource management is not limited to drinking water supply. For example, in arid regions an important focus of induced recharge is irrigation, given the prospective shortage of groundwater resources for this activity in these regions. Another important issue concerning the management of aquifer recharge relates to the engagement of academic, governmental and catchment-level institutions in the problem, because the academics are capable of studying the aquifer systems, including their resource availability, while the governments and water resource planners are empowered to regulate the construction and operation of dams even when they are small. The study of Lydia Kwoyiga and Catalin Stefan (Contribution 3) was developed in Northern Ghana and assessed the institutional feasibility of managed aquifer recharge methods in the Atankwidi catchment, where dry-season farmers may lose their source of livelihood due to limited access to groundwater. The results indicated a favorable political and administrative environment for the implementation of managed aquifer recharge. However, some essential backgrounds were lacking, making it difficult to execute or accelerate the desired task: firstly, quantitative information on groundwater demand for irrigation; secondly, scientific studies on the spatio-temporal distribution of groundwater flow depths, which are essential to assess the aquifer's storage capacity.

The value of detailed descriptions about groundwater flow depths for efficient groundwater resources management, either focused on drinking water supply or the irrigation of cropland, has been recognized by the authors of Contribution 4 and Contribution 5. In the first case, Dianlong Wang, Guanghui Zhang, Huimin Feng, Jinzhe Wang and Yanliang Tian developed a method to study temporal scale effects on groundwater levels and tested it in the Hufu Plain (China), namely in the city of Shijiazhuang, which is the capital of Hebei province. In this area, the population is larger than 17 million. To ensure the people's living, industry and irrigation, huge abstractions of groundwater take place from a water table aquifer for a long period of time. In the period 1960–2010, these abstractions led to the development of a constantly enlarging cone of depression, covering an area of approximately 20 km^2 in 1960 and $\approx$500 km^2 in 2010. At the center of this cone, the groundwater level varied from nearly 10 to 50 meters in the same timeframe. In spite of being progressive, the depression cone expansion was not regular overtime. Using their method, the aforementioned authors could distinguish five groundwater depth stages representing an equal number of exploitation stages: natural flow field, mild overexploitation, depression cone formation, serious overexploitation, and exploitation reduction. More important than the retrospective analysis, the set-up of a relationship between groundwater depth stages and exploitation stages provided essential knowledge for an eventual long-term monitoring program aiming at the recovery of groundwater levels in the region. In the second case, Ruiyan Wang, Simon Huston, Yuhuan Li, Huiping Ma, Yang Peng and Lihua Ding investigated the spatio-temporal stability of groundwater depth in the Yellow River Delta (North China Plain), with a specific concern for irrigation practices. The results of their research revealed that groundwater depth is highly nonlinear because irrigation, as well as surface water, topography, seawater intrusion and drainage, are heterogeneously distributed in the Delta. The outcomes also showed that a detailed assessment of groundwater depth fields and their stability patterns over time, would help control soil salinization, and hence be used to sustainably exploit the Delta's agricultural potential without saline intrusion.

The concerns about available freshwater for agriculture were extended in this special issue to the assessment of water footprints for agriculture and livestock products. The study by Ik Kim and Kyung-shin Kim conducted in the Republic of Korea (Contribution 6) quantified the water footprints of crops, open field vegetables, and vegetables in facility and fruits, as well as of beef, pork and chicken meat, in 2014. The estimates were presented on the cultivated area (hectare) and production (ton) bases and differed considerably among the products. For the per hectare case, the vegetables in facility contributed the most to the footprint and the open field vegetables contributed the least, representing water consumptions larger than 30,000 m^3/ha in the cases of cherry tomato or pepper, for example, and smaller than 5000 m^3/ha regardless of the type of vegetable produced in the open field. As regards the per ton case, the larger water footprints were estimated for the grapes and strawberries, which represented consumptions larger than 6000 m^3/ton in both cases. Among the livestock products, beef was the most stressful product with a consumption close to 20,000 m^3/ton, whereas chicken meat was the least demanding good with a consumption of <5000 m^3/ton. Overall, the footprint of total agricultural and livestock products in 2014 was approximately 27.9% of the total domestic water resources consumed in Korea.

The sustainable use of freshwater bears on the conservation of quality, besides all the issues related to quantity brought into the discussion so far. In the present special issue, one study was concerned with water quality and the nexus agriculture plus livestock activities > surface and groundwater pollution > compliance with the Nitrates Directive. The study was co-authored by Ewa Szalińska, Paulina Orlińska-Woźniak and Paweł Wilk (Contribution 7), and corresponded to a modeling exercise taking place in the Słupia River Basin (Poland) between 2002 and 2016. This European country is a main contributor to the excessive nutrient loads into the Baltic sea catchment, because it occupies a large portion of this catchment (ca. 18%) with a very large share in agricultural land area (ca. 50%). To reverse the situation, new rules are being proposed by the Polish authorities whereby the periods for which the application of nitrogen fertilizers is allowed are redefined (restricted). The modeling results make evident the decline in total nitrate released from farmlands within the Slupia basin, when a 20 day reduction is imposed on the fertilization period, maintaining the current application doses. On average, the reductions were 762 kg/month during the summer season, 7171 kg/month during the winter season, and 3966 kg/month for the entire period. Despite the promising results, the co-authors highlighted the financial and organizational costs resulting from the proposed action program, which are likely to meet discontentment from farmers.

Before being leached to surface water via runoff or to groundwater through infiltration, contaminants loaded by natural processes and/or anthropogenic activities stay in the soil compartment. Therefore, a full analysis of contamination, involving land use, soil and water, requires an understanding of soil contamination. In the present special issue, the examination of soil contamination covered the important theme of metal contamination (Contribution 8). The co-authors, Shixin Ren, Erling Li, Qingqing Deng, Haishan He and Sijie Li, while working in Lankao County (China), showed that heavy metal contents in arable soil (Cr, Ni, Cu, Zn, Cd, and Pb) depend on the planting mode, generally decreasing from vegetable greenhouse to garlic land and then to traditional crop farmland. Moreover, these metal concentrations were classified as "slight pollution" in ≈5% and "moderate pollution" in ≈1% of the studied soil samples. The areas where these samples were collected would benefit from mitigation measures, namely land circulation and the aggregation of smaller patches into larger ones, as proposed by the co-authors.

The sustainability of soil and water resources is inherently connected to the natural use of land, which means use that respects land capability, under predefined climatic conditions. Numerous studies have reported the negative impacts of land use and cover changes on the hydrology of catchments that inevitably alter the fertility of soils through amplified erosion, as well as the availability and quality of surface and groundwater through contaminant transport and infiltration. To document land use and cover changes, and the consequent hydrologic response, Contribution 9, co-authored by Xin Jin, Yanxiang Jin and Xufeng Mao, presented the case of Heihe River Basin located in Northwest China's arid region. The authors used a modified version of SWAT model capable of updating land use and occupation during the tested period (1990–2009), according to the historical evolution portrayed in the available maps. The results indicated the spreading of farmland, forest, and urban areas, while grassland and bare land areas showed decreasing trends. It was also made clear that key land use changes in the studied area were from grassland to farmland and from bare land to forest. The hydrologic response could be characterized by decreasing trends in surface runoff, groundwater runoff, and total water yield, while lateral flow and ET volume showed increasing trends under dry, wet, and normal conditions.

Another publication from this special issue (Contribution 10) studied the role of land use changes under climate change, based on land use, runoff and climatic data from the 1995–2015 period. This study was co-authored by Qun Liu, Zhaoping Yang, Cuirong Wang and Fang Han, and was developed in the Xinjiang region located in the Northwest of China. This is an arid region, where the water sources comprise river, lake and underground water provided by the melting of glaciers and snow. During the studied period and before, runoff significantly increased in response to the regional warming that accelerated this process. In the sequel, the cropland expanded rapidly. However, with the glacier's storage decreasing, the glaciers and snow melt water will also decrease, with inevitable consequences for runoff, that will drop proportionally. In some catchments within the studied area (Kai-Kong River Basin), the runoff has been decreasing in recent decades, while the cropland area is still large. This is placing a large pressure on groundwater for irrigation, with negative consequences for water levels (lowering), as already noted in other contributions.

The negative consequences of land use changes for soil were also addressed in this special issue in two studies. A first-order concern refers to the competing interests for land, which in some countries, such as Slovakia, resulted in preoccupying withdrawals of agricultural land, which means the conversion of these areas for non-agricultural purposes. The case of Slovakia was the focus of Contribution 11, co-authored by Lucia Palšová, Katarína Melichová and Ina Melišková. In this country, when the demand for investment activities and the convergence to the EU standards of living occurred (after 2004), many landowners who were receiving an agricultural land fund from the government abdicated from this income, but received a larger profit when they released (sold) their lands for other uses, such as residential. The most notable extreme in the amount of withdrawn agricultural land happened in 2008, which was the last year when the contributions for withdrawal were not levied on those that withdrew agricultural land from the top four quality categories. Besides the withdrawals for housing, conversions were also important for industry, transport and mining development purposes, especially in the western part of Slovakia. In keeping with the authors' opinion, the state should adopt a long-term conceptual document where the areas for agricultural land use are defined, taking into account the impact of developmental factors on land protection.

A second-order concern about the impact of land use changes on soil refers to inappropriate agricultural practices that are causing soil degradation worldwide. The study of Sameh Kotb Abd-Elmabod, Noura Bakr, Miriam Muñoz-Rojas, Paulo Pereira, Zhenhua Zhang, Artemi Cerdà, Antonio Jordán, Hani Mansour, Diego De la Rosa and Laurence Jones (Contribution 12), allowed the mapping of soil suitability classes in the El-Fayoum depression (Egypt) for a diversity of Mediterranean crops (wheat, corn, melon, potato, soybean, cotton, sunflower sugar beat, alfafa, peach, citrus fruits and olive). The maps were drawn for the current situation of soil factors and projected this for a scenario of improved manageable soil factors such as salinity, sodium saturation and drainage, using the Almagra model. The results were clear about the current situation of soil factors, for which the dominant soil suitability classes were the moderate and marginal classes. On the other hand, the highly suitable class was dominant under the projected scenario of improved manageable soil factors. Moreover, under this later scenario, the suitability for all crops improved except for the perennial crops, but one should recall that the most limiting factors for these crops are soil texture, depth, soil profile development and carbonate content, which are inflexible to modification. Overall, the maps of soil suitability classes published in this study are unequivocally valuable to decision makers for appropriate land-use planning and sustainable development in the El-Fayoum depression.

In some regions, land use changes related to anthropogenic pressures such as population or industrial growth are not the main driving force of a changing landscape. Instead, the observed changes are mostly the result of natural factors. This is particularly evident in regions that are prone to natural hazards such debris flows, landslides or earthquakes. In these areas, land use changes frequently occur where the consequences of a disaster were the most severe and/or extensive, involving the destruction of houses, roads and fields used for agriculture or livestock production. In general, the option in these cases is to move the uses or occupations in the affected areas to other places. A comprehensive study addressing this particular type of land use change has been published in the present special issue as Contribution 13. The study described land use changes and their driving forces in a debris flow active area of Wudu District, Gansu Province, China, and has been co-authored by Songtang He, Daojie Wang, Yong Li and Peng Zhao. This region is characterized by heavy torrential rains, loose entisols and moderate magnitude earthquakes, and as it is periodically affected by debris flow hazards, land use changes have occurred in the sequence of these natural events. The case of Houba village is given as an example. This village has been affected by five large-scale mudslides in the period 1956–2005, which destroyed private residences and various acres of farmland, burying livestock and damaging a stretch of a state highway. For security reasons, these lands were developed into woodland.

The last four papers published in the present Special Issue present panoramic views of many topics discussed thus far. These papers describe balances between resources, the environment and the socio-economic development, based on various concepts. For example, Contribution 14 discusses the concept of land carrying capacity, which can be defined as the number of people with a certain level of activity that can be sustained in a region without causing resource depletion or environment degradation. In their study, the co-authors Guangming Cui, Xuliang Zhang, Zhaohui Zhang, Yinghui Cao and Xiujun Liu used an index composed of cost-benefit indicators related to water and soil resources, eco-environment, society, economy and technology, to measure the land carrying capacity of eastern China, namely of seven cities from the Shandong Peninsula Blue Economic Zone, between 2007–2014. The results uncovered declines in carrying capacities for water and soil resources in most cities within the studied area, and increases in the eco-environment and social resources, as well as for in economy and technology of all cities. To revert the declines and improve the overall index, the authors proposed several mitigation measures, such as decreasing the proportion of added value of the primary industry to total Gross Domestic Product, promoting energy saving and emission reductions.

In Contribution 15, the co-authors Yuhan Yu, Mengmeng Yu, Lu Lin, Jiaxin Chen, Dongjie Li, Wenting Zhang and Kai Cao were interested in the Green Gross Domestic Product (GGDP), which reproduces the trade-off between ecosystem and economic systems. In their study, they were able to calculate and map the GGDP for China in the 1990–2015 period, based on historical data about Ecosystem Services Values (ESV) and Gross Domestic Product (GDP). They were also able to predict the evolution of ESV, GDP and GGDP until 2050 based on land use/land cover changes simulated by a Cellular Automata (CA) Markov model. The results indicated huge increases in GGDP (78%) from 1990 to 2015, which were also forecasted for the 2020–2050 period. However, the spatial patterns point to huge rises in the GGDP solely in the eastern territory, while the western part remained relatively unchanged in this regard. Moreover, in the regions where the GGDP was larger, the contribution of GDP to GGDP was substantial ($\approx$ 90%), while, in the other regions, this contribution was < 45%. Therefore, in the latter case, the green GDP is dependent on the preservation of the ESV to a large extent, while in the former case the regions sustain the green GDP with socio-economic development, eventually depressing the importance of preserving the ESV. The huge spatial differences detected for the green GDP, GDP and ESV among the regions, are likely to become the cause of social problems, both now and in the future. The co-authors finally advise decision makers to determine the hotspot regions where the problems may become more severe and make policies accordingly to preserve GDP or ESV.

The co-authors of Contribution 16 assessed the balance between societal development and the preservation of environmental values through the eco-security concept, which has been defined by the Insurance Accounting and Systems Association in 1989 as the "ecologically sustainable development that meets the environmental and ecological needs of the present generation without compromising the ability of future generations to meet their own environmental and ecological needs". In their study, Qiuyan Liu, Mingwu Wang, Xiao Wang, Fengqiang Shen and Juliang Jin developed a novel, multi-dimensional connection cloud model to determine land eco-security, and tested the method in the Wanjiang region, Anhui Province, eastern China. This region experienced extensive economic growth in the past 50 years, namely through the introduction of chemical plants, cast iron and forging plants, and building materials plants. This has caused a significant increase in population density and unbalanced the land eco-security in the region, because the cultivated area and forest coverage declined concurrently. The modeling results show that land eco-security in the studied area behaves unfavorably because the marginally unsafe and unsafe levels are dominant. The results are explained by the excessive use of chemical fertilizers and pesticides, excessive industrial gas and solid wastes, and ineffective cleaning, which contribute substantially to the negative impact of economic and social industry development on land eco-security. The co-authors indicated some measures for improving land eco-security through environmental protection, reasonable employment of resources, proper adjustment of industrial structure, and possible abatement of pollution.

The last published article, Contribution 17, co-authored by Zita Izakovičová, László Miklós and Viktória Miklósová, used an integrated analysis of three geosystems to identify competing interests for land among local activity sectors in the Trnava district, Slovakia, as well as the environmental problems derived therefrom. The geosystems comprise the space and the natural resources potentially used by the sectors (called primary landscape structure), the current use of these resources (secondary landscape structure), and the regulations and legal provisions for use of these resources (tertiary landscape structure). The intersections among the geosystems expose potential environmental problems that include endangering the ecological stability of the landscape, the natural resources and the immediate human environment. The identified competing interests and associated environmental problems could be summarized as follows: (*a*) the sectors "fight" for land to broaden their own territory. It is not uncommon to observe the use of high quality soils for non-agricultural purposes, such as industry, transport and residential areas, because developers find ways to change existing territorial plans and gain land even at the cost of compensation for the occupied land; (*b*) the sectors of forestry, agriculture, water management, nature conservation and recreation utilize large areas where their activities overlap, and sometimes are at conflict. This is the case where high quality soils overlap water protective zones, because the use of agrichemicals to increase productivity potentially endangers and pollutes the water resources; (*c*) there are "internal" conflicts within the sectors, exemplified by the competing occupation of shallow soils or steep slopes, with agriculture ignoring the severe consequences for erosion; (*d*) air and water pollution, with a specific concern about the presence of Jaslovské Bohunice nuclear power plant. Overall, the integrated analysis proved efficient to identify and map the areas of competing and conflicting interests, as well as the most preoccupying environmental problems. To increase efficiency, it would be necessary to bring the results from this study into official land use management plans.

List of Contributions:

1. de Salis, H.H.C.; da Costa, A.M.; Künne, A.; Sanches Fernandes, L.F.; Leal Pacheco, F.A. Conjunctive Water Resources Management in Densely Urbanized Karst Areas: A Study in the Sete Lagoas Region, State of Minas Gerais, Brazil. *Sustainability* **2019**, *11*, 3944.
2. da Costa, A.M.; de Salis, H.H.C.; Viana, J.H.M.; Leal Pacheco, F.A. Groundwater Recharge Potential for Sustainable Water Use in Urban Areas of the Jequitiba River Basin, Brazil. *Sustainability* **2019**, *11*, 2955.
3. Kwoyiga, L.; Stefan, C. Institutional Feasibility of Managed Aquifer Recharge in Northeast Ghana. *Sustainability* **2019**, *11*, 379.
4. Wang, D.; Zhang, G.; Feng, H.; Wang, J.; Tian, Y. An Approach to Study Groundwater Flow Field Evolution Time Scale Effects and Mechanisms. *Sustainability* **2018**, *10*, 2972.
5. Wang, R.; Huston, S.; Li, Y.; Ma, H.; Peng, Y.; Ding, L. Temporal Stability of Groundwater Depth in the Contemporary Yellow River Delta, Eastern China. *Sustainability* **2018**, *10*, 2224.
6. Kim, I.; Kim, K. Estimation of Water Footprint for Major Agricultural and Livestock Products in Korea. *Sustainability* **2019**, *11*, 2980.
7. Szalińska, E.; Orlińska-Woźniak, P.; Wilk, P. Nitrate Vulnerable Zones Revision in Poland—Assessment of Environmental Impact and Land Use Conflicts. *Sustainability* **2018**, *10*, 3297.
8. Ren, S.; Li, E.; Deng, Q.; He, H.; Li, S. Analysis of the Impact of Rural Households' Behaviors on Heavy Metal Pollution of Arable Soil: Taking Lankao County as an Example. *Sustainability* **2018**, *10*, 4368.
9. Jin, X.; Jin, Y.; Mao, X. Land Use/Cover Change Effects on River Basin Hydrological Processes Based on a Modified Soil and Water Assessment Tool: A Case Study of the Heihe River Basin in Northwest China's Arid Region. *Sustainability* **2019**, *11*, 1072.
10. Liu, Q.; Yang, Z.; Wang, C.; Han, F. Temporal-Spatial Variations and Influencing Factor of Land Use Change in Xinjiang, Central Asia, from 1995 to 2015. *Sustainability* **2019**, *11*, 696.
11. Palšová, L.; Melichová, K.; Melišková, I. Modelling Development, Territorial and Legislative Factors Impacting the Changes in Use of Agricultural Land in Slovakia. *Sustainability* **2019**, *11*, 3893.
12. Abd-Elmabod, S.; Bakr, N.; Muñoz-Rojas, M.; Pereira, P.; Zhang, Z.; Cerdà, A.; Jordán, A.; Mansour, H.; De la Rosa, D.; Jones, L. Assessment of Soil Suitability for Improvement of Soil Factors and Agricultural Management. *Sustainability* **2019**, *11*, 1588.
13. He, S.; Wang, D.; Li, Y.; Zhao, P. Land Use Changes and Their Driving Forces in a Debris Flow Active Area of Gansu Province, China. *Sustainability* **2018**, *10*, 2759.
14. Cui, G.; Zhang, X.; Zhang, Z.; Cao, Y.; Liu, X. Comprehensive land carrying capacities of the Cities in the Shandong Peninsula blue economic zone and their spatio-temporal variations. *Sustainability* **2019**, *11*, 439.
15. Yu, Y.; Yu, M.; Lin, L.; Chen, J.; Li, D., Zhang, W.; Cao, K National Green GDP Assessment and Prediction for China Based on a CA-Markov Land Use Simulation Model. *Sustainability* **2019**, *11*, 576.
16. Liu, Q.; Wang, M.; Wang, X.; Shen, F.; Jin, J. Land Eco-Security Assessment Based on the Multi-Dimensional Connection Cloud Model. *Sustainability* **2018**, *10*, 2096.
17. Izakovičová, Z.; Miklós, L.; Miklósová, V. Integrative Assessment of Land Use Conflicts. *Sustainability* **2018**, *10*, 3270.

Funding: This research was financed by National Funds of FCT—Portuguese Foundation for Science and Technology, under the project UID/QUI/00616/2019.

Conflicts of Interest: The author declares no conflict of interest.

References

1. Pacheco, F.A.L.; Varandas, S.G.P.; Sanches Fernandes, L.F.; Valle Junior, R.F. Soil losses in rural watersheds with environmental land use conflicts. *Sci. Total Environ.* **2014**, *485–486*, 110–120. [CrossRef] [PubMed]
2. Valle Junior, R.F.; Varandas, S.G.P.; Sanches Fernandes, L.F.; Pacheco, F.A.L. Environmental land use conflicts: A threat to soil conservation. *Land Use Policy* **2014**, *41*, 172–185. [CrossRef]
3. Valera, C.A.; Valle Junior, R.F.; Varandas, S.G.P.; Sanches Fernandes, L.F.; Pacheco, F.A.L. The role of environmental land use conflicts in soil fertility: A study on the Uberaba River basin, Brazil. *Sci. Total Environ.* **2016**, *562*, 463–473. [CrossRef] [PubMed]
4. Pacheco, F.A.L.; Sanches Fernandes, L.F. Environmental land use conflicts in catchments: A major cause of amplified nitrate in river water. *Sci. Total Environ.* **2016**, *548–549*, 173–188. [CrossRef] [PubMed]
5. Valle Junior, R.F.; Varandas, S.G.P.; Sanches Fernandes, L.F.; Pacheco, F.A.L. Groundwater quality in rural watersheds with environmental land use conflicts. *Sci. Total Environ.* **2014**, *493*, 812–827. [CrossRef]
6. Valle Junior, R.F.; Varandas, S.G.P.; Pacheco, F.A.L.; Pereira, V.R.; Santos, C.F.; Cortes, R.M.V.; Sanches Fernandes, L.F. Impacts of land use conflicts on riverine ecosystems. *Land Use Policy* **2015**, *43*, 48–62. [CrossRef]
7. Caldas, A.; Pissarra, T.; Costa, R.; Neto, F.; Zanata, M.; Parahyba, R.; Sanches Fernandes, L.; Pacheco, F. Flood Vulnerability, Environmental Land Use Conflicts, and Conservation of Soil and Water: A Study in the Batatais SP Municipality, Brazil. *Water* **2018**, *10*, 1357. [CrossRef]
8. Pacheco, F.A.L.; Sanches Fernandes, L.F.; Valle Junior, R.F.; Valera, C.A.; Pissarra, T.C.T. Land degradation: Multiple environmental consequences and routes to neutrality. *Curr. Opin. Environ. Sci. Heal.* **2018**, *5*, 79–86. [CrossRef]
9. Valera, C.A.; Pissarra, T.C.T.; Martins Filho, M.V.; Valle Junior, R.F.; Sanches Fernandes, L.F.; Pacheco, F.A.L. A legal framework with scientific basis for applying the 'polluter pays principle' to soil conservation in rural watersheds in Brazil. *Land Use Policy* **2017**, *66*, 61–71. [CrossRef]

Article

Conjunctive Water Resources Management in Densely Urbanized Karst Areas: A Study in the Sete Lagoas Region, State of Minas Gerais, Brazil

Hugo Henrique Cardoso de Salis [1], Adriana Monteiro da Costa [1], Annika Künne [2], Luís Filipe Sanches Fernandes [3] and Fernando António Leal Pacheco [4,*]

[1] Departamento de Geografia, Universidade Federal de Minas Gerais, Av. Antônio Carlos, 6.627-Pampulha-CEP, Belo Horizonte 31270-901, Minas Gerais, Brazil

[2] Geographic Information Science Group, Institute of Geography, Friedrich Schiller University, 07749 Jena, Germany

[3] Centro de Investigação e Tecnologias Agroambientais e Biológicas, Universidade de Trás-os-Montes e Alto Douro, Ap 1013, 5001–801 Vila Real, Portugal

[4] Centro de Química de Vila Real, Universidade de Trás-os-Montes e Alto Douro, Ap 1013, 5001–801 Vila Real, Portugal

* Correspondence: fpacheco@utad.pt

Received: 31 May 2019; Accepted: 17 July 2019; Published: 19 July 2019

Abstract: Headwater catchments store valuable resources of quality water, but their hydraulic response is difficult to assess (model) because they are usually deprived of monitoring stations, namely hydrometric stations. This issue becomes even more pertinent because headwater catchments are ideal for the practice of conjunctive water resources management involving the supply of towns with groundwater and surface water, a solution that can be used to mitigate overexploitation of groundwater resources in densely urbanized and populated areas. In this study, a stepwise approach is presented whereby, in a first stage, a gauged basin was modeled for stream flow using the JAMS J2000 framework, with the purpose to obtain calibrated hydraulic parameters and ecological simulated stream flow records. Having validated the model through a comparison of simulated and measured flows, the simulated record was adjusted to the scale of an ungauged sub-basin, based on a new run of JAMS J2000 using the same hydraulic parameters. At this stage, a second validation of modeled data was accomplished through comparison of the downscaled flow rates with discharge rates assessed by field measurements of flow velocity and water column height. The modeled basin was a portion of Jequitiba River basin, while the enclosed sub-basin was the Marinheiro catchment (state of Minas Gerais, Brazil). The latter is a peri-urban watershed located in the vicinity of Sete Lagoas town, a densely urbanized and populated area. This town uses 15.5 hm^3 year^{-1} of karst groundwater for public water supply, but the renewable resources were estimated to be 6.3 hm^3 year^{-1} The impairment between abstraction and renewable resources lasts for decades, and for that reason the town experiences systemic water table declines and sinkhole development. The present study claims that the storage of quality water in the Marinheiro catchment, in a dam reservoir, would help alleviate the depletion of groundwater resources in the karst aquifer because this catchment could deliver 4.73 hm^3 year^{-1} of quality surface water to the municipality without endangering ecologic flows. The construction of a small dam at the outlet of Marinheiro catchment could also improve aquifer recharge. Presently, the annual recharge in this catchment approaches 1.47 hm^3 but could be much larger if the small dam was installed in the water course and the captured stream water managed properly.

Keywords: hydrologic modeling; ungauged catchment; stream flow downscaling; karst aquifer; urban area; conjunctive water resources management; recharge; overexploitation; geo hazards

---

1. Introduction

Karst aquifers are valuable groundwater resources and therefore managed for public and private water supply in many regions around the globe [1–3]. In densely urbanized areas, the annual abstraction through pumping may exceed the renewable resources provided by recharge, leading the karst system to water table declines and subsidiary development of geo hazards (sinkholes). The relationship between overexploitation, water table declines, and sinkhole development has been reported in numerous studies [4–8] and described as follows: Excessive abstractions generate steep cones of depression that accelerate groundwater flow toward them, while slow phreatic recharge is replaced by more rapid downward percolation favoring suffosion, especially when the water table is lowered below the rock head. These two processes are accompanied by increases in the effective weight of the sediments due to loss of buoyant support that ultimately lead to sinkhole formation. Overcoming the impairment between abstraction and resource renewability could be accomplished through storm water infiltration. This is practiced in many urbanized areas to promote recharge and reestablish hydrology to pre-urbanization patterns [9,10]. However, the implementation of storm water infiltration in urban areas overlying karst aquifers is not recommended. Usually, urban storm water is significantly loaded with sediments and transports dissolved contaminants (metals, hydrocarbons), which can rapidly move through preferential flow paths and high permeability zones, reaching the karst aquifer in a short period of time [11–16]. In addition to groundwater contamination, storm water infiltration can also promote the development of new suffosional sinkholes.

A route to mitigate recharge declines and consequent groundwater resource reductions can be through the conjunctive management of water resources. This concept involves the combined use of groundwater and surface water to achieve public policy and management goals, and a few other cases consider the use of other sources such as recycled water [17]. Conjunctive water resources management enables greater water supply security and stability, helps adaptation to climate variation and uncertainty, and reduces depletion and degradation of water resources [18,19]. The implementation of conjunctive water resources management requires the selection of small catchments to store quality water in dam reservoirs. Forested headwater catchments can effectively accomplish the task [20]. Besides, if they overlay an overexploited karst aquifer, they can be used to promote diffuse recharge and improve water table recovery in the long term.

A successful selection of catchments to install dam reservoirs requires an evaluation of water resources usually accomplished through hydrologic modeling [21,22]. In general, hydrologic models simulate flow and routing processes at catchment and sub-catchment scales, based on spatially distributed relief, soils, and land use data, as well as on precipitation, temperature, and other climate records, and are validated for stream flow using available discharge records measured in hydrometric stations [23–26]. Hydrometric monitoring of rivers is practiced at the national scale in many countries and financially supported by public authorities but is usually restricted to a relatively small number of medium to large catchments because maintenance costs are large. Headwater catchments are barely included in the monitoring networks because their associated drainage basins are small. So, despite the importance of these basins as quality water sources ideal for the practice of conjunctive water resources management, they cannot be robustly characterized for hydrologic response, unless a stream discharge record is obtained in the field using portable velocity meters, while the hydrologic model is validated on the basis of that record. Validation of hydrologic models based on hydrometric station data are widespread in the literature [27,28], but that is not the case when the studies are developed in ungauged streams. The general purpose of this study is therefore to help improving hydrologic modeling of ungauged watersheds. The specific purposes include (1) use a spatially distributed hydrologic model (JAMS J2000 framework) to estimate stream discharge in a monitored catchment, (2) calibrate and validate the model using stream flow data obtained from a hydrometric station, (3) use the calibrated parameters to model stream discharges in an ungauged headwater sub-catchment, (4) validate the model using discharges measured in the field with a portable velocity meter, and (5) evaluate water resources and recharge capacity in the headwater catchment. The rationale underlying the proposed research is portrayed in Figure 1.

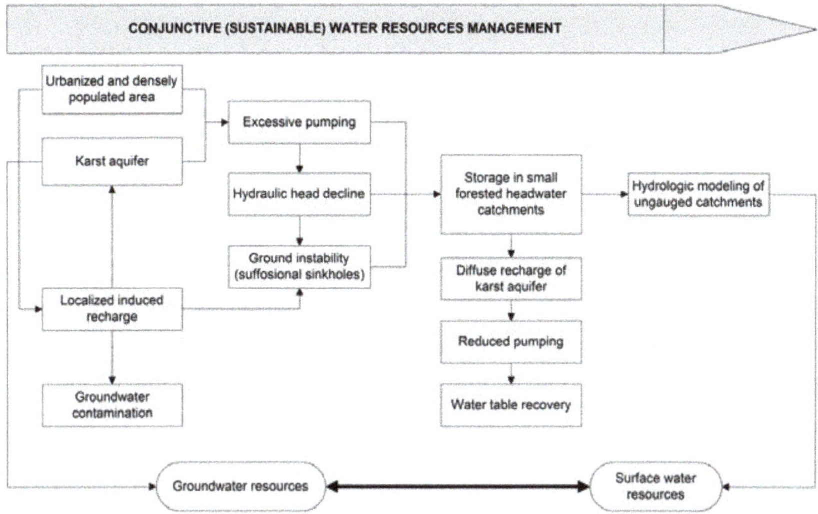

Figure 1. Conceptual approach to conjunctive water resources management in a karst aquifer.

The study area is the Marinheiro catchment, a headwater tributary catchment of Jequitiba River basin where the Sete Lagoas town is installed (state of Minas Gerais, Brazil). This town grew from 150.000 to 220.000 habitants in a period of 25 years (1993–2008). In that period, these people consumed 200 L habitant^{-1} day^{-1} of groundwater that was withdrawn from a karst aquifer. Groundwater was pumped from the aquifer using a significant number ($\approx$ 100) of large yield ($\approx$ 8.0 L s^{-1}) drilled wells that worked 16 h every day [29,30]. In 2014, Galvão et al. [31] evaluated the long standing effects of pumping in the geometry of hydraulic heads. A depressed area was delineated around the older wells (1942), where depths to the water have ranged from 14 m post drilling to 62 m in 2012 (48 m drawdown in 70 years). The study also suggested the link of this hydraulic head depression to the development of various suffosional sinkholes. It is therefore pertinent to study this area and help the Sete Lagoas Municipality to resolve the problem, suggesting the alleviation of pumping through the conjunctive use of karst groundwater and surface water from the Marinheiro stream in the public water supply. In this context, the Marinheiro catchment is a peri-urban basin of Sete Lagoas town whose recent master plan directs the region toward urban sprawl. These political guidelines make the Marinheiro catchment vulnerable to real estate speculation and soil sealing by the construction of houses and asphalt paving, as well as other impacts. It is also worth recalling that the Marinheiro main watercourse is ecologically relevant, acting as thinner of polluted waters flowing into the Jequitibá River.

2. Materials and Methods

2.1. Study Area

The Jequitiba River basin (area: 57,148 he) is located in the central region of Minas Gerais state (Brazil; Figure 2), and is characterized by rural environment, except in Sete Lagoas town, which is densely urbanized and populated. The study area comprises the Marinheiro catchment (1480 he), which a small tributary headwater catchment of the Jequitiba River basin located in the neighborhood of Sete Lagoas. The Jequitiba basin is located between the geographic coordinates X = 573,198 to 594,872 m and Y = 7,859,607 to 7,836,875 m, referred to the SIRGAS 2000 geodetic datum and UTM 23

South projection, while the Marinheiro catchment fits within the coordinates X = 581,100.3 to 587,493.5 m and Y = 7,841,747.5 to 7,847,420.3 m.

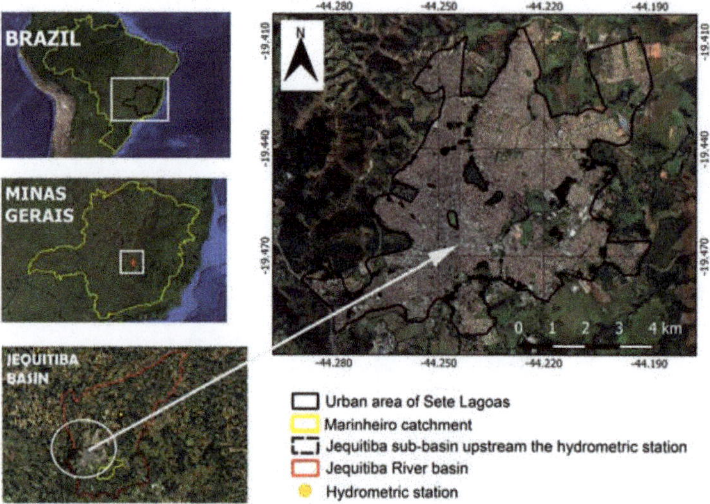

Figure 2. Location of study area.

According to Koppen's classification, the climate is subtropical (Cwa), characterized by dry winters and hot summers. The mean annual rainfall was 1291.2 mm in the 2000–2018 period, while the mean temperatures ranged between 18 °C and 24 °C from July to January–February, and showed a mean annual value of 21.8 °C. The altitudes range from 629 to 932 m in the Jequitiba River basin and between 717 and 918 m in the Marinheiro catchment (Figure 3a). The geologic substratum (Figure 3b) of the Jequitiba River basin comprises an Archean crystalline basement composed of orthogneisses, granites, and migmatites (24.3%) that were overlaid by Neoproterozoic carbonate rocks of the Bambuí Group, namely calcite and dolomite limestones from the Sete Lagoas Formation (42.7%), and pelitic rocks with interlayered carbonates from the Serra de Santa Helena Formation (28.7%). Later, this sequence was covered by alluvium, colluvium, and terrace sediments (4.4%) along and lateral to the main water courses [32,33]. The Marinheiro catchment is dominated by carbonate rocks from the Sete Lagoas formation, with a small area of Archean rocks cropping out around the catchment headwaters. The comparison of Figure 3b with Figure 3c (soil map [34]) reveals that Archean rocks as well as pelitic rocks cropping out in the catchment lowlands are overlaid with cambisols (36.7% coverage within the Jequitiba River basin), pelitic rocks cropping out in the catchment highlands are overlaid with neosols (12.2%), and limestones and terrigenous rocks are overlaid with latosols (49.6%). Accordingly, the Marinheiro catchment is mostly covered with latosols and cambisols (along the headwaters). Land in the Jequitiba River basin is mostly used for anthropogenic activities, such as livestock pasturing or agriculture, which are distributed along the drainage network (Figure 3d). Forests occupy 15.8% of the area, and urban areas 14.4% [35]. Agriculture is also dominant in the Marinheiro catchment, but native vegetation is kept dense in some areas, especially along the headwaters.

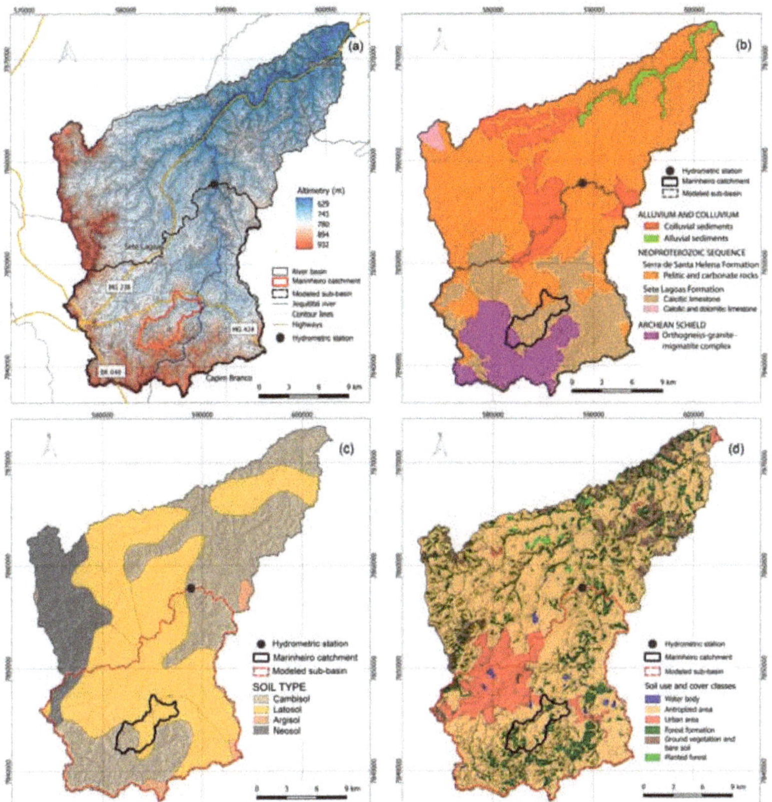

Figure 3. (**a**) Topographic map of Jequitiba River basin, with indication of towns, drainage network, and main road network; (**b**) lithologic map of Jequitiba River basin; (**c**) soil map of Jequitiba River basin; (**d**) land use and cover map of Jequitiba River basin. The geographic reference for the maps is the Universal Transverse Mercator (UTM) projection system, SIRGAS 2000 datum, 23 south time zone. The modeled sub-basin and Marinheiro catchment are also indicated in all panels.

The hydrogeology is characterized by fractured aquifers, developed in the Archean rocks. Fractured-karst aquifers developed in the pelitic rocks interlayered with carbonates from the Serra de Santa Helena formation, karst aquifers developed in the limestones from the Sete Lagoas formation, and porous aquifers developed in the terrigenous rocks and soil layer [36]. Recently, Monteiro et al. [37] estimated the recharge within the Jequitiba River basin (Figure 4) and derived average values for the various aquifer types (m^3 ha^{-1} year^{-1}): 829.3–1357.0 (porous), 829.3 (karst and fractured-karst), and 539.3 (fractured crystalline). According to Figure 4, average recharge within the Marinheiro catchment is 340 m^3 ha^{-1} year^{-1} within the fractured aquifer ($\approx$ 30% of the catchment) and 1883 m^3 ha^{-1} year^{-1} within the karst aquifer ($\approx$ 70%), which means 1420 m^3 ha^{-1} year^{-1} on average.

The town of Sete Lagoas has been supplied with groundwater from the karst aquifer since the 1950s [29], when the population was less than 25,000 inhabitants and groundwater was extracted from cisterns. These sites became obsolete as the population grew, being progressively replaced by drilled wells to satisfy the public water demand. In 1993, 65 drilled wells were pumped every day for 16 h, extracting 8.0 L s^{-1} each to satisfy the consumption of 150,000 people living in Sete Lagoas [29,38]. The situation was re-evaluated in 2008 by Botelho [30], who counted 94 drilled wells

(44% increase), keeping a similar mean abstraction (7.8 L s^{-1}) to satisfy the consumption of 220,000 habitants (47% increase).

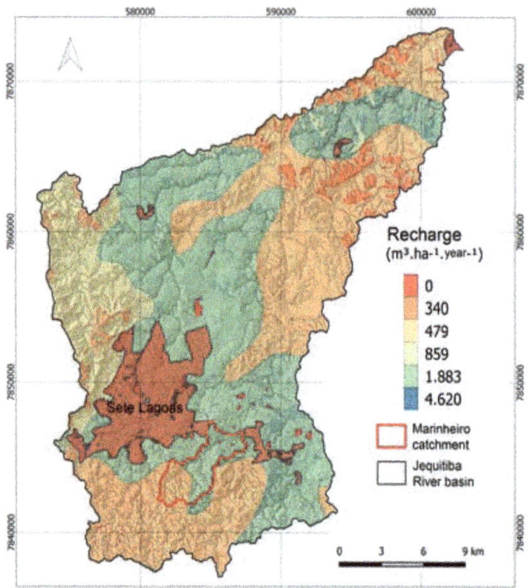

Figure 4. Recharge within the Jequitiba River basin and Marinheiro catchment. Adapted from Monteiro et al. [37].

2.2. Digital Data, Numerical Records and Software

The materials used in this study are indicated in Table 1 and comprised (a) a Digital Elevation Model (DEM) ALOS PALSAR with a spatial resolution of 12.5 m [39]; (b) Sentinel-2 satellite images with a spatial resolution of 10 m [40]; (c) the soil map of Minas Gerais state at scale of 1:650,000 and corresponding data on hydraulic conductivity [34]; (d) the geological map of Minas Gerais state at scale 1:1,000,000 [33]; (e) Climatic data from weather stations in the municipalities of Belo Horizonte (BH), Sete Lagoas (SL), Conceição do Mato Dentro (CMD), and Florestal (FLT) [41]; and (f) Hydrometric data of station 41410000 [42].

Table 1. Materials used in the JAMS J2000 hydrologic model, namely spatial data and climatic and stream flow records, and URLs of websites used for downloading the data.

Data Type	Use in the Hydrologic Model	URL of Website
Digital elevation model	Hydrologic Response Units (HRU)	https://www.asf.alaska.edu
Satellite images	Land use mapping and HRU	https://earthexplorer.usgs.gov/
Soil map and hydraulic conductivity data	HRU and data parameterization	http://www.dps.ufv.br
Geologic map	HRU and data parameterization	www.portaldageologia.com.br
Climatic data	Data for JAMS J2000 hydrologic model	http://www.inmet.gov.br
Stream flow data	Calibration/validation procedure	http://www.snirh.gov.br/hidroweb

The software Hydrus 1D (https://www.pc-progress.com) was used to estimate hydraulic conductivity of soils based on soil texture data released with the soil map. The SPRING software, version 4.3.5 (http://www.dpi.inpe.br/spring/english/), was used to interpret the satellite images and delineate land uses and occupations. The JAMS J2000 framework (http://jams.uni-jena.de) was used to perform the hydrologic modeling, including the calibration and validation procedures.

The Quantum GIS (https://www.qgis.org) and ArcMap (https://www.esri.com/) were used to produce the thematic maps (e.g., Figure 3), given its visualization capabilities exposed in numerous environmental studies [43–63].

2.3. Hydrologic Modeling

The hydrologic modeling was developed in the following four main steps (Figure 5): (1) pre-processing of climatic, stream flow, and cartographic data, followed by delineation of hydrologic response units as detailed below, based on the overlay analysis of relief, soils, and land use maps. The data records comprised the 2003–2016 period; (2) hydrologic modeling of a catchment monitored for stream discharge, namely a sub-basin of Jequitiba River basin monitored at the hydrometric station 41410000. The modeling was based on the JAMS J2000 framework developed by Krause [64] and comprehended a calibration (2003–2011) and a validation (2012–2016) period. This step was preceded by parameterization of input data and followed by the performance analysis of calibrated models based on goodness-of-fit indices; (3) "Downscaling" of stream flow discharges to the scale of an ungauged catchment enclosed in the monitored sub-basin, namely the Marinheiro catchment (study area). The "downscaling" was based on a second run of JAMS J2000, keeping the previous parameterization but adjusting the hydrologic analysis to the area, associated hydrologic response units, and climatic characterization of the Marinheiro catchment. The "downscaling" step was validated through comparison of discharge rates simulated by the JAMS J2000 framework and rates assessed in the field at the catchment outlet using a portable flow meter, as well as on the basis of a goodness-of-fit index; (4) Evaluation of reference and ecological stream flows aiming the assessment of surface water resources within the Marinheiro catchment. The four steps are detailed in the next paragraphs.

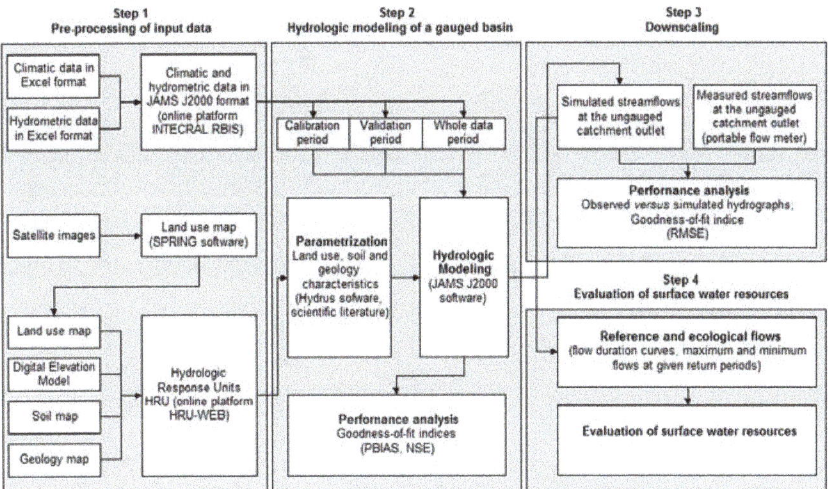

Figure 5. Workflow used to perform the hydrologic modeling.

In the first step, the variables precipitation, temperature, relative humidity, hours of sunshine, wind speed, and daily stream flow data, compiled from the weather and hydrometric stations, were organized in a series of Excel spreadsheets, namely listed in ascending order of time. In turn, these spreadsheets were submitted to the online platform INTECRAL RBIS—river Basin Information System (http://leutra.geogr.uni-jena.de/intecralRBIS)—developed by Zander and Kralisch [65], for the conversion of the Excel spreadsheets into files in the specific format of the JAMS J2000 framework.

The cartographic data were cropped to the area of interest and verified for geometric and geographic consistency. In addition, the satellite images were submitted to supervised classification of land use or cover using the SPRING software. The classification comprised classes for cultivated area, water bodies, pastures, forest and exposed soil.

The online platform HRU-WEB (http://intecral.uni-jena.de/hruweb-qs/) was used to delineate the homogeneous hydrologic response units (HRUs). The HRUs are used as modeling entities that have the same pedological, lithological, topographical, and land use/land cover characterizations and are heterogeneous from each other. They are connected by a topological routing scheme [66]. The lateral water flow is simulated allowing a fully explicit spatial discretization of hydrologic response within the modeled catchment. However, the delineation of HRUs may not account for karst heterogeneity. Nevertheless, the model can be calibrated so that the observed and modeled stream flows match well. This HRU-WEB platform is embedded with tools to intersect the digital elevation model with soil, geology and land use and cover maps producing the HRUs as output. Each HRU holds a specific identification number, centroid coordinates, and connection codes to adjacent HRUs or specific water lines [67].

In the second step, the JAMS J2000 framework model was applied to simulate hydrological processes at the scale of a sub-basin of the Jequitiba River basin located upstream of hydrometric station 41410000, which encloses the Marinheiro catchment. The J2000 is a process-based spatially explicit hydrological model, which simulates eco-hydrological processes on the river basin scale. The method is a part of a modular and object-oriented modeling system called JAMS. Altogether, the JAMS J2000 framework contains different environmental models and a plethora of components to create custom tailored models for different research questions. In addition, JAMS possesses modules for data preparation, analysis, and visualization. The full inventory of hydrologic modules used in this study is listed in Appendix A. It is beyond the scope of this paper to describe them all in detail, especially for their mechanics. Readers are invited to consult the original works.

The JAMS J2000 framework simulated discharge rates, on a daily time step, between the calendar years 2003 and 2016. Using water balances and routing algorithms, the model also simulated flow components (surface flow, percolation in soil layers, shallow and deep underground flow). The simulated discharge rates were compared with the homologous rates measured at the hydrometric station for the sake of calibration and validation. Calibration encompassed the 2003–2011 period while validation the 2012–2016 period. Besides the measured discharge rates, other parameters were used in the hydrologic modeling, which are listed in Tables A2–A4 of Appendix B. These parameters were processed in the various JAMS J2000 modules (Appendix A), which were run until optimized values (calibrated) were obtained. The calibration procedure was based on the NSIN II algorithm (Genetic Multi-objective II) with daily time step, while adopting 5000 iterations as the stopping rule [68]. The calibrated parameters are indicated in Appendix C. Following the calibration procedure, the hydrological model was tested for performance through assessment of goodness-of-fit indices (a) percentage of bias (PBIAS) [69] and (b) Nash-Sutcliffe (NSE) efficiency coefficient [70]. The performance intervals are provided as Appendix D. According to Gupta et al. [69], the PBIAS estimates the percentage trend of simulated data to be higher or lower than the observed data and can be described by the following equation:

$$PBIAS = \left[\frac{\sum_{t=1}^{n}(y_i - o_i)}{\sum_{t=1}^{n} o_i} \right] \times 100 \tag{1}$$

where PBIAS is the percentage of bias (%), y_i is the simulated flow (m^3 s^{-1}), and o_i is the observed flow (m^3 s^{-1}). A PBIAS = 0 occurs for a hydrological model with optimal performance. Positive or negative values indicate, respectively, that the model overestimates or underestimates the simulated flows. The NSE and LNSE coefficients are equated as follows [70]:

$$NSE = 1 - \frac{\sum_{i=1}^{n}(o_i - y_i)^2}{\sum_{i=1}^{n}(o_i - O^{med})^2} \tag{2}$$

$$LNSE \;=\; 1 - \frac{\sum_{i=1}^{n}(lno_i - lny_i)^2}{\sum_{i=1}^{n}(lno_i - lnO^{med})^2} \tag{3}$$

where "ln" represents the natural logarithm and o_i, o_{med}, y_i, and y_{med} represent, respectively, the observed flow, the average observed flow, the simulated flow, and the average simulated flow ($m^3\ s^{-1}$). The values of NSE and LNSE (dimensionless) can vary from $-\infty$ to 1. The closer to 1, the greater the adjustment between the simulated and observed values. Results below 0 indicate that the mean observed values are more representative than the values predicted by the model.

In the third step, the calibrated stream flows (2003–2016 series) were "downscaled" to the Marinheiro catchment scale and estimated at the corresponding outlet. To accomplish the "downscaling," the JAMS J2000 framework was run using the previously calibrated parameters (Appendix C) and the HRUs enclosed in the Marinheiro catchment. Then, the discharge rates estimated at the Marinheiro catchment's outlet using the JAMS J2000 framework were compared with homologous rates estimated on the basis of flow velocities and water column heights, measured with a portable flow meter in the period June of 2015 and February of 2018 (142 evaluations).

$$Q \;=\; Av \tag{4}$$

where Q ($m^3\ s^{-1}$) is the discharge rate, A (m^2) is the wet area of a circular stream section and v ($m\ s^{-1}$) is the measured flow velocity. The value of A was estimated as follows:

$$A \;=\; \frac{1}{8}(\theta - sen\,(\theta))\,D^2 \tag{5}$$

with

$$\theta \;=\; 2\,cos^{-1}\!\left(1 - \frac{2h}{D}\right) \tag{6}$$

where h (m) is the water column height measured at the stream section's center, while D (m) and θ (radians) are the stream section's diameter and the angle between the stream section central point and the stream water surface, respectively.

The performance of "downscaling" was checked through comparison of measured and simulated hydrographs and calculation of the RMSE estimator.

$$RMSE \;=\; \sqrt{\frac{\sum_{i=1}^{n}(O_i - E_i)^2}{n-1}} \tag{7}$$

where O_i and E_i represent the observed (measured Q) and estimated (with JAMS J2000 framework) values of a target variable, while n represents the number of measurements and corresponding estimations (142 in the present case).

In the fourth (and last) step, reference stream flows were evaluated within the Marinheiro catchment, based on the simulated series of discharge (JAMS J2000 framework, period 2003–2016). A reference stream flow launches the upper limit of water uses in a water course. According to Harris et al. [71], the legal application of a reference flow favors the protection of rivers, because the grants for stream water diversion in that context are based on low risk base flows. However, the setup of reference flows is also an obstacle to the implementation of a grant system [72]. The most common reference values are [73] $Q_{7,10}$, which is the minimum flow of seven days duration and 10 years return period, with a 10% risk of not being reached, and Q_{90}, which corresponds to a flow with 90% probability of being exceeded in time. They are both seen as ecological flows, meaning that any granted stream water diversion needs to permanently ensure these or larger flows in the target river [74,75]. In Brazil, states have adopted different criteria for setting up reference flows for granting but did not present justifications for the adoption of specific values. In Minas Gerais state, the Mining Institute of Water Management (IGAM; http://www.igam.mg.gov.br/) defined threshold values for the Marinheiro catchment, namely $Q_{7,10} = 0.029$ m^3 s^{-1} and $Q90 = 0.075$ m^3 s^{-1}. The comparison of IGAM values with counterparts determined by the JAMS J2000 framework represents additional performance criterion.

Besides the use in model performance assessment, the reference stream flows were used in the evaluation of water resources aiming the proposed conjunctive use of karst groundwater and surface water in the supply to Sete Lagoas town. Their calculation involved the definition of a flow duration curve for the Marinheiro stream that was based on a probabilistic model [76,77],

$$p_i = \frac{i}{n+1} \tag{8}$$

where p_i is the probability of reaching or exceeding a stream flow, i is the order of the ith sorted flow (descending order), and n is the number of ordered data. It also involved the definition of maximum and minimum flow values at different return periods (2, 5, 10, 50, and 100 years) using the log-normal probability distribution [78].

3. Results

The modeled sub-basin (Step 2; Figure 5) of Jequitibá River basin is located upstream of hydrometric station 41.410000. This sub-basin is represented in Figure 3 over the topographic, geologic, soil, and land use maps and corresponds to the upper part of Jequitiba River basin. During the hydrologic modeling, the overlay analysis of these geographic data generated 22,920 hydrologic response units characterized by similar soil types, relief classes, and land uses/occupations. These attributes, together with the climate data, were submitted to 5000 iterations in JAMS J2000 modules, which returned values for specific hydrologic parameters as listed the Appendix C. The hydrologic model also returned simulated stream flows that were compared with real discharge rates from 41.410000 station. The analysis of model performance was based on the PBIAS and NSE indices. The calculated PBIAS were −9.50 (calibration period), −3.65 (validation period), and 3.80 (whole period), indicating very good performance within the whole period (see reference levels in Appendix D). As regards NSE, the homologous results were 0.58, 0.67, and 0.64, indicating good to fair performance.

The spatial distribution of flow components is displayed in Figure 6. It is clear from this figure that direct flows dominate in the northwestern sector of the modeled sub-catchment where the Sete Lagoas town is located. This is obviously explained by surface waterproofing in this urban area. Concomitantly, the other flow components are lower in this sector, especially the deep groundwater flow. In the Marinheiro catchment, the flow components are, on average, direct flow (DF) = 36.95 mm year^{-1}, percolation in soil layer (S) = 69.06 mm year^{-1}, GF (upper) underground flow in the upper aquifer layer (GF) (upper) = 49.32 mm year^{-1}, and underground flow in the lower aquifer layer (GF) (lower) = 17.3 mm year^{-1}. If the underground components S, GF (upper), and GF (lower) are summed and converted in the m^3 ha^{-1} year^{-1} scale, the result (1356.8 m^3 ha^{-1} year^{-1}) is similar to the recharge estimated by Monteiro et al. [37] (1420 m^3 ha^{-1} year^{-1}).

The stream flows at the outlet of Marinheiro catchment were obtained through "downscaling" of JAMS J2000 flows estimated at the 41.410000 station (Step 3). The results are illustrated in the hydrograph of Figure 7. In the studied period, the daily average values of precipitation and discharge were 4.02 mm and 0.21 m^3 s^{-1}, respectively. When monthly averages were compiled from the hydrograph and compared with the homologous values assessed in the field with the portable flow meter, a fair approximation was observed between the two sets of values (Table 2). In the months of April to November, the simulated flows were lower than the observed flows. Conversely, in the rainy season the simulated flows were larger than the flows estimated in the field. The differences were larger in the months of January and March, indicating a slight overestimation of the simulated flows in these months. In the other months, the simulated flows approached the observed flows, especially during the dry period. Overall, the calculated RMSE index was 0.07 m^3 s^{-1}, which is indication of an acceptable "downscaling." Similar values of RMSE in hydrological models that presented good performance can be consulted in the works of Monte et al. [79] and Machado et al. [80].

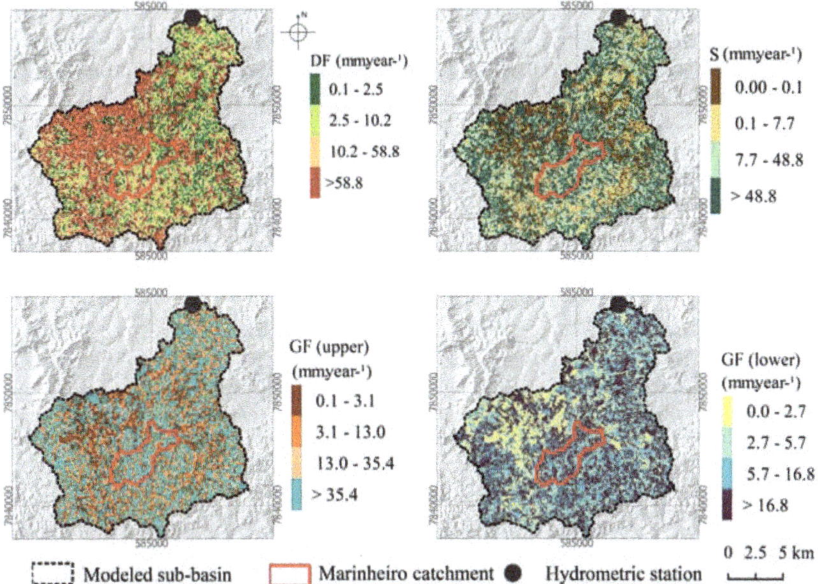

Figure 6. Flow components simulated by the JAMS J2000 framework within the Jequitiba River sub-basin located upstream the 41.410000 station. (DF)—direct (surface) flow; (S)—percolation in the soil layer; (GF upper)—underground flow in the upper aquifer layer; (GF lower)—underground flow in the lower aquifer layer.

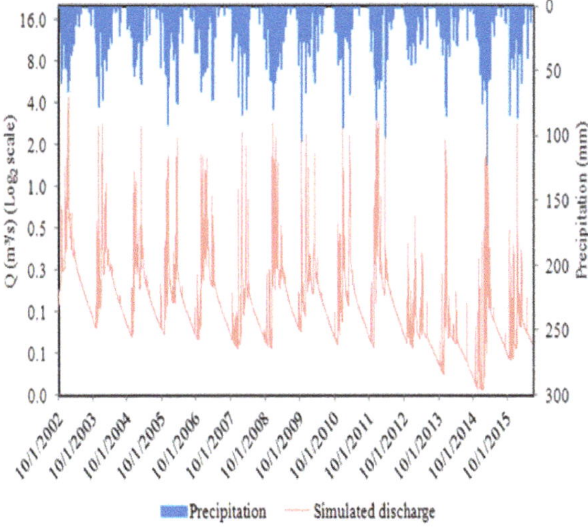

Figure 7. Discharge rates in the Marinheiro stream simulated with the JAMS J2000 model.

Table 2. Simulated (JAMS J2000), measured (portable flow meter), and residual (difference between simulated and measured) stream flows in the Marinheiro catchment, averaged on a monthly basis. The Root-mean-square deviation (RMSE) value indicates an acceptable performance of stream flow downscaling.

Month	Simulated Discharge ($m^3\ s^{-1}$)	Measured Discharge ($m^3\ s^{-1}$)	Residuals ($m^3\ s^{-1}$)
January	0.40	0.24	0.16
February	0.23	0.25	−0.02
March	0.30	0.17	0.13
April	0.16	0.17	−0.01
May	0.13	0.18	−0.05
Jun	0.11	0.16	−0.05
July	0.10	0.14	−0.04
Augus	0.09	0.12	−0.03
September	0.08	0.12	−0.04
October	0.09	0.14	−0.05
November	0.13	0.16	−0.03
December	0.36	0.30	0.06
	RMSE = 0.07		

The flow duration curve of Marinheiro catchment is illustrated in Figure 8. The figure shows that the flow was equal to or greater than 0.06 $m^3\ s^{-1}$ (base flow) in 90% of the time and equal to or greater than 0.36 $m^3\ s^{-1}$ (flood flows) in 10% of the time. The marked difference between flood flows and base flows may indicate low capacity to maintain the volume of water that is drained into the stream bed during the year, compromising the supply of water in the driest months unless the stream water is stored in a dam reservoir. Moreover, this low storage capacity can be explained both by the lithologic characteristics in which the river basin is located (karst) and by the abstractions that can occur in the watercourse. The minimum and maximum flows simulated for the Marinheiro catchment at various return periods are plotted in Figure 9. The maximum flows increase from 2.26 to 9.54 $m^3\ s^{-1}$ for increasing return periods (2 to 100 years). The minimum flows decrease from 0.14 to 0.06 $m^3\ s^{-1}$ for the same periods. The calculated ecological flows were $Q_{7,10}$ = 0.045 $m^3\ s^{-1}$ and Q_{90} = 0.071 $m^3\ s^{-1}$. These values were relatively close to the IGAM's values ($Q_{7,10}$ = 0.029 $m^3\ s^{-1}$ and Q_{90} = 0.075 $m^3\ s^{-1}$), especially the Q_{90}.

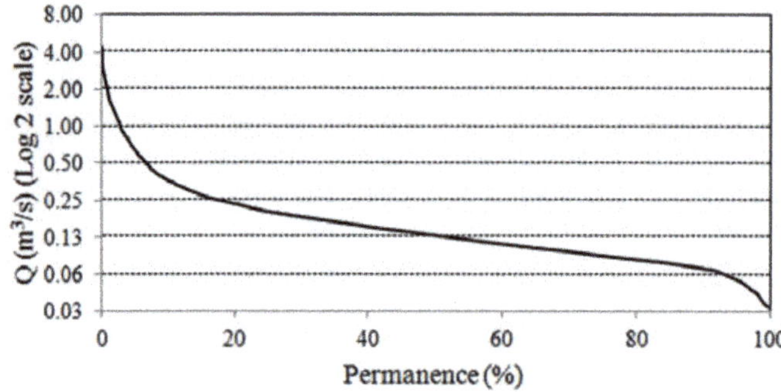

Figure 8. Flow duration curve of the Marinheiro catchment.

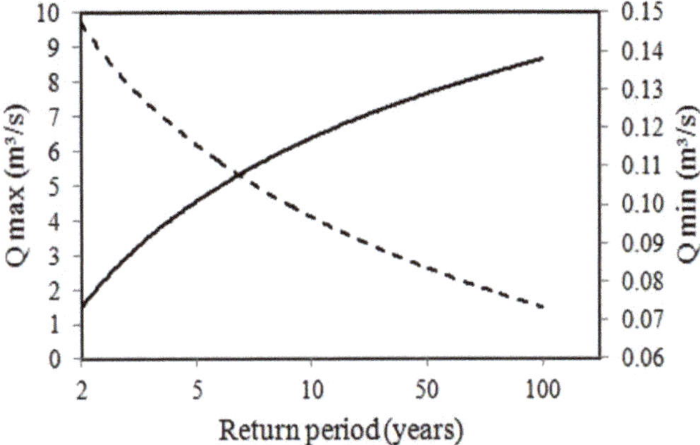

Figure 9. Return period of maximum (Qmax) and minimum (Qmin) flows in the Marinheiro catchment, estimated at return periods of 2, 5, 10, 50, and 100 years.

4. Discussion

The modeling of the Jequitiba River basin and the Marinheiro catchment based on the stepwise workflow of Figure 5 proved efficient, given the close match between simulated and measured stream flow discharges (Table 2) and generally good performance indicators (PBIAS, NSE, RMSE). The approximation between ecological indicators $Q_{7,10}$ and Q_{90}, derived from the hydrologic model and by the public institution IGAM, is also an indication of efficient modeling and robust modeling results. This was a key step and provides confidence in the interpretation and comments to follow.

In 2008, the ca. 220,000 inhabitants of Sete Lagoas were consuming approximately 200 L habitant^{-1} day^{-1} of karst aquifer water pumped from 94 drilled wells at an average rate of 7.8 L s^{-1} for 16 h every day [29,30]. This represented an annual abstraction of approximately 15.5 hm^3 of groundwater. Based on the hydrologic modeling, the annual groundwater resources (Figure 6) within the Sete Lagoas karst (14.4% of the modeled catchment) represented in the 2003–2016 period 6.3 hm^3 year^{-1}. The difference between availability and demand is therefore negative and approached –9.2 hm^3 year^{-1} in 2008. In fact, this difference has been negative for a long time and was the reason for the development of systemic water table declines and suffosional sinkholes. The duration curve of the Marinheiro catchment indicates that base flows are permanently (90% of the time) equal or greater than 0.06 m^3 s^{-1} (1.89 hm^3 year^{-1}). These flows are ecologic and therefore not grantable for uses such as public water supply. However, the flows above 0.06 m^3 s^{-1} can be considered for this use. For example, flows are equal to or larger than 0.13 m^3 s^{-1} for six months (Figure 8) and even larger for shorter periods of time. The relationship between flow and duration expressed in Figure 8 assumes the free flow of stream water. If a small dam or weir is installed in the water course, then the durations can be extended longer, namely one year, because the stream water is kept in the reservoir. In that case, the average flow c (0.21 m^3 s^{-1}) can be a good proxy for a permanent flow. The difference between this value and the ecologic flow represents 4.73 hm^3 year^{-1} of quality water that could be proposed for the public water supply in the context of conjunctive water resources management involving groundwater from the karst and surface water from the Marinheiro catchment. This conjunctive water resources management is suggested for other regions [18,81–85] and could help reduce pumping in the Sete Lagoas drilled wells to mitigate water table decline and suffosional sinkhole formation, as proposed by Galvão [31]. It is worth noting that the proposed discharge rate (0.21 m^3 s^{-1}) has a return period <2 years (Figure 9) and therefore is suited for water planning. We recall, however, the well-known limitations of return period assessments and the risk of failure of water resources evaluations based on this hydrologic indicator [86,87].

According to Monteiro et al. [37], recharge in the Marinheiro catchment approaches 1420 m^3 ha^{-1} $year^{-1}$ (142 mm $year^{-1}$). This recharge refills the Archean basement (30% of the catchment) but mostly the karst aquifer (70%), representing an input of 1.47 hm^3 $year^{-1}$ of quality water to the karst aquifer. It can be considered a high value, when compared with a range of values determined by Galvão [88] within the Serra de Santa Helena Environmental Protection Area next to the urbanized area of Sete Lagoas. The creation of a dam reservoir in the Marinheiro catchment could increase aquifer recharge to even larger values and hence improve productivity of boreholes downward. In a study based on stable isotopes, Galvão [89] related groundwater origin directly to local precipitation, with a limited recharge period, and locally receiving surface water contributions. It is therefore acceptable to expect the possibility of local recharge from a dam reservoir and to assume that recharge could be extended in time with the presence of that infrastructure. Studies all over the globe claim recharge improvements triggered by dam installation. For example, a study in AlKhod, Oman [90], claimed that controlled releases of water captured by a dam optimized water percolation and enhanced artificial recharge, which almost doubled in a period of eight years. Many other studies refer stream damming as measure to manage aquifer recharge [91,92], while the problem has been addressed from multiple stand points. For example, in a study developed in Ranchi (India), Mahto studied the feasibility of artificial groundwater recharge structures for urban and rural environment, using geospatial technology [93]. Other authors compared recharge improvements obtained with single dams and cascades of dams (check dams) [94]. The next step would be to look at the Marinheiro catchment as an opportunity to manage recharge expecting an improvement of hydraulic heads around Sete Lagoas.

From a practical standpoint, the modeling results confirmed the ability of the Marinheiro catchment to be used as source of quality surface water to the municipality of Sete Lagoas in the context of conjunctive water resources management. More importantly, the present research proved the efficiency of the modeling of ungauged catchments. Usually, the hydrologic modeling of gauged basins can be robustly validated with measured stream flows, but the scarcity of monitoring stations hampers validation of hydrologic models in headwater catchments. In the present study, a stepwise approach was implemented whereby a gauged basin is modeled for stream flow in a process involving calibration of hydraulic parameters and validation of simulated stream flows with measured counterparts. Subsequently, the calibrated parameters were used to model the stream flows of an enclosed sub-basin, in a process referred to as "downscaling." Finally, the "downscaled" stream flows were validated in the field by measurements of flow velocity and water column weight. To our knowledge, this sequential approach involving a double validation of modeling results is uncommon and can be viewed as relevant improvement of ungauged catchment hydrologic modeling.

5. Conclusions

A novel hydrologic approach was accomplished through a double run of the JAMS J2000 framework, whereby calibrated hydraulic parameters and simulated stream records were obtained at the scale of Jequitiba River basin (first run), and then the simulated flows were adjusted to the scale of the Marinheiro catchment (second run), which is a Jequitiba tributary. The Jequitiba basin also encloses the town of Sete Lagoas (state of Minas Gerais, Brazil), which has for decades experienced severe problems of water table declines and depletion of groundwater resources related to overexploitation of a karst aquifer used for public supply. The double run of JAMS J2000 was accompanied by a double validation of modeled stream flows. In the first run, the simulated stream flows could be validated by comparison with stream flows measured in a hydrometric station installed in a section of the Jequitiba section for years. The Marinheiro catchment is not gauged, and therefore that type of validation was not possible. Instead, the simulated (downscaled) stream flows were validated by discharge rates assessed in the field through measurements of flow velocity and water column heights. Having finished the modeling, flow duration and return periods of stream flow were assessed using probabilistic models. Altogether, the possibility to deliver 4.73 hm^3 $year^{-1}$ of quality surface water to Sete Lagoas to alleviate the pressure over the karst aquifer was considered. This water would be stored in a small dam that would also improve aquifer recharge

Author Contributions: Conceptualization, H.H.C.d.S., F.A.L.P., and A.M.d.C.; methodology, H.H.C.d.S., A.K. and A.M.d.C.; software, H.H.C.d.S. and A.K., validation, H.H.C.d.S. and F.A.L.P.; resources, A.M.d.C.; writing—original draft preparation, H.H.C.d.S.; writing—review and editing, A.M.d.C., L.F.S.F., and F.A.L.P.; supervision, A.M.d.C.; funding acquisition, A.M.d.C.

Funding: For authors integrated in the CITAB research center, this study was further financed by the FEDER/COMPETE/POCI—Operational Competitiveness and Internationalization Programme, under Project POCI-01-0145-FEDER-006958, and by National Funds of FCT—Portuguese Foundation for Science and Technology, under project UID/AGR/04033/2019. For the author integrated in the CQVR, the research was additionally supported by National Funds of FCT—Portuguese Foundation for Science and Technology, under project UID/QUI/00616/2019.

Conflicts of Interest: The authors declare no conflict of interest. The funders had no role in the design of the study; in the collection, analyses, or interpretation of data; in the writing of the manuscript or in the decision to publish the results.

Appendix A

Table A1. JAMS J2000 modules used in the hydrologic modeling.

Module	Parameter	Description
Start up	mFCa	Multiplier of field capacity
	mACa	Multiplier of air capacity
	initRG1	Initial capacity in the upper underground reservoir
	initRG2	Initial capacity in the lower underground reservoir
Interception	α, rain	Maximum interception capacity of leaf area
Water in soil	soiMaxDPS	Maximum storage capacity in the surface
	soilPolRed	Polynomial reduction factor of potential evapotranspiration
	soilLinRed	Linear reduction factor of potential evapotranspiration
	soilMaxInf1	Maximum infiltration in the April–September period
	soilMaxInf2	Maximum infiltration in the October–March period
	soilImpGT80	Relative infiltration capacity in areas with waterproofing larger than 80%
	soilImpLT80	Relative infiltration capacity in areas with waterproofing smaller than 80%
	soilDistMPSLPS	Coefficient of infiltration distribution between medium and large pores
	soilDiffMPSLPS	Diffusion coefficient from large to medium pores
	soilOutLPS	Output coefficient from large pores
	soilLatVertLPS	Distribution coefficient between interflow and percolation
	soilMaxPerc	Maximum percolation capacity
	soilConcRD1	Retention coefficient of surface flow
	soilConcRD2	Retention coefficient of interflow
Groundwater	gwRG1RG2dist	Distribution coefficient between storage in the upper and lower groundwater reservoirs
	gwRG1fact	Dynamic flow factor in the upper reservoir
	gwRG2fact	Dynamic flow factor in the lower reservoir
	gwCapRise	Capillary factor
Routing	flowRouteTA	Time of concentration

Appendix B

Parameterization of JAMS J2000 hydrologic model. The target watershed was the sub-basin of Jequitiba River basin located upstream hydrometric station 41410000. The modeling also holds for all enclosed catchments, namely the Marinheiro stream catchment. The parameterization of input data resorted to computer programs, research articles, and technical studies where lithologic, soil and land use properties were calculated or indicated. The selected values of all input parameters, as well as the corresponding sources of information, are listed in Tables A2–A4.

Table A2. Land use and occupation parameters used in the hydrologic model.

Land Use or Occupation	Albedo (%)	Superficial Resistance (s m^{-1})	Leaf Area Index (Dimensionless)	Effective Growth (m)	Root Depth (dm)
Cultivated area	20.0	70.0	0.6	1.1	2.0
Urbanized area	16.4	70.0	0.01	0.0	0.0
Cerrado biome	14.2	70.0	0.8	20.0	12.0
Water bodies	4.0	70.0	0.0	0.0	0.0
Forest	15.0	70.0	0.9	30.0	30.0
Bare land	20.0	70.0	0.0	0.0	0.0
Reference(s)	[95,96]	[97]	[98]	[99]	[97]

Table A3. Soil parameters used in the hydrologic model.

Soil Type	Depth (cm)	Minimum Permeability Coefficient (mm d^{-1})	Air Capacity (mm)	Field Capacity (mm)
Red-yellow argisol	170	1	40	600
Haplic cambisols	230	1	37	1150
Red-yellow latossols	250	1	38	1500
Tholic Litholic	50	1	13	125
Reference	Hydrus 1D software (https://www.pc-progress.com)			

Table A4. Lithologic parameters used in the hydrologic model.

Lithologic Type	Maximum Storage Capacity in the Upper Aquifer (mm)	Maximum Storage Capacity in the Lower Aquifer (mm)	Storage Coefficient in the Upper Groundwater Reservoir (d)	Storage Coefficient in the Lower Groundwater Reservoir (d)
Orthogneiss	50	900	13	365
Clastic sediments	50	800	16	365
Limestone	70	1000	17	365
Silstone	60	900	14	365
Reference	[100]			

Appendix C

Table A5. Intervals and values of JAMS J2000 module parameters obtained after calibration.

Module	Parameters	Interval	Unit	Calibrated Value
Start up	mFCa	0–5	–	4.99
	mACa	0–5	–	4.98
	initRG1	0–1	–	0.40
	initRG2	0–1	–	0.72
Interception	α, rain	0–10	mm	5.80
Water in soil	soiMaxDPS	0–10	mm	3.49
	soilPolRed	0–10	–	6.78
	soilLinRed	0–10	–	1.57
	soilMaxInf1	0–200	mm	129.97
	soilMaxInf2	1–200	mm	75.99
	soilImpGT80	0–1	–	0.07
	soilImpLT80	1–1	–	0.31
	soilDistMPSLPS	0–10	–	0.13
	soilDiffMPSLPS	0–10	–	0.34
	soilOutLPS	0–10	–	2.27
	soilLatVertLPS	0–10	–	0.70
	soilMaxPerc	0–20	mm	5.10
	soilConcRD1	0–10	–	1.49
	soilConcRD2	1–10	–	9.99

Table A5. *Cont.*

Module	Parameters	Interval	Unit	Calibrated Value
Groundwater	gwRG1RG2dist	0–1	–	0.31
	gwRG1fact	0–10	–	3.40
	gwRG2fact	0–10	–	1.27
	gwCapRise	0–1	–	0.41
Routing	flowRouteTA	0–100	h	46.80

Appendix D

Table A6. Reference values of PBIAS and NSE and their relation to hydrologic model performance. The evaluation of performance was done separately for the calibration period (2003 to 2011), validation period (2012 to 2016), and full data period (2003 to 2016), allowing us to verify the replication of parameters.

PBIAS (%)	NSE	Performance
0 a 10	0.75 a 1	Very good
10 a 15	0.65 a 0.75	Good
15 a 25	0.50 a 0.65	Fair
>25	<0.50	Inadequate

References

1. Kazakis, N.; Chalikakis, K.; Mazzilli, N.; Ollivier, C.; Manakos, A.; Voudouris, K. Management and research strategies of karst aquifers in Greece: Literature overview and exemplification based on hydrodynamic modelling and vulnerability assessment of a strategic karst aquifer. *Sci. Total Environ.* **2018**, *643*, 592–609. [CrossRef] [PubMed]
2. Fleury, P.; Ladouche, B.; Conroux, Y.; Jourde, H.; Dörfliger, N. Modelling the hydrologic functions of a karst aquifer under active water management—The Lez spring. *J. Hydrol.* **2009**, *365*, 235–243. [CrossRef]
3. Ladouche, B.; Marechal, J.-C.; Dorfliger, N. Semi-distributed lumped model of a karst system under active management. *J. Hydrol.* **2014**, *509*, 215–230. [CrossRef]
4. Kemmerly, P.R. A time-distribution study of doline collapse: Framework for prediction. *Environ. Geol.* **1980**, *3*, 123–130. [CrossRef]
5. Tihansky, A.B. Sinkholes, west-central Florida. *Land Subsid. USA* **1999**, *1182*, 121–140.
6. Kaufmann, O.; Quinif, Y. Geohazard map of cover-collapse sinkholes in the 'Tournaisis' area, southern Belgium. *Eng. Geol.* **2002**, *65*, 117–124. [CrossRef]
7. García-Moreno, I.; Mateos, R.M. Sinkholes related to discontinuous pumping: Susceptibility mapping based on geophysical studies. The case of Crestatx (Majorca, Spain). *Environ. Earth Sci.* **2011**, *64*, 523–537. [CrossRef]
8. Doğan, U.; Yılmaz, M. Natural and induced sinkholes of the Obruk Plateau and Karapınar-Hotamış Plain, Turkey. *J. Asian Earth Sci.* **2011**, *40*, 496–508. [CrossRef]
9. Bonneau, J.; Fletcher, T.D.; Costelloe, J.F.; Burns, M.J. Stormwater infiltration and the 'urban karst'—A review. *J. Hydrol.* **2017**, *552*, 141–150. [CrossRef]
10. Locatelli, L.; Mark, O.; Mikkelsen, P.S.; Arnbjerg-Nielsen, K.; Deletic, A.; Roldin, M.; Binning, P.J. Hydrologic impact of urbanization with extensive stormwater infiltration. *J. Hydrol.* **2017**, *544*, 524–537. [CrossRef]
11. Kalhor, K.; Ghasemizadeh, R.; Rajic, L.; Alshawabkeh, A. Assessment of groundwater quality and remediation in karst aquifers: A review. *Groundw. Sustain. Dev.* **2019**, *8*, 104–121. [CrossRef]
12. Ghasemizadeh, R.; Hellweger, F.; Butscher, C.; Padilla, I.; Vesper, D.; Field, M.; Alshawabkeh, A. Review: Groundwater flow and transport modeling of karst aquifers, with particular reference to the North Coast Limestone aquifer system of Puerto Rico. *Hydrogeol. J.* **2012**, *20*, 1441–1461. [CrossRef]
13. Groves, C.G.; Howard, A.D. Early development of karst systems: 1. Preferential flow path enlargement under laminar flow. *Water Resour. Res.* **1994**, *30*, 2837–2846. [CrossRef]

14. Malard, A.; Jeannin, P.-Y.; Vouillamoz, J.; Weber, E. An integrated approach for catchment delineation and conduit-network modeling in karst aquifers: Application to a site in the Swiss tabular Jura. *Hydrogeol. J.* **2015**, *23*, 1341–1357. [CrossRef]

15. Perrin, J.; Luetscher, M. Inference of the structure of karst conduits using quantitative tracer tests and geological information: Example of the Swiss Jura. *Hydrogeol. J.* **2008**, *16*, 951–967. [CrossRef]

16. White, E.L.; Aron, G.; White, W.B. The influence of urbanization of sinkhole development in central Pennsylvania. *Environ. Geol. Water Sci.* **1986**, *8*, 91–97. [CrossRef]

17. Furlong, C.; Jegatheesan, J.; Currell, M.; Iyer-Raniga, U.; Khan, T.; Ball, A.S. Is the global public willing to drink recycled water? A review for researchers and practitioners. *Util. Policy* **2019**, *56*, 53–61. [CrossRef]

18. Ross, A. Speeding the transition towards integrated groundwater and surface water management in Australia. *J. Hydrol.* **2018**, *567*, e1–e10. [CrossRef]

19. Soares, S.; Terêncio, D.; Fernandes, L.; Machado, J.; Pacheco, F. The Potential of Small Dams for Conjunctive Water Management in Rural Municipalities. *Int. J. Environ. Res. Public Health* **2019**, *16*, 1239. [CrossRef]

20. Křeček, J.; Haigh, M. Land use policy in headwater catchments. *Land Use Policy* **2019**, *80*, 410–414. [CrossRef]

21. Santos, R.M.B.; Sanches Fernandes, L.F.; Moura, J.P.; Pereira, M.G.; Pacheco, F.A.L. The impact of climate change, human interference, scale and modeling uncertainties on the estimation of aquifer properties and river flow components. *J. Hydrol.* **2014**, *519*, 1297–1314. [CrossRef]

22. Santos, R.M.B.; Sanches Fernandes, L.F.; Pereira, M.G.; Cortes, R.M.V.; Pacheco, F.A.L. Water resources planning for a river basin with recurrent wildfires. *Sci. Total Environ.* **2015**, *526*, 1–13. [CrossRef]

23. Beven, K.J.; Kirkby, M.J. A physically based, variable contributing area model of basin hydrology/Un modèle à base physique de zone d'appel variable de l'hydrologie du bassin versant. *Hydrol. Sci. Bull.* **1979**, *24*, 43–69. [CrossRef]

24. Abbott, M.B.; Bathurst, J.C.; Cunge, J.A.; O'Connell, P.E.; Rasmussen, J. An introduction to the European Hydrological System—Systeme Hydrologique Europeen, "SHE", 1: History and philosophy of a physically-based, distributed modelling system. *J. Hydrol.* **1986**, *87*, 45–59. [CrossRef]

25. Arnold, J.G.; Srinivasan, R.; Muttiah, R.S.; Williams, J.R. Large area hydrologic modeling and assessment part I: Model development. *J. Am. Water Resour. Assoc.* **1998**, *34*, 73–89. [CrossRef]

26. Coe, M.T.; Costa, M.H.; Howard, E.A. Simulating the surface waters of the Amazon River basin: Impacts of new river geomorphic and flow parameterizations. *Hydrol. Process.* **2008**, *22*, 2542–2553. [CrossRef]

27. Santhi, C.; Kannan, N.; Arnold, J.G.; Di Luzio, M. Spatial Calibration and Temporal Validation of Flow for Regional Scale Hydrologic Modeling. *JAWRA J. Am. Water Resour. Assoc.* **2008**, *44*, 829–846. [CrossRef]

28. Grayson, R.B.; Moore, I.D.; McMahon, T.A. Physically based hydrologic modeling: 1. A terrain-based model for investigative purposes. *Water Resour. Res.* **1992**, *28*, 2639–2658. [CrossRef]

29. Pessoa, P.F.P. Caracterização Hidrogeológica da Região Cárstica de Sete Lagoas-MG: Potencialidades e Riscos. MSc Thesis, São Paulo University, São Paulo, Brazil, 1996.

30. Botelho, L.A.L.A. Gestão dos Recursos Hídricos em Sete Lagoas/MG: Uma Abordagem a Partir da Evolução Espaçotemporal da Demanda e da Captação de Agua. MSc Thesis, Federal University of Minas Gerais, Belo Horizonte, Brazil, 2008.

31. Galvão, P.; Halihan, T.; Hirata, R. Evaluating karst geotechnical risk in the urbanized area of Sete Lagoas, Minas Gerais, Brazil. *Hydrogeol. J.* **2015**, *23*, 1499–1513. [CrossRef]

32. Iglesias, M.; Uhlein, A. Estratigrafia do Grupo Bambuí e coberturas fanerozóicas no vale do rio São Francisco, norte de Minas Gerais. *Rev. Bras. Geociências* **2009**, *39*, 256–266. [CrossRef]

33. CPRM/CODEMIG—Companhia de Pesquisa de Recursos Minerais/Companhia de Desenvolvimento Econômico de Minas Gerais. Mapa Geológico do Estado de Minas Gerais. Available online: www.portaldageologia.com.br (accessed on 1 October 2018).

34. Universidade Federal De Lavras; Fundação Estadual Do Meio Ambiente De Minas Gerais. Mapa de Solos do Estado de Minas Gerais. Available online: http://www.dps.ufv.br/?page_id=742 (accessed on 1 November 2018).

35. FBDS—Fundação Brasileira para o Desenvolvimento Sustentável. Mapeamento em Alta Resolução dos Biomas Brasileiros. Available online: http://geo.fbds.org.br/ (accessed on 1 November 2018).

36. Batista, R.C.R. Caracterização Hidrogeológica do Entorno do Centro Nacional de Pesquisa de Milho e Sorgo (CNPMS), em Sete Lagoas, MG. MSc Thesis, Universidade Federal de Minas Gerais, Belo Horizonte, Brazil, 2009.

37. Da Costa, A.M.; de Salis, H.H.C.; Viana, J.H.M.; Leal Pacheco, F.A. Groundwater Recharge Potential for Sustainable Water Use in Urban Areas of the Jequitiba River Basin, Brazil. *Sustainability* **2019**, *11*, 2955. [CrossRef]

38. IBGE Sistema IBGE de Recuperação Automática (SIDRA). Available online: http://www.sidra.ibge.gov.br/bda/orcfam/default.asp?t=2&z=t&o=23&u1=1&u2=1&u3=1&u4=1&u5=1&u6=1 (accessed on 1 November 2018).

39. UAF-NASA Alaska Satellite Facility—Making Remote-Sensing Data Accessible Since 1991. Available online: https://www.asf.alaska.edu/ (accessed on 1 November 2018).

40. The USGS Earth Explorer (USGS). USGS EarthExplorer. Available online: http://earthexplorer.usgs.gov/ (accessed on 28 November 2018).

41. Instituto Nacional de Meteorologia, Ministério da Agricultura, Dados Meteorológicos. Available online: http://www.inmet.gov.br/portal/ (accessed on 1 November 2018).

42. Agência Nacional de Águas. HidroWEB: Acervo de Dados Hidrológicos. Available online: http://www.snirh.gov.br/hidroweb/publico/apresentacao.jsf (accessed on 1 November 2018).

43. Pacheco, F.A.L.; Van der Weijden, C.H. Weathering of plagioclase across variable flow and solute transport regimes. *J. Hydrol.* **2012**, *420–421*, 46–58. [CrossRef]

44. Pacheco, F.A.L.; Szocs, T. "Dedolomitization reactions" driven by anthropogenic activity on loessy sediments, SW Hungary. *Appl. Geochem.* **2006**, *21*, 614–631. [CrossRef]

45. Fonseca, A.R.; Sanches Fernandes, L.F.; Fontainhas-Fernandes, A.; Monteiro, S.M.; Pacheco, F.A.L. The impact of freshwater metal concentrations on the severity of histopathological changes in fish gills: A statistical perspective. *Sci. Total Environ.* **2017**, *599–600*, 217–226. [CrossRef]

46. Fonseca, A.R.; Sanches Fernandes, L.F.; Fontainhas-Fernandes, A.; Monteiro, S.M.; Pacheco, F.A.L. From catchment to fish: Impact of anthropogenic pressures on gill histopathology. *Sci. Total Environ.* **2016**, *550*, 972–986. [CrossRef]

47. Valle Junior, R.F.; Varandas, S.G.P.; Sanches Fernandes, L.F.; Pacheco, F.A.L. Multi Criteria Analysis for the monitoring of aquifer vulnerability: A scientific tool in environmental policy. *Environ. Sci. Policy* **2015**, *48*, 250–264. [CrossRef]

48. Ferreira, A.R.L.; Sanches Fernandes, L.F.; Cortes, R.M.V.; Pacheco, F.A.L. Assessing anthropogenic impacts on riverine ecosystems using nested partial least squares regression. *Sci. Total Environ.* **2017**, *583*, 466–477. [CrossRef]

49. Santos, R.M.B.; Sanches Fernandes, L.F.; Cortes, R.M.V.; Varandas, S.G.P.; Jesus, J.J.B.; Pacheco, F.A.L. Integrative assessment of river damming impacts on aquatic fauna in a Portuguese reservoir. *Sci. Total Environ.* **2017**, *601–602*, 1108–1118. [CrossRef]

50. Sanches Fernandes, L.F.; Fernandes, A.C.P.; Ferreira, A.R.L.; Cortes, R.M.V.; Pacheco, F.A.L. A partial least squares—Path modeling analysis for the understanding of biodiversity loss in rural and urban watersheds in Portugal. *Sci. Total Environ.* **2018**, *626*, 1069–1085. [CrossRef]

51. Álvarez, X.; Valero, E.; Santos, R.M.B.; Varandas, S.G.P.; Sanches Fernandes, L.F.; Pacheco, F.A.L. Anthropogenic nutrients and eutrophication in multiple land use watersheds: Best management practices and policies for the protection of water resources. *Land Use Policy* **2017**, *69*, 1–11. [CrossRef]

52. Pacheco, F.A.L. Regional groundwater flow in hard rocks. *Sci. Total Environ.* **2015**, *506–507*, 182–195. [CrossRef]

53. Fernandes, L.F.S.; Marques, M.J.; Oliveira, P.C.; Moura, J.P. Decision support systems in water resources in the demarcated region of Douro—Case study in Pinhão river basin, Portugal. *Water Environ. J.* **2014**, *28*, 350–357. [CrossRef]

54. Fernandes, L.F.S.; dos Santos, C.M.M.; Pereira, A.P.; Moura, J.P. Model of management and decision support systems in the distribution of water for consumption. *Eur. J. Environ. Civ. Eng.* **2011**, *15*, 411–426. [CrossRef]

55. Pacheco, F.A.L. Application of Correspondence Analysis in the Assessment of Groundwater Chemistry. *Math. Geol.* **1998**, *30*, 129–161. [CrossRef]

56. Do Valle Júnior, R.F.; Siqueira, H.E.; Valera, C.A.; Oliveira, C.F.; Sanches Fernandes, L.F.; Moura, J.P.; Pacheco, F.A.L. Diagnosis of degraded pastures using an improved NDVI-based remote sensing approach: An application to the Environmental Protection Area of Uberaba River Basin (Minas Gerais, Brazil). *Remote Sens. Appl. Soc. Environ.* **2019**, *14*, 20–33. [CrossRef]

57. Pacheco, F.A.L.; Landim, P.M.B. Two-Way Regionalized Classification of Multivariate Datasets and its Application to the Assessment of Hydrodynamic Dispersion. *Math. Geol.* **2005**, *37*, 393–417. [CrossRef]

58. Pacheco, F.A.L.; Sousa Oliveira, A.; Van Der Weijden, A.J.; Van Der Weijden, C.H. Weathering, biomass production and groundwater chemistry in an area of dominant anthropogenic influence, the Chaves-Vila Pouca de Aguiar region, north of Portugal. *Water Air Soil Pollut.* **1999**, *115*, 481–512. [CrossRef]

59. Pacheco, F.A.L. Finding the number of natural clusters in groundwater data sets using the concept of equivalence class. *Comput. Geosci.* **1998**, *24*, 7–15. [CrossRef]

60. Valera, C.A.; Pissarra, T.C.T.; Martins Filho, M.V.; Valle Junior, R.F.; Sanches Fernandes, L.F.; Pacheco, F.A.L. A legal framework with scientific basis for applying the 'polluter pays principle' to soil conservation in rural watersheds in Brazil. *Land Use Policy* **2017**, *66*, 61–71. [CrossRef]

61. Hughes, S.J.; Cabecinha, E.; Andrade dos Santos, J.C.; Mendes Andrade, C.M.; Mendes Lopes, D.M.; da Fonseca Trindade, H.M.; dos Santos Cabral, J.A.F.A.; dos Santos, M.G.S.; Lourenço, J.M.M.; Marques Aranha, J.T.; et al. A predictive modelling tool for assessing climate, land use and hydrological change on reservoir physicochemical and biological properties. *Area* **2012**, *44*, 432–442. [CrossRef]

62. Modesto Gonzalez Pereira, M.J.; Sanches Fernandes, L.F.; Barros Macário, E.M.; Gaspar, S.M.; Pinto, J.G. Climate Change Impacts in the Design of Drainage Systems: Case Study of Portugal. *J. Irrig. Drain. Eng.* **2015**, *141*, 05014009. [CrossRef]

63. Siqueira, H.E.; Pissarra, T.C.T.; do Valle Junior, R.F.; Fernandes, L.F.S.; Pacheco, F.A.L. A multi criteria analog model for assessing the vulnerability of rural catchments to road spills of hazardous substances. *Environ. Impact Assess. Rev.* **2017**, *64*, 26–36. [CrossRef]

64. Krause, P.; Kralisch, S. The hydrological modelling system J2000—knowledge core for JAMS. In Proceedings of the MODSIM 2005 International Congress on Modelling and Simulation, Melbourne, Australia, 12–15 December 2005.

65. Zander, F.; Kralisch, S. River Basin Information System: Open Environmental Data Management for Research and Decision Making. *ISPRS Int. J. Geo-Inf.* **2016**, *5*, 123. [CrossRef]

66. Pfennig, B.; Kipka, H.; Wolf, M.; Fink, M.; Krause, P.; Flügel, W.A. Development of an extended routing scheme in reference to consideration of multi-dimensional flow relations between hydrological model entities. In Proceedings of the 18th World IMACS Congress and MODSIM09 International Congress on Modelling and Simulation, Cairns, Australia, 13–17 July 2009.

67. Flügel, W.A. Hydrological response units (HRU's) as modelling entities for hydrological river basin simulation and their methodological potential for modelling complex environmental process systems. Results from the Sieg catchment. *Erde* **1996**, *127*, 43–62.

68. Srinivas, N.; Deb, K. Muiltiobjective Optimization Using Nondominated Sorting in Genetic Algorithms. *Evol. Comput.* **1994**, *2*, 221–248. [CrossRef]

69. Gupta, H.V.; Sorooshian, S.; Yapo, P.O. Status of Automatic Calibration for Hydrologic Models: Comparison with Multilevel Expert Calibration. *J. Hydrol. Eng.* **2002**, *4*, 135–143. [CrossRef]

70. Nash, J.E.; Sutcliffe, J.V. River flow forecasting through conceptual models part I—A discussion of principles. *J. Hydrol.* **1970**, *10*, 282–290. [CrossRef]

71. Harris, N.M.; Gurnell, A.M.; Hannah, D.M.; Petts, G.E. Classification of river regimes: A context for hydroecology. *Hydrol. Process.* **2000**, *14*, 2831–2848. [CrossRef]

72. Ribeiro, M.M.R. No Alternativas para Outorga e a Cobrança Pelo Uso da Água: Simulação de um Caso. Ph.D. Thesis, Instituto de Pesquisas Hidráulicas, Universidade Federal de Rio Grande do Sul, Porto Alegre, Brazil, 2000.

73. Da Silva, A.M.; de Oliveira, P.M.; de Mello, C.R.; Pierangeli, C. Vazões mínimas e de referência para outorga na região do Alto Rio Grande, Minas Gerais. *Rev. Bras. Eng. Agríc. Ambient.* **2006**, *10*, 374–380. [CrossRef]

74. Longhi, E.H.; Formiga, K.T.M. Metodologias para determinar vazão ecológica em rios. *Rev. Bras. Ciências Ambient.* **2011**, *20*, 33–48.

75. Vestena, L.R.; Oliveira, E.D.; Cunha, M.C.; Thomaz, E.L. Vazão ecológica e disponibilidade hídrica na bacia das Pedras, Guarapuava-PR. *Rev. Ambiente Água* **2012**, *7*, 212–227. [CrossRef]

76. Vogel, R.M.; Fennessey, N.M. Flow-Duration Curves. I: New Interpretation and Confidence Intervals. *J. Water Resour. Plan. Manag.* **2006**, *120*, 485–504. [CrossRef]

77. Vogel, R.M.; Fennessey, N.M. Flow Duration Curves Ii: A Review of Applications in Water Resources Planning. *JAWRA J. Am. Water Resour. Assoc.* **1995**, *31*, 1029–1039. [CrossRef]

78. Naghettini, M.; Pinto, E.J.D.A. *Hidrologia Estatística*; CPRM: Belo Horizonte, Brazil, 2007; ISBN 978-85-7499-023-1.

79. Monte, B.; Costa, D.; Chaves, M.; Magalhães, L.; Uvo, C. Hydrological and hydraulic modelling applied to the mapping of flood-prone areas. *Rev. Bras. Recur. Hídricos* **2016**, *21*, 152–167. [CrossRef]

80. Machado, A.R.; Mello Junior, A.V.; Wendland, E.C. Avaliação do modelo J2000/JAMS para modelagem hidrológica em bacias hidrográficas brasileiras. *Eng. Sanit. Ambient.* **2017**, *22*, 327–340. [CrossRef]

81. Pulido-Velazquez, M.; Jenkins, M.W.; Lund, J.R. Economic values for conjunctive use and water banking in southern California. *Water Resour. Res.* **2004**, *40*. [CrossRef]

82. Fitch, P.; Brodaric, B.; Stenson, M.; Booth, N. Integrated groundwater data management. In *Integrated Groundwater Management: Concepts, Approaches and Challenges*; Springer International Publishing: Cham, Switzerland, 2016; pp. 667–692. ISBN 9783319235769.

83. Bazargan-Lari, M.R.; Kerachian, R.; Mansoori, A. A conflict-resolution model for the conjunctive use of surface and groundwater resources that considers water-quality issues: A case study. *Environ. Manag.* **2009**, *43*, 470–482. [CrossRef]

84. Zhu, T.; Marques, G.F.; Lund, J.R. Hydroeconomic optimization of integrated water management and transfers under stochastic surface water supply. *Water Resour. Res.* **2015**, *51*, 3568–3587. [CrossRef]

85. Blomquist, W.; Heikkila, T.; Schlager, E. Institutions and Conjunctive Water Management among Three Western States. In *Economics of Water Resources*; Routledge: London, UK, 2018; pp. 241–271.

86. Volpi, E.; Fiori, A.; Grimaldi, S.; Lombardo, F.; Koutsoyiannis, D. One hundred years of return period: Strengths and limitations. *Water Resour. Res.* **2015**, *51*, 8570–8585. [CrossRef]

87. Volpi, E. On return period and probability of failure in hydrology. *Wiley Interdiscip. Rev. Water* **2019**, *6*, e1340. [CrossRef]

88. Galvão, P.; Hirata, R.; Conicelli, B. Estimating groundwater recharge using GIS-based distributed water balance model in an environmental protection area in the city of Sete Lagoas (MG), Brazil. *Environ. Earth Sci.* **2018**, *77*, 398. [CrossRef]

89. Galvão, P.; Hirata, R.; Halihan, T.; Terada, R. Recharge sources and hydrochemical evolution of an urban karst aquifer, Sete Lagoas, MG, Brazil. *Environ. Earth Sci.* **2017**, *76*, 159. [CrossRef]

90. Abdalla, O.A.E.; Al-Rawahi, A.S. Groundwater recharge dams in arid areas as tools for aquifer replenishment and mitigating seawater intrusion: Example of AlKhod, Oman. *Environ. Earth Sci.* **2013**, *69*, 1951–1962. [CrossRef]

91. Morsy, K.M.; Morsy, A.M.; Hassan, A.E. Groundwater sustainability: Opportunity out of threat. *Groundw. Sustain. Dev.* **2018**, *7*, 277–285. [CrossRef]

92. Jaafar, H.H. Feasibility of groundwater recharge dam projects in arid environments. *J. Hydrol.* **2014**, *512*, 16–26. [CrossRef]

93. Mahto, S.; Kushwaha, A.; Siva Subramanian, M.; Nikita, N.; Singh, T.B.N. Feasibility of Artificial Groundwater Recharge Structures for Urban and Rural Environment of Ranchi in India using Geospatial Technology. *Hydrospat. Anal.* **2018**, *2*, 28–42. [CrossRef]

94. Parimalarenganayaki, S.; Elango, L. Assessment of effect of recharge from a check dam as a method of Managed Aquifer Recharge by hydrogeological investigations. *Environ. Earth Sci.* **2015**, *73*, 5349–5361. [CrossRef]

95. Braga, C.C.; Amanajás, J.C.; Alcântara, C.R.; Dantas, M.P. Avaliação do albedo nos diferentes tipos de cobertura do cerrado do Amapá—Brasil com imagens MODIS. *Territorium* **2018**, *25*, 129–134. [CrossRef]

96. Giongo, P.R.; Vettorazzi, C.A. Albedo da superfície por meio de imagens TM-Landsat 5 e modelo numérico do terreno. *Rev. Bras. Eng. Agrícola Ambient.* **2014**, *18*, 833–838. [CrossRef]

97. Testa, G.; Gresta, F.; Cosentino, S.L. Dry matter and qualitative characteristics of alfalfa as affected by harvest times and soil water content. *Eur. J. Agron.* **2011**, *34*, 144–152. [CrossRef]

98. Ribeiro, E.P.; Nóbrega, R.S.; Mota Filho, F.O.; Moreira, E.B. Estimativa dos índices de vegetação na detecção de mudanças ambientais na bacia hidrográfica do rio Pajeú. *Geosul* **2016**, *31*, 59–92.

99. Scolforo, J.R.; Oliveira, A.D.; Mello, J.M.; Silva, C.P.C.; Ferraz-Filho, A.C.; Andrade, I.S.; Abreu, E.C.R. Análise da estrutura fitossociológica dos fragmentos inventariados e dos grupos fisionômicos. In *Inventário Florestal de Minas Gerais: Floresta Estacional Decidual—Florística, Estrutura, Diversidade, Similaridade, Distribuição Diamétrica e de Altura, Volumetria, Tendências de Crescimento e Áreas Aptas para Manejo Florestal*; UFLA: Lavras, Brazil, 2008; pp. 99–114.

100. Schwarze, C.P.; Wollmann, H.A.; Binder, G.; Ranke, M.B. Short-term increments of insulin-like growth factor I (IGF-I) and IGF-binding protein-3 predict the growth response to growth hormone (GH) therapy in GH-sensitive children. *Acta Paediatr.* **1999**, *88*, 200–208. [CrossRef]

 *sustainability*

Article

Groundwater Recharge Potential for Sustainable Water Use in Urban Areas of the Jequitiba River Basin, Brazil

Adriana Monteiro da Costa [1], **Hugo Henrique Cardoso de Salis** [1], **João Hebert Moreira Viana** [2] and **Fernando António Leal Pacheco** [3,*]

[1] Federal University of Minas Gerais, 6620 Antônio Carlos Ave., Pampulha, Belo Horizonte, MG 31270-901, Brazil; drimonteiroc@gmail.com (A.M.d.C.); hugohcsalis@gmail.com (H.H.C.d.S.)

[2] Brazilian Agricultural Research Corporation (Embrapa Maize and Sorghum), Sete Lagoas, MG 35701-97, Brazil; joao.herbert@embrapa.br

[3] CQVR—Chemistry Research Centre, University of Trás-os-Montes and Alto Douro, Quinta de Prados Ap. 1013, 5001-801 Vila Real, Portugal

* Correspondence: fpacheco@utad.pt

Received: 25 March 2019; Accepted: 20 May 2019; Published: 24 May 2019

Abstract: The zoning of groundwater recharge potential would be attractive for water managers, but is lacking in many regions around the planet, including in the Jequitiba River basin, Minas Gerais, Brazil. In this study, a physically based spatially distributed method to evaluate groundwater recharge potential at catchment scale was developed and tested in the aforementioned Jequitiba River basin. The data for the test was compiled from institutional sources and implemented in a Geographic Information System. It comprised meteorological, hydrometric, relief, land use, and soil data. The average results resembled the annual recharge calculated by a hydrograph method, which worked as validation method. The spatial variation of recharge highlighted the predominant contribution of flat areas, porous aquifers, and forested regions to groundwater recharge. They also exposed the negative effect of urbanization. In combination, these factors elected the following sectors of the Jequitiba River basin as regions of high recharge potential: the south-southeast part of the headwaters in Prudente de Morais; Sete Lagoas towards the central part of the basin; and the region between Funilândia and Jequitiba, near the Jequitiba river mouth. Some management practices were suggested to improve groundwater recharge. The map of groundwater recharge potential produced in this study is valuable and is therefore proposed as tool for planners in the sustainable use of groundwater and protection of recharge areas.

Keywords: groundwater recharge; recharge zones; river basin; spatialization; relief; geology; forest; urbanization; water resource management; land use policy

1. Introduction

Water is a naturally circulating resource that is constantly recharged [1]. However, by 2025 it is estimated that around 5 billion people, out of a total population of around 8 billion, will be living in countries experiencing water stress [2]. Climate change has the potential to impose additional pressures in some regions. It is therefore urgent to chart our water future [3]. A river basin is the natural boundary where the relief directs the water to a common point known as river mouth. Incorrect management of river basins can seriously affect water availability, damaging both surface water and groundwater. One can refer, for example, the negative effect of pavement or other civil constructions on water infiltration in the soil [4,5]. In addition, other problems, such as intensification of erosive processes, can occur depending on the intensity level of soil waterproofing and compaction [6–10].

The percolation of water through the porous spaces of the soil and rocks is an important process to groundwater recharge, directly affecting the maintenance of various human activities, such as the water supply to urban centers, industrial parks, and agricultural activities [11–16]. The water stock in underground systems is connected to surface soil use conditions. Significant changes in the landscape can alter water regimes and groundwater quality [17]. Information on the relationships between surface water and groundwater allows a better water use through data that assists on the sustainable management of water resources [18–22]. Groundwater recharge zones are places where the ground surface allows water infiltration and percolation through the soil [23–27]. Water can be retained in the soil, reach the vadose zone or arrive at a geological system that can store and distribute it [15,28–31]. Recharge can be favored through storage in small dams or rainwater harvesting systems [32–37], which also aid the prevention of floods [38].

Despite the importance of recharge zones, no public policies are in force in Minas Gerais State (Brazil) to promote their sustainable use. The starting step comprises the identification and mapping of recharge areas, but this phase has not commenced. It is worth reinforcing that the inadequate use of recharge zones can reduce water infiltration in the soil through waterproofing and compaction; and low infiltration rates can concentrate and increase surface runoff, generating problems in urban centers such as flooding [39–43]. In addition, the intense exploitation of water resources in urban areas can generate water scarcity and even depletion [44–46]. At catchment scale, the inadequate managing of water sources and recharge areas can reduce water availability and increase the vulnerability of surface and groundwater to contamination [46–57]. It is therefore time for policy makers and water planners to start a robust assessment of groundwater recharge potential. Robustness relies on a thorough collection of information on physical-environmental factors, such as soil characteristics, geology, vegetation cover, and relief. Information on these issues allows the use of water volumes according to the natural capacity of the system [16].

Direct and indirect methods can be used to assess groundwater characteristics including recharge. Direct methods comprise geological and geophysical explorations, gravimetric and magnetic methods, and drilling tests. Indirect methods include hydrological modeling [58–66] using geographic information systems (GIS) combined with fieldwork, geochemical tracers [67–69], survey of specialized literature for standard values [15,16,70,71], among others. The choice for a method (direct or indirect) should consider the precision level needed, the project execution, and the resources available. Therefore, appropriate specific methodological approaches should be applied to each local condition that the river basin encompasses.

Significant amounts of groundwater are withdrawn from the Jequitiba river basin, because this catchment is located in a populous region of Minas Gerais State, which hosts a large population and several industries from different segments [72]. According to Pessoa [73], in 1993, the water for domestic use in the largest town in the basin (Sete Lagoas, representing 94.3% of the entire population) was supplied by groundwater resources, namely 65 drilled wells with an average yield of 8.0 L s^{-1} (520 L s^{-1} of total yield). In those days, the population of Sete Lagoas was nearly 150,000 and consumed approximately 200 L habitant^{-1} day^{-1}. Thus, the pressure over the drilled wells was evaluated in 16 h of pumping every day and considered preoccupying. Moreover, the quality of these resources was threatened because the domestic sewage system was lacking. The situation of Sete Lagoas was re-evaluated in 2008 by Botelho [74], with similar conclusions. 25 years after the evaluation of Pessoa, the number of drilled wells raised from 65 to 94 (44% increase), keeping a similar average yield (7.8 L s^{-1}), while the population of Sete Lagoas raised from 150,000 to 220,000 (47% increase). The sewage system was still lacking or incomplete. In 2014, Galvão et al. [75] evaluated the effects of pumping in the geometry of hydraulic heads within the area of Sete Lagoas where the number of drilled wells and pumping rates are larger. A hydraulically depressed area was delineated around the older wells (1942) where depths to the water have ranged from 14 m post drilling to 62 m in 2012 (48 m drawdown in 70 years). According to age versus drawdown data available in the study of Galvão, it is possible to estimate an average of 0.9 m year^{-1} of drawdown within the depressed area, caused by excessive pumping. The study of Galvão also suggested the link of this hydraulic head depression to the development of suffosional sinkholes. Moreover, the results of Botelho's interviews to SAAE (Autonomous Service of Water and Sewage) employees refer that " … as the urban space expands, the municipal authorities do not improve the supply and distribution systems, and do not prioritize studies and planning for the occupation of space … ". It is therefore urgent to help the municipality accomplishing the task, through provision of relevant information and data for planning such as the spatial distribution of groundwater recharge potential.

The general goal of this study was to estimate groundwater recharge potential within the Jequitiba River basin, to identify and delineate preferential areas for restoration, recovery, and protection. To accomplish this purpose, the following specific objectives had to be accomplished: (a) develop a physically based spatially distributed model to estimate groundwater recharge potential at catchment scale; (b) compile a diversified set of geospatial information, including meteorological, hydrometric, geologic, relief, land use, and soil data; (c) run the model using the compiled data and produce the final groundwater recharge potential map; (d) validate the model through calculation of groundwater recharge using an independent approach based on independent data, namely stream flow recession analysis.

2. Materials and Methods

2.1. Study Area

The Jequitiba River basin (JRB) crosses the municipalities of Sete Lagoas, Prudente de Morais, Funilândia, Jequitiba, and Capim Branco, in the state of Minas Gerais (MG), Brazil, and covers approximately 57,148 hectares (Figure 1a). The JRB is managed by the Water Resources Planning and Management Unit (UPGRH) Rio das Velhas (SF5). The main highways that cross the JRB are the BR040, MG024, and MG238. According to the 2014 Water Resources Plan for the Rio das Velhas River Basin, the Jequitiba River Strategic Territory Unit has a population of approximately 145,729 inhabitants, mostly concentrated in urban areas since 97.6% of them live in urban areas and 2.4% in rural areas. The largest municipality in the basin is Sete Lagoas, which accounts for 94.3% of the total population.

The climate, according to Koppen's classification, is the subtropical (Cwa), characterized by dry winter and hot summer. The rainfall regime of the area in 2000 to 2016 presented monthly precipitation of 3 mm to 319 mm, with a rainy season from October to March, and a dry season from April to September. The mean annual rainfall in the same period was 1291.2 mm, while the mean temperatures varied from 18 °C in July and 24 °C in January–February, with a mean value of 21.8 °C.

The soil classes found in the JRB (Figure 1b), according to the Brazilian Soil Classification System, are Latossolos (49.6%), Cambissolos (36.7%), Argissolos (1.3%), and Neossolos (12.2%) [76].

Considering the land use and cover map developed by the reference [77], almost 39,000 ha (68%) of the JRB is used for anthropogenic activities, comprising livestock pasturing or agriculture, which are distributed throughout the drainage basin domain (Figure 1c). Forest is the second most frequent class of land use in the JRB, with 15.8% of the area, followed by the urban area class, with 9.3% of the JRB area.

The geology is characterized by a stratigraphic sequence comprising an Archean crystalline basement made of orthogneisses, granites and migmatites, overlaid by Neoproterozoic carbonate rocks of Bambuí Group, namely calcite and dolomite limestones from Sete Lagoas Formation and pelitic rocks with interlayered carbonates from the Serra de Santa Helena Formation. Along the main water courses, the Neoproterozoic sequences are overlaid by terrigenous rocks composed of alluvium and colluvium sediments [78]. The spatial distribution of outcrops is illustrated in Figure 1d, indicating a predominance of pelitic rocks (57.6%), followed by limestones (20%), Archean basement (10%), colluvium (10%), and alluvium (8%) [79]. When the geologic map is compared with the soil map, a spatial association between lithotypes and soil types becomes apparent: the cambisols developed from the Archean rocks as well as from pelitic rocks cropping out in the catchment lowlands; the neosols developed from pelitic rocks cropping out in the catchment highlands; the latosols developed from limestones and terrigenous rocks.

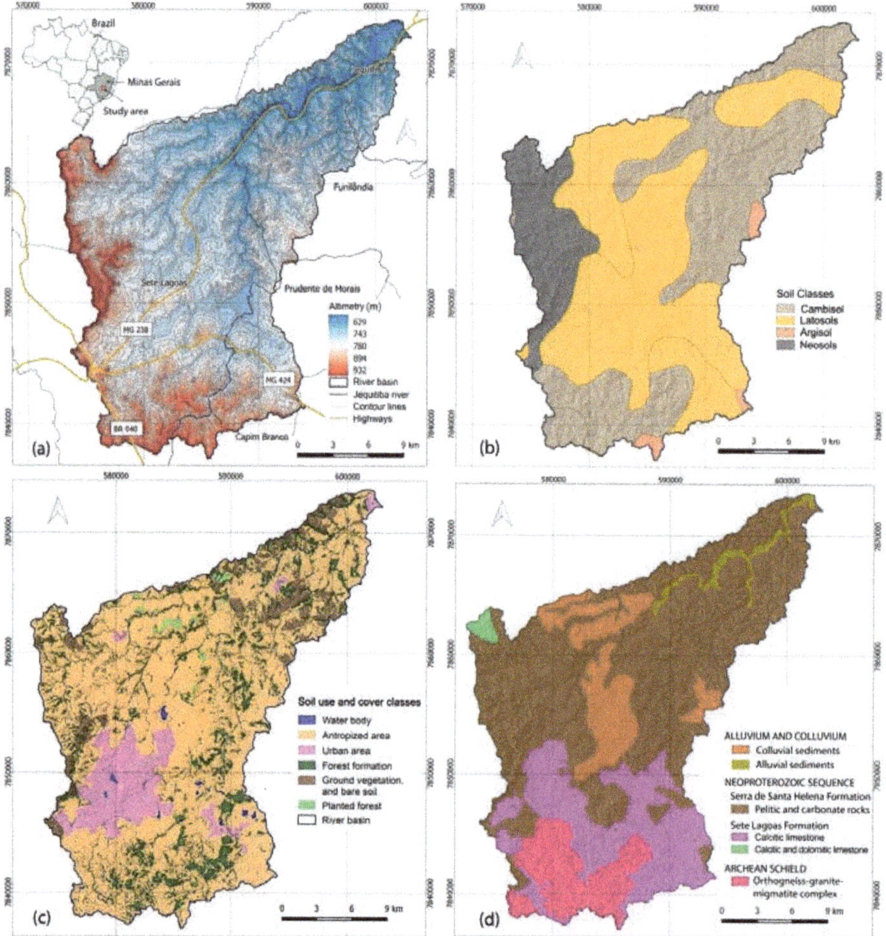

Figure 1. (**a**) Location of study area: Jequitiba River basin, Minas Gerais, Brazil; (**b**) soil map of Jequitiba River basin. (**c**) Land use and cover map of Jequitiba River basin; (**d**) lithologic map of Jequitiba River basin. The geographic reference for the maps is the Universal Transverse Mercator (UTM) projection system, Geocentric Reference System for America (SIRGAS) 2000 datum, 23 south time zone.

The hydrogeology is characterized by fractured aquifers composed of Archean rocks, fractured-karst aquifers composed of Serra de Santa Helena pelitic rocks interlayered with carbonates, karst aquifers composed of Sete Lagoas limestones, and porous aquifers composed of terrigenous rocks and the soil layer that can be thick [80]. Specific flows are very low in the fractured aquifers (average: 0.52 m^3 ha^{-1} m^{-1}), low in the fracture-karst aquifers (average: 20.84 m^3 ha^{-1} m^{-1}), and high in the karst aquifers (can reach 264 m^3 ha^{-1} m^{-1}) (http://www.cprm.gov.br). Given the thickness of soil and saprolite layers in large portions of the basin, aquifer recharge is largely conditioned by infiltration and storage in the soil. Latosols favor infiltration because soil particles in this soil type are mostly arranged as micro aggregates with high hydraulic conductivity. In the cambisols, water does not infiltrate easily, because soil particles are sometimes cemented or compacted and disposed in laminar layers [80].

2.2. Materials and Software

The materials used in this study are indicated in Table 1 and comprised: (a) a Digital Elevation Model (DEM) ALOS PALSAR with a spatial resolution of 12.5 meters [81]; (b) a land use and cover map at scale 1:25,000 [77]; (c) the soil map of Minas Gerais state at scale of 1:650,000 and corresponding data on total porosity and hydraulic conductivity [76]; (d) precipitation and real evapotranspiration data from weather stations in the municipalities of Belo Horizonte (BH), Sete Lagoas (SL), Conceição do Mato Dentro (CMD), and Florestal (FLT) [82]; (e) hydrometric data of station 41410000, in the mouth of Jequitiba River [83]; (f) the geological map of Minas Gerais state at scale 1:1,000,000 [79]; (g) data from the Rural Environmental Registry (CAR) of the municipalities in the JRB [84]; (h) Population data relative to the studied area [72]; (i) Quantum Geographic Information System (QGIS) program, version 2.18.4 [85].

Table 1. Materials (spatial data, climatic records, additional information), uses in the recharge potential evaluation model, and URLs of websites used for downloading the data.

Data Type	Use in the Recharge Evaluation Model	URL of Website
Digital elevation model	Calculation of slope length and steepness factor (Equation (1))	https://www.asf.alaska.edu
Land use and cover map	Calculation of runoff parameter (Equation (1))	http://geo.fbds.org.br
Soil map and associated porosity and hydraulic conductivity data	Calculation of percolation factor (Equation (2))	http://www.dps.ufv.br
Rainfall and evapotranspiration data	Calculation of recharge potential (Equation (3))	http://www.inmet.gov.br
Stream flow data	Validation of recharge potential using na independent method (recession analysis)	http://www.snirh.gov.br/hidroweb
Geologic map	Information for discussion	www.portaldageologia.com.br
Administrative data	Additional information	http://www.car.gov.br
Population data	Additional information	http://www.sidra.ibge.gov.br

2.3. Methods

The method proposed in this study to evaluate groundwater recharge is spatially distributed and based on water balance. The general workflow is illustrated in Figure 2 and was developed in five main stages: (i) acquisition of topographic, land use and soil maps as well as compilation of climatic records (precipitation and real evapotranspiration); (ii) calculation of the surface runoff factor (RF), based on evaluation of hillside lengths and slopes (LS factor) as well as on runoff coefficients (C); (iii) calculation of water percolation factor (PF), based on soil characteristics (total porosity—n; hydraulic conductivity—Ks); (iv) calculation of groundwater recharge in each point of a catchment and an average value for the entire catchment, using a geographic information system; (v) validation of results, through comparison of previously calculated average recharge with a counter value estimated by an independent method (e.g., the hydrograph recession method based on stream flow analysis). The five stages are described in detail in the following paragraphs.

In the first stage, the annual averages for precipitation and real evapotranspiration in the hydrological stations of municipalities near the study area (SL, BH, CMD, and FLT) were estimated using data from 2000 to 2018. Longer time series would be more adequate for recharge estimation in a changing climate, but they were not available. The information was interpolated by calculating the inverse distance weighting (IDW) raised to the power of two. This interpolator was used because it reduces the influence of the values recorded in the farthest stations from the JRB, with higher weights for the values of the nearest stations [86]. Subsequently, the data was spatialized and cut to the limits of the study area.

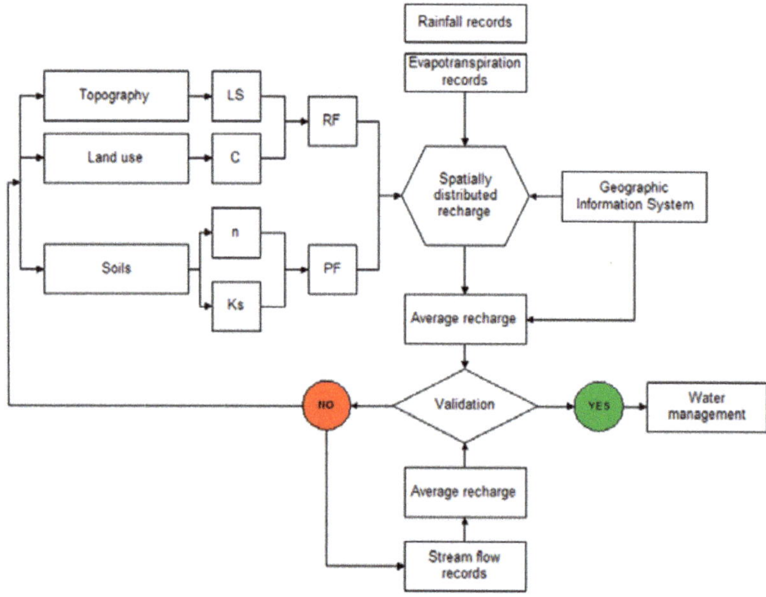

Figure 2. Workflow for the estimation of groundwater recharge potential.

In the second stage, the land use and cover map and topographic information derived from the digital elevation model (slope length and steepness factor, LS-factor) were used to calculate a surface runoff factor, based on the method proposed by Böhner and Selige [87]. The surface runoff factor was calculated according to Equation (1):

$$RF = 1 - (C + LS_{Fuzzy}) \tag{1}$$

where RF is the surface runoff factor (dimensionless), C is the coefficient of surface runoff (dimensionless; adopted values in Table 2), and LS_{Fuzzy} is the slope length and steepness factor estimated using the method of Desmet and Govers [88]. The LS_{Fuzzy} parameter was subsequently adjusted to a range of 0 to 1 by a fuzzy logic algorithm; the closer to 1, the steeper the slope is in the landscape.

Table 2. Coefficient of surface runoff adopted for each soil use and cover class in the Jequitiba River basin, Minas Gerais (MG), Brazil (Adapted from [89]).

Soil Use and Cover Classes	C
Anthropized area	0.5
Urban area	0.85
Forest formation	0.1
Planted forest	0.13
Ground vegetation, and bare soil	0.6

In the third stage, the soil map and information on soil total porosity (n) and hydraulic conductivity (Ks) were used to calculate a percolation factor (PF) for each soil class in the tested basin. The values of Ks and n were mostly compiled from field studies published by Pedron [90,91]. The values used in the modeling are depicted in Table 3. The percolation factor was evaluated through Equation (2):

$$PF = n \times Ks_{Fuzzy} \tag{2}$$

where PF is the water percolation factor (dimensionless), and Ks_{Fuzzy} (dimensionless) is the soil hydraulic conductivity adjusted to a range of 0 to 1 through fuzzy logic; the closer to 1, the greater the soil hydraulic conductivity in the specific class.

Table 3. Total porosity and hydraulic conductivity values adopted for each soil type in the Jequitiba River basin, MG, Brazil (main source: field studies of Pedron [90,91]).

Soil Type	Ks (mm/h)	Total Porosity
Cambisol	16	0.53
Latosol	52	0.69
Argisol	19	0.44
Neosol	32	0.43

In the fourth stage, values were calculated for the groundwater recharge potential in each point of the studied catchment, using Equation (3):

$$RPot = [(P - ETr) \times RF \times PF] \times 10 \tag{3}$$

where $RPot$ is the groundwater recharge potential (m^3 ha^{-1} year^{-1}), P is the annual average precipitation (mm year^{-1}), ETr is the average real evapotranspiration (mm year^{-1}), RF is the runoff factor; and PF is the percolation factor.

The results were validated in the fifth stage, by comparing the total recharge volume of the basin with the total estimated recharge volume by the recession curve analysis method, using the Jequitiba River hydrograph. Data from 2010 to 2012 was used because it was the most recent period in which water flow records had the greatest continuity in the historical series without failures or missing data. These data were used to calculate the Maillet equation (Equation (4)) [65,92]:

$$Q_t = Q_0 e^{-at} \tag{4}$$

wherein Q_t is flow at time t (m^3 s^{-1}), Q_0 is the flow at the beginning of the recession (m^3 s^{-1}), α is the coefficient of recession, t is the time (days) from the beginning to the end of the recession, and e is the basis of the Neperian logarithm (2.71828).

Thus, the coefficient of recession can be determined numerically, based on the logarithmic form expressed in Equation (5):

$$\alpha = \frac{LogQ_0 - LogQ_t}{0.4343t} \tag{5}$$

Subsequently, the groundwater recharge volume was calculated using Equation (6):

$$V = Q_0 \times t^* / \alpha \tag{6}$$

where V is the recharge volume (m^3), Q_0 is the flow at the beginning of the recession (m^3 s^{-1}), t^* is the unit converter (days to seconds; 86,400), and α is the recession coefficient (dimensionless).

3. Results

The interpolation of climatic data from the rainfall stations in the JRB region (period 2000–2018) showed a variation in average annual precipitation from 1296 mm year^{-1} in Sete Lagoas to 1340 mm year^{-1} in regions near the river mouth, in the municipality of Jequitiba (Figure 3a). The spatial distribution of the real evapotranspiration was inversely related to the precipitation, presenting 634 mm year^{-1} to 649 mm year^{-1}, with higher values in Sete Lagoas, and lower values near the headwaters of the drainage area and in the river mouth (Figure 3b). The highest real evapotranspiration in Sete Lagoas can be attributed to the urban area, which waterproofs the soil making the surface water exposed for a longer time, favoring the evaporation rather than the infiltration process.

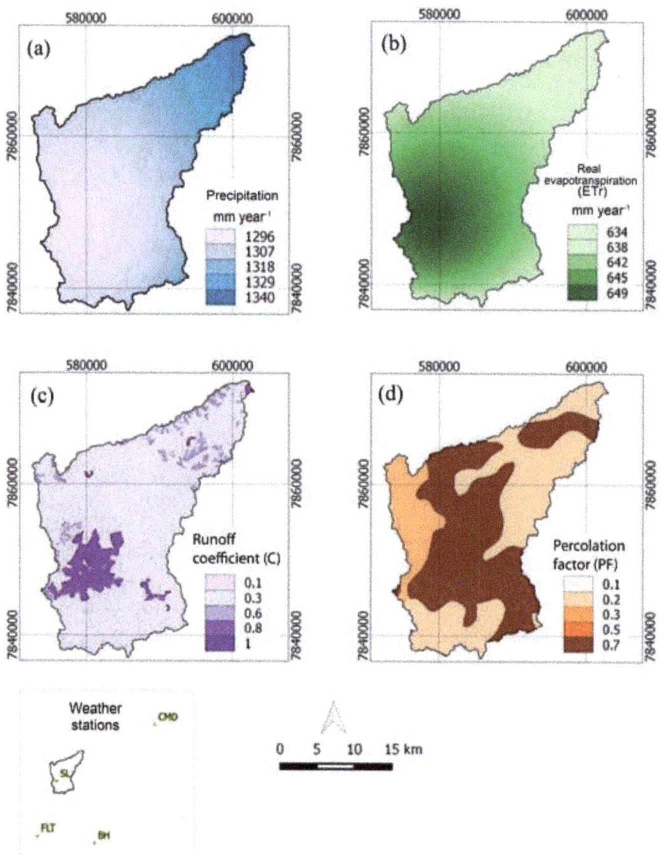

Figure 3. Spatialization of hydrological components used to estimate the groundwater recharge potential (Equation (3)) of the Jequitiba River basin, MG, Brazil: (**a**) Precipitation; (**b**) real evapotranspiration; (**c**) surface runoff factor—RF; Equation (1); (**d**) percolation factor—PF; Equation (2).

The surface runoff (RF) calculated by Equation (1) took into account the land uses (Figure 1c) and associated surface runoff coefficients (C; Table 2), as well as the slope length and steepness factor estimated for each point in the Jequitiba catchment (LS_{fuzzy}). The calculation of LS_{fuzzy} resorted to tools embedded in the GIS platform (QGIS), which implement methods that are proper for topographically complex landscape units [88]. The results reveal a strong influence of C in the values of RF because the largest values of RF (0.8–1.0) occur in the Sete lagoas town where the C value is the highest (0.85; urban area). The urban densification in Sete Lagoas contributed to a higher surface runoff, because urban drainage systems prevent storm water from infiltrating into unsaturated parts of the soil (Figure 3c).

The percolation factor (*PF*) calculated by Equation (2) took into account the spatial distribution of soil types (Figure 1b) as well as the corresponding porosities and hydraulic conductivities (Table 3). The percolation factor ranged from 0.1 to 0.7 (dimensionless), expressing the combined variation of soil permeability and total porosity in the area (Figure 3d). The PF values are over estimated in the urban area of Sete Lagoas, because the figure does not account for the effects on PF caused by compaction and cementation. These overestimated PF values produce little impact on the recharge potential (see Figure 4) because the corresponding RF values are very low ($\approx$0).

The mapping of recharge potential (Figure 4) resulted from a combination of maps using the proper tools of QGIS. The input maps were spatial distributions of precipitation, real evapotranspiration, RF factor and PF factor, illustrated in Figure 3, which were combined on a pixel basis according to Equation (3). The results showed that the groundwater recharge potential of the JRB ranges from 0 to 4626.4 m^3 ha^{-1} year^{-1}, with a mean of 953.72 m^3 ha^{-1} year^{-1}. The areas with the highest recharge potential are in regions with dense arboreous vegetation cover, flat or slightly undulated relief, and developed and structured soils, whose porosity and hydraulic conductivity allow water percolation to the water table. These areas are concentrated in the south-southeast of the headwaters in Prudente de Morais, in Sete Lagoas towards the central part of the basin, and between Funilândia and Jequitiba, near the Jequitiba river mouth. The concentrated areas with lower groundwater recharge potential are mainly in Sete Lagoas, due to its urban area, which causes waterproofing of the soil through roofs and pavements (Figure 5).

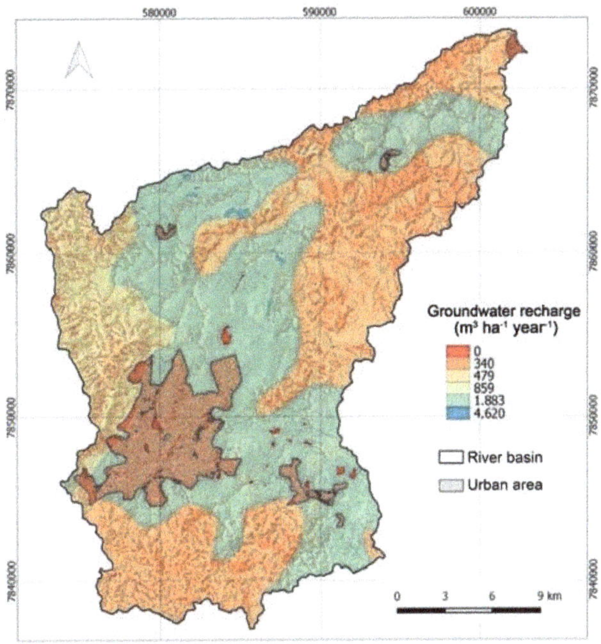

Figure 4. Groundwater recharge potential of the Jequitiba River basin, MG, Brazil.

Figure 5. Example of waterproofed area in the Jequitiba River basin, MG, Brazil, due to urban areas in the municipality of Sete Lagoas.

The validation of the results indicated that the groundwater recharge volumes obtained by the spatialization of the groundwater recharge potential and that obtained by the analysis of the hydrograph recession were similar, with a relative volumetric difference of 14.4% (Table 4). The hydrograph used to calculate recharge based on stream flow recession is displayed in Figure 6. The recession constant (α; Equation (5)) was calculated three times, considering the three recession periods represented in the figure and the corresponding initial discharge (Q_0), final discharge (Q_1), and recession time (t_1-t_0). Then, an average value was used in Equation (6), coupled with an average Q_0, to obtain the mean recharge potential.

Table 4. Groundwater recharge volumes in the Jequitiba River basin, MG, Brazil, estimated by two methods.

Method	Volume (m^3)	ΔV	ΔV (%)
Spatialization	3.63×10^9		
Recession analysis	3.11×10^9	5.22×10^8	14.4%

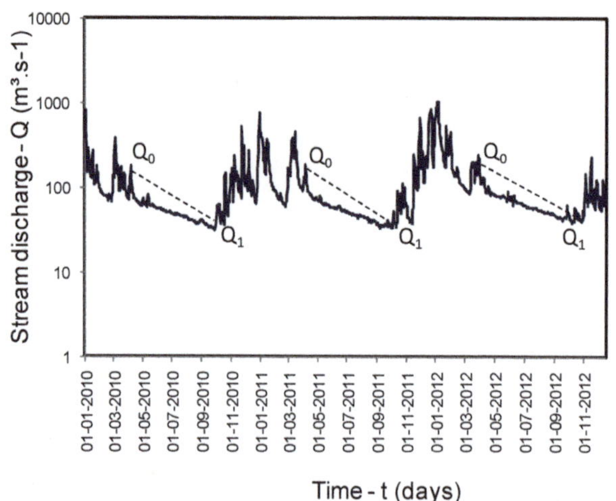

Figure 6. Hydrograph of Jequitiba River with indication of Q_1 and Q_0 values used to calculate the recession constant (Equation (5)).

The small difference between the calculated recharge volumes are likely to be related with differences between the two recharge estimation methods. In general, comparison of recharge estimates is difficult because of different levels of unavoidable inherent uncertainty associated with each method [93]. In the present study, the spatialization approach estimates recharge using a water balance equation based on precipitation and evapotranspiration data as well as on spatially distributed topographic, land use, and soil data, while the hydrograph approach estimates recharge using a storage variation equation based on stream flows. The spatialization approach is likely to incorporate larger uncertainty in the recharge values because the number of parameters involved in the calculations is larger. However, the hydrograph approach provides a single (average) recharge estimate while the spatialization approach provides an estimate for each point in the catchment.

It is worth to note that estimation of recharge did not account for differences among aquifer types. It is unlikely that these differences significantly affect the recession based results, because measured stream discharges represent an average flow through all the aquifers. Eventually, aquifers with markedly different characteristics located in specific places within the catchment could influence the water balance based results, because recharge estimation in this case depends on location. The small difference between recession and water balance based recharge values suggest that geology has limited influence on recharge estimation in the studied catchment, and that recharge is likely to be predominantly controlled by infiltration capacity and thickness of soils. The impact of unsaturated zone thickness on recharge has been recently reported in a study in China [94].

4. Discussion

4.1. Appreciation of Model Results

Information on the characteristics, potential and limitations of each part of a given river basin allows an adequate management in accordance with the demands of the local natural environmental system [95–99]. Thus, although Sete Lagoas has the highest area of the JRB, and soil and relief that favor groundwater recharge, it was behind Prudente de Morais, with an annual average of 1013.8 m^3 ha^{-1} $year^{-1}$ (Table 5, upper panel). This result was mainly due to its dense population, which leads to waterproofing of the soil. Prudente de Morais presented the highest average annual groundwater recharge potential, with approximately 1350.8 m^3 ha^{-1} $year^{-1}$ because it is in the part of the basin that concentrates areas with high recharge potential. Capim Branco had the lowest average annual groundwater recharge potential because it has a lower relative area and is in a region with predominance of surface runoff due to steep slopes, and areas with pastures and agriculture.

The average groundwater recharge potentials under the rock classes in the JRB were higher in regions with alluvial sediments (1357 m^3 ha^{-1} $year^{-1}$), colluvial sediments (1061 m^3 ha^{-1} $year^{-1}$), and carbonate and politic rocks (829.5 m^3 ha^{-1} $year^{-1}$) (Table 5, middle panel). These rocks form porous or fractured-karst aquifers, which present great water storage potentials [100]. Lower average annual recharge potentials were found in regions with igneous rocks (539 m^3 ha^{-1} $year^{-1}$), because they are hard, massive, and crystalline rocks, and the water infiltrates mainly through their fractures [101]. The water volume stored in these aquifers depends on the characteristics of cracks in each lithologic type, thus requiring detailed databases that characterize rock fractures in each region.

The average groundwater recharge potential was estimated through the land use and cover map and recharge potential map. According to the results, areas with planted forest had the highest average (1899.4 m^3 ha^{-1} $year^{-1}$) due to their topographic conditions and pedological characteristics (deep and well-structured soils), which are more suitable for forest management (Table 5, lower panel). The highest standard deviation (778.5 m^3 ha^{-1} $year^{-1}$) was also found in these areas, denoting their greater variability. Various studies analyzed the influence of land cover and the benefits of forests for groundwater recharge [102–110].

Table 5. Average groundwater recharge potential of each municipality (upper panel), lithologic type (middle panel), and land use (lower panel) of the Jequitiba River basin.

Parameter	Average Recharge Potential (m^3 ha^{-1} $year^{-1}$)
Municipality	
Prudente de Morais	1350.8
Sete Lagoas	1013.8
Jequitiba	807.5
Funilândia	633.2
Capim Branco	480.8
Lithologic types	
Alluvial sediments	1357.0
Colluvial sediments	1061.5
Carbonate and pelitic rocks	829.3
Igneous rocks	539.3
Land use and cover	
Forest	1899.4
Agriculture and pasture	888.4
Native vegetation	672.3
Ground vegetation and bare soil	484.0
Urban area	217.1

The average annual groundwater recharge potential of planted forest areas may be different when considering the specific values of maximum evapotranspiration of the crop, physiological characteristics, and management practices. Tree density, root system efficiency to remove soil water, and patterns of stomata opening and closing at certain periods of the year (dry and rainy seasons) and day (morning, afternoon, and night) of planted forests should be considered in further studies to obtain more detailed estimates [111,112]. For example, Leite et al. [113] evaluated water relations in eucalyptus stands and found different patterns for each spacing used and that evapotranspiration of eucalyptus crops with greater spacing between plants is about 0.91 mm.day^{-1} in the dry season, and 4.48 mm.day^{-1} in the rainy season. Other studies related recharge reductions with increasing root depths [114,115].

The average groundwater recharge potential of anthropogenic areas reached 884.4 m^3 ha^{-1} $year^{-1}$, with a standard deviation of 554.1 m^3 ha^{-1} $year^{-1}$. The areas with natural forest formation had the third highest average annual groundwater recharge potential, reaching more than 670 m^3 ha^{-1} $year^{-1}$, with standard deviation of 481 m^3 ha^{-1} $year^{-1}$. This difference was due to the topographical and pedological characteristics of the areas. Areas with natural forest formations are different from planted forested areas, and anthropized areas with agriculture or pasture; they have naturally better adaptation to undulated terrain, with steep slopes and shallow soils.

Undulate to mountainous reliefs present native vegetation, which favors groundwater recharge more than areas without well-managed vegetation cover, because the forest reduces runoff and favors water percolation. Thus, forest formations prevent soil losses due to runoff and wind action, maintain soil physical and mechanical stability, and, consequently, assist in water storage and in groundwater supply.

4.2. Controls of Groundwater Recharge and the Need to Delineate Zones of Groundwater Recharge Potential

The karst aquifer of Jequitiba River basin is probably undergoing over-exploitation due to increasing water demand, especially around the Sete Lagoas town where water table decline and suffosional sinkholes are reported for decades. This is enough reason to start an evaluation of groundwater recharge potential, as noted by Arulbalaji et al. [116] while conducting a study for delineation of groundwater potential zones in Southern Western Ghats, India. The methodological approach used in that study was also based on geographic information system (GIS), because it is rapid and provides first-hand information on the studied topic for further developments. However, the GIS was used to calculate a groundwater potential index as result of an overlay weight analysis, while the approach used in the Jequitiba River basin was physically based. Another study [117] based on a groundwater potential index and involving slope, drainage density, land use, geology, lineament density and geomorphological features from a mountainous region in Nalgonda district, Telangana, India, elected as best suitable areas the pediplains and valleys with minimum slope and sediment filling. Similar results were obtained by Singh et al. [118] in the Deccan Volcanic Province of Maharashtra, India, who further highlighted the predominance of low to medium groundwater potential within the studied basaltic area. Fractured igneous rocks were also the less suited areas as regards groundwater recharge potential in the Jequitiba River basin. Slope was not a major control factor of groundwater recharge, because the Jequitiba catchment is mostly shaped on a region of low slope (is monotonous in this regard). As in our case, geomorphologic parameters such as slope were not able to explain groundwater occurrence within the basement terrain of Keffi Area, North-Central Nigeria [119]. A study developed in the coastal part of Arani and Koratalai River basin, Southern India [120], also based on weighted overlay analysis but validated with hydraulic head and borehole distribution analyses, exposed the negative effects of urbanization on groundwater potential, as we also evidenced in the Jequitiba River basin. The most suited areas in the Arani and Koratalai River basin were used for agriculture, but this was also the predominant land use (>80%). The study of Dar et al. [121] carried out in the Mamundiyar basin (India) and still based on overlay analysis, attributed large weights to geomorphology and drainage density using data from previous works, and consequently identified pediment and pediplains with low drainage density as most suited areas. Land use was given the lowest weight. Overall, the current survey of literature identified the weighted overlay analysis as predominant technique for evaluating groundwater potential. Although easy to apply, this technique relies on the rating and weighting of layers, which is dependent on personal evaluations and therefore subjective. The method used in this study is physically based, and therefore objective. Validation was also lacking in many studies. The study conducted in the Arani and Koratalai River basin validated groundwater potential with hydraulic head data, relating the high potential zones with regions where the hydraulic head surface is flat. In the present study, the average groundwater recharge potential estimated by the spatially distributed method was compared with an average value obtained by the recession flow analysis, with very promising results (14% difference). We are therefore confident on the reliability of our physically based and spatially distributed model.

4.3. Management Considerations

Various recent studies estimated aquifer recharge using spatially distributed methods [122–124]. In most cases, the aim was merely to apply techniques, but other studies used the results as tools for water management. For example, Hund et al. [115], while working in Costa Rica, developed a recharge indicator based on a specific relationship between groundwater recharge and rainfall, which allowed estimating total groundwater recharge for a wet season from previously measured cumulative rainfall. The indicator permitted water managers to assess if a specific year was likely fall into a low recharge category prior to the end of the wet season. This information could then be used to trigger short-term adaptation strategies with the goal to "bank" groundwater while surface water sources are still available in the wet season. In the present study, the spatial information on recharge can also be used to identify areas for water "banking", namely the areas characterized by lowest recharge potentials.

The recharge potential estimated in the present study refers to natural recharge. The calculated values are likely to be underestimated in the urban area of Sete Lagoas, because the so-called urban component of recharge has not been forecasted and can be much larger than the natural component. In urban areas, there are various new components that must be considered in addition to the natural recharge from precipitation. These include leakages from water supply network and storm water drainage systems, and were reported to be 10 times greater than the natural recharge in a study in the city of Hyderabad, India [125]. Besides the implications for groundwater resource evaluations, the urban recharge issue is relevant for water quality management because the aforementioned leakages may contain contaminants such as metals or hydrocarbons [126–131], as well as for the geotechnical management of the territory because concentrated infiltration through storm water drainage systems can lead to suffosional sinkhole development in karst areas [132]. Urban and water planners should be made aware of these issues and work together to avoid aquifer recharge through storm water drainage systems, especially in the Sete Lagoas town that is laid over a karst.

The areas used for agriculture and pasture are predominant in the studied catchment and were considered the second most favorable areas for groundwater recharge. It is however necessary to recall that recharge in agricultural areas is largely influenced by irrigation practices. Porhemmat et al. [133] investigated the effects of irrigation methods on potential groundwater recharge and concluded that annual potential groundwater recharge under furrow irrigation was estimated in the range of 19–228 mm, with an average of 111.3 mm. However, mean annual potential groundwater recharge under the sprinkler (4.1 mm) and drip (0.7 mm) was an order of magnitude lower than furrow irrigation. These results raise questions about the sustainability of water-saving irrigation methods, and should be attended by farmers and water planners from the Jequitiba watershed. On the other hand, the practice of furrow irrigation in karst areas is more likely to trigger or accelerate the development of suffosional sinkholes [134,135] with negative consequences for the practice of agriculture.

5. Conclusions

The proposed objective was to develop and apply a physically based spatially distributed recharge estimation model in a catchment largely influenced by anthropogenic activities (Jequitiba River, Minas Gerais, Brazil), to serve as water management tool for water planners and policy makers. The results showed areas with greatest potential concentrated in south-southeast regions of the basin and near the river mouth. The groundwater potential was highest in the porous aquifers overlaid by planted forests and lowest in the igneous rocks overlaid by urban areas. Waterproofing was considered a key factor of reduced recharge in the urban area of Sete Lagoas. The adoption of management practices are expected to improve natural groundwater recharge, and hence, significantly increase the available water volume to the local population. Water management initiatives may include seasonal storage of surface water in areas of low recharge potential or adjustment of irrigation methods. Preservation of forest vegetation is also recommended because these areas were considered the most favorable for groundwater recharge. The use of a physically based approaches may be more adequate in the evaluation of groundwater recharge potential, relative to more frequently used weighted overlay analyses, because the former methods are potentially free from subjective steps.

Author Contributions: Conceptualization, A.M.d.C., H.H.C.d.S., and J.H.M.V.; methodology, A.M.d.C., H.H.C.d.S., and J.H.M.V.; validation, H.H.C.d.S and F.A.L.P.; resources, A.M.d.C.; writing—original draft preparation, H.H.C.d.S.; writing—review and editing, A.M.d.C., J.H.M.V., and F.A.L.P.; supervision, A.M.d.C.; funding acquisition, A.M.d.C.

Funding: This research was funded by the World Wildlife Fund WWF-Brasil (https://www.wwf.org.br/wwf_brasil/), project number "CPT 001299-2018", and A.M.C and H.H.C.S received a grant by WWF. For the author integrated in the Vila Real Chemistry Research Centre—CQVR (http://cqvr.purpleprofile.pt/), the research was additionally supported by National Funds of the Portuguese Foundation for Science and Technology (FCT), under the project UID/QUI/00616/2019.

Acknowledgments: We thank the support of the team of Soil and Environment Laboratory of the Federal University of Minas Gerais for the help in the analyses made for the project.

Conflicts of Interest: The authors declare no conflict of interest. The funders had no role in the design of the study; in the collection, analyses, or interpretation of data; in the writing of the manuscript or in the decision to publish the results.

References

1. Oki, T.; Kanae, S. Global hydrological cycles and world water resources. *Science* **2006**, *313*, 1068–1072. [CrossRef]
2. Arnell, N.W. Climate change and global water resources. *Glob. Environ. Chang.* **1999**, *9*, S31–S49. [CrossRef]
3. Water Resources Group. *Charting Our Water Future: Economic Frameworks to Inform Decision-Making*; McKinsey & Company: New York, NY, USA, 2009.
4. Lerner, D.N. Groundwater recharge in urban areas. *Atmos. Environ. Part B Urban Atmos.* **1990**, *24*, 29–33. [CrossRef]
5. Lerner, D.N. Identifying and quantifying urban recharge: A review. *Hydrogeol. J.* **2002**, *10*, 143–152. [CrossRef]
6. Pacheco, F.A.L.; Varandas, S.G.P.; Sanches Fernandes, L.F.; Valle Junior, R.F. Soil losses in rural watersheds with environmental land use conflicts. *Sci. Total Environ.* **2014**, *485–486*, 110–120. [CrossRef] [PubMed]
7. Valle Junior, R.F.; Varandas, S.G.P.; Sanches Fernandes, L.F.; Pacheco, F.A.L. Environmental land use conflicts: A threat to soil conservation. *Land Use Policy* **2014**, *41*, 172–185. [CrossRef]
8. Botelho, M.H.C. *Água de Chuva: Engenharia das Águas Pluviais nas Cidades*; Blucher: São Paulo, Brazil, 2012; ISBN 978-85-212-0596-8.
9. De Barros, L.C. *Barraginhas: Água de Chuva para Todos*; Empresa Brasileira de Pesquisa Agropecuária, Embrapa Milho e Sorgo: Brasília, Brazil, 2009; ISBN 978-85-7383-447-5.
10. Timm, L.C.; Reichardt, K. *Solo, Planta e Atmosfera: Conceitos, Processos e Aplicações*; Manole On-line publisher: Barueri, Brazil, 2008; ISBN 8520417736.
11. Scanlon, B.R.; Faunt, C.C.; Longuevergne, L.; Reedy, R.C.; Alley, W.M.; Mcguire, V.L.; Mcmahon, P.B. Groundwater depletion and sustainability of irrigation in the US High Plains and Central Valley. *Proc. Natl. Acad. Sci. USA* **2012**, *109*, 9320–9325. [CrossRef]
12. Siebert, S.; Burke, J.; Faures, J.M.; Frenken, K.; Hoogeveen, J.; Döll, P.; Portmann, F.T. Groundwater use for irrigation—A global inventory. *Hydrol. Earth Syst. Sci.* **2010**, *14*, 1863–1880. [CrossRef]
13. Wada, Y.; Van Beek, L.P.H.; Bierkens, M.F.P. Nonsustainable groundwater sustaining irrigation: A global assessment. *Water Resour. Res.* **2012**, *48*. [CrossRef]
14. Sophocleous, M. Groundwater recharge and sustainability in the High Plains aquifer in Kansas, USA. *Hydrogeol. J.* **2005**, *13*, 351–365. [CrossRef]
15. Todd, D.K. *Hidrologia de Águas Subterrâneas*; Edgard Blucher: São Paulo, Brazil, 1967; ISBN 1000213223042.
16. Gribbin, J.B. *Introdução à Hidráulica, Hidrologia e Gestão de Águas Pluviais*; Cengage Learning: São Paulo, Brazil, 2014; ISBN 8522116342.
17. Valle, R.F.; Varandas, S.G.P.; Sanches Fernandes, L.F.; Pacheco, F.A.L. Groundwater quality in rural watersheds with environmental land use conflicts. *Sci. Total Environ.* **2014**, *493*, 812–827. [CrossRef] [PubMed]
18. Kalbus, E.; Reinstorf, F.; Schirmer, M. Measuring methods for groundwater—Surface water interactions: A review. *Hydrol. Earth Syst. Sci.* **2006**, *10*, 873–887. [CrossRef]
19. Winter, T.C. Relation of streams, lakes, and wetlands to groundwater flow systems. *Hydrogeol. J.* **1999**, *7*, 28–45. [CrossRef]
20. Sophocleous, M. Interactions between groundwater and surface water: The state of the science. *Hydrogeol. J.* **2002**, *10*, 52–67. [CrossRef]
21. Kollet, S.J.; Maxwell, R.M. Integrated surface-groundwater flow modeling: A free-surface overland flow boundary condition in a parallel groundwater flow model. *Adv. Water Resour.* **2006**, *29*, 945–958. [CrossRef]
22. Wisler, C.O.; Brater, E.F. *Hidrologia*; Ao Livro Técnico: Rio de Janeiro, Brazil, 1964.
23. Chowdhury, A.; Jha, M.K.; Chowdary, V.M. Delineation of groundwater recharge zones and identification of artificial recharge sites in West Medinipur district, West Bengal, using RS, GIS and MCDM techniques. *Environ. Earth Sci.* **2009**. [CrossRef]
24. Zaidi, F.K.; Nazzal, Y.; Ahmed, I.; Naeem, M.; Jafri, M.K. Identification of potential artificial groundwater recharge zones in Northwestern Saudi Arabia using GIS and Boolean logic. *J. African Earth Sci.* **2015**, *111*, 156–169. [CrossRef]
25. Saraf, A.K.; Choudhury, P.R.; Roy, B.; Sarma, B.; Vijay, S.; Choudhury, S. GIS based surface hydrological modelling in identification of groundwater recharge zones. *Int. J. Remote Sens.* **2004**, *25*, 5759–5770. [CrossRef]
26. Yeh, H.F.; Lee, C.H.; Hsu, K.C.; Chang, P.H. GIS for the assessment of the groundwater recharge potential zone. *Environ. Geol.* **2009**, *58*, 185–195. [CrossRef]

27. Yeh, H.F.; Cheng, Y.S.; Lin, H.I.; Lee, C.H. Mapping groundwater recharge potential zone using a GIS approach in Hualian River, Taiwan. *Sustain. Environ. Res.* **2016**, *26*, 33–43. [CrossRef]

28. Nimmo, J.R.; Healy, R.W.; Stonestrom, D.A. Aquifer Recharge. In *Encyclopedia of Hydrological Sciences*; Wiley: Hoboken, NJ, USA, 2006; ISBN 0470848944.

29. Allison, G.B.; Gee, G.W.; Tyler, S.W. Vadose-Zone Techniques for Estimating Groundwater Recharge in Arid and Semiarid Regions. *Soil Sci. Soc. Am. J.* **2010**, *58*, 6–14. [CrossRef]

30. Jiménez-Martínez, J.; Skaggs, T.H.; van Genuchten, M.T.; Candela, L. A root zone modelling approach to estimating groundwater recharge from irrigated areas. *J. Hydrol.* **2009**, *367*, 138–149. [CrossRef]

31. Delin, G.N.; Healy, R.W.; Lorenz, D.L.; Nimmo, J.R. Comparison of local- to regional-scale estimates of ground-water recharge in Minnesota, USA. *J. Hydrol.* **2007**, *334*, 231–249. [CrossRef]

32. Jódar-Abellán, A.; Albaladejo-García, J.A.; Prats-Rico, D. Artificial groundwater recharge. Review of the current knowledge of the technique. *Revista de la Sociedad Geológica de España* **2017**, *30*, 138–149.

33. Bouwer, H. Artificial recharge of groundwater: Hydrogeology and engineering. *Hydrogeol. J.* **2002**, *10*, 121–142. [CrossRef]

34. Stiefel, J.M.; Melesse, A.M.; McClain, M.E.; Price, R.M.; Anderson, E.P.; Chauhan, N.K. Effects of rainwater-harvesting-induced artificial recharge on the groundwater of wells in Rajasthan, India. *Hydrogeol. J.* **2009**, *17*, 2061. [CrossRef]

35. Singh, A.; Panda, S.N.; Kumar, K.S.; Sharma, C.S. Artificial groundwater recharge zones mapping using remote sensing and gis: A case study in Indian Punjab. *Environ. Manag.* **2013**, *52*, 61–71. [CrossRef]

36. Terêncio, D.P.S.; Sanches Fernandes, L.F.; Cortes, R.M.V.; Pacheco, F.A.L. Improved framework model to allocate optimal rainwater harvesting sites in small watersheds for agro-forestry uses. *J. Hydrol.* **2017**, *550*, 318–330. [CrossRef]

37. Terêncio, D.P.S.; Sanches Fernandes, L.F.; Cortes, R.M.V.; Moura, J.P.; Pacheco, F.A.L. Rainwater harvesting in catchments for agro-forestry uses: A study focused on the balance between sustainability values and storage capacity. *Sci. Total Environ.* **2018**, *613–614*, 1079–1092. [CrossRef]

38. Bellu, A.; Sanches Fernandes, L.F.; Cortes, R.M.V.; Pacheco, F.A.L. A framework model for the dimensioning and allocation of a detention basin system: The case of a flood-prone mountainous watershed. *J. Hydrol.* **2016**, *533*, 567–580. [CrossRef]

39. Tucci, C.E.M.; Collischonn, W. *Drenagem Urbana E Controle de Erosão*, 2nd ed.; Oficina de Textos: São Paulo, Brazil, 1998; ISBN 978-85-7975-160-8.

40. Mignot, E.; Li, X.; Dewals, B. Experimental modelling of urban flooding: A review. *J. Hydrol.* **2019**, *568*, 334–342. [CrossRef]

41. De Roo, A.; Odijk, M.; Schmuck, G.; Koster, E.; Lucieer, A. Assessing the effects of land use changes on floods in the meuse and oder catchment. *Phys. Chem. Earth Part B Hydrol. Ocean. Atmos.* **2001**, *26*, 593–599. [CrossRef]

42. Acreman, M.; Holden, J. How wetlands affect floods. *Wetlands* **2013**, *33*, 773–786. [CrossRef]

43. Caldas, A.M.; Pissarra, T.C.T.; Costa, R.C.A.; Neto, F.C.R.; Zanata, M.; Parahyba, R.d.B.V.; Fernandes, L.F.S.; Pacheco, F.A.L. Flood vulnerability, environmental land use conflicts, and conservation of soil and water: A study in the Batatais SP municipality, Brazil. *Water* **2018**, *10*, 1357. [CrossRef]

44. Minderhoud, P.S.J.; Erkens, G.; Pham, V.H.; Bui, V.T.; Erban, L.; Kooi, H.; Stouthamer, E. Impacts of 25 years of groundwater extraction on subsidence in the Mekong delta, Vietnam. *Environ. Res. Lett.* **2017**. [CrossRef]

45. Sušnik, J.; Vamvakeridou-Lyroudia, L.S.; Savić, D.A.; Kapelan, Z. Integrated System Dynamics Modelling for water scarcity assessment: Case study of the Kairouan region. *Sci. Total Environ.* **2012**, *440*, 290–306. [CrossRef]

46. Custodio, E. Aquifer overexploitation: What does it mean? *Hydrogeol. J.* **2002**, *10*, 254–277. [CrossRef]

47. Álvarez, X.; Valero, E.; Santos, R.M.B.; Varandas, S.G.P.; Sanches Fernandes, L.F.; Pacheco, F.A.L. Anthropogenic nutrients and eutrophication in multiple land use watersheds: Best management practices and policies for the protection of water resources. *Land Use Policy* **2017**, *69*, 1–11. [CrossRef]

48. Santos, R.M.B.; Sanches Fernandes, L.F.; Pereira, M.G.; Cortes, R.M.V.; Pacheco, F.A.L. Water resources planning for a river basin with recurrent wildfires. *Sci. Total Environ.* **2015**, *526*, 1–13. [CrossRef]

49. Ferreira, A.R.L.; Sanches Fernandes, L.F.; Cortes, R.M.V.; Pacheco, F.A.L. Assessing anthropogenic impacts on riverine ecosystems using nested partial least squares regression. *Sci. Total Environ.* **2017**, *583*, 466–477. [CrossRef]

50. Pacheco, F.A.L.; Santos, R.M.B.; Sanches Fernandes, L.F.; Pereira, M.G.; Cortes, R.M.V. Controls and forecasts of nitrate yields in forested watersheds: A view over mainland Portugal. *Sci. Total Environ.* **2015**, *537*, 421–440. [CrossRef]

51. Santos, R.M.B.; Sanches Fernandes, L.F.; Pereira, M.G.; Cortes, R.M.V.; Pacheco, F.A.L. A framework model for investigating the export of phosphorus to surface waters in forested watersheds: Implications to management. *Sci. Total Environ.* **2015**, *536*, 295–305. [CrossRef]

52. Pacheco, F.A.L.; Pires, L.M.G.R.; Santos, R.M.B.; Sanches Fernandes, L.F. Factor weighting in DRASTIC modeling. *Sci. Total Environ.* **2015**, *505*, 474–486. [CrossRef]

53. Valle Junior, R.F.; Varandas, S.G.P.; Sanches Fernandes, L.F.; Pacheco, F.A.L. Multi criteria analysis for the monitoring of aquifer vulnerability: A scientific tool in environmental policy. *Environ. Sci. Policy* **2015**, *48*, 250–264. [CrossRef]

54. Pacheco, F.A.L.; Sanches Fernandes, L.F. The multivariate statistical structure of DRASTIC model. *J. Hydrol.* **2013**, *476*, 442–459. [CrossRef]

55. Pacheco, F.A.L.; Sanches Fernandes, L.F. Environmental land use conflicts in catchments: A major cause of amplified nitrate in river water. *Sci. Total Environ.* **2016**, *548–549*, 173–188. [CrossRef]

56. Rocha, C.I.O. Plano Diretor de Drenagem Urbana em Cidades Planejadas: Premissa de Zoneamento Baseado no Risco de Contaminação da Água Subterrânea. Ph.D. Thesis, Escola Politécnica da Universidade de São Paulo, São Paulo, Brazil, 2014.

57. Sanches Fernandes, L.F.; Pacheco, F.A.L.; Cortes, R.M.V.; Jesus, J.J.B.; Varandas, S.G.P.; Santos, R.M.B. Integrative assessment of river damming impacts on aquatic fauna in a Portuguese reservoir. *Sci. Total Environ.* **2017**, *601–602*, 1108–1118.

58. Pacheco, F.A.L. Regional groundwater flow in hard rocks. *Sci. Total Environ.* **2015**, *506–507*, 182–195. [CrossRef]

59. Pacheco, F.A.L.; van der Weijden, C.H. Weathering of plagioclase across variable flow and solute transport regimes. *J. Hydrol.* **2012**, *420–421*, 46–58. [CrossRef]

60. Van Der Weijden, C.H.; Pacheco, F.A.L. Hydrochemistry, weathering and weathering rates on Madeira island. *J. Hydrol.* **2003**, *283*, 122–145. [CrossRef]

61. Pacheco, F.A.L.; Van Der Weijden, C.H. Mineral weathering rates calculated from spring water data: A case study in an area with intensive agriculture, the Morais Massif, northeast Portugal. *Appl. Geochem.* **2002**, *17*, 583–603. [CrossRef]

62. Pacheco, F.A.L.; Alencoão, A.M.P. Role of fractures in weathering of solid rocks: Narrowing the gap between laboratory and field weathering rates. *J. Hydrol.* **2006**, *316*, 248–265. [CrossRef]

63. Pacheco, F.A.L.; Van der Weijden, C.H. Role of hydraulic diffusivity in the decrease of weathering rates over time. *J. Hydrol.* **2014**, *512*, 87–106. [CrossRef]

64. Pacheco, F.A.L.; Van der Weijden, C.H. Modeling rock weathering in small watersheds. *J. Hydrol.* **2014**, *513*, 13–27. [CrossRef]

65. Pacheco, F.A.L.; Van der Weijden, C.H. Integrating topography, hydrology and rock structure in weathering rate models of spring watersheds. *J. Hydrol.* **2012**, *428–429*, 32–50. [CrossRef]

66. De Luca, D.A.; Lasagna, M.; Gisolo, A.; Morelli di Popolo e Ticineto, A.; Falco, M.; Cuzzi, C. Potential recharge areas of deep aquifers: an application to the Vercelli–Biella Plain (NW Italy). *Rend. Lincei* **2019**, *30*, 137–153. [CrossRef]

67. Blasch, K.W.; Bryson, J.R. Distinguishing sources of ground water recharge by using $\delta 2H$ and $\delta 18O$. *Ground Water* **2007**, *45*, 294–308. [CrossRef] [PubMed]

68. Ingram, R.G.S.; Hiscock, K.M.; Dennis, P.F. Noble gas excess air applied to distinguish groundwater recharge conditions. *Environ. Sci. Technol.* **2007**, *41*, 1949–1955. [CrossRef]

69. Sukhija, B.S.; Reddy, D.V.; Nagabhushanam, P.; Hussain, S.N.; Giri, V.Y.; Patil, D.J. Environmental and injected tracers methodology to estimate direct precipitation recharge to a confined aquifer. *J. Hydrol.* **1996**, *177*, 77–97. [CrossRef]

70. Sanches Fernandes, L.F.; Santos, C.; Pereira, A.; Moura, J. Model of management and decision support systems in the distribution of water for consumption: case study in North Portugal. *Eur. J. Environ. Civ. Eng.* **2011**, *15*, 411–426. [CrossRef]

71. Fernandes, L.F.S.; Marques, M.J.; Oliveira, P.C.; Moura, J.P. Decision support systems in water resources in the demarcated region of Douro—Case study in Pinhão river basin, Portugal. *Water Environ. J.* **2014**, *28*, 350–357. [CrossRef]

72. Instituto Brasileiro de Geografia E Estatística—IBGE Sistema IBGE de Recuperação Eletrônica (SIDRA). Available online: http://www.sidra.ibge.gov.br (accessed on 1 November 2018).

73. Pessoa, P.F.P. Caracterização Hidrogeológica da Região Cárstica de Sete Lagoas-MG: Potencialidades e Riscos. Ph.D. Thesis, São Paulo University, São Paulo, Brazil, 1996.

74. Botelho, L.A.L.A. *Gestão dos Recursos Hídricos em Sete Lagoas/MG: UMA Abordagem a Partir da Evolução Espaçotemporal da Demanda e da Captação de Água;* Federal University of Minas Gerais: Belo Horizonte, Brazil, 2008.

75. Galvão, P.; Halihan, T.; Hirata, R. Evaluating karst geotechnical risk in the urbanized area of Sete Lagoas, Minas Gerais, Brazil. *Hydrogeol. J.* **2015**, *23*, 1499–1513. [CrossRef]

76. Universidade Federal De Lavras; Fundação Estadual Do Meio Ambiente De Minas Gerais. Mapa de Solos do Estado de Minas Gerais. Available online: http://www.dps.ufv.br/?page_id=742 (accessed on 1 November 2018).

77. FBDS—Fundação Brasileira para o Desenvolvimento Sustentável. Mapeamento em Alta Resolução dos Biomas Brasileiros. Available online: http://geo.fbds.org.br/ (accessed on 1 November 2018).

78. Iglesias, M.; Uhlein, A. Estratigrafia do Grupo Bambuí e coberturas fanerozóicas no vale do rio São Francisco, norte de Minas Gerais. *Rev. Bras. Geociências* **2018**, *39*, 256–266. [CrossRef]

79. CPRM/CODEMIG—Companhia de Pesquisa de Recursos Minerais/Companhia de Desenvolvimento Econômico de Minas Gerais. Mapa Geológico do Estado de Minas Gerais. Available online: www.portaldageologia.com.br (accessed on 1 October 2018).

80. Batista, R.C.R. Caracterização Hidrogeológica do Entorno do Centro Nacional de Pesquisa de Milho e Sorgo (CNPMS), em Sete Lagoas, MG. MSc Thesis in Sanitation, Environment and Water Resources, Federal University of Minas Gerais, Belo Horizonte, Brazil, 2009.

81. UAF-NASA Alaska Satellite Facility—Making Remote-Sensing Data Accessible Since 1991. Available online: https://www.asf.alaska.edu/ (accessed on 1 November 2018).

82. Instituto Nacional de Meteorologia, Ministério da Agricultura, P. e A. Dados Meteorológicos. Available online: http://www.inmet.gov.br/portal/ (accessed on 1 November 2018).

83. Agência Nacional de Águas. HidroWEB: Acervo de Dados Hidrológicos. Available online: http://www.snirh.gov.br/hidroweb/publico/apresentacao.jsf (accessed on 1 November 2018).

84. Sistema Nacional de Cadastro Ambiental Rural. Consulta Pública dos Dados do Cadastro Ambiental Rural. Available online: http://www.car.gov.br/publico/municipios/downloads (accessed on 1 November 2018).

85. QGIS a Free and Open Source Geographic Information System. Available online: https://www.qgis.org/en/site/ (accessed on 1 November 2018).

86. Mazzini, P.L.F.; Schettini, C.A.F. Avaliação de metodologias de interpolação espacial aplicadas a dados hidrográficos costeiros quase-sinóticos. *Braz. J. Aquat. Sci. Technol.* **2013**, *13*, 53–64. [CrossRef]

87. Böhner, J.; Selige, T. Spatial prediction of soil attributes using terrain analysis and climate regionalisation. *Göttinger Geogr. Abhandlungen* **2002**, *115*, 13–27.

88. Desmet, P.J.J.; Govers, G. A GIS procedure for automatically calculating the USLE LS factor on topographically complex landscape units. *J. Soil Water Conserv.* **1996**, *51*, 427–433.

89. American Society of Civil Engineers (ASCE). *Design and Construction of Sanitary and Storm Sewers;* ASCE Manual of Practice No 37, WPCF Manual of Practice No 9; American Society of Civil Engineers: New York, NY, USA, 1969; ISBN 0686304098.

90. de Pedron, F.A.; Fink, J.R.; Rodrigues, M.F.; de Azevedo, A.C. Condutividade e retenção de água em Neossolos e saprolitos derivados de arenito. *Rev. Bras. Cienc. do Solo* **2011**, *35*, 1253–1262.

91. de Araújo Pedron, F.; Fink, J.R.; Rodrigues, M.F.; de Azevedo, A.C. Hydraulic conductivity and water retention in leptosols-regosols and saprolite derived from sandstone, Brazil. *Rev. Bras. Ciência do Solo* **2011**, *35*, 1253–1262. [CrossRef]

92. Santos, R.M.B.; Sanches Fernandes, L.F.; Moura, J.P.; Pereira, M.G.; Pacheco, F.A.L. The impact of climate change, human interference, scale and modeling uncertainties on the estimation of aquifer properties and river flow components. *J. Hydrol.* **2014**, *519*, 1297–1314. [CrossRef]

93. Scanlon, B.R.; Healy, R.W.; Cook, P.G. Choosing appropriate techniques for quantifying groundwater recharge. *Hydrogeol. J.* **2002**, *10*, 18–39. [CrossRef]

94. Cao, G.; Scanlon, B.R.; Han, D.; Zheng, C. Impacts of thickening unsaturated zone on groundwater recharge in the North China Plain. *J. Hydrol.* **2016**, *537*, 260–270. [CrossRef]

95. Koch, H.; Vögele, S. Dynamic modelling of water demand, water availability and adaptation strategies for power plants to global change. *Ecol. Econ.* **2009**, *68*, 2031–2039. [CrossRef]

96. Fyles, H.; Madramootoo, C. Water Management. In *Emerging Technologies for Promoting Food Security: Overcoming the World Food Crisis*; Woodhead Publishing: Sawston, Cambridge, UK, 2015; ISBN 9781782423539.

97. Butler, D.; Memon, F. *Water Demand Management*; IWA Publishing: London, UK, 2006; p. 380. ISBN 1843390787.

98. Russell, S.; Fielding, K. Water demand management research: A psychological perspective. *Water Resour. Res.* **2010**. [CrossRef]

99. Dessu, S.B.; Melesse, A.M.; Bhat, M.G.; McClain, M.E. Assessment of water resources availability and demand in the Mara River Basin. *Catena* **2014**, *115*, 104–114. [CrossRef]

100. McMahon, P.B.; Plummer, L.N.; Böhlke, J.K.; Shapiro, S.D.; Hinkle, S.R. A comparison of recharge rates in aquifers of the United States based on groundwater-age data. *Hydrogeol. J.* **2011**. [CrossRef]

101. Cook, P.G. *A Guide to Regional Groundwater Flow in Fractured Rock Aquifers*; CSIRO Land and Water: Henley Beach, Australia, 2003; p. 115, ISBN 1 74008 233 8.

102. Scanlon, B.R.; Reedy, R.C.; Stonestrom, D.A.; Prudic, D.E.; Dennehy, K.F. Impact of land use and land cover change on groundwater recharge and quality in the southwestern US. *Glob. Chang. Biol.* **2005**, *11*, 1577–1593. [CrossRef]

103. Brauman, K.A.; Freyberg, D.L.; Daily, G.C. Land cover effects on groundwater recharge in the tropics: Ecohydrologic mechanisms. *Ecohydrology* **2012**, *5*, 435–444. [CrossRef]

104. Ilstedt, U.; Bargués Tobella, A.; Bazié, H.R.; Bayala, J.; Verbeeten, E.; Nyberg, G.; Sanou, J.; Benegas, L.; Murdiyarso, D.; Laudon, H.; et al. Intermediate tree cover can maximize groundwater recharge in the seasonally dry tropics. *Sci. Rep.* **2016**. [CrossRef]

105. Zomlot, Z.; Verbeiren, B.; Huysmans, M.; Batelaan, O. Spatial distribution of groundwater recharge and base flow: Assessment of controlling factors. *J. Hydrol. Reg. Stud.* **2015**, *4*, 349–368. [CrossRef]

106. Maréchal, J.C.; Varma, M.R.R.; Riotte, J.; Vouillamoz, J.M.; Kumar, M.S.M.; Ruiz, L.; Sekhar, M.; Braun, J.J. Indirect and direct recharges in a tropical forested watershed: Mule Hole, India. *J. Hydrol.* **2009**, *364*, 272–284. [CrossRef]

107. Le Maitre, D.C.; Scott, D.F.; Colvin, C. A review of information on interactions between vegetation and groundwater. *Water SA* **1999**, *25*, 137–152.

108. Zhang, H.; Hiscock, K.M. Modelling the impact of forest cover on groundwater resources: A case study of the Sherwood Sandstone aquifer in the East Midlands, UK. *J. Hydrol.* **2010**, *392*, 136–149. [CrossRef]

109. Zhang, Y.K.; Schilling, K.E. Effects of land cover on water table, soil moisture, evapotranspiration, and groundwater recharge: A Field observation and analysis. *J. Hydrol.* **2006**, *319*, 328–338. [CrossRef]

110. Allen, A.; Chapman, D. Impacts of afforestation on groundwater resources and quality. *Hydrogeol. J.* **2001**, *9*, 390–400. [CrossRef]

111. Schlesinger, W.H.; Jasechko, S. Transpiration in the global water cycle. *Agric. For. Meteorol.* **2014**, *189–190*, 115–117. [CrossRef]

112. Zhang, L.; Dawes, W.R.; Walker, G.R. Response of mean annual evapotranspiration to vegetation changes at catchment scale. *Water Resour. Res.* **2001**, *37*, 701–708. [CrossRef]

113. Leite, F.P.; Barros, N.F.; Novais, R.F.; Sans, L.M.A.; Fabres, A.S. Relações hídricas em povoamento de eucalipto com diferentes densidades populacionais. *Rev. Bras. Ciência do Solo* **1999**, *23*, 9–16. [CrossRef]

114. Consensa, C.O.B. Balanço Hídrico em Povoamento de Eucalipti com Diferentes Densidades de Plantas em Argissolo. Master's Thesis, Federal University of Santa Maria, Santa Maria, RS, Brazil, 2015.

115. Hund, S.V.; Allen, D.M.; Morillas, L.; Johnson, M.S. Groundwater recharge indicator as tool for decision makers to increase socio-hydrological resilience to seasonal drought. *J. Hydrol.* **2018**, *563*, 1119–1134. [CrossRef]

116. Arulbalaji, P.; Padmalal, D.; Sreelash, K. GIS and AHP Techniques Based Delineation of Groundwater Potential Zones: a case study from Southern Western Ghats, India. *Sci. Rep.* **2019**. [CrossRef]

117. Rajaveni, S.P.; Brindha, K.; Elango, L. Geological and geomorphological controls on groundwater occurrence in a hard rock region. *Appl. Water Sci.* **2015**, *7*, 1377–1389. [CrossRef]

118. Singh, P.; Thakur, J.K.; Kumar, S. Delineating groundwater potential zones in a hard-rock terrain using geospatial tool. *Hydrol. Sci. J.* **2013**, *58*, 213–223. [CrossRef]

119. Kudamnya, E.A.; Edet, A.E.; Ekwere, A.S. Geomorphological Control on Groundwater Occurrence within the Basement Terrain of Keffi Area, North-Central Nigeria. In *Advances in Sustainable and Environmental Hydrology, Hydrogeology, Hydrochemistry and Water Resources*; Springer: Cham, Switherland, 2019.

120. Suganthi, S.; Elango, L.; Subramanian, S.K. Groundwater potential zonation by remote sensing and GIS techniques and its relation to the groundwater level in the coastal part of the Arani and Koratalai river basin, Southern India. *Earth Sci. Res. J.* **2013**, *17*, 87–95.

121. Dar, I.A.; Sankar, K.; Dar, M.A. Deciphering groundwater potential zones in hard rock terrain using geospatial technology. *Environ. Monit. Assess.* **2011**, *173*, 597–610. [CrossRef]

122. Yenehun, A.; Walraevens, K.; Batelaan, O. Spatial and temporal variability of groundwater recharge in Geba basin, Northern Ethiopia. *J. African Earth Sci.* **2017**, *134*, 198–212. [CrossRef]

123. Cheo, A.E.; Voigt, H.J.; Wendland, F. Modeling groundwater recharge through rainfall in the Far-North region of Cameroon. *Groundw. Sustain. Dev.* **2017**, *5*, 118–130. [CrossRef]

124. Sandoval, J.A.; Tiburan, C.L. Identification of potential artificial groundwater recharge sites in Mount Makiling Forest Reserve, Philippines using GIS and Analytical Hierarchy Process. *Appl. Geogr.* **2019**, *105*, 73–85. [CrossRef]

125. Wakode, H.B.; Baier, K.; Jha, R.; Azzam, R. Impact of urbanization on groundwater recharge and urban water balance for the city of Hyderabad, India. *Int. Soil Water Conserv. Res.* **2018**, *6*, 51–62. [CrossRef]

126. Kalhor, K.; Ghasemizadeh, R.; Rajic, L.; Alshawabkeh, A. Assessment of groundwater quality and remediation in karst aquifers: A review. *Groundw. Sustain. Dev.* **2019**, *8*, 104–121. [CrossRef]

127. Ghasemizadeh, R.; Hellweger, F.; Butscher, C.; Padilla, I.; Vesper, D.; Field, M.; Alshawabkeh, A. Review: Groundwater flow and transport modeling of karst aquifers, with particular reference to the North Coast Limestone aquifer system of Puerto Rico. *Hydrogeol. J.* **2012**, *20*, 1441–1461. [CrossRef]

128. White, W.B. Conceptual models for karstic aquifers. *Speleogenes. Evol. Karst Aquifers* **2003**, *1*, 1–6.

129. Groves, C.G.; Howard, A.D. Early development of karst systems: 1. Preferential flow path enlargement under laminar flow. *Water Resour. Res.* **1994**, *30*, 2837–2846. [CrossRef]

130. Malard, A.; Jeannin, P.Y.; Vouillamoz, J.; Weber, E. An integrated approach for catchment delineation and conduit-network modeling in karst aquifers: Application to a site in the Swiss tabular Jura. *Hydrogeol. J.* **2015**, *23*, 1341–1357. [CrossRef]

131. Perrin, J.; Luetscher, M. Inference of the structure of karst conduits using quantitative tracer tests and geological information: Example of the Swiss Jura. *Hydrogeol. J.* **2008**, *16*, 951–967. [CrossRef]

132. White, E.L.; Aron, G.; White, W.B. The influence of urbanization of sinkhole development in central Pennsylvania. *Environ. Geol. Water Sci.* **1986**, *8*, 91–97. [CrossRef]

133. Porhemmat, J.; Nakhaei, M.; Altafi Dadgar, M.; Biswas, A. Investigating the effects of irrigation methods on potential groundwater recharge: A case study of semiarid regions in Iran. *J. Hydrol.* **2018**, *565*, 455–466. [CrossRef]

134. Atapour, H.; Aftabi, A. Geomorphological, geochemical and geo-environmental aspects of karstification in the urban areas of Kerman city, southeastern, Iran. *Environ. Geol.* **2002**, *42*, 783–792.

135. Gutiérrez, F.; Galve, J.P.; Guerrero, J.; Lucha, P.; Cendrero, A.; Remondo, J.; Bonachea, J.; Gutiérrez, M.; Sánchez, J.A. The origin, typology, spatial distribution and detrimental effects of the sinkholes developed in the alluvial evaporite karst of the Ebro River valley downstream of Zaragoza city (NE Spain). *Earth Surf. Process. Landforms* **2007**, *32*, 912–928. [CrossRef]

 *sustainability*

Article

Institutional Feasibility of Managed Aquifer Recharge in Northeast Ghana

Lydia Kwoyiga *[ID] and **Catalin Stefan**[ID]

Department of Hydrosciences, TU Dresden, 01069 Dresden, Germany; catalin.stefan@tu-dresden.de
* Correspondence: lydia.kwoyiga@tu-dresden.de

Received: 4 December 2018; Accepted: 9 January 2019; Published: 13 January 2019

Abstract: As part of global efforts to address the challenges that are confronting groundwater for various purposes (including irrigation), engineering methods such as Managed Aquifer Recharge (MAR) have been adopted. This wave of MAR has engulfed some parts of Northern Ghana, characterized by insufficient groundwater for dry-season irrigation. Inspired by the strides of these schemes, the paper assesses the institutional feasibility of MAR methods in the Atankwidi catchment where dry-season farmers may lose their source of livelihood due to limited access to groundwater. We used both primary and secondary data, together with policy documents, to address the following questions: (i) What provisions and impacts formal government institutions had for MAR, and; (ii) what catchment-level institutions exist which may influence MAR. The results show that formal government institutions do not prohibit the adoption of MAR in the country. Among these institutions, it is realized that laws/legislative instruments provide sufficient information and support for MAR than policies and administrative agencies. Moreover, catchment-level institutions which are informal in the form of taboos, rules, norms, traditions, and practices, together with local knowledge play a significant role as far as groundwater issues in the catchment are concerned, and are important for the adoption of MAR methods.

Keywords: Managed Aquifer; Recharge; Groundwater; Institutions; Ghana

---

1. Introduction

Sustainability of water resources for any livelihood activity can be achieved through groundwater engineering methods, such as Managed Aquifer Recharge (MAR). This is because over the years, MAR has proved its ability to replenish aquifers and augment groundwater supplies. This is seen in its wider application where a global inventory puts MAR applications at 1200 schemes; a solution to addressing water scarcity under different climatic, geographic, and socioeconomic conditions [1]. Citing some examples, Page et al. [2] demonstrated that MAR is contributing to achieving sustainable water management, especially in urban areas, as it is also able to store water from different sources. It is, therefore, not surprising that MAR is considered to be a tool capable of promoting groundwater adaptation to climate change and its impacts [3].

Besieged by limited groundwater in some parts of northern Ghana, particularly in the dry season for farming, artificial methods of boosting groundwater resources which include MAR have been adopted [4]. Among the various techniques of MAR implemented are Aquifer Storage and Recovery (ASR) and Pit Infiltrations; both referring primarily to techniques of getting water infiltrated, as classified by International Groundwater Resources Assessment Centre IGRAC [5]. These schemes have offered rural farmers in this part of the country the opportunity to undertake dry-season irrigation in the face of challenging water conditions. Discussions relating to similar schemes are well-documented in places like India [6].

In the Atankwidi catchment in northern Ghana, the population depends almost entirely on groundwater resources. Most smallholder irrigators dig wells in the river bed and adjacent low-lying areas near rivers to exploit groundwater to meet their irrigation water needs. As part of the farming conditions, farmers are expected to refill these wells after every dry-season farming session, which makes the activity difficult. Also, Ghana designed a Riparian Buffer Zone Policy in 2011 [7], which was meant to create vegetative buffers for the preservation and functioning of the country's water bodies and vital ecosystems. When this policy is implemented, these famers will be without any livelihood activity. At the peak of the dry season, these farmers experience water scarcity, for which Barry et al. [8] noted that they will need better technology in order to access groundwater at deeper depths. These farmers' contributions to irrigation in the Upper East region is significant in that Dittoh et al. [9] noted them to be more than those engaged in surface water irrigation. Unfortunately, in recent times, the available groundwater for this activity have been insufficient.

Associated factors, such as increasing population and what Laube et al. [10] discussed as favourable conditions for the dry season (e.g., better infrastructure, the presence of a tomato factory etc.) are further increasing the demand for groundwater in general in the catchment. There are also reports of the incidence of fallen groundwater tables already in the northern part of Ghana, including Atankwidi [11,12] and these also have implications for groundwater availability for dry-season irrigation.

Apart from the aforementioned water challenges identified in the Atankwidi catchment, it has been documented that Ghana will become water-stressed by 2025. Climate change will further exacerbate the situation, as it will bring a reduction in groundwater recharge of 5–22% for 2020 and 30–40% for 2050 [13]. In a study on the Volta Basin about the potential impacts of climate change on subsurface and base flow for groundwater resources, the results show a reduction in recharges in the year 2020 of 17%, 5%, and 22% for Pra, Ayensu, and the White Volta, respectively, while for 2050 these values will increase to 29%, 36%, and 40% for the representative basins [14].

In view of the prevailing limited water conditions and those anticipated in the catchment, some farmers have abandoned farming or adopted measures which are either expensive or have little returns. For instance, Kwoyiga and Stefan [15] brought to the fore that in order to continue with dry-season farming, some irrigators have adopted strategies such as adopting crops that depend less on groundwater as a way of coping with limited groundwater. Unfortunately, some of these strategies do not yield better returns to the farmers.

Considering the importance of dry-season groundwater irrigation, especially in the Atankwidi catchment [16] and guided by the prospects of the emerging MAR schemes in the same northern Ghana, it will be prudent to assess the feasibility of MAR to boost groundwater for irrigation in the dry season. To make this feasible, it is realised that institutions play a crucial role in achieving MAR implementation. It is, therefore, not surprising that Gale [17] stated that even when the hydrological and hydrogeological parameters of a recharge scheme are favourable, consideration also needs to be given to institutions. Asano and Cotruvo [18] lamented that "the lack of guidelines governing artificial recharge of groundwater is currently hampering the implementation . . . of groundwater recharge operations".

Nonetheless, it is observed that while MAR activities are taking shape in Ghana, research relating to it at the moment focus on Geographic Information System GIS tools for mapping and technologies [4]. To the best of our knowledge, the institutional aspect of MAR is yet to receive attention. Even at the global level, it is realized that limited studies focus on the institutional aspect of MAR. For instance, Dillon et al. [3] provided a catalogue of eleven papers, and unfortunately, of their contributions to the increase of MAR application, only one paper discussed the role of institutions. Specifically in regard to Africa, though artificial recharge methods date back to the 1900s, existing literature largely focus on aquifer characterization and maps production [19]. This paper, therefore, assesses the scope of the existing formal government institutions and their implications on the adoption of MAR for irrigational purposes in the Atankwidi catchment.

It is realized that institutions define the kind of requirements or considerations (e.g., entitlement to share a source of water, a permit to construct a well, approval to recharge water to an aquifer, etc.) that need to be met for MAR projects to be approved. These institutions also define regulatory frameworks to cover issues like water rights. They further define agencies or organizations that may be responsible for the management and operations of the project [20]. Apart from legal and regulatory issues, land rights, demand management of groundwater, and the conjunctive use of surface and groundwater are also given attention by institutions [17]. Mechlem [21] exemplified the situation by stating that prior authorization is a precondition for the success of MAR projects. Groundwater instruments like licenses are also needed to abstract, store, and recharge water with legislation dovetailing the conditions for the operation of such a project. Drawing examples from Arizona, Megdal and Dillon [22] note *inter alia* that the success of MAR can be attributed to these arrangements where proposed projects are subjected to rigorous scrutiny based on permit requirements, which include specifications of the kind of MAR projects that should be introduced. The focus here is, therefore, on the technical aspect, approval conditions, regulatory structures, and agencies/organizations that may affect MAR activities in the Atankwidi catchment.

The second objective of this paper looks at the informal institutions at the catchment level and how they may influence the implementation and management of MAR projects for dry-season farming. Page et al. [2] noted that apart from the hydrogeology, topography, and, hydrology, sociocultural and regulatory factors are crucial when choosing a suitable MAR site. Gale [17] opined this by admitting that the success of MAR was also dependent on the role of the local or rural people. Information from such people and their local environment could facilitate planning in terms of the area/community selection, participation responsibilities (construction), and technical options. Dillon [23] shared a similar view that the social environment is a factor that may largely determine the extent to which MAR can achieve its potential for water supplies. The Atankwidi catchment is a rural and traditional catchment where informal institutions or customary water practices encapsulating culture, social relations, and networks predominantly regulate groundwater irrigation. Local knowledge is the main driver of groundwater irrigation in the catchment. Guided by this, assessing the informal institutions may provide important information on the planning, implementing, managing, and operating of MAR projects in a better way.

The overall aim of this study is thus to provide information about institutional guidelines that need consideration by proponents of MAR who may wish to undertake MAR for irrigational purposes in a developing country such as Ghana. The paper starts by looking at issues surrounding groundwater dry-season irrigation in the Atankwidi catchment of northern Ghana. We further review the literature about MAR globally and in Ghana, and offer an insight into the characteristics of the Atankwidi catchment and the source of data for the study. A presentation on both formal government- and catchment-level traditional/informal institutions constitutes the next part of the discussion. The paper concludes by recommending a comprehensive study on the storage capacity of all the aquifers in the catchment. There is also a need for a detailed quantitative study on the actual groundwater demand for dry-season irrigation in the catchment.

2. Review of the Existing Literature

2.1. Managed Aquifer Recharge

According to Dillon [23], MAR refers to the purposeful recharge of water to aquifers for subsequent recovery or environmental benefits. According to Gale [17], MAR projects enable the storage of water in aquifers for subsequent use, to increase groundwater levels, improve water quality, address saline intrusion in waters, etc. Asano and Cotruvo [18] revealed that artificial recharge methods (including MAR) are able to reverse a decline in groundwater levels, enhance groundwater quality by preventing saltwater intrusion, and enable the storage of reclaimed municipal wastewater for future use. Dillon [23] argues that MAR can harvest and reuse water, citing the City of Mount Gambier in Australia where drainage wells have replenished a karstic aquifer for 120 years without indications of poor water quality. MAR also offers an opportunity to build groundwater mounds which block the inflow of contaminated water from areas upstream [24]. Bouwer [25] explained that artificial recharge can address problems of land subsidence, store water, and enhance water quality through soil aquifer treatment or geo-purification, among many other benefits. It is further realized that ponded infiltration (during percolation through the vadose zone and passage through aquifer) can improve the quality of treated sewage effluent or enhance the quality of surface water for irrigation purposes [26]. There is extensive literature on other classifications of MAR [2,3,17,23]. Table 1 provides some highlights of MAR [5,27].

Table 1. Managed Aquifer Recharge (MAR) methods.

Main MAR Methods	Specific MAR Methods	
Techniques referring primarily to getting water infiltrated	Well, shaft, and borehole recharge	Aquifer Storage and Recovery (ASR)/Aquifer Storage, Transfer and Recovery (ASTR) shallow wells/shaft/pit infiltration
	Spreading methods	Infiltration ponds and basin Flooding, Ditch, furrow, drains irrigation
	Induced bank infiltration	River/lake bank filtration Dune filtration
Techniques referring primarily to intercepting the water	In-channel modifications	Recharge dams, Subsurface dams, Sand dams, Channel spreading
	Runoff harvesting	Rooftop rainwater harvesting Barriers and bunds Trenches

Source: Adapted from International Groundwater Research Assessment Centre (IGRAC) [5] and Ringleb et al. [27].

2.2. MAR in Northern Ghana

MAR schemes are noted to be emerging in the northern part of the country. The primary goal of these schemes is usually to make water available for agricultural purposes in the dry season. Some of the communities where these schemes are located are flood-prone areas that usually experience water scarcity, particularly in the dry season. The beneficiary communities are also poor in terms of socio-economic development. A major characteristic of these MAR schemes is that they are usually combined with or augmented by other technologies in developing groundwater for use. These schemes have also been implemented by either individuals or agencies outside the domain of formal government agencies. The Kpaloworgu Sand Dam in the Upper West region, the first of its kind, was initiated by non-Ghanaian individuals. The Aquifer Storage and Recharge (ASR) and Pit Infiltration MAR schemes were implemented by institutes/agencies, not directly connected to formal government institutions.

Regarding the ASR method, its schemes have been implemented by the International Water Management Institute (IWMI) as part of the "Securing Water for Improved Seed and High-Value Vegetable Production in Flood-Prone Areas of Northern Ghana" (Secure Water) scheme. The ASR method is supported by the Bhungroo technology (an an Indian indigenous water-harvesting technique) which, according to Owusu et al. [28], allows for the harvesting of excess floodwater for agricultural use during the dry season. The technology encompasses the harvesting, storage, and abstraction of water. The technology comes with the lifting of water from the Bhungroo and putting it into an overhead tank for on-site water storage and distribution, and a drip or sprinkler irrigation system for applying water to crops [28]. The Jagsi and Kpasenkpe communities in the West Mamprusi District in the Northern Region and the Weisi community in Builsa South District in the Upper East region are the beneficiary communities. These are flood-prone communities that experience floods annually.

The Pit Infiltration with PAVE technology is found only in the northern region of Ghana and adopted by Conservation Alliance International (CA). PAVE Irrigation Technology is a German-originated rainwater harvesting and aquifer recharge irrigation system that injects excess water underground during periods of rainy days and floods. The technology captures flood water and filters and injects into the aquifer and unsaturated fractures. The water is stored underground, from which farmers can use it for about six months. Simple pumps are used to tap the water from the pits. Water is extracted from the injection pipe, or through an alternative pipe. Savelugu Nanton

Municipal Assembly, Tolon-Kumbungu, and the West Mamprusi districts constitute the beneficiary districts. There are thirteen projects.

2.3. Institutions and their Nature

Definitions of institutions come from different sources, each reflecting the backgrounds or perspectives of the theorists. A commonly agreed definition of institutions is thus lacking [29,30]. As a result, one finds a plethora of them in existence, with new ones still emerging. Common among these definitions is the one offered by North [31] which states that institutions are "the rules of the game in a society". Saleth [32] added that institutions are the rules, norms, and strategies which guide the activities and behaviour of individuals. Within the domain of water resources, institutions are the "rules that define action situations, delineate action sets, provide incentives, and determine outcomes ... in the context of water development, allocation, use, and management" [32].

In discussing their nature, Helmke and Levitsky [33] explained that institutions are both formal and informal rules and procedures that structure social interactions. To simply this, they defined informal institutions as "socially shared rules, usually unwritten, that are created, communicated, and enforced outside of officially sanctioned channels. By contrast, formal institutions are rules and procedures that are created, communicated, and enforced through channels widely accepted as official" [33]. Rauf [34] is of a similar view that institutions are both formal and informal, with the informal institutions in the form of norms, customs, and traditions having the additional advantage of generating social capital and influencing resource utilization among people. Institutions are both formal and informal structures; they are often multi-purpose, intermittent, and semi-opaque in operation, and not consciously designed [35]. Focusing on formal institutions, Saleth and Dinar [36] decomposed formal institutions with regard to water resources as laws, policies, and administrative structures.

Concerning their place within the domain of water resources, Kemper [37] discussed that institutions define groundwater instruments, which deal with user rights, abstraction permits or concessions, groundwater tariffs, and subsidies, as they even create groundwater markets. Vatn [30] also noted that institutions define access or ownership of given natural resources. Gale [17] suggested that in examining institutions in relation to MAR, attention should be given to water rights, land ownership, legal and regulatory issues, etc.

In Ghana, Fuest et al. [38] noted that institutions are in the form of statutory laws, legal instruments and regulations, national policies, by-laws, local laws, and project laws. In tracing their source, as far as groundwater use for irrigation is concerned, one sees them emanating from two levels: the national and local levels. As a country where decentralization is practised, the Local Government Act 1993 Act 462 mandates District/Municipal/Metropolitan Assemblies to make by-laws and take certain decisions for the purpose of a function conferred on them. Therefore, local entities like the Assemblies, based on their local conditions, make by-laws to complement those at the national level. Nonetheless, groundwater instruments have not been designed and used within the domain of groundwater irrigation in the country. Occasions that call for securing permits for the use of groundwater for irrigation by smallholder farmers are also uncommon.

In view of all these things, this paper chooses to define institutions as both formal and informal structures where the former comprises policies, laws/legislative instruments, and administrative structures, while the latter encompasses norms, taboos, traditions, etc. which operate outside officially sanctioned channels. These institutions serve multiple purposes and are consciously or unconsciously developed.

3. Materials and Methods

3.1. Study Area

The study area of choice was the Atankwidi catchment, a tributary of the White Volta Basin (Figure 1). It is transboundary in nature and covers an area of about 286 km^2. The portion of the catchment in Ghana is about 159km^2 [39]. Its population in 2010 was 45,841 [40]. The catchment in Ghana is located in the Upper East region, of which four Districts/Municipalities—namely, the Kasena/Nankana Municipality, Kasena/Nankana East District, Bolgatanga Municipality, and Bongo District—are its local political units. The catchment covers six communities in Ghana: Kandiga, Sirigu, Yuwa, Zorko, parts of Sumbrugu, and Mirigu (Figure 1). The people in these communities speak the same language, with similar ethnic characteristics. Agriculture is their major economic activity. Sumbrungu is the most populous and more urbanized community among them.

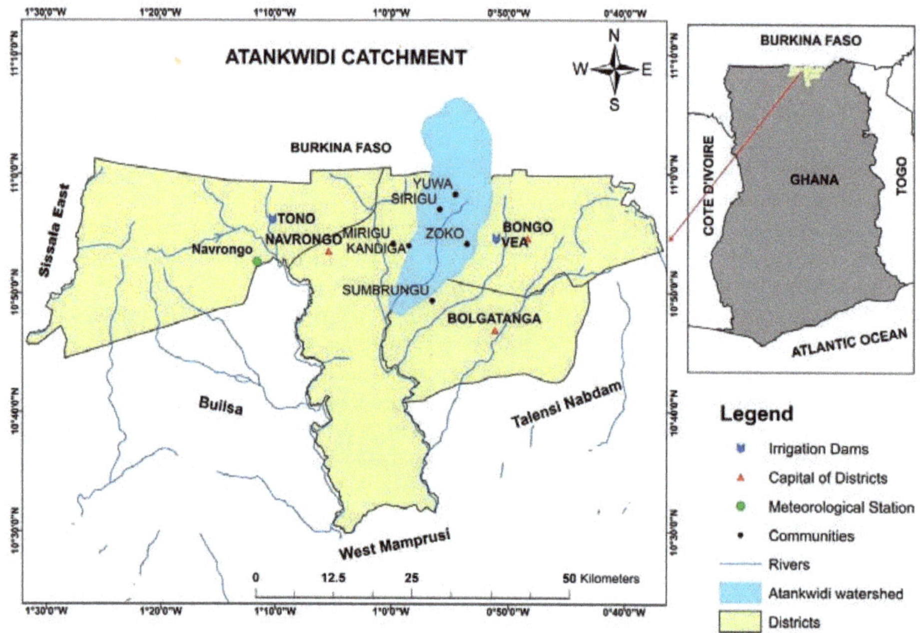

Figure 1. Atankwidi catchment.

The catchment was chosen because the people here rely almost entirely on groundwater for their livelihood activities. It remains a major catchment in the northern part of the country, where groundwater irrigation is significant in the dry season.

The climatic conditions of the catchment, according to Barry et al. [8] are that of the Sudan-Savanna zone, associated with high temperatures and a mono-modal rainfall distribution, with a distinct rainy season lasting from approximately May to September. In examining the entire Volta Basin of which the Atankwidi catchment forms part of, Amisigo [41] stated that the basin is characterized by two main geological systems: the Precambrian platform and a sediment layer, and the Voltain system (which covers about 45% of Ghana). The basin is also noted for the absence of primary porosity; therefore, groundwater occurrence in most of the basin is through the development of secondary porosity. The aquifer systems in the basin (including its portion in Ghana) are highly discontinuous with groundwater, occurring mostly under semi-confined or leaky conditions [42]. Specifically, on the Atankwidi catchment, Martin [11] noted that a larger part of the catchment is associated with the presence of the Paleoproterozoic granitoids. Also, faulting activities have resulted in two smaller shearing faults which are found in the northern tip of the area in the west-east direction.

There are three aquifers characterizing the catchment [11]. These are the discontinuous, shallow, perched aquifer, the regolith aquifer, and the fractured aquifer (Figure 2). Among these three, the regolith aquifer constitutes the principal aquifer in the weathered mantle, resulting in a continuous aquifer whose average saturated thickness is 25 m and hydraulic conductivity being from 2.5E-6 to 2.5-5E m/s, which supplies the yield of most boreholes. The discontinuous shallow perched aquifer is characterized by coarse soils of 0.5 m to 1.5 m thickness and covered by a less permeable clayey or lateritic layer. Nonetheless, this aquifer provides water to traditional wells at very shallow depth, even though they dry up in the dry season.

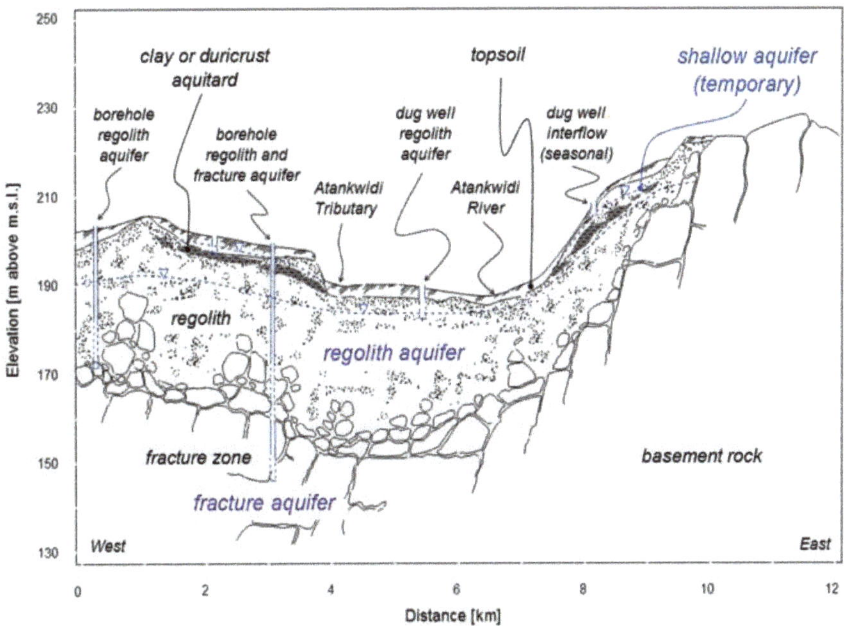

Figure 2. Hydrogeological cross-section of the Atankwidi catchment (Source: Martin [11]).

Focusing on the Volta basin (of which the Atankwidi catchment forms part of), Amisigo [41] observed that the mean monthly potential evapotranspiration exceeds the mean monthly rainfall for most of the year in the basin. Regarding recharge, Namara et al. [16] stated that recharge in the Volta Basin is highly variable, both spatially and temporally, and also low when compared with annual rainfall and evapotranspiration. According to Martin [11], groundwater recharge specifically in the Atankwidi catchment is between 1 and 13% of the mean annual rainfall, which is about 990 mm.

Groundwater quality in the entire Volta basin presents health concerns, especially with regard to the high level of fluoride concentration found in the central part of the basin in northern Ghana [16]. Nonetheless, the water quality is considered suitable for irrigation [43]. The total irrigable area by groundwater, as documented by Barry et al. [8] is about 387 ha.

According to Kwoyiga and Stefan [15], groundwater irrigation is largely undertaken by men. These farmers draw from local knowledge to explore and exploit groundwater for irrigation. Farmers rely on themselves, family members, and friends for financial support and to provide security on their farms. The sizes of the farms are different, and a farm may depend on 2–3 wells for water. Wells are either shallow as in the riverine or deep when on the field. Water is extracted using buckets, ropes, and pumping machines. The crops grown are mostly vegetables like onion, spinach, pepper, tomatoes, cabbage, lettuce, etc. Farmers market their crops either on their farms or in the nearby markets.

3.2. Data for the Study

In order to have an in-depth knowledge and understanding, and also to support proponents of MAR with information about the institutions that need consideration when implementing MAR for irrigation purposes in a developing country such as Ghana, a qualitative approach was employed. In each of the six study communities, two focus-group discussions (each group comprised six members) were held with farmers. Four key informants, such as chiefs, elders, and *tindana* (earth priest) in each community were contacted to obtain data about institutions that regulate groundwater development, as well as use in the catchment. Two officials from the Water Resources Commission (WRC) and one official from the Environmental Protection Agency (EPA) in the Upper East region were also contacted to obtain information about the institutions and management support that may aid MAR implementation.

All questions were open-ended. This was to give respondents the opportunity to freely respond to questions without being limited or influenced. Moreover, there was no pre-existing information about the informal institutions in the catchment—hence the need to avoid close-ended questions. Interviews with the people at the catchment were done using interview guides because most of the people there were illiterate. Questionnaires were used to get responses from the officials at the WRC in the Upper East region. An interview guide was used during the interview with the official at the EPA, which was a decision taken by the official.

All respondents in this category were purposefully selected. Thus, a four-month field trip was undertaken in Ghana in 2017 and a two-month field trip in 2018.

Formal discussions with three purposefully selected members of the Innovative web-based Decision Support System for Water Sustainability under a Changing Climate (INOWAS) Junior Research Group in Technische Universität Dresden, TU Dresden, Germany, together with secondary data from this group about MAR methods, applications, and projects yielded supporting data. A review of the policy documents and legal and institutional frameworks for providing groundwater in Ghana was carried out. These documents were accessed from the internet, as well as from the offices of the WRC and EPA in Bolgatanga.

The first author who collected the data relied on an interpreter who doubled up as a research assistant during the interviews. The data collection tools employed were interviews, questionnaires, observation, conversations, and informal discussions. Data were first recorded, transcribed, and then analyzed manually.

4. Results and Discussion

Drawing from the empirical data and extensive analysis of existing literature and policy documents from Ghana, the ensuing presentations look at the various institutions in the country that need to be considered prior to and during the implementation of MAR methods for irrigational purposes in the Atankwidi catchment.

4.1. Formal Government Institutions

A review of Ghana's official documents and secondary data, together with interview responses of officials of the WRC and the EPA provided the following results. The results here show the various relevant formal government institutions designed at both national and District/Municipal (local) governments. However, the presentation here about the list of relevant institutions is inexhaustible, since new ones are emerging while some existing ones are being modified.

4.1.1. Laws/Legislative Instruments

Currently, there is no legal provision/rule in Ghana that prohibits MAR implementation, and neither is there any single legislation which states that MAR should be implemented to boost groundwater resources. However, some existing legislation or legal instruments appear more explicit and relevant, which proponents of MAR projects need to consider.

- The 1992 constitution of the Republic of Ghana

Land is an important factor as far as MAR is concerned. In Ghana, land and groundwater appear inseparable (an interpretation of the local people). Land ownership in the country has changed over the years; however, a final decision was taken in the 1990s. The 1992 constitution (Article 36(8)) vests all customary lands in the appropriate stool, skin, or landowning family on behalf of and in trust for their people, to be managed as pertained in the duties of the traditional authorities based on customary law. In the Atankwidi catchment, the land is considered skin land. The land is owned by the community, families, or clans. It is entrusted in the various earth priests called *tindana,* who act as the mediator between the gods and the people. This is important for engineers and proponents for MAR when deciding on the source and location of MAR project. Land for such a purpose must be negotiated with and obtained from the local people.

- The Water Resources Commission Act 1996 (Act 522)

The act stipulates that regardless of geographic regions, all water resources belong to the state to be held in trust by the president of the Republic of Ghana. This connotes that groundwater resources associated with MAR form part of the country's pool of water resources, whose ownership is vested in the President of the Republic. This is important for understanding groundwater rights and their enforcement in the catchment, as far as MAR schemes are concerned.

- The Drilling Licence and Groundwater Development Legislative Instrument (L.I) 1827

This Act mandates engineers to obtain drilling licences from the Water Resources Commission. Before the project commences, there is a need for written notification to the Water Resources Commission for permission. It is thus required of MAR engineers to acquire drilling licences beforehand. Wells drillers, as part of the MAR process, are also cautioned by this act to make sure that wells drilled do not pose threats in the form of pollution or contamination to groundwater aquifers.

- Water Use Regulations (L. I) 1692: 2001

MAR goes beyond some of the exemptions of the Water Use Regulations (L. I) 1692: 2001. The regulations make exemptions to people who: apply manual means to lift groundwater from wells; persons who intend to abstract water through mechanical means where the abstraction level is below 5 L/s; and farmers who use the water to cultivate areas of land that do not exceed 1 h. These water users are, however, obliged to register their activities with the District Assembly. Depending on the scale of the MAR scheme, it is important for users of the water to obtain permits, especially if is for irrigation on a large scale.

- The Rivers Act, 1903

This Act states that "a person shall not, without a licence from the Minister, pump, divert, or by any means cause water to flow from a river (a) for purposes of irrigation, or for mines, factories, or any other commercial or industrial purposes, or (b) to generate power". The source of water for MAR includes rivers and other surface water bodies. It is, therefore, compulsory for MAR engineers or proponents to get permits or licences in a situation where the source of water for MAR (river bank infiltration method) is from a river, like the Atankwidi River.

- The Dam Safety Regulations, 2016

These regulations state that licence is required to construct, alter, operate, conduct, or decommission a dam. Therefore, should any MAR methods (subsurface dams or sand dams), be desired, registration of the project is deemed necessary, after which the Dam Safety Licence is issued by the Dam Safety Commission before the project kick-starts.

- Environmental Assessment Regulations, 1999

Regardless of the scale of MAR, an Environmental Impact Assessment may be required, as stipulated in the Environmental Assessment Regulations, 1999. These regulations stipulate that the Environmental Impact Assessment is mandatory for groundwater development for industrial, agricultural, or urban purposes.

From the results, it is realized that there exist a plethora of laws/legislation relating to MAR in Ghana. However, there are no applicable by-laws of the four Districts/Municipalities in this regard. A critical look at the existing legislation shows that they are designed to serve multiple purposes, which confirms the argument of Cleaver [44] about the nature of institutions in Africa. These institutions, in view of MAR, touched on issues such as the source of water for recharge, land, and water rights, construction/drilling licences, environmental permits, and use of the water. They are legally binding and may hinder MAR implementation if the approval requirements are not met. However, there is missing information concerning the specific guidelines for water treatment (quality) for recharge, so MAR operators may, therefore, need to adopt certain international guidelines. There is also no information on the preferred MAR methods in the country. Groundwater instruments, as far as irrigation is concerned, are also not designed. Even though local farmers, based on local knowledge, conjunctively use surface water and groundwater, this has not been spelt out by these institutions. Demand management is only applicable to groundwater for domestic purposes. These issues are important, and may negatively affect the implementation and operations of MAR if not properly considered. Therefore, it behooves MAR operators to consider these before initiating MAR projects. Like Casanova et al. [45] noted of France, and Megdal et al.'s [46] discussion of the USA state of Arizona, Ghana's laws/legislation, to some extent, are explicit on the requirements that need to be met in order for MAR to be accepted. When compared with policies and administrative agencies/organisations, it can be concluded that the relevant existing laws provide better and more comprehensive information for MAR activities in the country, and failure to comply with them may result in the rejection of MAR projects.

4.1.2. Policies

Specific policy provisions for MAR as part of groundwater resource development are lacking. Nonetheless, the following policies provide information relating to MAR implementation.

- Groundwater Development Strategy, 2011

This strategy, though yet to be implemented, may contribute to the realization of MAR objectives. The strategy intends to boost data and information on groundwater, strengthen capacity in terms of technical and organisational aspects, and encourage stakeholder participation, among other things. When implemented, this strategy may reduce the burden of engineers or proponents of MAR schemes who will need to build the capacities of the organization or agencies to manage and operate the schemes.

- The National Rainwater Harvesting Strategy, 2011

Formulated by the then Ministry of Water Resources, Works, and Housing (MWRWH) as a roadmap to augment water service delivery in both rural and urban areas of the country, the strategy spans a period of 2012–2025. It concerns rainwater harvesting, which may be relevant regarding the source of water for MAR projects.

- The Ghana Climate Change Strategy, 2012

This strategy advocates for methods of improving water resources for agricultural activities. It implies that in this era of climate change and its impacts, MAR may be considered a part of the country's adaptation measures.

- The Ghana National Climate Change Policy, 2014

This policy promotes the constructions of water storage systems through rainwater harvesting. Water harvesting in this source could boost the availability of water sources for MAR.

- The Ghana Water Policy, 2007

This policy also encourages Ghanaians to harvest rainwater at the household level and that of the community level. This is to boost water availability in general by broadening water sources, and is also a way of containing excess water on the ground surface. This is a plus towards the realization of MAR, as rainwater harvesting could serve as the source of water for recharge.

- The National Environmental Policy, 2012

This policy lends credence to the need to subject water resources development projects to an Environmental Impact Assessment. As per this policy, MAR projects are therefore supposed to undergo impact assessment to attain the nature of impacts that MAR may have on the environment.

In view of policies, issues such as the source of water for MAR and organizational management of the water, as well as awareness/education about MAR activities have been captured. However, these are only general and are not action-oriented, as far as MAR is concerned. None of these policies specifically made provisions for groundwater development through recharge. A critical look at the climate change adaptation policies reveals that the aim of harvesting surface water in this regard is to avert hydrological disasters (flood control) and not to capture water to deliberately boost recharge. The National Rainwater Harvesting Strategy and the Ghana Water Policy, 2007 have been put in place as part of measures to boost household and municipal water supply, but not for groundwater recharge. One would have expected that the other related policies in the country, such as the Food and Agricultural Sector Development Policy (FASDEP I& II) of 2003 and 2007, respectively; the Ghana Water Policy, 2007; the National Climate Change Adaptation Strategy, 2012; and National Climate Change Adaptation Policy, 2014 would have prioritized and included groundwater resources development (including recharge) in them—unfortunately, none of these did.

In that same direction, the Ghana Irrigation Development Policy 2011, in section 5.3.1, skeletally stated that efforts shall be made to promote access to safer groundwater or safer irrigation practices where only marginal-quality water is available. The state of the Water Policy 2007 regarding groundwater irrigation is but vague. This is because the policy in section 2.2.3 (Water for Food Security) only touches on supporting micro-irrigation schemes among rural areas without specifying the source of water for these schemes. Generally, water resource policies outline the roles of the government and other stakeholders, define monitoring and controlling measures, and state how to build capacity for management. Unfortunately, a review of the existing policies shows that groundwater development through MAR is completely absent within the policy framework of the country. This may negatively affect the adoption of MAR in terms of resources in Ghana. Proponents of MAR will need to identify the ways and means of addressing these constraints when planning MAR schemes.

4.1.3. Administration

There are some administrative bodies in the country that deal with water resources in general on one hand, and with irrigation on the other. The activities of the following agencies nonetheless relate to MAR and groundwater.

- Water Resources Commission (WRC)

This agency has representative offices at the regional level. The commission processes all water rights and permits. It is the lead regulator of all water resources in Ghana. It is therefore important for MAR proponents to involve this commission, especially at the planning stage.

- Environmental Protection Agency (EPA)

This agency is the leading environmental agency of the country's government, with offices in all the regions. It is responsible for conducting Environmental Impact Assessments (EIA). Prior to MAR implementation, notification to this agency is required.

- Water Research Institute (WRI)

This institute, unfortunately, has no offices at the regional level, as it is only based in Accra, the capital. It is relevant for MAR projects because this outfit has conducted research and collaborated with other research organisations concerning information-gathering pertaining to the hydro-geology, hydrochemistry, and other issues about groundwater for the entire country.

- District Assemblies (DA)

The assemblies are decentralized agencies or local government entities who play both political and administrative roles at the local level. The assemblies, as part of their responsibilities, are supposed to monitor drilling and wells construction activities in the district. MAR wells constructors will, therefore, need to register with the assembly.

- River Basin Management Boards

These boards have been purposefully constituted to manage river basins in the country. Unfortunately, their focus is more on surface water than groundwater resources. Despite this, they are regarded as important stakeholders whose platform enables the management of water resources, as well as for conflict resolution.

The critical issues raised here include responsibilities about groundwater regulation, pollution, and hydrogeological knowledge/information. The WRC only regulates, rather than develops [47], and from the interviews, it is realized that its responsibilities, especially at the regional level, focus more on surface water than on groundwater resources. At the moment, MAR implementation may be constrained due to the absence of technical, financial, and management support from these institutions. For instance, it is difficult to identify people with technical skills and resources available to design, construct, and operate MAR projects.

Groundwater development and management, which should have been a prerogative of the Irrigation Development Authority according to the Irrigation Development Authority Act, 1977, has not been carried out (see Ministry of Food and Agriculture [48]). The District/Municipal Assemblies only support the management of groundwater facilities for domestic purposes. This implies that currently, and as far as groundwater development for irrigation is concerned, MAR projects cannot be associated with any administrative agency in the country. It is not surprising that the MAR projects in the northern part of the country are spearheaded and championed largely by agencies such as the Conservation Alliance and the International Water Management Institute, together with the local communities. As such, MAR proponents and engineers should be willing to develop groundwater with limited support from these agencies. Proponents should also identify individuals or agencies that may be responsible for managing the MAR projects when implemented. This can be done through collaboration with the WRC, other water agencies in the country, or through the creation of an independent organization for such a purpose.

4.2. Catchment-Level Institutions

From the interviews conducted in the six communities in Atankwidi, it has been revealed that informal institutions remain strong and influential as far as groundwater irrigation is concerned in the catchment.

Nature of Institutions

The Atankwidi catchment, as noted already, is a rural catchment where informal institutions are in full operation. These are limited largely to the catchment and the neighbouring communities who are connected through kinship or the social network. These informal institutions are in the form of:

- Taboos
- Rules
- Customs/norms/traditions
- Groundwater leaders

In order to highlight the relevance and provisions of these institutions in respect of MAR, a detailed presentation is shown in Table 2. This takes into consideration the regulatory aspect, approval requirements, and agencies/organizations in regard to MAR.

Table 2. Catchment-level institutions, and their provisions for MAR.

Institution	Provision for MAR
Rules 1. Land and groundwater are inseparable gifts from the gods/nature and therefore belong to every member of the catchment; thus, access is free. 2. Extraction of groundwater is mostly the responsibility of individual farmers. 3. Farmers exploit groundwater through the construction and maintenance of individual wells. No one is supposed to trespass on another irrigator's land/groundwater. Irrigators, however, sometimes help one another by granting free access to water in their wells. 4. Wells can be constructed at any time of the year. 5. Information about the spiritual component of irrigation is kept secret among irrigators.	1. Land/groundwater here is owned by the community; thus, land for siting the MAR scheme must be acquired from the local people. 2. MAR projects can be individually or communally owned. This is important for managing and operating MAR projects. 3. Farmers may be able to operate MAR projects (small-scale) individually; however, social network/capital may influence the process, depending on the nature of the MAR project. 4. MAR projects can be implemented at any time of the year. 5. Local knowledge of farmers about groundwater offers a quicker way to understanding and managing the MAR projects.
Taboos 1. There shall be no construction of wells in sacred groves or places considered sacred. 2. There shall be no fetching of water at night, especially near places considered to be the abodes of the gods.	1. The physical location of the MAR scheme must be accepted and approved by the local people. 2. As a rural catchment where there are no wastewater treatment plants, the source of water will obviously not come from a wastewater treatment plant. Therefore, where the source of water for the MAR scheme is from a river, detailed community consultations (chiefs/earth priests) are required prior to implementation.
Customs/Norms/Traditions 1. All water resources are sacred. 2. Rituals and sacrifices are made for abundant water (rains to recharge).	1. Water from MAR projects is equally considered sacred and treated diligently. 2. This is important for conserving and protecting MAR projects, due to the importance attached to all water resources in the catchment.
Groundwater leaders Chiefs, elders, heads of clans/families, farmer groups/associations, and youth groups.	These are local but traditional political leaders of the catchment, gate keepers, and major decision-makers. They will contribute to managing and operating MAR projects. Their participation in decisions regarding the scheme must always be considered.

The catchment-level institutions, as far as groundwater irrigation is concerned, are informal. They are undocumented and enshrined largely in the belief systems and traditions of the people, as described by Helmke and Levitsky [33]. They are in the form of taboos, traditions, and norms, and many others are embedded in the local knowledge of the people, which is applied in groundwater irrigation in the catchment. They are also local in nature in that they are only limited to the catchment. These institutions have not been consciously developed for groundwater irrigation purposes only. They have historical connotations, are enmeshed in the people's culture, and are thus not easily altered. Social relations and networks penetrate these institutions. The satisfaction derived from the application of these institutions is not necessarily expressed in economic gains, but is more about social welfare. Their relevance has been noted by Obeng-Odoom [49] in the entire water sector of Ghana in the past.

Fortunately, they have some provisions for MAR (Table 2), especially in relation to planning and operating MAR projects. These institutions provide information on authorization, the approval of sites, and the location of MAR projects. They also define water rights. These institutions further spell out the leadership structure in the catchment that may be crucial for the governance, management, and operation of MAR schemes. Every farmer already possesses knowledge of groundwater, which is significant for building/improving human resource capacities to operate MAR projects.

It can be said that these institutions do not prohibit, but rather favour the adoption of MAR in the catchment. However, choices/decisions of the local people must be respected. For instance, any scheme that fails to take into consideration the taboos, customs, and traditions of the local people will be resisted and may not be implemented.

5. Conclusions

The increasing growth rate of the population, fallen groundwater tables, impacts of climate change on water resources, and the booming of groundwater irrigation may constitute the basis for more MAR activities in Ghana. Moreover, although developing countries like Ghana have no specific institutions regarding artificial methods of groundwater recharge, the existing ones do not prohibit it.

Formal government institutions, like laws or legislative instruments, provide sufficient but relevant information on the requirements and regulatory structures for MAR schemes in Ghana, even though policy formulations failed to significantly capture groundwater development through artificial methods. Policy support at the moment may negatively impact the adoption and operation of MAR projects in the country. Proponents of MAR will, therefore, need to mobilize a lot of resources (technical, financial, and managerial) in order to achieve MAR goals in the country. Catchment-level institutions, in the form of rules, taboos, customs, and practices, favor MAR adoption through local knowledge, planning, and management and operations. All institutions must thus be given significant attention to facilitate the approval, easy acceptance, implementation, and operation of MAR projects.

For MAR to be effectively adopted in Atankwidi, this paper recommends a comprehensive study of the storage capacity to be done of all the aquifers in the catchment. Currently, there is no available study about the storage capacity of all the aquifers in the catchment; the only study at the moment is by Barry et al. [8] who studied only one of the aquifers. There is also the need for a detailed quantitative study about the actual groundwater demand for dry-season irrigation in the catchment, since the information at the moment is largely qualitative. Land use is another issue that needs consideration. Environmental degradation, population growth, and agricultural activities continuously alter the environment. Land-use studies have been done already, but this information needs to be updated so as to map out areas that may be suitable for MAR.

Author Contributions: Conceptualization; Data curation; Formal analysis; Funding acquisition; Methodology and Writing–original draft were by L.K.; while Resources, Supervision; Writing–review & editing, were done by C.S.

Funding: The research received funding from the Government of Ghana through DAAD (Germany) and the Graduate Academy, TU Dresden for the doctoral studies/field work in Ghana. The University for Development Studies (UDS), Ghana also supported this study financially.

Acknowledgments: The authors acknowledge the contributions of Paul Alagidede through informal discussions to the article. The INOWAS Junior Research Group, TU Dresden is also acknowledged for its support in terms of literature/materials for the article. We further acknowledge the assistance of Mr Francis Anafo during the data collection in Ghana. We thank the people of the Atankwidi catchment and the officials of the various organisations/agencies in Ghana for their responses and materials. The authors again, wish to thank the anonymous reviewers for carefully reviewing the article.

Conflicts of Interest: The authors declare that there is no conflict of interest.

References

1. Stefan, C.; Ansems, N. Web-based global inventory of managed aquifer recharge applications. *Sustain. Water Resour. Manag.* **2018**, *4*, 153–162. [CrossRef]
2. Page, D.; Bekele, E.; Vanderzalm, J.; Sidhu, J. Managed Aquifer Recharge (MAR) in Sustainable Urban Water Management. *Water* **2018**, *10*, 239. [CrossRef]
3. Dillon, P.; Pavelic, P.; Nava, A.P.; Weiping, W. Advances in multi-stage planning and implementing managed aquifer recharge for integrated water management. *Sustain. Water Resour. Manag.* **2018**, *4*, 145–151. [CrossRef]

4. Owusu, S.; Mul, M.L.; Ghansah, B.; Osei-Owusu, P.K.; Awotwe-Pratt, V.; Kadyampakeni, D. Assessing land suitability for aquifer storage and recharge in northern Ghana using remote sensing and GIS multi-criteria decision analysis technique. *Model. Earth Syst. Environ.* **2017**, *3*, 1383–1393. [CrossRef]

5. International Groundwater Resources Assessment Centre (IGRAC). Artificial Recharge of Groundwater in the World, Report. Available online: https://www.un-igrac.org/sites/default/files/resources/files/2008_IGRAC_Global%20MAR%20Inventory%20Report.pdf (accessed on 10 April 2018).

6. Christoff, S.P.; Sommer, M.J. Women's Empowerment and Climate Change Adaptation in Gujarat, India: A Case-Study Analysis of the Local Impact of Transnational Advocacy Networks. *Sustainability* **2018**, *10*, 1920. [CrossRef]

7. Ministry of Water Resources, Works and Housing. *Riparian Buffer Zone Policy for Managing Freshwater Bodies in Ghana*; Government of Ghana: Accra, Ghana, 2011.

8. Barry, B.; Kortatsi, B.; Forkuor, G.; Gumma, M.K.; Namara, R.E.; Rebelo, L.-M.; van den Berg, J.; Laube, W. *Shallow Groundwater in the Atankwidi Catchment of the White Volta Basin: Current Status and Future Sustainability*; IWMI: Colombo, Sri Lanka, 2010; Volume 139.

9. Dittoh, S.; Awuni, J.A.; Akuriba, M.A. Small pumps and the poor: A field survey in the Upper East Region of Ghana. *Water Int.* **2013**, *38*, 449–464. [CrossRef]

10. Laube, W.; Schraven, B.; Awo, M. Smallholder adaptation to climate change: Dynamics and limits in Northern Ghana. *Clim. Chang.* **2012**, *111*, 753–774. [CrossRef]

11. Martin, N. Development of a Water Balance for the Atankwidi Catchment, West Africa: A Case Study of Groundwater Recharge in a Semi-Arid Climate. Ph.D. Thesis, University of Goettingen, Goettingen, Germany, 2006.

12. Johnston, R.M.; McCartney, M. *Inventory of Water Storage Types in the Blue Nile and Volta River Basins*; IWMI Working Paper 140, International Water Management Institute (IWMI): Colombo, Sri Lanka, 2010; 40p.

13. Kankam-Yeboah, K.; Obuobie, E.; Amisigo, B. *Climate Change Impacts on Water Resources in Ghana*; Ghana National Commission for UNESCO: Accra, Ghana, 2009; pp. 65–69.

14. Water Resources Commission. White Volta River Basin—Integrated Water Resources Management Plan. Available online: http://webcache.googleusercontent.com/search?q=cache:vkxrZJOB6eIJ:www.wrc-gh.org/dmsdocument/19+&cd=2&hl=en&ct=clnk&gl=de (accessed on 20 March 2016).

15. Kwoyiga, L.; Stefan, C. Groundwater Development for Dry Season Irrigation in North East Ghana: The Place of Local Knowledge. *Water* **2018**, *10*, 1724. [CrossRef]

16. Namara, R.E.; Horowitz, L.; Nyamadi, B.; Barry, B. *Irrigation Development in Ghana: Past Experiences, Emerging Opportunities, and Future Directions*; GSSP Working Papers; International Food Policy Research Institute: Accra, Ghana, 2011.

17. Gale, I. *Strategies for Managed Aquifer Recharge (MAR) in Semi-Arid Areas*; UNESCO IHP: Pari, France, 2005.

18. Asano, T.; Cotruvo, J.A. Groundwater recharge with reclaimed municipal wastewater: Health and regulatory considerations. *Water Res.* **2004**, *38*, 1941–1951. [CrossRef]

19. MacDonald, A.M.; Bonsor, H.C.; Dochartaigh, B.É.Ó.; Taylor, R.G. Quantitative maps of groundwater resources in Africa. *Environ. Res. Lett.* **2012**, *7*, 024009. [CrossRef]

20. Dillon, P.; Pavelic, P.; Page, D.; Beringen, H.; Ward, J. Managed aquifer recharge: An introduction. *Waterlines Rep. Ser.* **2009**, *13*, 86.

21. Mechlem, K. Groundwater governance: The role of legal frameworks at the local and national level—Established practice and emerging trends. *Water* **2016**, *8*, 347. [CrossRef]

22. Megdal, S.B.; Dillon, P. Policy and economics of managed aquifer recharge and water banking. *Water* **2015**, *7*, 592–598. [CrossRef]

23. Dillon, P. Future management of aquifer recharge. *Hydrogeol. J.* **2005**, *13*, 313–316. [CrossRef]

24. Moeck, C.; Radny, D.; Popp, A.; Brennwald, M.; Stoll, S.; Auckenthaler, A.; Berg, M.; Schirmer, M. Characterization of a managed aquifer recharge system using multiple tracers. *Sci. Total Environ.* **2017**, *609*, 701–714. [CrossRef] [PubMed]

25. Bouwer, H. Artificial recharge of groundwater: Hydrogeology and engineering. *Hydrogeol. J.* **2002**, *10*, 121–142. [CrossRef]

26. Greskowiak, J.; Prommer, H.; Massmann, G.; Johnston, C.D.; Nützmann, G.; Pekdeger, A. The impact of variably saturated conditions on hydrogeochemical changes during artificial recharge of groundwater. *Appl. Geochem.* **2005**, *20*, 1409–1426. [CrossRef]

27. Ringleb, J.; Sallwey, J.; Stefan, C. Assessment of Managed Aquifer Recharge through Modeling—A Review. *Water* **2016**, *8*, 579. [CrossRef]

28. Owusu, S.; Cofie, O.O.; Osei-Owusu, P.; Awotwe-Pratt, V.; Mul, M.L. *Adapting Aquifer Storage and Recovery Technology to the Flood-Prone Areas of Northern Ghana for Dry-Season Irrigation*; International Water Management Institute (IWMI): Colombo, Sri Lanka, 2017; Volume 176.

29. De Koning, J. Reshaping Institutions: Bricolage Processes in Smallholder Forestry in the Amazon. Ph.D. Thesis, Wageningen University, Wageningen, The Netherlands, 2011.

30. Vatn, A. *Institutions and the Environment*; Edward Elgar Publishing: Cheltenham, UK, 2005; ISBN 978-1-84542-574-6.

31. North, D. *Institutions, Institutional Change and Economic Performance*; Cambridge University Press: Cambridge, UK, 1990.

32. Saleth, R.M. Understanding water institutions: Structure, environment and change process. In *Water Governance for Sustainable Development*; Perret, S., Farolfi, S., Hassan, R., Eds.; Earthscan: London, UK, 2006.

33. Helmke, G.; Levitsky, S. Informal institutions and comparative politics: A research agenda. *Perspect. Politics* **2004**, *2*, 725–740. [CrossRef]

34. Rauf, M. Innovations and informal institutions: An institutionalist approach to the role of social capital for innovation. *J. Acad. Res. Econ.* **2009**, *1*, 25–33.

35. Cleaver, F. *Development through Bricolage: Rethinking Institutions for Natural Resource Management*; Routledge: London, UK, 2017.

36. Saleth, R.M.; Dinar, A. *The Institutional Economics of Water: A Cross Country Analysis of Institutions and Performance*; Edward Elgar: Cheltenham, UK, 2004.

37. Kemper, K.E. Instruments and institutions for groundwater management. In *The Agricultural Groundwater Revolution: Opportunities and Threats to Development*; Giordano, M., Villholth, G.K., Eds.; CAB International: London, UK, 2007; pp. 153–172.

38. Fuest, V.; Ampomah, B.; Haffner, S.A.; Tweneboah, E. *Mapping the Water Sector of Ghana: An Inventory of Institutions and Actors*; Center for Development Research: Bonn, Germany, 2005.

39. Salifu, T.; Agyare, W.A. Distinguishing different land use types using surface albedo and normalized difference vegetation index derived from the SEBAL for the Atankwidi and a farm sub catchments in Ghana. *J. Eng. Appl. Sci.* **2012**, *7*, 69–80.

40. Ghana Statistical Service. *2010 Population and Housing Census: Summary Report of Final Results*; GSS, Government of Ghana: Accra, Ghana, 2012.

41. Amisigo, A.B. Modelling Riverflow in the Volta Basin of West Africa: A Data-Driven Framework. Ph.D. Thesis, University of Bonn, Bonn, German, 2006.

42. Kortatsi, B. Groundwater utilization in Ghana. *IAHS Publ. Ser. Proc. Rep. Intern Assoc. Hydrol. Sci.* **1994**, *222*, 149–156.

43. Barnie, S.; Anornu, G.; Kortatsi, B. Assessment of the Quality of Shallow Groundwater for Irrigation in the Atankwidi Sub-Basin of the White Volta Basin, Ghana. *J. Nat. Sci. Res.* **2014**, *4*, 1–11.

44. Cleaver, F. Reinventing institutions: Bricolage and the social embeddedness of natural resource management. *Eur. J. Dev. Res.* **2002**, *14*, 11–30. [CrossRef]

45. Casanova, J.; Devau, N.; Pettenati, M. Managed Aquifer Recharge: An Overview of Issues and Options. In *Integrated Groundwater Management: Concepts, Approaches and Challenges*; Jakeman, A.J., Barreteau, O., Hunt, J., Rinaudo, J., Ross, A., Eds.; Springer: Cham, Switzerland, 2016; pp. 413–434. ISBN 978-3-319-23575-2.

46. Megdal, S.B.; Dillon, P.; Seasholes, K. Water banks: Using managed aquifer recharge to meet water policy objectives. *Water* **2014**, *6*, 1500–1514. [CrossRef]

47. Chokkakula, S.; Giordano, M. Do policy and institutional factors explain the low levels of smallholder groundwater use in Sub-Saharan Africa? *Water Int.* **2013**, *36*, 790–808. [CrossRef]

48. Ministry of Food and Agriculture. *National Irrigation Policy, Strategies and Regulartory Measures*; Ministry of Food and Agriculture: Accra, Ghana, 2011.

49. Obeng-Odoom, F. Marketising the commons in Africa: The case of Ghana. *Rev. Soc. Econ.* **2016**, *74*, 390–419. [CrossRef]

 *sustainability*

Article

An Approach to Study Groundwater Flow Field Evolution Time Scale Effects and Mechanisms

Dianlong Wang [1,2,*], Guanghui Zhang [2], Huimin Feng [3], Jinzhe Wang [2] and Yanliang Tian [2,*]

[1] State Key Laboratory of Simulation and Regulation of Water Cycle in River Basin, China Institute of Water Resources and Hydropower Research, No. 20 Chegongzhuang Road, Beijing 100048, China

[2] Institute of Hydrogeology and Environmental Geology, Chinese Academy of Geological Sciences, No. 258 Zhonghua Street, Shijiazhuang 050800, China; huanjing@heinfo.net (G.Z.); 5885970@sina.com (J.W.)

[3] College of Urban and Rural Construction, Shanxi Agricultural University, No. 1 Mingxian Road, Taigu 030801, China; fenghuimin1997@163.com

* Correspondence: sxndwdl@163.com (D.W.); yanliang209@163.com (Y.T.); Tel.: +86-0351-466-6415 (D.W.); +86-0311-6750-5832 (Y.T.)

Received: 24 July 2018; Accepted: 20 August 2018; Published: 21 August 2018

Abstract: The temporal scale effect is an important issue for groundwater system evolution research. The selection of an appropriate time scale will enhance the understanding of the characteristics and mechanisms of groundwater flow field evolution. In this study, a methodology was provided to analyze the groundwater system evolution, focusing on the choice of the suitable time step for identifying the distinct stages of evolution, characterized by different behavior linked to the management of the groundwater system. The evolution trend of the groundwater level in the center of the cone of depression at different time scales, combined with the F test and the groundwater system balance index (R_e) categories, were used for the choice of the time step and the division of the evolution stages. Based on the transformed groundwater level time series using the selected best time step, the main factors controlling the groundwater evolution were assessed for the different stages. Our results show that the methodology can exactly identify the different important stages of the evolution, and they can be used to individually study these stages, which can help to reveal the mechanisms of the groundwater evolution more easily. Therefore, it is useful to obtain an increased knowledge of the regional groundwater dynamics.

Keywords: groundwater flow field; scale effects; discrete wavelet transform; time series analysis; multiple stresses

---

1. Introduction

Groundwater is becoming increasingly important, because it can be used to support the public water supply and ecosystem services, especially during longer drought periods [1]. Therefore, understanding the mechanisms and characteristics of the groundwater system evolution is critical for the sustainable development and utilization of groundwater resources [2,3]. The time scale of the groundwater level used in the research affects our understanding of the characteristics and mechanisms of the groundwater field evolution, and affects the recognition of temporal and spatial dimension characteristics. If the time scale is too brief, there will be too many evolution stage divisions, and the understanding of the evolution and long-term trends of the groundwater flow field will be reduced. On the contrary, if the time scale is excessively lengthy, the threshold characteristics may be masked [4,5]. Therefore, a suitable time scale must be identified in order to obtain a comprehensive understanding of the mechanisms and characteristics of the groundwater system evolution, assessing of the main influencing factors of the groundwater system evolution.

In recent years, a number of numerical and time series analyses have been in use for the study of groundwater evolution. Firstly, the groundwater flow model is a frequently used tool to study the groundwater system. Many researchers have studied the groundwater level signal variation of specific areas through establishing the groundwater numerical model [6–9]. Secondly, the time series statistical analysis is another common method for the analysis of groundwater level variation. Lafare et al. [10] used the seasonal trend decomposition to analyze the groundwater variation, which can decompose the groundwater level fluctuations signal into the following three components: (i) the trend component; (ii) the seasonal or repeated component; and (iii) the remainder, residual, or noise component. Asmuth et al. [11] decomposed the time series of the groundwater head fluctuations related to multiple stresses. Autocorrelation and cross-correlation are usually used to assist in the identification of the main factors influencing the groundwater evolution, and to evaluate the potential delay between the application of the factor and the response within the signal [12,13].

Nevertheless, the methods mentioned above cannot achieve the selection of an appropriate time step for identifying the distinct stages of groundwater system evolution. Discrete wavelet transform (DWT) analysis is a useful tool for such an application, because it requires no assumptions of statistical stationarity. We can therefore use this method to investigate the scale-dependent variations and co-variations of environmental properties that change spatially or contain transient features. This allows us to identify the important stages of evolution within a time series, and to study these stages individually. Although the DWT technique has been widely used in various fields [4,14,15], there was no research on the identification of a distinct component within the evolution of a groundwater system with different stages characterized by different behaviors using this approach.

In this study, we took the Hufu Plain, in North China, as a case-study area, and proposed a new methodology applicable to study the groundwater flow evolution. With this methodology, we can obtain increased knowledge of the regional groundwater dynamics, as follows: (i) select a suitable time step for identifying the distinct stages of groundwater system evolution, (ii) assess the main factors controlling the groundwater evolution for the different stages by using the transformed groundwater level time series using the selection best time step, and (iii) analyze the mechanisms of the groundwater system evolution based on the division of the different stages.

The groundwater flow field in the Hufu Plain has changed considerably over the past 50 years. Since the severe regional drought of 1973, the abstraction of the Hufu Plain groundwater has drastically increased compared with 1961–1972; the groundwater level below the surface has declined from <5 m, during the 1960s and 1970s, to 5–50 m today. In particular, the horizontal water flow has continuously slowed, while the vertical water flux has continuously increased, and the direction of the groundwater flow has changed from a west to east movement under the natural state to the current state of flowing from the peripheral regions of the overexploited area to the center of the groundwater depression cone [16]. Several studies have examined the aspects of groundwater system evolution in the Hufu Plain using time series analysis [17,18], groundwater modeling, and numerical modeling [19–22]. There is less information, however, on the time scale effect of the groundwater flow field in this area.

2. Materials and Methods

2.1. Study Areas

The Hufu Plain is located in the middle of the North China Plain (Figure 1a). It is bounded in the west by Mount Taihang and on the east, north, and south by gently undulating plains. The Hufu Plain has an area of about 8205 km^2. The Hutuo River and Fuyang River are the main rivers in the Hufu Plain, and the annual runoff has decreased by over 80% [23] since the 1970s, because of the wholesale construction of the water storage projects upstream and the severe regional drought. The elevation of the Hufu Plain ranges from 15 to 87 m above sea level, and the annual precipitation ranges from 200 to 1000 mm/year, which is concentrated mostly in the summer.

The Hufu Plain aquifer can be divided into the following four zones [24]: the Holocene aquifer (I), the upper Pleistocene aquifer (II), the middle Pleistocene aquifer (III), and the lower Pleistocene aquifer (IV). Given the close connection between I and II, the two aquifer zones are hereafter described as the shallow groundwater aquifer (I + II). The three aquifer zones of I + II, III, and IV, as shown in Figure 1h, are separated by a relatively thick layer of clay, which acts as a confining unit over the lower zones.

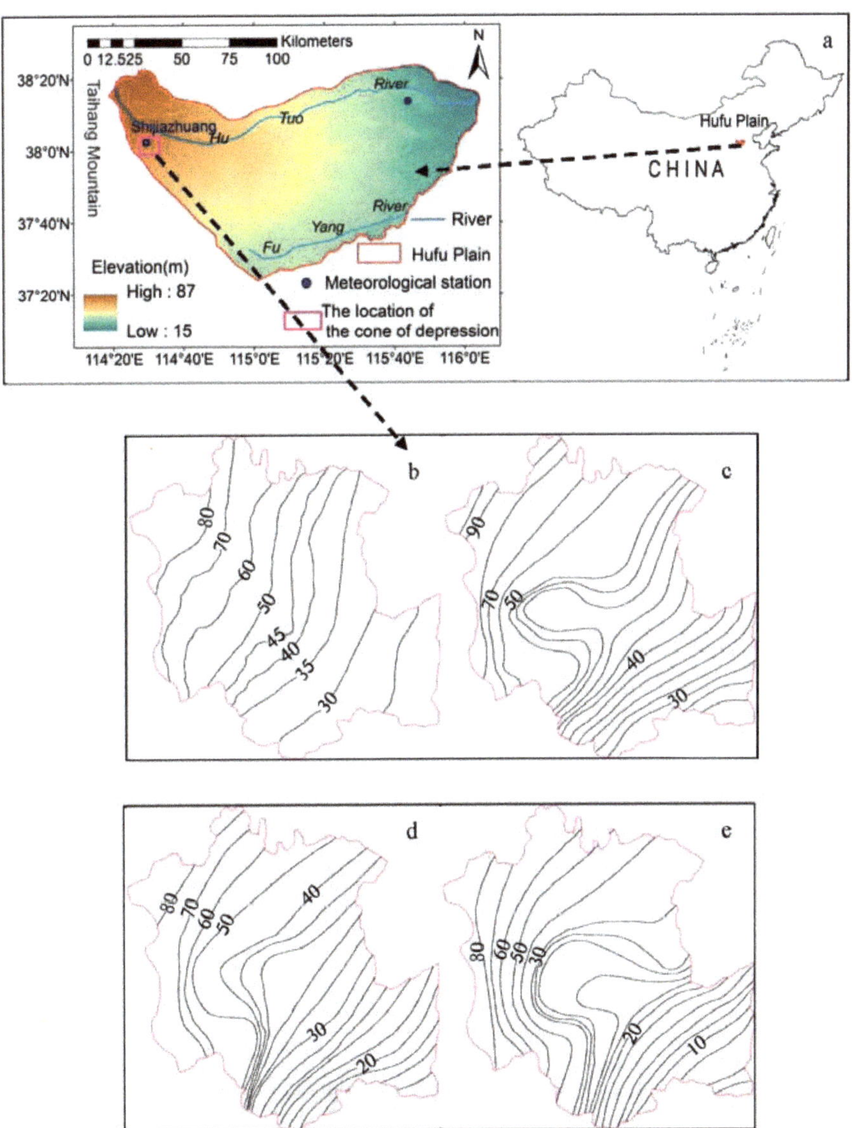

Figure 1. *Cont.*

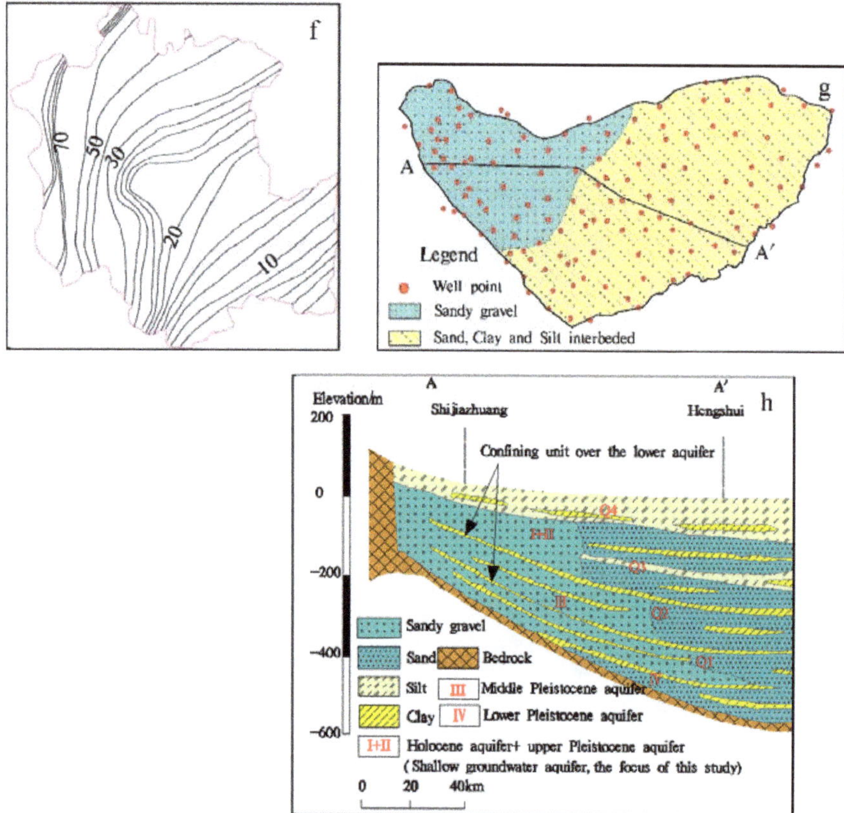

Figure 1. (**a**) The Hufu Plain research area, the groundwater level depression cone area; (**b**) groundwater elevation in the area of the cone of depression for 1961, (**c**) 1980, (**d**) 1995, (**e**) 2005, and (**f**) 2010; (**g**) lithological and monitoring wells distribution of the shallow groundwater aquifer; and (**h**) hydrostratigraphy along the cross-section line marked AA′ in (**g**).

The shallow groundwater aquifer (I + II) is unconfined, and the main geological deposit of this aquifer is sandy gravel with a depth of 20–40 m in the upper parts of the plain. In the middle parts of the plain, the aquifer is composed of alternate layers of sand, clay, sand, silt, and sand. The middle Pleistocene aquifer (III) and the lower Pleistocene aquifer (IV) are confined with a depth of over 20 m and consist of quaternary sediments of sandy gravel, sand, and clay, respectively.

The shallow groundwater aquifer is the main productive aquifer for the Hufu Plain, and is what this paper mainly deals with. Over 100 monitoring wells, mainly located in shallow groundwater aquifers, were selected (Figure 1g), which provided the groundwater level and drawdown records on monthly and yearly scales, and the depth of the wells varied from 9 m to 150 m [25,26].

Groundwater is an important water source for regional development and accounts for >80% of the total water supply, with irrigation as the main water use. Groundwater from the shallow, unconfined, aquifer layers is seriously overexploited, and the groundwater level declines by about 1 m each year [27,28]. In Shijiazhuang, the provincial capital of Hebei province, where the population is more than 17 million and the cropland accounts for over 50% of the total city region, the severe abstraction of the groundwater for living, industry, and irrigation has led to the formation of a constantly enlarging groundwater depression cone. The cumulative gross groundwater abstraction now has exceeded 18 billion m^3 over the past 50 years in the groundwater depression cone [25].

According to the literature and site observations, there have been many climatic and environmental changes in the study area during the past few decades. For example, the annual precipitation significantly decreased in 1971–2010, compared with the 1950–1960s (Table 1). Since the 1970s, there has been a drastic decrease in the directly available surface water, and an increased demand for water in industrial and agricultural development due to the severe regional drought. The groundwater abstractions increased, leading to a consistent annual expansion in the area of the cone of depression of 8.63 km^2/year, and a continued decline in the groundwater level in the center of the cone of depression, with a rate of 1.02 m/year [29]. Moreover, the water balance of the Hufu Plain has changed considerably over time. During 1961–1967, the total recharge was far larger than the discharge, and the surface water was an important water supply source; whereas during 1968–2010, with the construction of the large and medium-sized reservoirs in the upper reaches, and especially with the seepage treatment of the reservoir dam [30,31], the discharge volumes of the watercourse in the lower reaches and a lateral inflow from the mountain ranges in the west have reduced over 90% and 60%, respectively, and the recharge was less than the discharge [25].

Table 1. The variation characteristics of average precipitation in different decades.

Decades	Precipitation/mm	Range of Variation/%
1951–1960	614.2	0.00
1961–1970	525.5	14.4
1971–1980	467.0	−24.0
1981–1990	477.5	−22.3
1991–2000	470.4	−23.4
2001–2010	472.0	−23.2

Notes: the annual precipitation range of variation in 1961–1970, 1971–1980, 1981–1990, 1991–2000, and 2011–2010 were calculated relative to the annual precipitation in 1951–1960.

2.2. Data

The data used in this study include meteorological data, input fluxes and output fluxes of the groundwater flow field, the groundwater level in the center of the cone of depression, and the area of the cone of depression (Figure 2); the average groundwater level of the Hufu Plain; and the groundwater level distribution area of the Hufu Plain over the period of 1961–2010. We used the groundwater level in the center of the cone of depression, the area of the cone of depression, and the average groundwater level of the Hufu Plain to characterize the groundwater flow field evolution as follows: (i) we used the groundwater level in the center of the cone of depression to implement the provided methodology of this study, (ii) the average groundwater level of the whole plain, the area of the cone of depression to test the rationality of the results of the step (i), if these two indexes evolution characterizes are obviously different in each division stages of a specific time scale, we consider the step (i) is suitable; if on the contrary, it is unsuitable.

The meteorological data in this study include the annual precipitation over 1961–2010. These data were obtained from a monthly precipitation dataset and a 0.5° × 0.5° resolution produced by the China Meteorological Administration (http://data.cma.cn), from two national meteorological stations (Figure 1a). The average groundwater level data of the Hufu Plain, the groundwater level in the center of the cone of depression, and the area of the cone of depression over 1961–1975 were provided by Liu [29], or can be obtained from the Yearbook of the Shijiazhuang Groundwater Level Statistical [26]. This resource determined the depression cone area based on whether the groundwater elevation contour normal direction changed, relative to the normal direction in the last year. The groundwater level distribution area data of the Hufu Plain were produced by Wang et al. [27], using an inverse distance weighted method through ArcGIS 10.0 (ESRI, Redlands, CA, USA). The curve of the ratio of groundwater recharge volume to exploitation volume (R_e) over 1961–1975 was produced by Zhang et al. [23] using the groundwater numerical model.

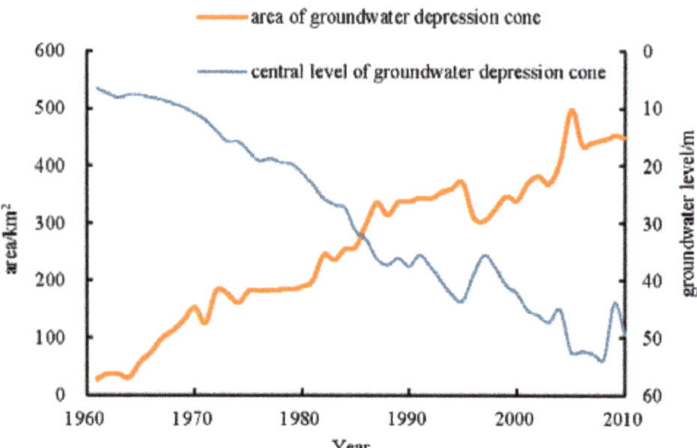

Figure 2. Variation of the original groundwater level in the center of the cone of depression and the area of the cone of depression from 1961–2010.

The input fluxes (infiltration from precipitation, leakage of watercourses, infiltration from canal systems, infiltration from irrigation, and lateral influx from the mountain in the west) and output fluxes (exploitation, lateral out flux, and phreatic water evaporation, which was set as 0 according to Zhang [24]) of the groundwater flow field, the mean groundwater level of the Hufu Plain, the groundwater level in the center of the cone of depression and the area of the cone of depression from 1976 to 2010 were obtained from the Report on Geological Environmental Monitoring for Shijiazhuang [25], in which the input and output fluxes were estimated using the corresponding hydrogeology parameters and groundwater model. The curve of the ratio of the groundwater recharge volume to the exploitation volume (R_e) over 1976–2010 was calculated using the data of the input and output fluxes, which fully considered the possible delay of the recharge. Digital elevation data with a resolution of 90 m were downloaded from the shuttle radar topography mission [32].

2.3. Research on Scale Effect

The DWT method was used to carry out the multi-scale time effect analysis for the identification of the characteristics of the groundwater system evolution. The DWT method is effective for multi-scale analysis. It can filter the research sequences at various time scales, eliminate high-frequency signals and noise, and then reconstruct the low-frequency signals obtained at a single scale [4]. Low frequency wavelet coefficients are reconstructed on various time scales into an identification quantity sequence to obtain an evolution trend and transition the time nodes of the research sequence at each time scale.

Firstly, Equation (1) is used to carry out DWT for the groundwater level in the center of the cone of depression, from 1961 to 2010, at the different time scales, and the low frequency wavelet coefficients and the high frequency wavelet coefficients are obtained for the different time scales, as follows:

$$w_f(a,b) = |a|^{-1/2} \Delta t \sum_{k=1}^{N} \left(f(k\Delta t)\overline{\phi}\left(\frac{k\Delta t - b}{a}\right) \right), \tag{1}$$

where $W_f(a, b)$ denotes the wavelet transform coefficient; $\phi(t)$ denotes the mother wavelet or basic wavelet function, according to the study by Sang et al. [33] and Wang et al. [4], 'db3' was chosen as the mother wavelet function in this study; $\overline{\phi}(t)$ is the conjugate functions of the $\phi(t)$; a denotes expansion and contraction factor, reflecting the cycle length of wavelet; b denotes the time parameter, which is the shift factor relative to time, t; N signifies that the time, t, is divided into N equal parts, $\Delta t = t/N$; k denotes the time step; and $f(t)$ is the groundwater level in the center of the cone of depression sequence over the period 1961–2010.

Secondly, Equation (2) was used to carry out single-scale reconstruction for each low-frequency signal of each scale, as follows:

$$f(k\Delta t) = \sum_{a,b} w_f(a,b)\overline{\varphi_{a,b}}(k\Delta t), \tag{2}$$

Thirdly, based on the reconstruction low-frequency curve of the groundwater level in the center of the cone of depression in different time scales, the transition points of the groundwater system were identified, and the evolution trend of the curve between the two adjacent points was clarified using the method of linear regression analysis. Furthermore, in order to test the rationality of the transition point, an F test was used to compare the differences in the slope between the regression models. If $p < 0.05$, the transition points were considered to be significant. The SPSS11.5 software (IBM, New York, NY, USA) was used to carry out the regression analysis and F test.

Finally, the method of comparing and analyzing the evolution trend of the groundwater level in the center of the cone of depression at different time scales and the groundwater recharge–exploitation balance was used to determine the time nodes of the evolution stages of the groundwater flow fields in the research area. The ratio of groundwater recharge volume to exploitation volume (R_e) can clearly explain the balance state of the groundwater system, where $R_e > 1$ implies that the recharge of the groundwater system is larger than the discharge, while the groundwater system is in a state of increasing storage; $R_e < 1$ indicates that the groundwater system is in a state of decreasing storage. The following five balance index categories can be distinguished: unbalanced toward to increasing storage state ($R_e \geq 2.0$), and the amount of the exploitation was less than one half of the recharge volumes; slightly unbalanced to increasing storage state ($1 < R_e < 2$), and the amount of the exploitation was one half to 1.0 times that of the recharge volumes; balanced ($R_e = 1$); slightly unbalanced toward to decreasing storage state ($0.5 \leq R_e < 1$), and the amount of the exploitation was one to two times that of the recharge volumes; and unbalanced toward to decreasing storage state ($R_e < 0.5$), and the amount of the exploitation was more than 2.0 times that of the recharge volumes.

2.4. Dominant Factor Analysis

The following method was used to study the dominant factors at different evolutionary stages of the groundwater flow field. The dynamic model of groundwater system in the research area was defined as follows:

$$Q_{rre} + Q_{pre} + Q_{fre} + Q_{lre} + Q_{wre} - Q_{ldi} - Q_{ev} - Q_{ex} = \Delta H \mu F, \tag{3}$$

where ΔH denotes the groundwater level amplitude, m; μ denotes the specific yield of the aquifer in the groundwater level amplitude zone, a dimensionless parameter; F denotes the area of the Hufu Plain, km^2; Q_{rre} denotes the input flux for the leakage of watercourse, m^3; Q_{pre} denotes the input flux of the precipitation infiltration, m^3; Q_{fre} denotes the input flux for the leakage of the canal irrigation field, m^3; Q_{lre} denotes the input flux of lateral influx, m^3; Q_{wre} denotes the input flux of well irrigation regression, m^3; Q_{ldi} denotes lateral outfluxes, m^3; Q_{ex} denotes the volume of groundwater exploitation, m^3; and Q_{ev} denotes the volume of evaporation discharge of phreatic water, m^3. The following equation is obtained by transforming Equation (3):

$$\Delta H = \frac{1}{\mu F}\left(Q_{rre} + Q_{pre} + Q_{fre} + Q_{lre} + Q_{wre} - Q_{ldi} - Q_{ev} - Q_{ex}\right), \tag{4}$$

Through the analysis of Equation (4), it is clear that the input flux of the regional groundwater system increases the groundwater level, while the output flux decrease the groundwater level; the imbalance between input and output fluxes causes the ΔH change, which denotes the shift of the regional groundwater level; when the input flux is greater than the output flux, the groundwater level rises and $\Delta H > 0$; when the groundwater level drops, $\Delta H < 0$; and when the input flux is equal to the output flux, the groundwater level remains unchanged. When considering the distance of the movement of the groundwater level, in an evolution stage, the total distance of the movement of the groundwater level was calculated as follows:

$$D = \frac{1}{\mu F}\left(Q_{rre} + Q_{pre} + Q_{fre} + Q_{lre} + Q_{wre} + Q_{ldi} + Q_{ev} + Q_{ex}\right),$$
(5)

Therefore, the effect of the intensity of each input flux and output flux of the groundwater system on the evolution of the groundwater flow field at a certain evolutionary stage was calculated using the following formula where the input flux is positive and the output flux is negative, as follows:

$$\alpha_i = \pm\frac{Q_i}{D\mu F},$$
(6)

where i denotes the serial number of input flux or output flux, Q_i denotes the ith input or output flux, α_i denotes the effect intensity of the ith input or output flux on groundwater flow field, and D is the distance of the groundwater level movement.

The fuzzy coincident matrix method [34] was used to calculate the dominant factors for the years with missing input and output flux data. Firstly, the main factors affecting the changes in the groundwater flow field were ordered based on what the groundwater scientist qualitatively decides is important, according to their experience. Then, according to the order of 'importance', a binary comparison between every two pairs was carried out separately, based on the experiences of the scientist. For example, a binary comparison was made between the 'importance' of groundwater exploitation quantity (a_i) and that of the precipitation infiltration (a_j). If a_i is more important than a_j, $r_{ij} = (0.6–1.0)$, $r_{ji} = (0–0.5)$, and $r_{ij} + r_{ji} = 1.0$. If a_i and a_j are equally important, $r_{ij} = 0.5$, $r_{ji} = 0.5$; if a_j is more important than a_i, $r_{ij} = (0.6–1.0)$, $r_{ji} = (0–0.5)$, and $r_{ij} + r_{ji} = 1.0$ ($i = 1, 2, \dots, n; j = 1, 2, \dots, n$). Correspondingly, a priority matrix ($R = [r_{ij}] n \times n$) of each evolution stage was obtained. The uniform transformation (Equations [7] and [8]) was carried out for the priority matrix to obtain the fuzzy coincident matrix ($B = [b_i] n \times 1$), and square root normalization (Equation (9)) was carried out for the fuzzy coincident matrix, generating the weight matrix ($W = (\omega_i) n \times 1$), as follows:

$$b_{ij} = \frac{b_i - b_j}{2n} + 0.5,$$
(7)

$$B_i = \sqrt[n]{\sum_j^n b_{ij}},$$
(8)

$$w_i = \frac{B_i}{\sum_i^n B_i},$$
(9)

2.5. Quantitative Analysis

A regression analysis was used to evaluate the quantitative relationships between the groundwater flow field evolution, precipitation changes, and groundwater exploitation. A regression analysis on the groundwater level series, precipitation series, and groundwater exploitation volume series were performed to establish the regression equations, and a t-test was used to test the significance of the regression models. The statistical software SPSS11.5 was used to conduct these analyses.

3. Results

3.1. Scale Effect and Stage Division of the Evolution of the Groundwater Flow Field

Using the groundwater level in the center of the cone of depression at different time scales of the Hufu Plain combined with the F test, the balance index (R_e) categories of the groundwater system to demonstrate how a particular choice of the suitable time step and the division of the evolution stage was made. Firstly, the methods described in Section 2.3 were used to obtain the 1–5-year and 12-year (Figure 3) time scale maps of the evolution characteristics of the groundwater flow field in the research area. As shown in Figure 3, there was a significant difference in the groundwater level among the six time scales, because the high-frequency signals and noise of the original time series were eliminated gradually with the time scales' increase, retaining only the interesting part of the original signal, depending on the application. We can therefore use these data to investigate the scale-dependent variations of the groundwater level signals that contain transient features, allowing us to identify different important stages of evolution within a time scale series, and to study these stages individually. Based on Figure 3, we can roughly estimate the transition points of the groundwater system evolution in each time scale.

Secondly, we used the F test to compare the evolution slope differences between the linear regression model for each division in each time scale, eliminated the transition points that are statistically not significant, and then got the exact time nodes of the groundwater system evolution of different time scales. There were five transition points on the two-year time scale (1968, 1980, 1988, 1995, and 2005; Figure 3a), dividing the whole sequence into six divisions. There were seven transition points in the evolution of the groundwater flow field at the one-year time scale (1967, 1976, 1988, 1991, 1994, 1998, and 2006; Figure 3b), dividing the whole sequence into eight divisions. There were five transition points on the three-year time scale (1964, 1980, 1988, 1995, and 2005; Figure 3c), dividing the whole sequence into six divisions. There were two transition points in the four-year and five-year time scale (1964 and 1995; Figure 3d,e), dividing the whole sequence into three divisions. There were no transition points on the 12-year time scale (Figure 3f).

Thirdly, use of the balance index (R_e) categories of the groundwater system to test the rationality of the stages divisions of the groundwater system evolution of different time scales, if the balance state (balance index (R_e) categories) of the groundwater system in the different division stages of a specific time scale is obviously different, we consider this time scale as suitable; if on the contrary, it is unsuitable. There were five transition points on the two-year time scale (Figure 3a), dividing the evolution of the groundwater flow field into the following six stages: 1961–1967, 1968–1980, 1981–1988, 1989–1995, 1996–2005, and 2006–2010. Compared with the curve, R_e, during 1961–1967, R_e was greater than 2.0, the flow field was in a state of unbalanced toward to increasing storage, and the groundwater level remained basically unchanged. During 1968–1980, R_e was in 1.0–2.0 (except for 1973 and 1976), on average, $R_e = 1.32$, the groundwater flow field was in a state of slightly unbalanced toward increasing storage. During 1981–1995, the average value of R_e was 0.46, which implies the groundwater system was generally in a state of unbalance toward to the decreasing storage. In 1988, the $R_e = 0.38$, and several years before and after the nodes it was less than 0.5, so the time nodes of 1988 should be eliminated. During 1996–2004, R_e increased during the catastrophic flood period (1996 when the precipitation was 1096 mm), but it was still less than 1.0 in the other years, on average, $R_e = 0.70$, the flow field was in a state of slightly unbalanced towards the decreasing storage, and the groundwater level continued to decline. From 2006–2010, $R_e = 0.40$, the groundwater flow field was in a state of unbalance towards the decreasing storage. Overall, the transition points on a two-year time scale can clearly express the evolution stages of the groundwater flow field.

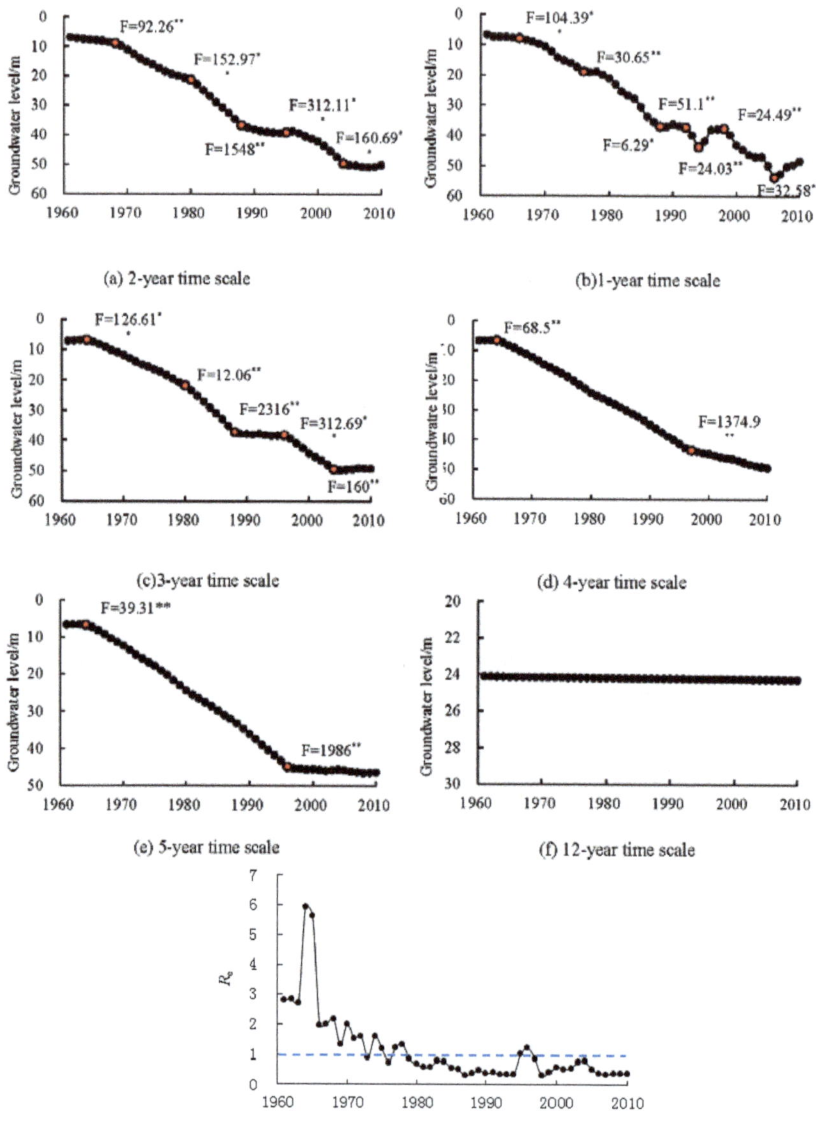

(g) The curve of the ratio of groundwater recharges volume to exploitation volume

Figure 3. Evolution characteristics of the groundwater flow field in the research area using (**a**) a two-year time scale, (**b**) a one-year time scale, (**c**) a three-year time scale, (**d**) a four-year time scale, (**e**) a five-year time scale, (**f**) a twelve-year time scale. (**g**) The curve of the ratio of groundwater recharge volume to exploitation volume, the symbol ⊙ indicates the transition points, and the labels provide the F test values between the division sequences linear regression model slope, * indicates significance levels of 0.05, ** indicates significance levels of 0.01.

Figure 3b shows that the time nodes of a one-year time scale did not coincide with curve R_e, especially the nodes of 1991 and 1994. The R_e of several years before and after the nodes was less than

0.5. During this period, the groundwater flow field was generally in a state of unbalance toward the decreasing storage, and it could not be divided into an evolution stage. On the four-year time scale and five-year time scale, there were no transition points during 1965–1997. Considering the evolutionary sequence of the groundwater recharge–exploitation balance (R_e), R_e rapidly changed from >1.0 to <1.0 during 1965–1997. The groundwater flow field changed from a state of increasing storage to a state of decreasing storage. However, these stages and their characteristics were eliminated. On the 12-year scale (Figure 3f), the evolution characteristics of the groundwater flow field formed a straight line with all of the evolution characteristics being generalized, and all of the features of each stage lost. The groundwater flow field evolution shows no stage characteristics on a large scale such as this.

Compared with the two-year time scale, the one-year time scale reflects more transition points during groundwater flow field evolution, and the stage divisions are not sufficiently clear. At the three-year time scale, the first transition point, 1964, is not accurate, because the R_e was generally greater than 2.0 during 1961–1967, and the groundwater flow field was in a state of unbalance toward increasing storage. At the 4-, 5-, and 12-year time scales, the characteristics of the stages were lost. Some transition points during the evolution process were over-homogenized and obscured. Therefore, we adopted the two-year time scale to divide groundwater flow field evolution into five stages. During 1961–1967, the groundwater was in a natural state, so the field was a natural flow field. During 1968–1980, the annual average exploitation volumes were 2.13 billion m^3/year, but due to several drought years (e.g., the precipitation was 224 and 304 mm in 1973 and 1975, respectively), the groundwater system was subject to mild overexploitation. During 1981–1995, the annual average exploitation volume was 2.56 billion m^3/year, and the accumulated overexploitation volume was 9.4 billion m^3 [25], forming a stable groundwater depression cone. During 1996–2005, the exploitation volumes were 2.29 billion m^3/year, the overexploitation volumes increased to 15.4 billion m^3 [25], which was serious overexploitation. During 2006–2010, the exploitation volume was 1.98 billion m^3/year [25], making the period an exploitation reduction stage.

In the process of the choice of the time-step and the division of the different stages, we used the *F* test to reduce the subjectivity of the selection of the transition points, using the transition of the state of the groundwater flow field to test the rationality of the division of the evolution stages; however, a few transition points may need to be eliminated in accordance to the state of the groundwater system, which may still rely on the experience of the groundwater scientist, so the experience of the scientist should always be valued.

3.2. Characteristics of Groundwater Flow Field Evolution

The Figure 1b–f shows the groundwater elevation in the area of the cone of depression for the natural flow field, mild overexploitation, depression cone formation, and exploitation reduction stage. In the natural flow field stage (1961–1967), the groundwater level in the research area was characterized by a change from a deep to shallow groundwater level in the area from the West Piedmont to the Eastern Plain. The groundwater level in this area was less than 10 m and the distribution areas mainly had groundwater levels of <5 m (Table 2). No obvious groundwater depression cone was formed.

Table 2. The groundwater level distribution area of the Hufu Plain in each evolution stage (km^2).

Year	<5 m	5–10 m	10–15 m	15–20 m	20–25 m	25–30 m	30–35 m	35–40 m	40–45 m	>45 m
1965	7945	260								
1975	791	5525	1832	57						
1985	239	1932	3788	1904	252	89				
1995	62	293	2011	1209	2851	1386	315	76	2	
2005	18	257	758	1277	893	908	2024	1742	296	30

In the mild overexploitation stage (1968–1980), the area with groundwater levels of <5 m was significantly reduced. The distribution areas mainly have groundwater levels of 5–10 m. The distribution areas with groundwater levels >10 m appear, with an area reaching 1889 km^2, accounting for 23.1% of the total regional area. The analysis of the linear regression ($p < 0.05$) showed that the rate of decline of the groundwater level was 0.38 m/year, the groundwater level in the center of the cone of depression descended at a speed of 1.01 m/year, while the area of the cone of depression expanded at 6.37 km^2/year.

In the depression cone formation stage (1981–1995), the ratio of the distribution area with groundwater level of less than 10 m to the Hufu Plain shrank drastically. Distribution areas mainly have a groundwater level of 10–25 m, and the areas with groundwater levels of more than 35 m appear, accounting for 4.2% the area of the total region. The rate of decline of the average groundwater level in the region was 0.69 m/year. The average annual rate of descent in the groundwater level in the center of the cone of depression was 1.35 m/year, and the area of the cone of depression was increased by 11.35 km^2/year.

In the serious overexploitation stage (1996–2005), the distribution area, with groundwater levels of <20 m, shrank drastically and mainly had groundwater levels of 30–40 m. The distribution areas with groundwater levels of >45 m occurred, accounting for 4.7% of the total area. The rate of descent of the average groundwater level in the region was 0.69 m/year. The average rate of descent of the groundwater level in the region was increased by 0.34 m/year, compared with the rate in the depression cone formation stage, and the spread rate of the area of the cone of depression was accelerated to 12.39 km^2/year.

In the exploitation reduction stage (2006–2010), the exploitation volume in the serious overexploitation area decreased each year. The rate of descent of the regional average groundwater level was 0.74 m/year. The expansion of the area of the cone of depression slowed during this stage.

3.3. Analysis of Dominant Factors

During 1961–1967, the groundwater flow field showed natural stage characteristics. The fuzzy coincident matrix method was used to calculate the magnitude of the main controlling factors, because of a lack of input and output flux data. The ranking of the relative importance of each input and output flux was as follows: infiltration from precipitation > exploitation > leakage of watercourse > infiltration from irrigation > infiltration from canal system > lateral influx > lateral outflux. A binary comparison was made between each two, according to the order of exploitation, infiltration from precipitation, leakage of watercourse, infiltration from canal system, infiltration from irrigation, lateral influx, and lateral outflux, and a priority matrix was obtained (Equation (10)). The following weight matrix was obtained by uniform transformation (Equations (7) and (8)) and square root normalization (Equation (9)) of the priority matrix, as follows: (0.20, 0.23, 0.15, 0.11, 0.12, 0.10, and 0.08).

$$
\begin{bmatrix}
0.50 & 0.30 & 0.70 & 0.80 & 0.80 & 0.90 & 0.90 \\
0.70 & 0.50 & 0.80 & 0.90 & 0.90 & 0.90 & 1.00 \\
0.30 & 0.20 & 0.50 & 0.60 & 0.60 & 0.70 & 0.70 \\
0.20 & 0.10 & 0.40 & 0.50 & 0.40 & 0.50 & 0.60 \\
0.20 & 0.10 & 0.40 & 0.60 & 0.50 & 0.60 & 0.60 \\
0.10 & 0.10 & 0.30 & 0.50 & 0.40 & 0.50 & 0.50 \\
0.10 & 0.00 & 0.30 & 0.40 & 0.40 & 0.40 & 0.50
\end{bmatrix}
\tag{10}
$$

For the mild overexploitation, depression cone formation, serious overexploitation, and exploitation reduction stages, Equations (3)–(6) were used for calculation. The results are shown in Table 3. Infiltration from precipitation had the largest effect in the natural flow field stage and was the main controlling factor, followed by exploitation. In the mild overexploitation, depression cone formation, serious overexploitation, and exploitation reduction stages, exploitation had the largest effect and was the main controlling factor, followed by infiltration from precipitation.

Table 3. Intensity levels of each input flux and output flux of the groundwater flow field (unit: %).

Equilibrium Elements		Natural Flow Field Stage	Mild Overexploitation Stage	Depression Cone Formation Stage	Serious Overexploitation Stage	Exploitation Reduction Stage
				Evolution Stage		
Input fluxes	Infiltration from precipitation	23.00	19.19	20.67	20.06	24.30
	Leakage of watercourse	15.00	2.82	3.12	4.68	0.81
	Infiltration from canal system	11.00	9.48	3.24	2.40	1.82
	Infiltration from irrigation	12.00	11.39	8.12	7.72	7.71
	Lateral influx	10.00	8.28	8.77	6.73	5.88
	Subtotal	71.00	51.15	43.92	41.58	40.52
Output fluxes	Exploitation	20.00	43.45	52.24	55.96	53.47
	Lateral outflux	8.00	5.39	3.84	2.47	6.01
	Subtotal	28.00	48.85	56.08	58.43	59.48

3.4. Quantitative Analysis of Influence of Precipitation and Exploitation on Groundwater Levels

Exploitation was the main factor causing the decline of the groundwater level, and precipitation is an important factor influencing the changes in the groundwater level. In the natural flow field and the mild overexploitation stages, of only a few years in duration, the groundwater system was in a state of decreasing storage, and the groundwater levels responded sensitively to the changes in the exploitation quantity. In the natural flow field and mild overexploitation stages, there was a clear correlation between the exploitation quantity and the groundwater level. A regression analysis suggested that when the exploitation volumes were increased by 100 million m^3, the mean groundwater level in the research areas decreased by 0.15 m in the natural flow field stage, and decreased by 0.34 m in the mild overexploitation stage (Figure 4).

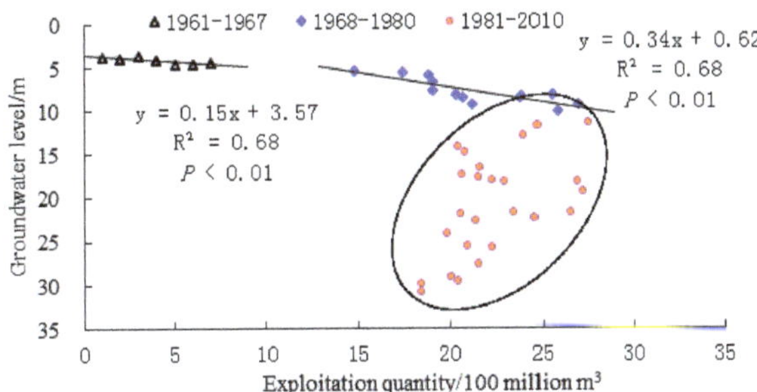

Figure 4. Relationship between exploitation quantity and groundwater level.

During the depression cone formation, serious overexploitation, and exploitation reduction stages, the volumes of the groundwater abstraction were far greater than the recharge volume. In Figure 4, the data from 1981–2010 are shown as scattered dots, and the relationship between the groundwater level and exploitation volumes is not obvious. Moreover, according to Equation (4), under the condition of the total output fluxes being larger than the input fluxes, even if the exploitation declined, the groundwater level would continue to descend. However, the increased precipitation can slow the decline of the groundwater level. Figure 5 shows that when the precipitation amount was increased by 100 mm, the descent amplitude of the groundwater level dropped by 0.37 m. In some years, with ample precipitation, the groundwater level even rose. For example, the precipitation was 1096 mm in

1995–1996. During this period, the groundwater exploitation reduced, the recharge increased, the total input fluxes were larger than output fluxes, and the groundwater levels correspondingly rose.

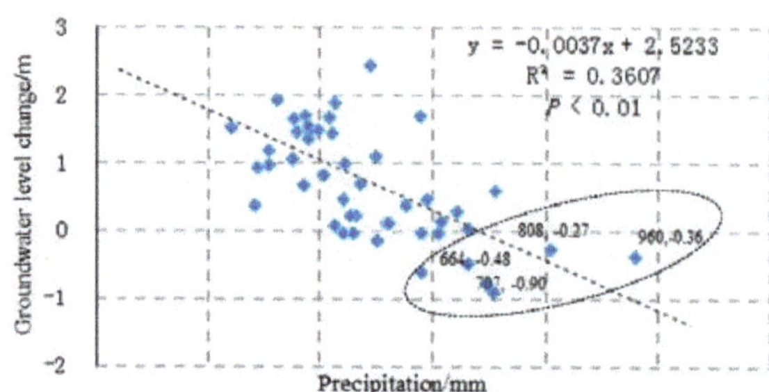

Figure 5. Relationship between precipitation and change in groundwater level.

4. Discussion

In this paper, we use the groundwater level in the center of the cone of depression at different time scales of the Hufu Plain, combined with the F test and the balance index (R_e) categories of the groundwater system, so as to study the shallow groundwater aquifer system evolution time scale effects and mechanisms, and the applicability of the method to other types of aquifers needs to be tested in future studies.

The DWT method was used to carry out the multi-scale time effects analysis for the groundwater level in the center of the cone of depression from 1961 to 2010, and get the 1–5-year and 12-year time scale maps of the evolution characteristics of the groundwater flow field in the research area (Figure 3, Figure 6); based on the maps, we can roughly estimate the transition points of the groundwater system evolution in each time scale. Compared with the original groundwater level time series (Figure 2), the one-year time scale time series is similar to the original series, and with the time scale increasing, more and more details of the time series were filtered; as the time scale increased to 12 years, the evolution of the groundwater level formed a straight line and all of the evolution characters were generalized, so it is necessary for us to identify a suitable time scale for more clearly expressing the groundwater system evolution. Although our methodology can be used in any regions where groundwater is the main water supply, the time scale effects may be significantly different for various study areas, because of the difference of the hydrogeology conditions. In the same time scale, the characters of the groundwater system evolution for the different study areas may be different, and the time series evolution characters as the time scales increase are also varied, so you probably choose different time scales to study the groundwater evolution for the different regions.

In order to reduce the subjectivity of the choice of the suitable time step and the division of the different evolution stages, we quantified the balance index (R_e) of the groundwater system. The following five balance index categories can be distinguished: unbalanced toward to increasing storage state ($R_e \geq 2.0$); slightly unbalanced to increasing storage state ($1 < R_e < 2$); balanced ($R_e = 1$); slightly unbalanced toward to decreasing storage state ($0.5 \leq R_e < 1$); and unbalanced toward to decreasing storage state ($R_e < 0.5$), which can help us discriminate the stage of the divisions of the groundwater system evolution of the different time scales, but in the process of the division of the evolution stages, the experiences of the groundwater scientists are required, and thus the researchers who are not specialists in groundwater science may meet some difficulties when utilizing the methods.

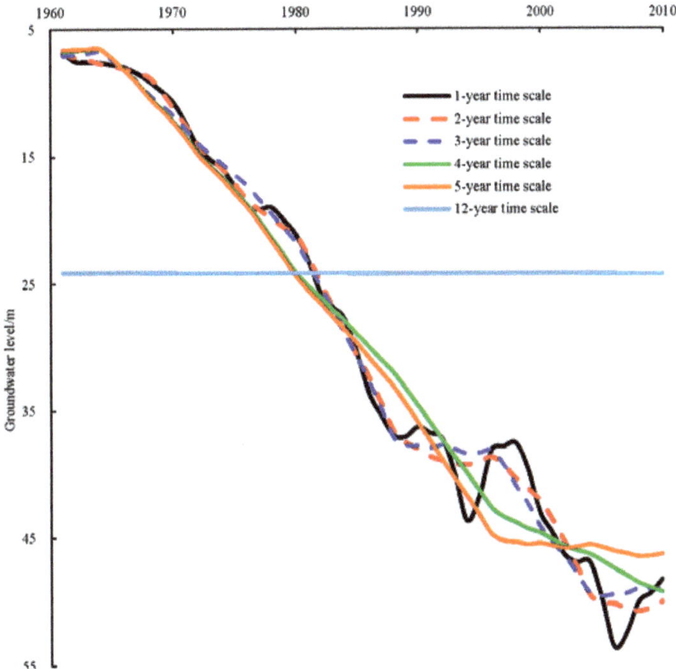

Figure 6. The groundwater level evolution characteristics with the time scale increasing.

5. Conclusions

The paper provides a new method to study the groundwater flow field temporal evolution characteristics, focusing on the choice of the correct time step for identifying the distinct stages of evolution characterized by different characteristics linked to the management of groundwater system. The Hufu Plain was taken as a case study to demonstrate the methodology. The results showed that a two-year time scale was the most suitable for studying the groundwater flow field evolution in the Hufu Plain. According to the groundwater level on two-year time scale, we can divide the groundwater flow field evolution into the following five stages: natural flow field, mild overexploitation, depression cone formation, serious overexploitation, and exploitation reduction.

The main factors controlling the groundwater evolution were then assessed for the different stages, still using the transformed two-year time scale groundwater level time series. The infiltration from precipitation was the main factor controlling groundwater level changes during the natural flow field and the effect intensity was 23%. During the mild overexploitation, depression cone formation, serious overexploitation, and exploitation reduction stages, the effect of the intensity of the exploitation was 43.5%, 52.25%, 55.96%, and 53.47%, respectively, and it was the main controlling factor. With the methodology, after careful tests on a number of different systems, we can obtain an increased knowledge regarding regional groundwater dynamics.

Author Contributions: D.W., G.Z., H.F. and J.W. conceived, designed, and conducted the research, but also collected and analyzed the data. Y.T. reviewed and edited the paper; He was also the corresponding author.

Funding: This research was funded by the National Natural Science Foundation of China (NSFC) (Project No. 41172214, 41502253, 41702263, 51822907, 51479210); Research and demonstration of water quality guarantee and ecological restoration technology in the World Horticultural Exposition and winter Olympics in Gui river basin (2017ZX07101004).

Acknowledgments: We thank LetPub (www.letpub.com) for its linguistic assistance during the preparation of this manuscript.

Conflicts of Interest: The authors declare no conflict of interest.

References

1. Murphy, J.M.; Sexton, D.M.H.; Jenkins, G.J. *UK Climate Projections Science Report: Climate Change Projections*; Met Office Hadley Centre: Exeter, UK, 2009.
2. Xu, Y.Q.; Mo, X.G.; Cai, Y.L.; Li, X.B. Analysis on groundwater table drawdown by land use and the quest for sustainable water use in the Hebei Plain in China. *Agric. Water Manag.* **2005**, *75*, 38–53. [CrossRef]
3. Perrina, J.; Ferrant, S.; Massuel, S.; Dewandel, B.; Maréchal, C.; Aulong, S.; Ahmed, S. Assessing water availability in a semi-arid watershed of southern India using a semi-distributed model. *J. Hydrol.* **2012**, *460–461*, 143–145. [CrossRef]
4. Wang, W.S.; Ding, J.; Li, Y.Q. *Hydrogeology Wavelet Analysis*; Chemical Industry Press: Beijing, China, 2005.
5. Nalley, D.; Adamowski, J.; Khalil, B. Using discrete wavelet transforms to analyze trends in stream flow and precipitation in Quebec and Ontario (1954–2008). *J. Hydrol.* **2012**, *475*, 204–228. [CrossRef]
6. Parkin, G.; Birkinshaw, S.J.; Younger, P.L.; Rao, Z.; Kirk, S. A numerical modeling and neural network approach to estimate the impact of groundwater abstractions on river flows. *J. Hydrol.* **2007**, *339*, 15–28. [CrossRef]
7. Padilla, F.; Méndez, A.; Fernández, R.; Vellando, P.R. Numerical modeling of surface water/groundwater flows for freshwater/saltwater hydrology: The case of the alluvial coastal aquifer of the Low Guadalhorce River, Malaga, Spain. *Environ. Geol.* **2008**, *55*, 215–226. [CrossRef]
8. Yidana, S.M.; Alfa, B.; Banoeng-Yakubo, B.; Addai, M.O. Simulation of groundwater flow in a crystalline rock aquifer system in Southern Ghana—An evaluation of the effects of increased groundwater abstraction on the aquifers using a transient groundwater flow model. *Hydrol. Process.* **2012**, *28*, 1084–1094. [CrossRef]
9. Iwasaki, Y.; Nakamura, K.; Horino, H.; Kawashima, S. Assessment of factors influencing groundwater—Level change using groundwater flow simulation, considering vertical infiltration from rice—Planted and crop-rotated paddy fields in Japan. *Hydrogeol. J.* **2014**, *22*, 1184–1855. [CrossRef]
10. Lafare, A.E.A.; Peach, D.E.; Hughes, A.G. Use of seasonal trend decomposition to understand groundwater behavior in the Permo—Triassic Sandstone aquifer, Eden Valley, UK. *Hydrogeol. J.* **2016**, *24*, 141–158. [CrossRef]
11. Asmuth, J.R.; Maas, K.; Bakker, M.; Petersen, J. Modeling Time Series of Groundwater Head Fluctuations Subjected to Multiple Stresses. *Groundwater* **2008**, *46*, 30–40. [CrossRef]
12. Bloomfield, J.P.; Marchant, B.P. Analysis of groundwater drought building on the standardized precipitation index approach. *Hydrol. Earth Syst. Sci.* **2013**, *17*, 4769–4787. [CrossRef]
13. Chae, G.T.; Yun, S.T.; Kim, D.S.; Kim, K.H.; Joo, Y. Time—Series analysis of three years of groundwater level data (Seoul, South Korea) to characterize urban groundwater recharge. *Q. J. Eng. Geol. Hydrogeol.* **2010**, *43*, 117–127. [CrossRef]
14. Wang, W.S.; Yuan, P.; Ding, J. Wavelet analysis and its application to stochastic simulation of daily flow. *J. Hydraul. Eng.* **2000**, *11*, 43–48. (In Chinese)
15. Kang, L.; Liu, S.R.; Liu, X.Z. Multiresolution and periodicity analysis of hydrological and meteorological factors in upper reaches of Minjiang River. *Acta Ecol. Sin.* **2016**, *36*, 1253–1262. (In Chinese)
16. Zhang, G.H.; Fei, Y.H.; Zhang, X.N.; Yan, M.J. Abnormal variation of groundwater flow field in plain area of Hutuo River basin and analysis on its cause. *J. Hydraul. Eng.* **2008**, *39*, 747–752. (In Chinese)
17. Yang, Y.H.; Watanabe, M.; Sakura, Y.; Tang, C.Y.; Hayashi, S. Groundwater table and recharge changes in the Piedmont region of Taihang Mountain in Gaocheng City and its relation to agricultural water use. *Water SA* **2002**, *28*, 171–178. [CrossRef]
18. Kendy, E.; Zhang, Y.Q.; Liu, C.M.; Wang, J.X.; Steenhuis, T. Groundwater recharge from irrigated cropland in the North China Plain: Case study of Luancheng County, Hebei Province, 1949–2000. *Hydrol. Process.* **2004**, *18*, 2289–2302. [CrossRef]
19. Shu, Y.Q.; Villholth, K.G.; Jensen, K.H.; Stisen, S.; Lei, Y. Integrated hydrological modeling of the North China Plain: Options for sustainable groundwater use in the alluvial plain of Mt. Taihang. *J. Hydrol.* **2012**, *464–465*, 79–93. [CrossRef]

20. Du, S.H.; Su, X.S.; Lu, H. The Artificial Recharge Effects of Groundwater Reservoir under Different Precipitation Plentiful Scanty Encounter in Hutuo River. *J. Jilin Univ. (Earth Sci. Ed.)* **2010**, *40*, 1090–1097. (In Chinese)
21. Nakayama, T.; Yang, Y.H.; Watanabe, M.; Zhang, X.Y. Simulation of groundwater dynamics in the North China Plain by coupled hydrology and agricultural models. *Hydrol. Process.* **2006**, *20*, 3441–3466. [CrossRef]
22. Hu, Y.K.; Moiwo, J.P.; Yang, Y.H.; Han, S.M.; Yang, Y.M. Agricultural water saving and sustainable groundwater management in Shijiazhuang Irrigation District, North China Plain. *J. Hydrol.* **2010**, *393*, 219–232. [CrossRef]
23. Zhang, G.H.; Fei, Y.H.; Nie, Z.L.; Yan, M.J. *Theory and Method of Regional Groundwater Flow Field Evolution and Assessment*; Sciences Press: Beijing, China, 2014.
24. Zhang, Z.J. *Investigation and Assessment of Sustainable Utilization of Groundwater Resources in North China Plain*; Geological Publishing House: Beijing, China, 2009; pp. 157–180, ISBN 978-7-116-06105-7.
25. Geological Environmental Monitoring Institute of Hebei Province (GEMIHP). Report on Geological Environmental Monitoring of Shijiazhuang of 1981–1985, 1986–1990, 1991–1995, 1996–2000, 2001–2005, 2006–2010. Available online: www.ngac.cn/125cms/c/qggnew/index.htm (accessed on 17 August 2018).
26. GEMIHP. Yearbook of Shijiazhuang Groundwater Level Statistical of 1961–1980. Available online: www.ngac.cn/125cms/c/qggnew/index.htm (accessed on 17 August 2018).
27. Wang, J.Z.; Zhang, G.H.; Yan, M.J.; Nie, Z.L. Research on the Plot Groundwater spatial-temporal Evolvement Rule in Hutuo Valley. *J. Arid Land Resour. Environ.* **2009**, *23*, 5–11. (In Chinese)
28. Mao, X.S.; Liu, C.M. Groundwater changing trends and agriculture sustainable development in Taihang mountain foot Plain of North China. *Res. Soil Water Conserv.* **2001**, *8*, 147–149. (In Chinese)
29. Liu, Z.P. Impact of Agricultural Activities on Regional Groundwater Variation—A Case of Shijiazhuang Plain. Ph.D. Thesis, Chinese Academy of Geological Sciences, Beijing, China, 2007.
30. Fei, Y.H. The influence of Huangbizhuang reservoir dam foundation seepage cut-off on the downstream development and counter measures. *Site Investig. Sci. Technol.* **1996**, *6*, 13–15. (In Chinese)
31. He, P. Primary Analysis of Groundwater Circumstances Influence of Huangbizhuang Village Reservoir Water proof Project to the Hutuo River Alluvium. *Ground Water* **2009**, *31*, 121–123. (In Chinese)
32. Jarvis, A.; Reuter, H.I.; Nelson, A.; Guevara, E. Hole-Filled Seamless SRTM DataV4. International Center for Tropical Agriculture. Available online: http://srtm.csi.cgiar.org (accessed on 17 August 2018).
33. Sang, Y.F.; Wang, D. Wavelet selection method in hydrologic series wavelet analysis. *J. Hydraul. Eng.* **2008**, *39*, 295–300. (In Chinese)
34. Zhang, J.; Zhang, L.C.; Xu, C.M.; Zhang, J.Z.; Li, X.B. Application of fuzzy consistent matrix method in multi-objective decision plain of slim hole drilling. *Nat. Gas Ind.* **2003**, *23*, 61–62. (In Chinese)

Article

Temporal Stability of Groundwater Depth in the Contemporary Yellow River Delta, Eastern China

Ruiyan Wang [1,2], Simon Huston [3,*], Yuhuan Li [2,*], Huiping Ma [2], Yang Peng [1] and Lihua Ding [1]

1   College of Resources and Environment, Shandong Agricultural University, Tai'an 271018, China; wry@sdau.edu.cn (R.W.); sqliu@sdau.edu.cn (Y.P.); gypsdau@sdau.edu.cn (L.D.)
2   National Engineering Laboratory for Efficient Utilization of Soil and Fertilizer Resources, Tai'an 271018, China; lcheng@sdau.edu.cn
3   Centre for Real Estate, Royal Agricultural University, Cirencester GL7 6JS, UK
*   Correspondence: simon.huston@rau.ac.uk (S.H.); yuhuan@sdau.edu.cn (Y.L.)

Received: 10 May 2018; Accepted: 20 June 2018; Published: 28 June 2018

Abstract: Sustainable development calls for the wise use of groundwater resources. Of particular concern is saline intrusion into productive agricultural land, which is contiguous with densely populated coastal settlements. To reverse saline intrusion in such coastal regions, information about the groundwater depth in terms of its spatio-temporal variability is essential. Using survey data from 2004 to 2007, the research revealed the temporal variation characteristics of groundwater depth in the Contemporary Yellow River Delta. It explored the temporal stability characteristics of groundwater depth by using the coefficient of variation, Spearman rank correlation coefficient, and average relative deviation and standard deviation, and confirmed that the representative point reflected the average groundwater depth of the study area. Results showed that spatial variation of the groundwater depth in the study area was medium, but the variation coefficient of groundwater depth showed the seasonal changes. The spatial variation coefficient was largest in the dry season; the other months were relatively stable. The groundwater depth in the study area had strong temporal stability. The correlation between the Spearman rank correlation coefficient and the time lags showed that the spatial pattern of groundwater depth in the study area was similar across two or three years but the similarity weakened beyond this period. The representative points of the whole area showed a good linear correlation, and were spatially concentrated. In different years or time periods, the representative points were not the same but belonged to the medium groundwater depth grade in the area. The study provides useful guidance for Yellow River irrigation, preventing saline intrusion and the restoration of saline-alkali soils. It offers a theoretical basis for identifying regional satellite groundwater depth monitoring points.

Keywords: sustainable development; water resources; Contemporary Yellow River Delta; groundwater depth; temporal stability

---

1. Introduction

Over the past couple of decades, China has undergone spectacular economic development, coupled with deteriorating environmental conditions [1]. From 1998 to 2012, the proportion of urban population increased from 30.4% to 52.6% but the urbanization and industrialization process consumed 7.93 million ha of farmland [2]. Aware of the danger, China is restructuring its growth towards quality. Specifically, to mitigate the pressure of farmland loss and guarantee national food security, the Chinese government has recently adopted a series of farmland exploitation policies.

The Yellow River Delta (YRD) is one of China's three major river delta systems. Geologically, it is relatively young land that was formed less than 150 million years ago. Its name reflects the 1 billion tons of sediment the Yellow River carries annually downstream into its estuary where the deposits increase land area by about 2400 hm^2 annually [3]. Engineering-focused national policy has sought to exploit the rich land resources of the YRD. For example, the YRD is the core demonstration area in the "Bohai granary project" that is known as the National Science and Technology Support Program. However, as a newly formed estuarine delta, YRD is naturally characterized by extensive coverage of primary salinization [4]. Over recent decades, secondary anthropogenic salinization has also been severe to the extent that it now threatens food production and the environment [5,6]. For these reasons, the Chinese central government plans to restore salinized land in the YRD and especially control secondary salinization. The YRD is dominated by rivers and oceans, so groundwater is shallow. Indeed, in most areas it is less than 3 m below the surface but has a high degree of salinity. These characteristics of groundwater, coupled with silt soil texture, lead to a strong soil capillary effect which brings groundwater salt to the soil surface. Therefore, in this region the salt transportation characteristics of groundwater is especially important with soil salinity strongly linked to groundwater depth (GD) [7,8]. For national food security and to sustainably exploit the YRD agricultural potential without saline intrusion, information on GD is urgently needed. However, GD is highly and nonlinearly variable in both time and space due to the heterogeneity of surface water, topography, seawater intrusion and human activities such as irrigation and drainage. Eltahir and Yeh demonstrated that seasonal variations of shallow groundwater are strongly influenced by climatologic variables, including precipitation and evapotranspiration [9]. Many authors, including Famiglietti et al., Brocca et al., and Li and Rodell, have investigated soil moisture spatio-temporal variability [10–12]. Western and Blöschl found it increased with increasing extent [13]. However, the study of GD spatio-temporal variability often requires a large amount of observation data and many monitoring points. Unfortunately, in most areas, observation wells are generally sparse. Therefore, few studies have been conducted on GD spatio-temporal variability owing to the scarcity of available, high-quality measurement time series at a regional scale. Satellite-based remote sensing methods can provide time series of GD estimates over large areas. For instance, at regional scales the Gravity Recovery and Climate Experiment (GRACE) satellite observations have proved useful for evaluating groundwater variations and trends [14,15], but the GRACE spatial resolution is about 15,000 km^2 at mid-latitudes [16,17], making it difficult to validate remote sensing data with well observations. The economic use of remote sensing to accurately estimate average GD for large areas calls for data validation. The first step to solve this validation problem is to identify representative groundwater monitoring points.

Temporal stability is the most important feature of soil property spatio-temporal variability. This concept was first described by Vachaud et al. who defined it as "the time-invariant association between spatial location and classical statistical parameters of a given soil property" [18]. The concept was later expanded by Kachanoski et al., who described the temporal stability of soil property as a reflection of the temporal persistence of the spatial structure. Temporal stability indicates the degree by which soil moisture content spatial patterns change over time; a strong temporal stability indicates a stable pattern over time even though the actual soil moisture content values may vary greatly within a given area [19]. Temporal stability analysis has been successfully applied to identify locations that represent the mean soil moisture content of an area [20]. Soil moisture content information can be up-scaled or downscaled and even resolve missing data sets [21–23]. GD temporal stability assessment helps resolve GD issues but also improves the design of groundwater monitoring networks. Knowledge of temporal variability helps determine the number of wells required to quantify the regional mean GD at a given time. Despite its importance in helping to identify stable spatio-temporal variability structures to control soil salinization, no GD study has been conducted.

The study had four main objectives: (1) to clarify the spatial pattern and the temporal stability characteristics of GD in the Contemporary Yellow River Delta; (2) to identify temporally stable locations to reduce time series sampling of average GD on a regional scale (approximately 3000 km^2); (3) to determine how long temporal stability can persist; and (4) to identify time-stable points for effective remote sensing data validation. To illustrate the benefits of the time-stability concept, the research compared the average GD determined from time-stable points with GD data from satellite remote sensing.

The research articulated optimization of a well observation network and illuminated how validated satellite data could be employed cost-effectively to monitor GD.

2. Materials and Methods

2.1. Study Area

The Yellow River Delta (117°31′–119°18′ E, 36°55′–38°16′ N) is situated in the northeast of the Shandong Province, China, on the southern bank of the Bohai Sea (Figure 1). The YRD is traversed by the Yellow River, which conveys heavy sediment discharge. The nutrient-rich waters support a classical river wetland ecosystem which sustains the most important population of migrating birds in the world. The YRD is composed by the ancient, modern and contemporary delta body. The contemporary delta is a fan shape resulting from a periodic swing of the mouth channel, with the Yuwa village as the vertex. Since 1934, the delta has grown to reach its current 3000 km^2 (Figure 1). In Figure 2, the yellow line displays the boundary of the contemporary delta. The study area was in the contemporary delta but sampling points (Figure 1) were limited to by the availability of observation wells which restricts interpolation of the results. The climate is characterized by continental monsoon in the North Temperate Zone with seasonal fluctuations in precipitation and temperature. The annual average precipitation and evaporation are approximately 600 mm and 1944 mm, respectively, with 70% of the total precipitation occurring between July and August [24]. Currently, large-scale irrigation system construction to introduce Yellow River water into reservoirs and channels proceeds apace with ongoing intensification of land use.

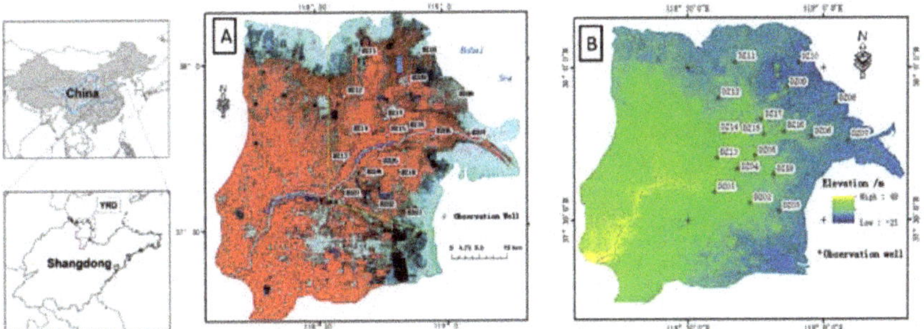

Figure 1. Spatial distribution of the observation wells in the study area. (**A**) The wells listed on the landsat8 color image composited by Bands 543; (**B**) The wells listed on the elevation map.

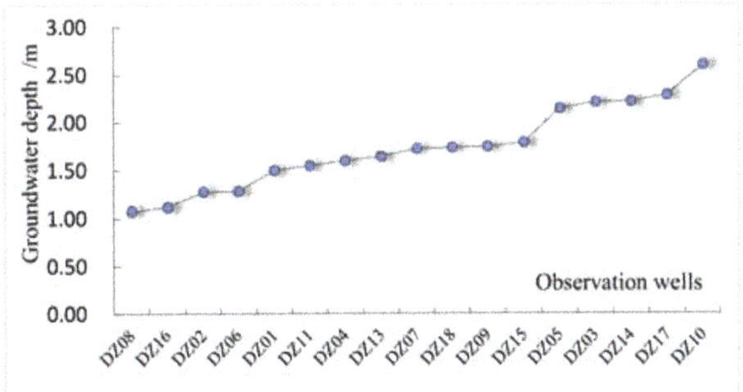

Figure 2. Sequence diagram of groundwater depth survey from 2004–2007.

2.2. Data and Method

2.2.1. Data Collection

(1)   Groundwater depth dataset

The GD dataset came mainly from 18 observation wells numbered DZ01 to DZ18, measured at regular intervals during the period of 2004–2007. Generally, the depth of the observation well was 5–6 m, with well tube diameter 50 mm. The material of well wall is PVC, and the lower part of the wall tube consists of some small holes, surrounded by a fine sand and other filter media. The groundwater monitoring device has a probe for detecting the GD and conductivity (see Figure 1). The location of observation wells was determined based on the accessibility of roads and the gradient of the GD, as shown in Figure 1. In the study area, groundwater monitoring wells DZ01, DZ04, DZ05, DZ06, DZ07, DZ13, DZ15 and DZ16 were arranged from inland to the coast, but also close to the Yellow River channel.DZ02, DZ03, DZ18, DZ14, DZ17, DZ08, DZ09, DZ12, DZ10, and DZ11 are distributed from south to north.

Measurements began on 5 May 2004. Initially, researchers measured only the ten wells located to the north of the Yellow River. On the following day, wells south of the Yellow River were measured. From then on, each point was measured once every 5 days, and all locations were sampled each time in the same time and order.

Researchers took three measurements and the truncated mean was computed by discarding the highest and the lowest value. Researchers also measured the geography coordinates and elevation data of the wells by GPS device. Researchers logged measurement time, irrigation, precipitation and other weather conditions in situ.

(2)   Elevation data

We used the elevation data from the processed SRTM (Shuttle Radar Topography Mission) DATA VERSION 4.1. The data are in Geotiff format with 90 m resolution, in decimal degrees and datum WGS84. They are derived from the USGS/NASA SRTM-4 data (http://srtm.csi.cgiar.org); the elevation map of the study area is shown as Figure 1B.

(3)   Terrestrial water storage (TWS) dataset

In this study, we used the Level-3 products from the GRACE project data products as the TWS. The Level-3 mass anomalies datasets are associated with the most up-to-date Level-2 data obtained from three data centers: Jet Propulsion Laboratory (JPL), Center for Space Research (CSR), and Geo Forschungs Zentrum Potsdam (GFZ). Monthly mean TWS equivalent thickness, $1° \times 1°$ gridded data files were retrieved from the GRACE Tellus website (https://grace.jpl.nasa.gov/data/get-data/monthly-massgrids-land/), CU GRACE website (http://geoid.colorado.edu/grace/dataportal.html) and ICGEM website (http://icgem.gfz-potsdam.de/ICGEM/ICGEM.html).

(4)   Surface water and soil moisture storage dataset

Surface water (SW) and soil moisture storage (SMS) were obtained from the outputs of the Community Land Model 2.0 (CLM 2.0) in Global Land Data Assimilation System (GLDAS-1) Products corresponding to the study period. Outputs of CLM are available since 1979 at the spatial resolution of 1 degree, and the study area only has one grid cell. The monthly SMS was computed by aggregating SMS of all the soil layers with a total soil thickness ranging from 0 cm to 82.9 cm.

2.2.2. Method of Deriving Maps of the Groundwater Depth

We calculated the groundwater tables for the wells by subtracting the GD from the elevation measured in situ. Then, a digital groundwater table model (DGTM) was produced by using Inverse

Distance Weight (IDW) interpolation. In discontinuous environments with weak soil moisture spatial correlation, linear regression models can capture the impact of features and outperform distance-based methods. Nevertheless, in complex terrains such as the YRD, IDW outperforms Kriging for interpolation. The resolution of the DGTM data was the same as the SRTM data, and the maps of the mean GD for each month were derived by subtracting the DGTM from the elevation data.

2.2.3. Satellite-Based Groundwater Storage Anomaly Estimation

To determine the anomaly in satellite-based groundwater storage (DGWS), anomalies of other water components of terrestrial water cycle, i.e., soil moisture (DSM), snow water equivalent (DSNW) and surface water (DSW) equivalents, were removed from terrestrial water storage (TWS). Correspondingly, monthly SMS changes were computed as the difference of SMS in two consecutive months. The spatial averages of monthly evapotranspiration and SMS changes were computed from 1980 to 2015. The monthly evapotranspiration (E) and changes in SMS (DSMS) were aggregated to obtain annual values.

2.2.4. Method of Analysis: Temporal Stability

To characterize the GD temporal stability, the research estimated standard soil moisture statistics.

Coefficient of Variation

The coefficient of variation captured the spatial variability of GD. The coefficient of variation (C_V) reflects the relative variability at sampling times; in other words, the degree of discretization of the random variable and is defined as:

$$C_V = \sigma/\mu \tag{1}$$

where: σ is the GD standard deviation in a given sampling time, μ is the average GD at sampling time.

In order to characterize the degree of variability, the coefficient of variation (C_V) values were analyzed as suggested by Nielsen and Bouma [25], C_V can quantitatively ascertain the magnitude of the spatial variability as weak when $C_V < 10\%$, moderate if $10\% < C_V < 100\%$ and strong when $C_V > 100\%$.

Relative Deviation

The temporal stability of the GD patterns was also assessed following the classic approach of the ranking stability analysis [18,19,22]. This encompassed the computation of the mean relative difference (MRD) and standard deviation of relative difference (SDRD) for all measurements. The relative difference was defined as:

$$\delta_{ij} = \frac{\left(h_{ij} - \overline{h_j}\right)}{\overline{h_j}} \tag{2}$$

where δ_{ij} is the GD measured at site i and time j.

$$\overline{h_j} = \frac{1}{n}\sum_{i=1}^{n} h_{ij} \tag{3}$$

Represents the mean value of GD at time j, and n is the number of sampling sites. Therefore, the mean relative difference for site i was defined as:

$$MRD_i = \frac{1}{m}\sum_{j=1}^{m} \delta_{ij} \tag{4}$$

where m represents the number of sampling times. Similarly, the standard deviation of the relative difference at site i was defined as:

$$\text{SDRD}_i = \sqrt{\frac{1}{m-1} \sum_{j=1}^{m} \left(\varepsilon_{ij} - \overline{\delta_i} \right)^2} \tag{5}$$

The mean relative deviation and its standard deviation reflect the temporal stability of the GD. The smaller the relative deviation and the standard deviation, the higher the temporal stability. Therefore, the principle of "when the average relative deviation is close to zero, the standard deviation is small" was employed to select the well locations on behalf of the averaged conditions.

Spearman Rank Correlation Coefficient

The Spearman rank correlation coefficient method is a non-parametric test method which can express the intensity and direction of the same variable at different times. In this study, we used this statistic to compare the similarity and persistence for the GD spatial pattern on different dates by means of autocorrelation analysis. The Spearman rank correlation coefficient was defined as:

$$r_s = 1 - \frac{6 \sum_{i=1}^{n} \left(R_{ij} - R_{ik} \right)^2}{n(n^2 - 1)} \tag{6}$$

where R_{ij} is the rank of GD measured at site i and time j, R_{ik} is the rank of the GD measured at site i but at time k, and n is the total number of observations. A Spearman r_s equal to 1 indicates a total agreement of the spatial patterns between the two dates and therefore perfect temporal stability of the GD. In other words, the closer is to one, the more stable the process involved. Spearman rank correlation coefficient was computed for each study site.

The statistical analyses and plots of the GD data used SPSS 16.0 and Excel 2013 software.

3. Results and Analysis

3.1. Group the Observation Wells

Figure 2 illustrates the observation wells, sorted according to the average of the GD during the study period.

As can be seen from Figure 2, the GD average in the area is high, roughly between 1.0 m and 3.0 m. the maximum and the minimum value of the mean GD in the study area appeared at DZ10 and DZ08 in Figure 2 respectively. Based on the curvature typology of the GD in Figure 2, the observation wells can be divided into three groups: wells with lower GD including DZ05 DZ03, DZ14, DZ17 and DZ10, wells with higher GD including DZ08, DZ16. DZ02 and DZ06, others are wells with medium GD.

3.2. Temporal Dynamics of Groundwater Depth

The statistical characteristics of the GD obtained from the monthly average of all the observation wells in the period from 2004 to 2007 are shown in Figure 3.

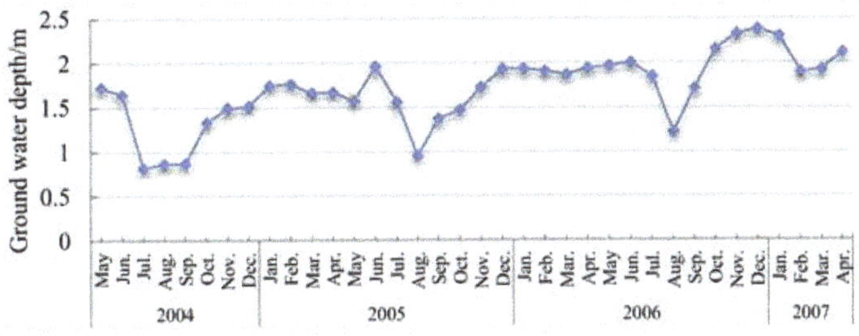

Figure 3. Seasonal distribution of groundwater depth.

From Figure 3, the GD in the Contemporary Yellow River Delta has obvious monthly variation, and the GD varied in response to precipitation and surface water in this area. A sharp decrease in GD occurred and reached the minimum value in months 7–9 each year which corresponds to the wet period and rainy season. After the rains, precipitation and water level decrease as the Yellow River enters its local dry season. Notwithstanding seasonal fluctuations, in general groundwater depth gradually increases to its maximum between December and February which corresponds to the usual regional dry season. Regional droughts in spring 2005 and 2006 explain the spike and sustained high GD reflecting the low water level.

3.3. Spatial Variation Characteristics over Time

The spatial variance of each month in the study area was calculated according to Equation (1), and the results are shown in Figure 4.

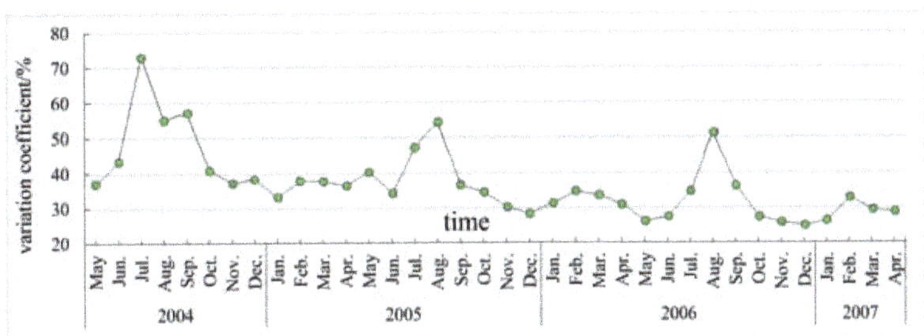

Figure 4. Spatial variation coefficient of groundwater depth of each month in 2004–2007.

Figure 4 shows that the monthly variation coefficient of GD in the study area fluctuated between 24.96% and 73.04% during the study period. According to the grouping rules described above for the degree of spatial variance in Equation (1), the degree of GD spatial variation is moderate in the Contemporary YRD.

The fluctuation of the spatial variation of GD illustrated distinct seasonal cycles. Generally, peaks in variation coefficient manifested in July–September 2004, July–August 2005 and August 2006, respectively. These periods corresponded to the wet season in the study area where groundwater was at shallow depth. In comparison, the fluctuation of the spatial variance in other months was gentle. In addition, the spatial variance showed a decreasing trend from 2004 to 2007, which is consistent with the characteristic for the temporal dynamics of the GD.

We visually demonstrated the spatial variance features of the GD and compared the possible switch in the spatial organization of GD with the season change. Taking 2005 as an example, interpolated maps of the GD were created for both wet periods and dry periods in the study area by the method listed in Section 2.2.2. The maps are presented in Figure 5.

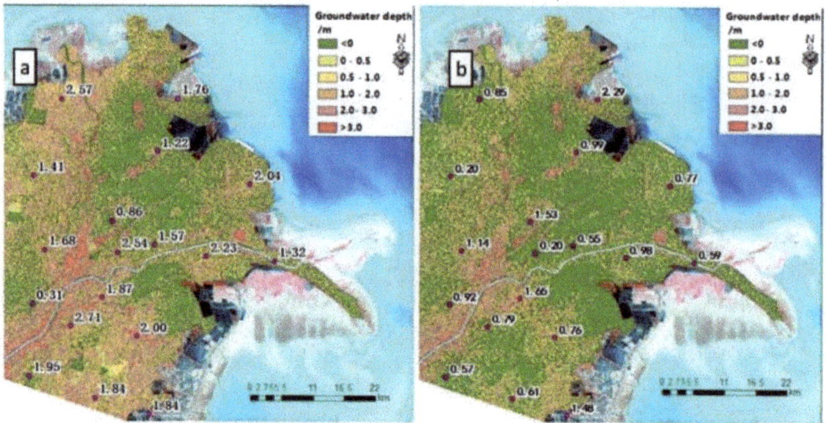

Figure 5. Interpolated map (IDW) of mean groundwater depth in January and August 2005. (**a**) Groundwater depth in January; (**b**) Groundwater depth in August.

The resulting maps (Figure 5a,b), despite a certain degree of uncertainty for some border areas due to extrapolation, confirm and make the spatial patterns during the two conditions visually appreciable. On the maps, the places where the buried depth is less than zero were covered with surface water. Figure 6 illustrates that, overall, the spatial distribution of GD in August was more dispersed and staggered in different grades than in January, indicating the spatial variance was more distinct in the wet period than other months during a season cycle.

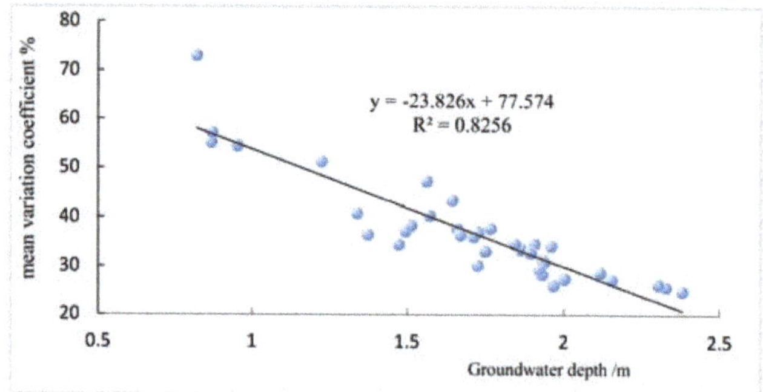

Figure 6. Relationship between groundwater depth and variation coefficient.

Furthermore, to make clear the relationship between spatial variability and GD, statistical regression analysis was performed on the monthly mean GD across the whole region and monthly spatial variation coefficients, as shown in Figure 6.

Figure 6 illustrates that the spatial variance coefficient has a good linear negative correlation with the GD, with the coefficient R^2 of the model being 0.83, which indicates that when the groundwater of the whole study area is higher, the regional spatial variation is larger. According to the meteorological data of the study region, in 2006, a dry year, the GD was lower than in 2004 and 2005, and the mean annual GD in these three years was 1.43 m, 1.68 m and 1.99 m with corresponding spatial variability coefficients of 43.98%, 36.39% and 30.93% respectively. The serial data of annual averages validates the relevance between the two variances, but also implies that, in this region, climate plays a more significant role in influencing the spatial variability of GD. The research confirmed spatial variability of groundwater depth: compared to wet periods, drought weakens it.

3.4. Time Stability of Groundwater Depth

3.4.1. Analysis for the Temporal Stability of Spatial Pattern

Spearman's rank correlation coefficients confirmed the temporal stability for spatial patterns of GD. The results for past months are presented in Table 1. The scatter plot of the mean GD and the Spearman rank correlation coefficient for each month is shown in Figure 7.

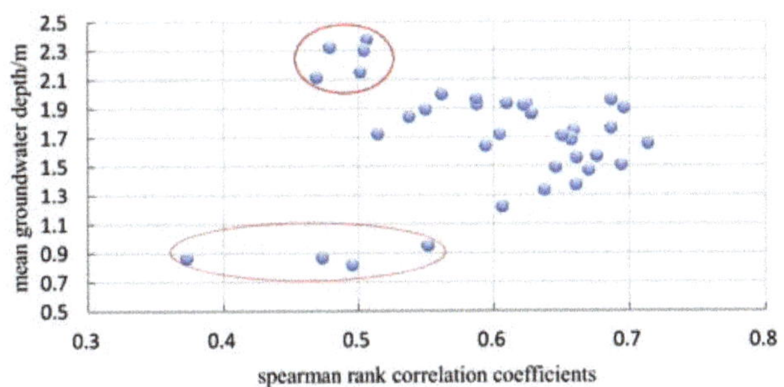

Figure 7. Scatter plot of Spearman rank correlation coefficient and groundwater depth.

The values in Table 1 show that the relationships were significant at $p < 0.01$ or $p < 0.05$ in most months, indicating strong time persistence in the spatial pattern of GD across the study area. These results are consistent with those in other areas [18,26–29]. Figure 8 shows that the Spearman rank correlation coefficient is lower at the observation wells when the monthly mean GD >2 m or <1 m (labeled with red circles) while, when the groundwater is in medium depth, the Spearman rank correlation coefficients are higher, implying the spatial patterns of GD in these months are similar to that of other times. These results indicate that the temporal stability for spatial patterns of GD is seasonal, in dry or wet season, it is poor, while in regular season it is stronger.

At the same time, Table 1 illustrates that the temporal stability presented a time-associated drift, with higher correlations between sampling series close in time and decreasing with increasing time lags. Furthermore, to analyze the relationship between spatial pattern and time lagging for GD, bivariate correlation analysis was performed for the Spearman rank correlation coefficients and the corresponding time lags of each month. The results are presented in Table 2.

From the table, the relationships were all significant at $p < 0.01$, indicating that the spatial patterns of GD can persist for short periods but cannot be maintained over longer periods. Moreover, the effect of time lags on the temporal patterns of GD was evident. As shown in Table 1, for example in May, the coefficients changes from significant correlation to irrelevant with the time lags increasing from 1 monthly lag to 18 monthly lags. Although only limited inferences can be drawn from three years of data, the continuity and concentration of significant correlation coefficients suggests spatial patterns endure for around 1.5–3 years, but further research is obviously needed to strengthen the preliminary result.

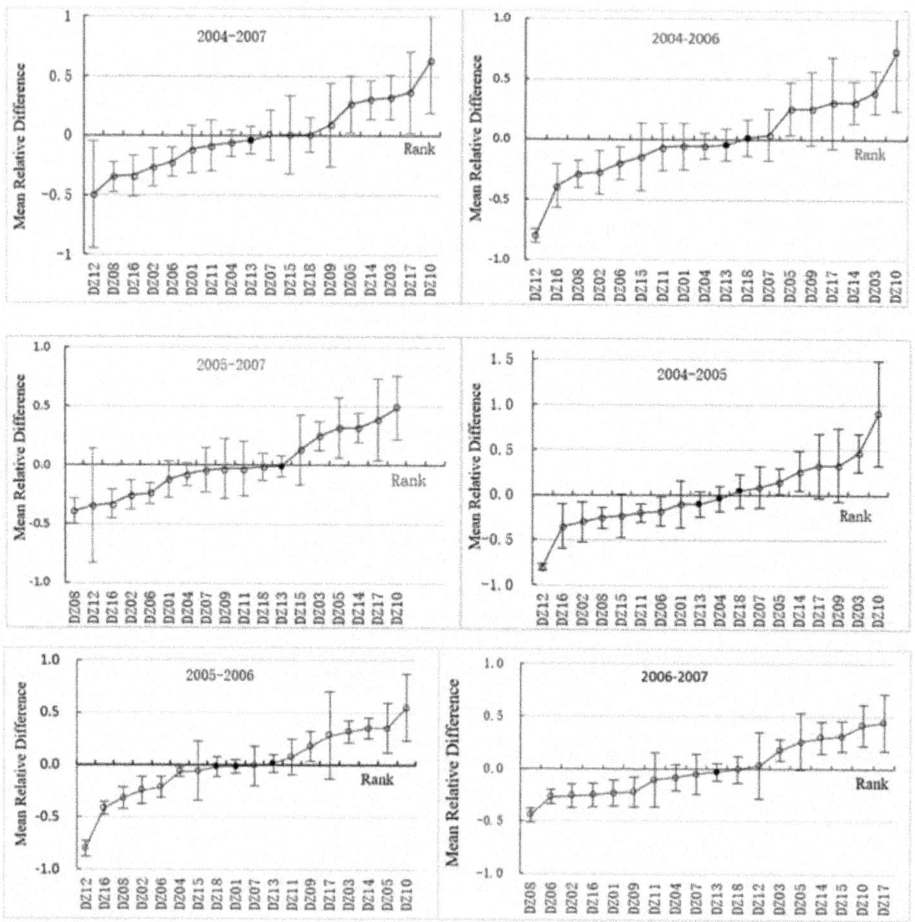

Figure 8. Mean relative difference and standard deviation of groundwater depth for different time part of each survey well. Note: Vertical bars represent standard deviation. Representative locations are marked in black. Vertical bars correspond to ±one standard deviation of the relative difference (%) over time.

Table 1. Spearman rank correlation coefficient of past months.

Month	2004–2005					2005–2006				2006–2007		
	May	August	December	February	April	August	December	February	April	August	December	February
May	1.000	0.521 *	0.676 **	0.616 **	0.732 **	0.470 *	0.299	0.443 *	0.389	0.378	0.294	0.298
August		1.000	0.520 **	0.550 **	0.556 **	0.613 **	0.464 *	0.342	0.461 *	0.087	−0.121	0.096
December			1.000	0.953 **	0.876 **	0.753 **	0.673 **	0.732 **	0.645 **	0.518 *	0.428 *	0.531 *
February				1.000	0.851 **	0.623 **	0.781 **	0.783 **	0.738 **	0.488 *	0.455 *	0.577 **
April					1.000	0.609 **	0.602 **	0.680 **	0.740 **	0.488 *	0.329	0.476 *
August						1.000	0.549 **	0.644 **	0.449 *	0.563 **	0.190	0.250
December							1.000	0.874 **	0.874 **	0.488 *	0.320	0.461 *
February								1.000	0.866 **	0.775 **	0.579 **	0.635 **
April									1.000	0.539 *	0.366	0.513 *
August										1.000	0.804 **	0.808 **
December											1.000	0.833 **
February												1.000
Sum/month	17	16	24	24	25	20	24	30	20	16	12	12

Note: Due to the table is too large, only a few months in the table; Sum: the time interval reached the significant level; ** Statistical significance at $p < 0.01$, * Statistical significance at $p < 0.05$.

Table 2. The determination coefficient of Spearman rank correlation coefficient and the time lags.

2004-2005	May	June	July	August	September	October	November	December	January	February	March	April
Pearson coefficient	−0.799 **	−0.610 **	−0.815 **	−0.856 **	−0.865 **	−0.855 **	−0.789 **	−0.789 **	−0.676 **	−0.716 **	−0.757 **	−0.749 **
2005–2006	May	June	July	August	September	October	November	December	January	February	March	April
Pearson coefficient	−0.781 **	−0.691 **	−0.778 **	−0.754 **	−0.722 **	−0.745 **	−0.676 **	−0.712 **	−0.740 **	−0.758 **	−0.540 **	−0.438 **
2006–2007	May	June	July	August	September	October	November	December	January	February	March	April
Pearson coefficient	−0.707 **	−0.735 **	−0.771 **	−0.816 **	−0.776 **	−0.801 **	−0.820 **	−0.816 **	−0.851 **	−0.839 **	−0.634 **	−0.574 **

** Statistical significance at $p < 0.01$.

3.4.2. Identification of Representative Locations

To further explore the dynamics of GD and its stability in different periods, the 36-month study period was split into three time slices—three years, two years and one year—for separate study. Temporal stability analysis was conducted for the 18 wells. According to Equations (2)–(5), the MRD and SDRD of each observation well in these periods were determined, the ascending order results of which are shown in Figure 8. The error lines of each sample are the SDRD, they are low (<35%) for many of the sites, indicating temporal stability for the GD in the study area.

The main purpose of measuring temporal stability was to precisely estimate field mean GD at identified locations; the most temporally stable locations have the lowest SDRD [18]. However, the disadvantage of only using SDRD to identify stable locations is that the locations with the lowest SDRD may not necessarily correspond to the locations that best represent the field mean (MRD closest to zero). To investigate the tradeoff between the two methods, we used the criteria "MRD is close to 0 and SDRD is lower", to help identify the temporally stable locations. Figure 9 shows that the mean relative deviation and standard deviation of GD were not identical on different time scales. Table 3 illustrates the representative locations and the stability in different periods.

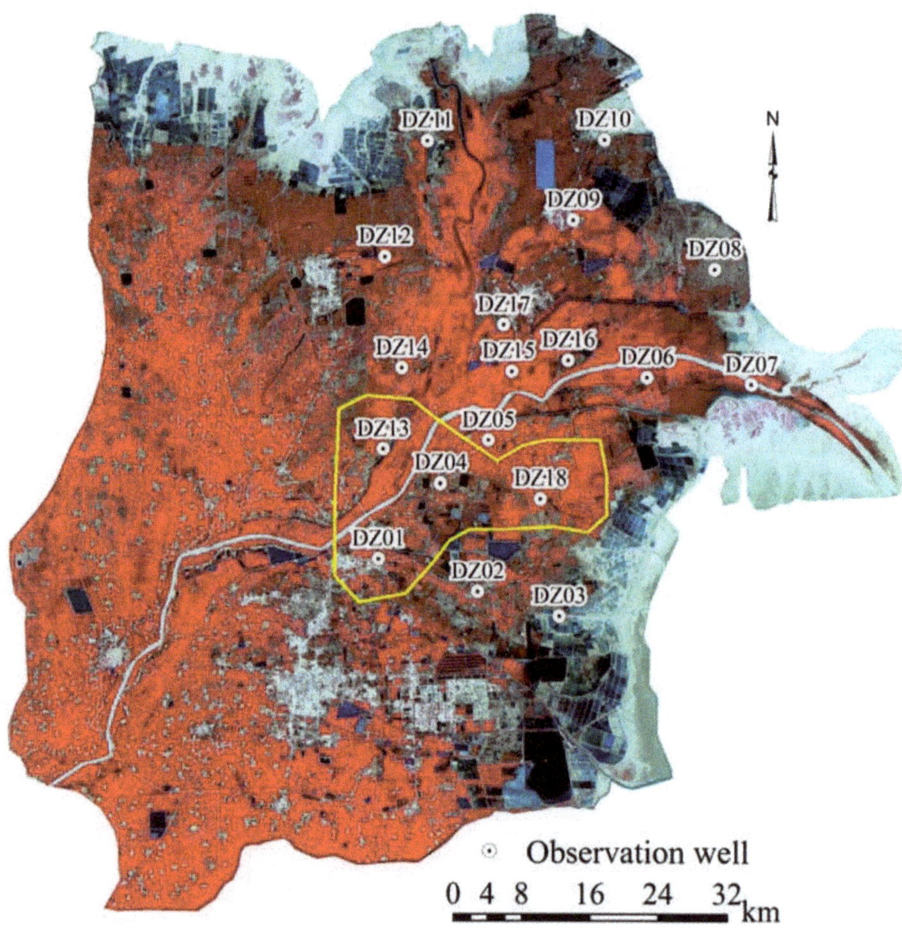

Figure 9. The locations of DZ01/DZ04/DZ13/DZ18.

Table 3. Representative locations in different time scales.

Time Scale	Representative Locations		Unstable Locations	
	Well Code	Groundwater Depth Grade	Well Code	Groundwater Depthm and Grade
2004–2007	DZ13	medium	DZ12 and DZ10	0.95 higher and 2.60 lower
2004–2006	DZ13, DZ18	medium	DZ12 and DZ10	0.95 higher and 2.60 lower
2005–2007	DZ13	medium	DZ12 and DZ10	0.95 higher and 2.60 lower
2004–2005	DZ04, DZ13 and DZ18	medium	DZ12 and DZ10	0.95 higher and 2.60 lower
2005–2006	DZ13, DZ01 and DZ18	medium	DZ12 and DZ10	0.95 higher and 2.60 lower
May 2006–24 December 2007	DZ13	medium	DZ08 and DZ10	1.08 higher and 2.60 lower

On the 3-year scale: the observation well DZ13, with medium grade GD, had a small mean relative difference values and standard deviations when compared to the other observation wells as shown in Figure 9. This well was considered the most representative site, which could be used on long time scales as accurate estimators of the study region average. Two wells, DZ10 and DZ12, with the maximum and minimum GD respectively, were the most unstable locations in the study area.

On the 1-year scale: using the same method as above to identify the representative locations on the one-year time scale, the observation wells DZ04/DZ13/DZ18, DZ01/DZ13/DZ18, and DZ13 were considered the most representative sites during the periods of May 2004–April 2005, May 2005–April 2006 and May 2006–April 2007, respectively, all with the medium grade GD. The observation wells DZ12, and DZ10 remained the most unstable locations in May 2004–April 2005 and May 2005–April 2006, respectively, which are same as on the 3-year scale. The most unstable GD locations were DZ08 and DZ10 in May 2006–April 2007; mean GD of DZ08 during the study period is 1.08 m, a little higher than the DZ12, and belongs to the grade of higher GD.

On the 2-year scale: still using the same method as above to identify the representative locations on the 2 years scale, during the periods of May 2004–April 2006 and May 2005–April 2007, the observation wells DZ18/DZ13 and DZ13 were the most representative sites, respectively, all still belonging to the medium grade of GD. The observation wells DZ12 and DZ10 remained the most unstable locations in May 2004–April 2006 and May 2005–April 2007, respectively, which are same as on the most of time scales.

Synthesizing the results, while groundwater depth temporal stability varies along the 3 timescales, all the most unstable stations belong to the grades of shallower or higher groundwater depths, and the most temporal stable wells all belong to the grades of medium GD (see Table 3 above for thresholds). This suggests that, for the study region, areas with medium GD have stronger temporal stability, while areas with lower or higher GD have weaker temporal stability. Furthermore, the representative stations are adjacent to each other in the spatial position (labeled with a yellow line in Figure 10), likely due to similar spatial GD variability. The illustrated area is the representative region of GD in the Contemporary YRD. The measurement of GD here will cost effectively achieve the dynamic monitoring of GD in the whole Contemporary YRD.

It is notable that DZ13 is included in the representative locations of each period. Therefore, considering the persistence in application, DZ13 was selected as the representative location of the area and can be used to estimate the average GD of the study region. Conjunction for these results with the previous results of Spearman rank correlation analysis is indicates that without a persistent pattern, there would not be temporally stable sites, and a persistent spatial pattern of GD can strengthen the application of temporal stability analysis.

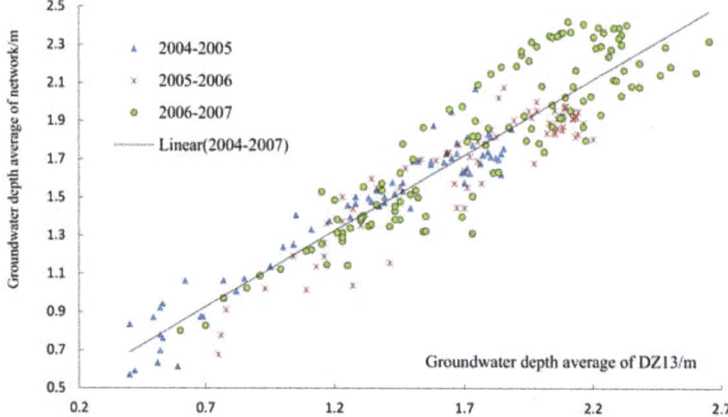

Figure 10. Correlation between the representative location and mean of the whole region for groundwater depth in 2004–2007 (259 samples).

3.4.3. The Rationality Test of Representative Locations

To determine if the representative locations accurately estimates the GD average, groundwater depth of the representative well measured every 5 days were compared to the GD average of all observation wells in the area at the corresponding date. Shown as in Figure 10.

Figure 10 and Table 4 contain the plots and statistics of the representative location in each period versus the whole observation network average for the entire data series. The GD of the representative station is slightly floating around the network average (Figure 10). There is a negligible difference (RMSE is 0.053 m/m). When we compare the mean between the representative station average and the network average of each period, the difference between the two estimates is about 0.022 m/m. The locations were identified, based on the GD (x) of which, the region mean (y) could be estimated by y = 0.795x + 0.371, as shown in Figure 10, and a paired t-test indicates they are equal at a 99% confidence level. This indicates that the DZ13 is a good representation of the study region with respect to groundwater depth during the experiment.

Table 4. Comparison between the representative location and mean of the whole region for groundwater depth in different period.

Period	Pearson Correlation	Mean of the DZ13/m	Mean of the Region/m	Difference/m/m
5 May 2004–28 April 2005	0.961 **	1.294	1.419	0.088
2 May 2005–27 April 2006	0.910 **	1.707	1.670	−0.022
2 May 2006–2 December 2007	0.898 **	1.782	1.812	0.017
5 May 2004–27 April 2006	0.943 **	1.501	1.544	0.028
2 May 2005–2 December 2007	0	1.754	1.759	0.003
5 May 2004–2 December 2007	0.924 **	1.629	1.665	0.022

** Significant at 0.01 level (two tails).

3.4.4. Validate the Satellite-Based DGWS

To test whether the representative point can adequately represent the entire area to validate the satellite-based DGWS, we compared three products of satellite-based DGWS (labeled Jpl, Gfz and Csr) with the DGWS of DZ13 and the average of all observation wells by the bivariate correlation analysis, respectively. The results are presented in Table 5. Frustratingly, their relevance is poor. However, when we delayed satellite-based DGWS by one month, their relevance was promoted to significant at the 0.01 level, and the results are shown in Table 5. The delayed products of satellite-based DGWS are labeled DJpl, DGfz and DCsr, respectively. The correlation between DZ13 and DGfz is the best, with a Pearson correlation value of 0.707. The time series of monthly GD change at representative point DZ13; Gfz and DGfz are presented as Figure 11.

Table 5. The Pearson Correlation of DGWS from satellite-based data, DZ13 and the average of all wells.

	Jpl	Gfz	Csr	DJpl	DGfz	DCsr
DZ13	0.081	−0.049	0.031	0.673 **	0.707 **	0.635 **
All wells	0.089	−0.033	0.044	0.634 **	0.644 **	0.566 **

** Significant at 0.01 level (two tails).

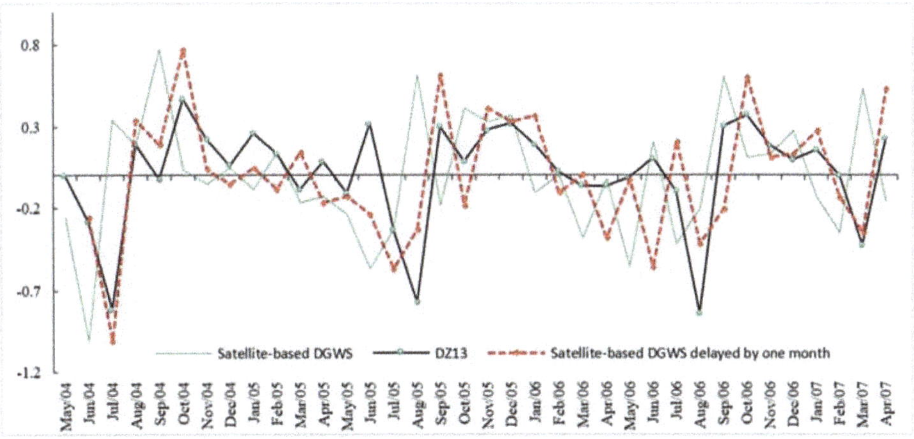

Figure 11. Time series of DZ13and the satellite-based DGWS for monthly groundwater depth change in 2004–2007 (Note: The present values of the satellite-based DGWS and the satellite-based DGWS delayed by one month are 0.1 times the original value).

Table 5 and Figure 11 shows that the correlation between DZ13 and satellite-based DGWS is similar to that of the average of all observation wells, indicating that DZ13 can replace the regional average to verify the satellite-based DGWS. There is no significant correlation between the observed data and the satellite-based data of DGWS in the current month, but significant correlation with the data delayed by one month. This phenomenon may be due to the hydrological characteristics of the study area. The study area is near the ocean and, due to the influence of the Yellow River and the sea, surface runoff is rich, and the soil moisture content is high. Therefore, the groundwater mainly comes from the recharge of the soil moisture and surface water, and the change of soil moisture and surface water strongly affects the change of GD. The more soil moisture and surface water, the smaller the groundwater depth. Generally, due to the recharge of the soil moisture to groundwater taking some time, and the satellite-based DGWS being derived from TWS change deducting soil moisture and surface water, there will be a time lag between the satellite-based DGWS and the measured GD. This also implies that the GRACE Level 3 products can be used to predict the GD in the area one month in advance.

4. Conclusions and Discussion

The following results can be summarized:

The groundwater depth in the Contemporary Yellow River Delta has obvious seasonal variation, and the GD is consistent with the change of surface water and precipitation in the area. The results of the spatial variability analysis found groundwater spatial depth variability in the study area was moderate. The monthly variation coefficient fluctuated from 24.96% to 73.04%. The variation coefficients of GD also have a seasonal aspect. In each year, the spatial coefficient of variation in the wet season was the largest while other months were relatively stable. The results of the correlation analysis showed that spatial variability of GD anomalies exhibits dependence on spatial mean of GD. The spatial variability of GD in wet season was generally stronger than in dry season. Research in other regions found similar results in soil moisture and water storage [20,29–32]. The results of this study are of great significance to the restoration of saline-alkali soil. In saline soil, the salt content of soil is not inseparable from the high mineral degree of groundwater, the smaller the spatial variability of groundwater depth, the more stable the salt content and the more favorable for plant growth. Therefore, the regulation of GD provides an important means to control the saline-alkali degree in the contemporary Yellow River Delta. A further possible implication is that the contribution of groundwater to local evapotranspiration and runoff is important under very wet and dry conditions. This may be a justification for incorporating groundwater into various type of water management systems.

The results of Spearman rank correlation analysis show that the GD demonstrates obvious temporal stability characteristics. The temporal stability of GD at different water levels will exhibit different temporal stability: more stable in time of normal water level than either in dry season or wet periods. Suggested reasons for differences in GD in dry and wet conditions include precipitation recharge, irrigation, vegetation root absorption, and strong surface evaporation. During the dry period, the evaporation of soil surface is strong, and the capillary movement of water in the soil cause groundwater level decline. At the same time, in areas where farms introduce the Yellow River water for irrigation, this artificially improves the groundwater level and thereby increases the temporal variability of GD. During wet and dry periods, precipitation and vegetation root uptake could differ. It is different with the soil moisture. For example, Martinez et al., using data from a network consisting of 23 soil moisture stations in 36 months (from June 1999 to May 2002) to study the soil moisture temporal stability in 1285 km^2 located at the central part of the Douro Basin in Spain, found that soil moisture temporal stability during dry periods was always stronger than in wet ones. These authors also observed that the period with the lowest temporal stability coincided with the transition from dry to wet [33]. Williams et al., who sampled continuously throughout the year, found that soil moisture temporal stability was lower during the relatively dry period [34]. The divergence of the conclusions between the two fields shows that groundwater depth variations did not strictly followed soil moisture dynamics, leading to a certain decoupling between soil moisture and groundwater patterns in temporal stability.

The correlation analysis between Spearman rank correlation coefficient and time lags shows that the spatial pattern of GD is similar in a certain period, which is about 1.5–3 years in the study area, and the similarity will be weakened beyond this period. Similar results have been found in soil moisture. For example, Schneider and Martinez-Fernandez et al. found that the spatial pattern of soil moisture content was similar across 2–3 years, and the similarity decreased with increasing time interval [30,33].

The results of the representative stations of GD indicate that all the selected representative stations belong to a moderate grade of groundwater depth, but low depth and high depth of the station tends to be more unstable, indicating that the stations with medium depth often have a stronger temporal stability. The representative location is DZ13, with highest prediction accuracy, which implied a single sampling location could be used to estimate, and thus represent, the region mean. However, the relatively small spatial domain of the study area could account for this single representative point. For larger study areas, one representative point may be insufficient to estimate the mean GD of the whole region. Notwithstanding, it is obvious that temporal stability analysis could benefit future groundwater satellite products because it can verify stable GD networks as well as identify representative and anomalous sites, thus improving any ground validation programs.

As the bulk of current temporal stability research concentrates on soil moisture, it is useful to compare our study results in this area. It found that, although the groundwater is an important source of soil moisture, its temporal stability differs, likely due to other soil moisture factors at play. Further research is needed to verify this. In addition, the current research results show that while spatial scale influences soil moisture temporal stability, the extent of the impact remains unclear. This finding contrasts with Coppola et al., who found surface soil moisture temporal stability around 225–8800 m^2 with low mean relative deviations (MRD), typically 20–60% [35]. On the coastal plain of the United States, Bosch et al. reported that MRD of the soil moisture in the scope of 3750 km^2 is more than 200% [36], while Broccain, who investigated two larger area (178 and 242 km^2), found that the range of MRD was only about 50% [11].

Finally, we compared the data serial of the GD from the representative location and the remote sensing data. The results indicated that the representative location can replace the regional average to verify the satellite-based DGWS. Satellites such as GRACE can forecast the groundwater depth at region scale, with the DGWS afforded by remote sensing providing valuable and convenient monitoring information. Nevertheless, the temporal stability of the GD at different spatial scale remains unresolved. Future studies are needed to answer this question.

Author Contributions: R.W., Y.L., H.M., Y.P. and L.D. conceived, designed and conducted the research but also collected and analyzed the data. S.H. reviewed and edited the paper. He also was the corresponding author.

Funding: National This research received no external funding from "Thirteen Five" Program (2017YFD0200702), Shandong "Double Tops" Program (SYL2017XTTD02), Shandong Key R & D Program (2017CXGC0306) and the Cultivate Plan Funds for Young Teacher and the Science and Technology Innovation Foundation for Youth of Shandong Agricultural University (No. 23694).

Acknowledgments: The data used in this study were acquired as part of the mission of NASA's Earth Science Division and archived and distributed by the Goddard Earth Sciences (GES) Data and Information Services Center (DISC). Sincere thanks are also extended to China's national science and technology infrastructure platform construction project: Earth system science data sharing platform to provide information support.

Conflicts of Interest: The authors declare no conflicts of interest.

References

1. Kuang, W.; Liu, J.; Dong, J.; Chi, W.; Zhang, C. The rapid and massive urban and industrial land expansions in China between 1990 and 2010: A CLUD-based analysis of their trajectories, patterns, and drivers. *Landsc. Urban Plan.* **2016**, *15*, 21–22. [CrossRef]

2. Liu, Y.; Fang, F.; Li, Y. Key issues of land use in china and implications for policy making. *Land Use Policy* **2014**, *40*, 6–12. [CrossRef]

3. International Research and Training Center on Erosion and Sedimentation (IRTCES). *Case Study on Utilization of Sediment Resource in the Lower Yellow River*; International Research and Training Center on Erosion and Sedimentation: Beijing, China, 2010.

4. Zhang, T.T.; Zeng, S.L.; Gao, Y.; Ouyang, Z.T.; Li, B.; Fang, C.M.; Zhao, B. Assessing impact of land uses on land salinization in the Yellow River Delta, China using an integrated and spatial statistical model. *Land Use Policy* **2011**, *28*, 857–866. [CrossRef]

5. Fang, H.; Liu, G.; Kearney, M. Georelational analysis of soil type, soil salt content, landform, and land use in the Yellow River Delta, China. *Environ. Manag.* **2005**, *35*, 72–83. [CrossRef]

6. Qinghua, Y.E.; Liu, G.; Tian, G.; Chen, S.; Huang, C.; Chen, S.; Liu, Q.; Chang, J.; Shi, Y. Geospatial-temporal analysis of land-use changes in the Yellow River Delta during the last 40 years. *Sci. China Ser. D Earth Sci.* **2004**, *47*, 1008–1024.

7. Fan, X.M.; Liu, G.H.; Tang, Z.P. Analysis on main contributors influencing soil salinization of Yellow River Delta. *J. Soil Water Conserv.* **2010**, *24*, 139–144.

8. Yao, R.; Yang, J. Quantitative analysis of spatial distribution pattern of soil salt accumulation in plough layer and shallow groundwater in the Yellow River Delta. *Trans. Chin. Soc. Agric. Eng.* **2007**. [CrossRef]

9. Eltahir, E.A.B.; Yeh, P.J.F. On the asymmetric response of aquifer water level to floods and droughts in Illinois. *Water Resour. Res.* **1999**, *35*, 1199–1217. [CrossRef]

10. Famiglietti, J.S.; Ryu, D.; Berg, A.A.; Rodell, M.; Jackson, T.J. Field observations of soil moisture variability across scales. *Water Resour. Res.* **2008**, *44*. [CrossRef]

11. Brocca, L.; Tullo, T.; Melone, F.; Moramarco, T.; Morbidelli, R. Catchment scale soil moisture spatial–temporal variability. *J. Hydrol.* **2012**, *422–423*, 63–75.

12. Li, B.; Rodell, M. Spatial variability and its scale dependency of observed and modeled soil moisture under different climate conditions. *Hydrol. Earth Syst. Sci.* **2013**, *17*, 1177–1188. [CrossRef]

13. Western, A.W.; Blöschl, G. On the spatial scaling of soil moisture. *J. Hydrol.* **1999**, *217*, 203–224. [CrossRef]

14. Rodell, M.; Chen, J.; Kato, H.; Famiglietti, J.S.; Nigro, J.; Wilson, C.R. Estimating groundwater storage changes in the Mississippi River basin (USA) using GRACE. *Hydrogeol. J.* **2007**, *15*, 159–166. [CrossRef]

15. Voss, K.A.; Famiglietti, J.S.; Lo, M.H.; Linage, C.D.; Rodell, M.; Swenson, S.C. Groundwater depletion in the middle east from GRACE with implications for transboundary water management in the Tigris-Euphrates-Western Iran region. *Water Resour. Res.* **2013**, *49*, 904–914. [CrossRef] [PubMed]

16. Rowlands, D.D.; Luthcke, S.B.; Klosko, S.M.; Lemoine, F.G.R.; Chinn, D.S.; Mccarthy, J.J.; Cox, C.M.; Anderson, O.B. Resolving mass flux at high spatial and temporal resolution using GRACE intersatellite measurements. *Geophys. Res. Lett.* **2005**, *32*, 319–325. [CrossRef]

17. Swenson, S.; Wahr, J. Post-processing removal of correlated errors in grace data. *Geophys. Res. Lett.* **2006**, *33*. [CrossRef]

18. Vachaud, G.; Silans, A.P.D.; Balabanis, P.; Vauclin, M. Temporal stability of spatially measured soil water probability density function. *Soil Sci. Soc. Am. J.* **1985**, *49*, 822–828. [CrossRef]

19. Kachanoski, R.G.; Jong, E.D. Scale dependence and the temporal persistence of spatial patterns of soil water storage. *Water Resour. Res.* **1988**, *24*, 85–91. [CrossRef]

20. Penna, D.; Brocca, L.; Borga, M.; Fontana, G.D. Soil moisture temporal stability at different depths on two alpine hillslopes during wet and dry periods. *J. Hydrol.* **2013**, *477*, 55–71. [CrossRef]

21. Blöschl, G.; Wagner, W. *Remote Sensing in Hydrology and Water Management*; Springer: Berlin, Germany, 2009.

22. Grayson, R.B.; Western, A.W. Towards areal estimation of soil water content from point measurements: Time and space stability of mean response. *J. Hydrol.* **1998**, *207*, 68–82. [CrossRef]

23. Dumedah, G.; Coulibaly, P. Evaluation of statistical methods for infilling missing values in high-resolution soil moisture data. *J. Hydrol.* **2011**, *400*, 95–102. [CrossRef]

24. Zhao, Y.M.; Song, C.S. *Scientific Survey of the Yellow River Delta National Nature Reserve*; China Forestry Publishing House: Beijing, China, 1995.

25. Nielsen, D.R.; Bouma, J. *Soil Spatial Variability*; Pudoc: Wageningen, The Netherlands, 1985.

26. Brocca, L.; Melone, F.; Moramarco, T.; Morbidelli, R. Soil moisture temporal stability over experimental areas in central Italy. *Geoderma* **2009**, *148*, 364–374. [CrossRef]

27. Heathman, G.C.; Larose, M.; Cosh, M.H.; Bindlish, R. Surface and profile soil moisture spatio-temporal analysis during an excessive rainfall period in the Southern Great Plains, USA. *Catena* **2009**, *78*, 159–169. [CrossRef]

28. Gao, L.; Shao, M. Temporal stability of shallow soil water content for three adjacent transects on a hillslope. *Agric. Water Manag.* **2012**, *110*, 41–54. [CrossRef]

29. Gao, L.; Shao, M. Temporal stability of soil water storage in diverse soil layers. *Catena* **2012**, *95*, 24–32. [CrossRef]

30. Schneider, K.; Huisman, J.A.; Breuer, L.; Zhao, Y.; Frede, H.G. Temporal stability of soil moisture in various semi-arid steppe ecosystems and its application in remote sensing. *J. Hydrol.* **2008**, *359*, 16–29. [CrossRef]

31. Biswas, A.; Si, B.C. Identifying scale specific controls of soil water storage in a hummocky landscape using wavelet coherency. *Geoderma* **2011**, *165*, 50–59. [CrossRef]

32. Jia, Y.H.; Shao, M.A. Temporal stability of soil water storage under four types of revegetation on the northern Loess Plateau of China. *Agric. Water Manag.* **2013**, *117*, 33–42. [CrossRef]

33. Martínezfernández, J.; Ceballos, A. Temporal stability of soil moisture in a large-field experiment in Spain. *Soil Sci. Soc. Am. J.* **2003**, *67*, 1647–1656. [CrossRef]

34. Williams, C.J.; Mcnamara, J.P.; Chandler, D.G. Controls on the temporal and spatial variability of soil moisture in a mountainous landscape: The signature of snow and complex terrain. *Hydrol. Earth Syst. Sci.* **2009**, *13*, 1325–1336. [CrossRef]

35. Coppola, A.; Comegna, A.; Dragonetti, G.; Lamaddalena, N.; Kader, A.M.; Comegna, V. Average moisture saturation effects on temporal stability of soil water spatial distribution at field scale. *Soil Tillage Res.* **2011**, *114*, 155–164. [CrossRef]
36. Bosch, D.D.; Lakshmi, V.; Jackson, T.J.; Choi, M.; Jacobs, J.M. Large scale measurements of soil moisture for validation of remotely sensed data: Georgia soil moisture experiment of 2003. *J. Hydrol.* **2006**, *323*, 120–137. [CrossRef]

 *sustainability*

Article

Estimation of Water Footprint for Major Agricultural and Livestock Products in Korea

Ik Kim [1],* and Kyung-shin Kim [2]

[1] SMaRT-Eco Consulting Firm, Seoul 06338, Korea
[2] Department of Environment & Energy Engineering, Sungshin University, Seoul 01133, Korea; kyskim@sungshin.ac.kr
*   Correspondence: kohung@smart-eco.co.kr; Tel.: +82-10-8233-0904

Received: 13 April 2019; Accepted: 20 May 2019; Published: 25 May 2019

Abstract: The Republic of Korea is the only country classified with severe water stress among the 34 Organization for Economic Co-operation and Development (OECD) member countries. Additionally, the self-sufficiency rate of grain in Korea is 27%, which is 1/3 the average of OECD member countries. Because food cannot be produced without water, demand-driven water management of agricultural and livestock products applying water footprints is needed for food security. For this, this study estimates the water footprints of 42 agricultural products and three livestock products. Based on the results, the water footprint of the vegetables grown in facility such as a greenhouse is 7.9 times larger per ton than the footprint of the vegetables cultivated in the open field. Furthermore, the water footprint per ton of beef is about 4.2 times the average water footprint per ton of vegetables grown in facility. Based on the water footprint data of 45 agricultural and livestock products, the footprint of total agricultural and livestock products in 2014 is approximately 27.9% of the total domestic water resources consumed in Korea.

Keywords: water footprint; agricultural and livestock products; Penman–Monteith equation; evapotranspiration; climate conditions

---

1. Introduction

The Republic of Korea has experienced rapid industrialization and urbanization since the 1970s. In 2015, the United Nations reported that Korea was the 23rd most densely populated country in the world with 509 people/km^2 [1]. Rapid industrialization and urbanization have led to a continual decline in the ground water level. Korea's average annual precipitation is 1274 mm, which is 1.6 times that of the average global precipitation. However, the precipitation per capita is 2660 m^3/year, which is 1/6 of the world's average because of the high population density [2]. Kim's study on the Sustainable Water Management Legislation for Climate Change Response published at Yonsei University revealed that the number of extreme drought events in Korea will double in the next 100 years, and the mean drought duration will increase six fold [3]. Additionally, in March 2012, the OECD published their Environmental Outlook to 2050, stating that Korea was classified as the only country with severe water stress among 34-member countries with a stress ratio of over 40% [4]. The ecological footprint is used as an indicator expressing the levels of human consumption. It is a metric of the biologically productive area needed to provide for everything that people use from nature (e.g., fruits and vegetables, fish, wood, fibers, absorption of carbon dioxide from fossil fuel use, and space for buildings and roads). The Global Footprint Network reported in August 2018 that Korea's ecological footprint stood at 8.5 and those of Japan, the UK, the USA, and France were 7.8, 4.0, 2.3, and 1.7, respectively [5]. Thus, the report showed that Korea's consumption was much higher than that of the other OECD member countries.

Foods such as vegetables and meat are among the most important items affecting the ecological footprint. Korea's cereal self-sufficiency rate, including feedstuffs, is 27%, which is 1/3 of the average

(83%) of OECD member countries [6,7]. Food production and processing requires a large amount of water. If Korea's food self-sufficiency rate is 100%, there will be a greater demand for water. In fact, Korea imports nearly 3/4 of its food consumed. Therefore, a large amount of water in foreign countries is being used to grow their food. This water called "virtual water" [8]. The importance of virtual water trading has been emphasized to solve the problem of global water depletion [9].

According to the Water Footprint Network (WFN), the concept of water footprint is defined as a measure of humans' appropriation of freshwater in volumes of water consumed and/or polluted [10]. Generally, water is required to manufacture most products. Additionally, water is required to produce agricultural and livestock products. In particular, imported food is produced by using water from the food production site. A food production area can be depleted by exporting large quantities. Thus, water footprint indicators should be used to measure and manage water consumption.

This study estimates the water footprint of 45 agricultural and livestock products in Korea. An extrapolation method is used to extend the water footprints per ton of 45 agricultural and livestock products to the water footprint of the total consumption of 45 products.

2. Materials and Methods

2.1. Characteristics of Korean Weather

Plants are affected by weather conditions during their growth via evapotranspiration. Weather conditions vary slightly depending on the area. Figure 1 shows a map of South Korea. It has nine Provinces, two special cities, and six metropolitan cities. Korea has a temperate climate with four distinct seasons: spring, summer, autumn, and winter. In spring and autumn, there are many sunny days caused by the migratory anticyclone. In summer, the North Pacific high-pressure system brings hot, humid weather. Winters are cold and dry because of the expanding Siberian high-pressure zone. The average temperature in Korea ranges from 10 to 15 °C with the exception of mountain and island areas. August is the hottest month with temperatures ranging from 23 to 26 °C, whereas the coldest month is January with temperatures between −6 and 3 °C. Annual precipitation per region is 1200–1500 mm in the central region and 1000–1800 mm in the southern region. Gyeongsangbuk Province is ~1000–1300 mm, some parts of Gyeongsangnam Province are about 1800 mm, and Jeju Province is ~1500–1900 mm. Fifty to sixty percent of the annual precipitation falls intensively in summer. Generally, the northwest wind in winter is relatively stronger, and in summer, the southwest wind is stronger [11].

Figure 1. Map of provinces of South Korea.

2.2. Major Agricultural and Livestock Products Studied

As shown in Table 1, this study selects 45 species that are most commonly consumed by Koreans as the target agricultural and livestock products for estimating the water footprint. The target products combine both the 2014 agricultural and livestock income collection published by the Rural Development Administration and the statistical data on agricultural and livestock products of the National Statistical Office [12,13]. Agricultural products are nine kinds of crops, 13 kinds of open field vegetables, 12 kinds of vegetables in facilities, and eight kinds of fruits. Fruits and vegetables are classified as both open field and in-facility, depending on the cultivation method livestock products such as beef, pork, and chicken are included. Here cultivation in facility means that agricultural products can be grown in agricultural structures such as greenhouses during the winter season.

Table 1. Lists of 42 agricultural products and three livestock products.

Agricultural Products (42)				Livestock Products (3)
Crops (9)	**Open Field Vegetables (13)**	**Vegetable in Facilities (12)**	**Fruits (8)**	
Rice plant Crest Barley Beer barley Corn Bean Sweet potato Spring potato Autumn potato	Spring radish Autumn radish Highland radish Carrot Spring cabbage Autumn cabbage Highland cabbage West cabbage Spinach Watermelon Yellow pepper Garlic Onion	Facility radish Facility cabbage Facility spinach Facility lettuce Facility watermelon Facility melon Cucumber Facility pumpkin Tomato Cherry tomato Strawberry Facility pepper	Apple (open field) Pear (open field) Peach (open field) Persimmon (open field) Grape (open field) Grape (in facility) Tangerine (open field) Tangerine (in facility)	Beef Pork Chicken

2.3. Water Footprint Assessment Model

International Organization for Standardization (ISO) 14046 (2014) and the WFN Water Footprint Assessment Manual are the two methodologies internationally accepted for estimating water footprint in a country, region, or product [10,14]. Both the methodologies present different water types for estimating water footprint. Thus, ISO 14046 classified water types as freshwater, brackish water, surface water, sea water, ground water, and fossil water, whereas the WFN Water Footprint Assessment Manual classifies water types as green water, blue water, and gray water. The water types of the former are identified according to the water intake point and are closely related to human life and industrial activities. Those of the latter have intimate connections with agricultural activities, which are distinguished by the use of water. It is common to generate and manage operations data in factories by the water types proposed in ISO 14046 in Korea. However, it is more appropriate for the agricultural sector to apply the water types of WFN. Considering the characteristics of the two mentioned methods, the method for estimating the water footprint for the major agricultural and livestock products proposed in this study is based on the methodological procedures and requirements of WFN. The major data categories to be collected, as shown in Table 2, are green water, blue water, and gray water, and the detailed water resource type for blue water is set to follow the requirements of ISO 14046. Additionally, the requirements of ISO 14044 are integrated with WFN's methods of calculating the amount of indirect water and estimating the environmental impacts throughout the entire life cycle of agricultural and livestock products [15,16]. Thus, the water footprint defined in this study integrates direct and indirect water footprints.

Table 2. Water types by data categories.

Activities		Data Category	Water Type		
			Green	Blue	Gray
Direct Water	Irrigation water	• Groundwater, surface water		○	
	Effective rainfall	• Precipitation	○		
	Waste water	• COD, GOD, SS, T-N, T-P			
Indirect Water	Raw material/energy	• Raw material: strain, seedling, fertilizer,	○	○	○
		• Pesticide, farm materials			
		• Energy: electricity, lubricant oil, heavy oil			
	Waste water	• COD, GOD, SS, T-N, T-P			○
	Solid waste	• Sludge, waste package		○	

According to the proposed method, this study estimates the water footprints for major agricultural and livestock products presented in Table 1. Furthermore, this study defines functional units, which quantifies the description of performance requirements fulfilled by the product system in different ways for agricultural and livestock products. The water footprint of agricultural products was set up in two ways: a ton basis (m³/ton) and a hectare basis (m³/ha). For livestock products, the functional unit was set to a ton basis. As shown in Figure 2, the system boundary for estimating the water footprint of agricultural products includes seeding, planting, cultivation, and harvesting. That for livestock encompasses feed production, feeding, grazing, slaughtering, and processing.

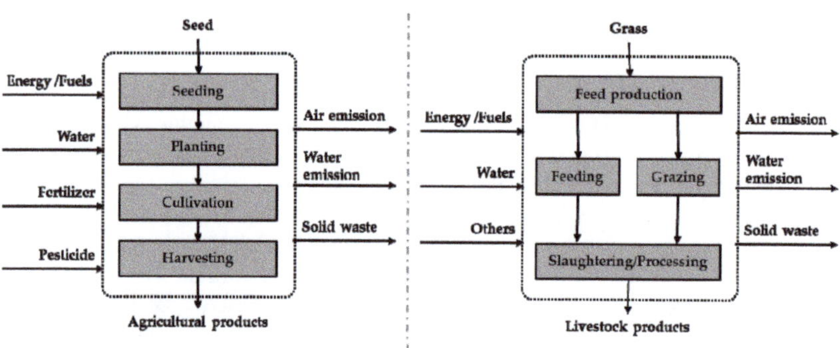

Figure 2. System boundaries of agricultural and livestock products.

Table 2 presents data categories collected throughout the life cycle of agricultural and livestock products in 2013. Here, green water includes the on-land precipitation that does not run off or recharge the ground water, but is instead stored in the soil, temporarily staying on top of the soil, or residing in vegetation. Blue water is fresh surface and ground water found in freshwater lakes, rivers, and aquifers [17]. The gray water footprint concept is the amount of fresh water required to assimilate pollutants to meet specific water quality standards. Direct water includes irrigation water (i.e., surface water and groundwater), precipitation, and wastewater, including water quality indicators, such as biochemical oxygen demand (BOD), chemical oxygen demand (COD), suspended solid (SS), total nitrogen (T-N), and total phosphorus (T-P) [18]. The amount of irrigation is calculated by excluding the value of effective rainfall and the cultivation water from the evapotranspiration calculated by the Food and Agriculture Organization (FAO) of the United Nations using the Penman–Monteith equation [17,19–21]. Additionally, because water is not actually wasted in agricultural fields, and agricultural wastewater is not generally found in Korea, this study does not estimate gray water separately. However, gray water estimated from industrial wastewater is considered. Indirect water is calculated by multiplying the collective activities by consumptive water use factors of individual activities. Consumptive water use factors are converted from the national lifecycle inventory database.

2.4. Estimation of Direct Irrigation Water

The amount of direct water caused by the evapotranspiration of agricultural products, including crops, is estimated using the FAO Penman–Monteith equation (Figure 3). The technical procedure for measuring direct water quantity for each crop comprises an estimation of evapotranspiration for each crop, a calculation of irrigation water needed, and a measurement of direct water quantity considering scarcity.

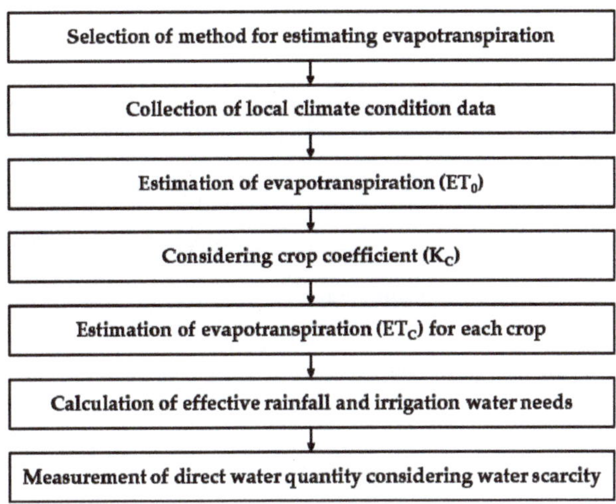

Figure 3. Technical procedure for measuring direct water quantity for each crop.

2.4.1. Estimation of evapotranspiration for each crop

Evapotranspiration for each crop (*ETc*) is estimated in two steps: net evapotranspiration of the plant (*ETo*) and evapotranspiration for each agricultural product. First, considering Korea's climate, the Penman–Monteith equation recommended by FAO is used to measure net evapotranspiration. Equation (1) shows the equation [22].

$$ET_0 = \frac{0.408\Delta(R_n - G) + \gamma\frac{900}{T+273}u_2(e_s - e_a)}{\Delta + \gamma(1 + 0.34u_2)} \tag{1}$$

where *ETo* is the reference evapotranspiration (mm/day), Rn is the net radiation at the crop surface (MJ/m^2·day), *G* is the soil heat flux density (MJ/m^2·day), *T* is the mean daily air temperature at 2 m height (°C), U_2 is the wind speed at 2 m height (m/s), e_s is the saturation vapor pressure (kPa), e_a is the actual vapor pressure (kPa), $e_s - e_a$ is the saturation vapor pressure deficit (kPa), Δ is the slope vapor pressure curve (kPa/°C), and γ is the psychrometric constant (kPa/°C).

Equation (1) is used to estimate the net evapotranspiration. The following information of daily weather is collected from 79 weather stations in Korea from 2003 to 2012: latitude, longitude, and altitude; maximum and minimum temperature; relative humidity; average vapor pressure; average wind speed; sunshine duration; solar radiation; precipitation; and soil information. Second, the evapotranspiration of each crop was calculated by multiplying the net evapotranspiration of the plant by the crop coefficient provided by Rural Development Administration (RDA) [23].

2.4.2. Calculation of Effective Rainfall and Irrigation Water Need

Effective rainfall is determined depending on the difference between total rainfall and actual evapotranspiration. It can be measured directly from the climatic parameters and the usable ground reserves. At ground level, water from effective rainfall is categorized as surface run-off and infiltration. Equation (2) calculates effective rainfall.

$$Re(t) = D(t) - D(T-1) - Req(t) + ETc(t) \tag{2}$$

where $Re(t)$ is the effective rainfall at t days (mm), $D(t)$ is the soil moisture content at t days (mm), $D(t-1)$ is the soil moisture content at $t-1$ days (mm), $Req(t)$ is the net irrigation at 1 day (mm), and $ETc(t)$ is the consumptive use (or evapotranspiration) by a crop at t dasy (mm).

Equations (3) and (4) show that if the minimum value of $D(t)$ is less than the sum of $D(t-1)$ and $Re(t)$, irrigation water is not required. However, if the minimum value of $D(t)$ is larger than the sum of $D(t-1)$ and $Re(t)$, then we subtract $ETc(t)$. Irrigation water is then calculated as the sum of the maximum value of $D(t)$ and $ETc(t)$, minus the sum of $D(t-1)$ and $Re(t)$.

$$If\ Dmin \leq D(t-1) + Re(t), Req(t) = 0 \tag{3}$$

$$If\ Dmin \geq D(t-1) + Re(t) - ETc(t), Req(t) = Dmax - D(t\,1) - Re(t) + ETc(t) \tag{4}$$

2.4.3. Measurement of Direct Water Consumption

The amount of irrigation water required is converted to the amount of surface and ground water considering the rate of consumption by the source of water consumed in each region. The converted surface water and ground water usage are changed into direct water consumption by multiplying the water scarcity index by the water source, developed using the water scarcity footprints method proposed by Tokyo University [24,25]. Table 3 shows the water scarcity index applied in this study.

Table 3. Water scarcity index by water source.

	Precipitation	Surface Water: River	Surface Water: Reservoir	Ground Water
Water scarcity index	1.0	2.5	6.9	35.1

3. Results and Discussion

3.1. Distribution of Weather Data

The distribution of six weather indices was analyzed to estimate the evapotranspiration of each agricultural product (Figure 4). These data were recorded from 79 weather stations in Korea by analyzing the daily weather conditions from 1 January 2003, to 31 December 2012. Approximately 288,000 data points were collected for each weather indicator [11]. Figure 4 shows that all the bar graphs were obtained by plotting the indicator values for the weather condition data on a monthly basis. The highest monthly temperature distribution was the highest in August and the lowest was in January. Additionally, the monthly lowest temperature distribution was lowest in January and highest in July. The duration of sunshine was the longest in July and August (about 14 h). Relative humidity exceeded 50% in the summer of July and August and was less than 20% from January to April. The average wind speed increased to 16 m/s in August and September when typhoons were frequent. June was the lowest. Finally, precipitation was concentrated from July to September.

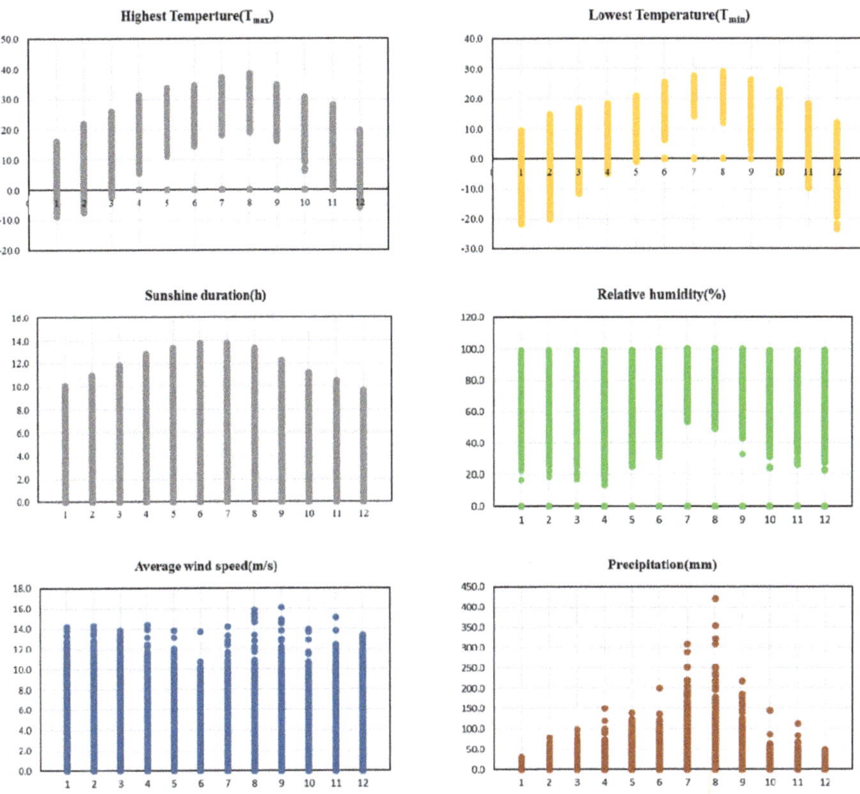

Figure 4. Monthly distribution chart of weather data (2003–2012).

Figure 5 presents schematic distribution results of six weather indicators per year. The annual distribution chart appears relatively constant compared to the monthly distribution chart of Figure 4. Maximum temperatures exceeding 35 °C were observed several times every year. The highest temperature in 2012 was estimated to be close to 38 °C and the temperature was the lowest in 2004 and 2007. The duration of sunshine averaged 13.5 h, but, in 2009, it was up to 14 h. The relative humidity was lowest in 2002, and the average wind speed was highest in 2005 and 2012. Precipitation was the lowest in 2008 and relatively small in 2003 and 2009 compared with other years.

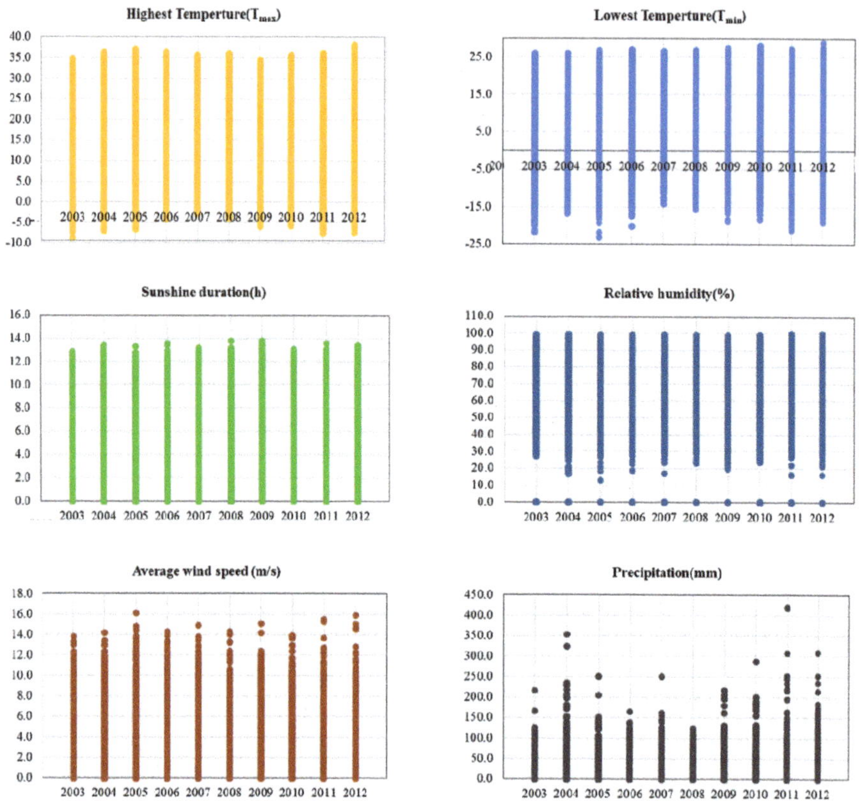

Figure 5. Annual distribution chart of weather data (2003–2012).

3.2. Water Footprint of Agricultural and Livestock Products

3.2.1. Direct Water Footprint of 43 Agricultural Products on a Hectare Basis

Figure 6 shows the direct water footprint per hectare of 43 agricultural products estimated using the water footprint assessment model developed for this study. Among the products, the water footprint of rice was the largest at 11,741 m^3/ha, and that of the autumn potato was the smallest at 2096 m^3/ha. In the case of open field, open field vegetables, the footprint of pepper was the highest at 4994 m^3/ha, and spinach was the lowest at 2132 m^3/ha. With regard to vegetables grown in facilities, the footprint of cucumber was the largest at 34,962 m^3/ha and that of spinach was the smallest at 3195 m^3/ha. The reason is that the crop coefficient developed considering the growth period of the crop of cucumber is bigger than that of spinach. Among the fruits, the footprint of the grape was the largest at 17,159 m^3/ha, and the least was 6813 m^3/ha. In total, fruits were considered to have a higher water footprint per unit area than crops and vegetables. The indirect water consumption of vegetables grown in facility was much higher than that of vegetables grown in an open field, because vegetables grown in facility consume more fuel, energy, and water than open field vegetables.

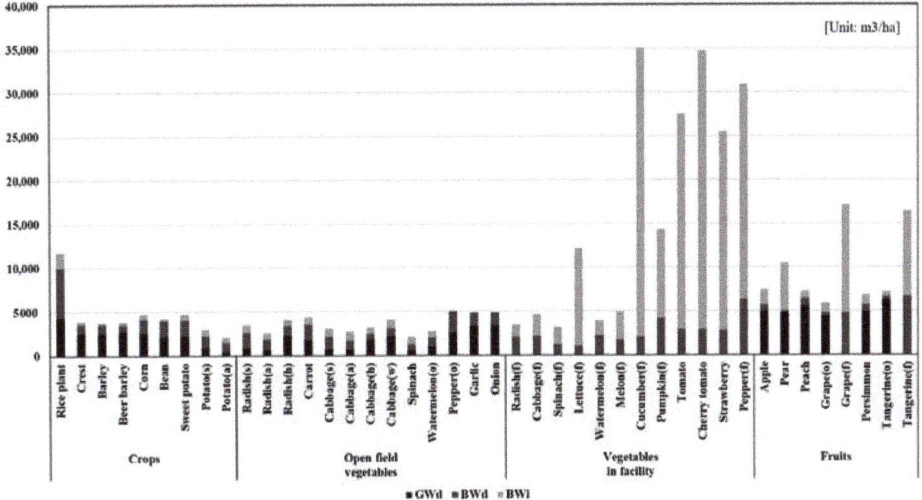

Figure 6. Comparison of direct water footprint per ha for 43 agricultural products. Black bar indicates green water as direct water (GW_d) and dark gray bar indicates blue water as direct water (BW_d), and finally, light gray bar indicates blue water as indirect water (BW_i).

3.2.2. Direct Water Footprint of 43 Agricultural Products in Ton Basis

Figure 7 illustrates the direct water footprint of 43 agricultural products on a ton basis. As shown in Figure 7, the water footprint of soybean among the crops was the highest at 3859 m³/ton, 2.9 times the average water footprint of crops: 1320 m³/ton. Among the open field, open field vegetables, the water footprint of spinach was the highest at 930 m³/ton, 3.2 times the average of open field vegetables: 287 m³/ton. Next, among the vegetables grown in the facility, the water footprint of strawberries was the largest at 6046 m³/ton: 2.7 times the average of vegetable, 2268 m³/ton. Finally, the water footprint of the grapes cultivated in facility was the highest at 7085 m³/ton: 3.5 times the average of 2027 m³/ton of fruits. Here, the reason why the water footprint was different for each agricultural product is because the crop coefficients were different, as mentioned.

From Figure 7, the average water footprint of the vegetables grown in facility was the highest among the four types of agricultural products, followed by fruits, crops, and vegetables grown in the open field. The average water footprint of vegetables grown in facilities was about 7.9 times the average water footprint of the vegetables grown in the open field.

The pattern of the bar graph shown in Figure 7 is different from the pattern of Figure 6, especially for soybeans, grapes, and cucumbers. For soybeans and grapes, the water footprints per hectare are low, but the water footprint per ton is relatively high. However, the water footprint of cucumber has the opposite pattern compared to those of soybeans and grapes. Thus, yields per hectare of soybean and grape were relatively low compared to other crops, and the cucumber yield per hectare was relatively higher than other agricultural products. Thus, it is not suitable for estimating the water footprint for agricultural products, because the result of water footprint per area did not reflect the production yield per area.

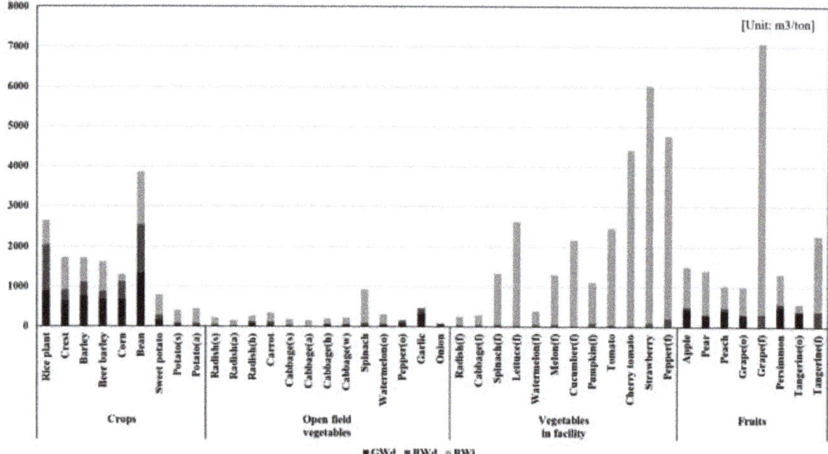

Figure 7. Comparison of water footprint per ton for 43 agricultural products. Black bar indicates green water as direct water (GW$_d$) and dark gray bar indicates blue water as direct water (BW$_d$), and finally, light gray bar indicates blue water as indirect water (BW$_i$).

3.2.3. Direct Water Footprint of Three Livestock Products on a Ton Basis

Figure 8 illustrates the results of the direct water footprint of three livestock products. The water footprint of beef, including green and blue water from direct and indirect sources, was 19,600 m³/ton. The water footprints of pork and chicken on a ton basis were 5272 m³/ton and 4008 m³/ton, respectively. The water footprint of beef was the largest, because the intake of feed consumed during breeding was higher than that of pigs and chickens. The result of the water footprints of three livestock products show that the water footprint of beef was 3.7 times and 4.9 times pork and chicken, respectively. The blue water footprints of three livestock products were approximately 66%. However, the water footprint per ton of beef was analyzed to be 8.6 times that of the average water footprint per ton of the vegetables grown in facilities. The water footprint per ton of pork and chicken was 2.3 times and 1.8 times larger than that of vegetables grown in facility, respectively. The average water footprint of meat per ton was 4.2 times higher than that of vegetables in facility.

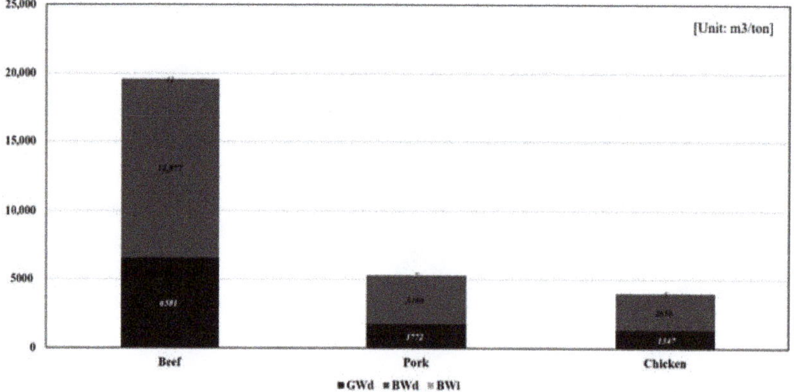

Figure 8. Comparison of water footprint per ton for 3 livestock products. Note: In this figure, the black bar means green water as direct water (GW$_d$) and the dark gray bar is blue water as direct water (BW$_d$), and finally, the light gray bar is blue water as indirect water (BW$_i$).

3.3. Comparison of Water Consumption per Region

Generally, the water footprint for agricultural products is affected by weather conditions. In Korea, the regional variation of climate in terms of temperature, wind speed, and precipitation is large between the northern and southern regions. Therefore, it is important that water footprints are calculated and compared at the regional level considering their weather characteristics. The subjective product for comparison is rice. The cultivation area for rice spreads nationwide, and the water footprint is larger than other crops.

Figure 9 depicts direct water footprint per region and year. Here, the purpose of analyzing the water footprint only for direct water is that indirect water consumption from agricultural materials such as fertilizer, pesticides, and mulching vinyl is not affected by domestic weather conditions. In 2003, it consumed 9600 m^3 of direct water to harvest rice of 1 ha, which is the least consumption in a year. However, the years with the highest direct water consumption were 2004 and 2012, more than 10,300 m^3 per 1 ha. The region-wise annual water consumption of Chungcheongnam Province was the largest, and that of Jeollanam Province was the least. Moreover, the central regions, including Gyeonggi Province, Gangwon Province, Chungcheongbuk Province, and Chungcheongnam Province, had higher direct consumption per 1 ha than the southern regions, including Gyeongsangnam Province, Gyeongsangbuk Province, Jeollanam Province, and Jeollabuk Province.

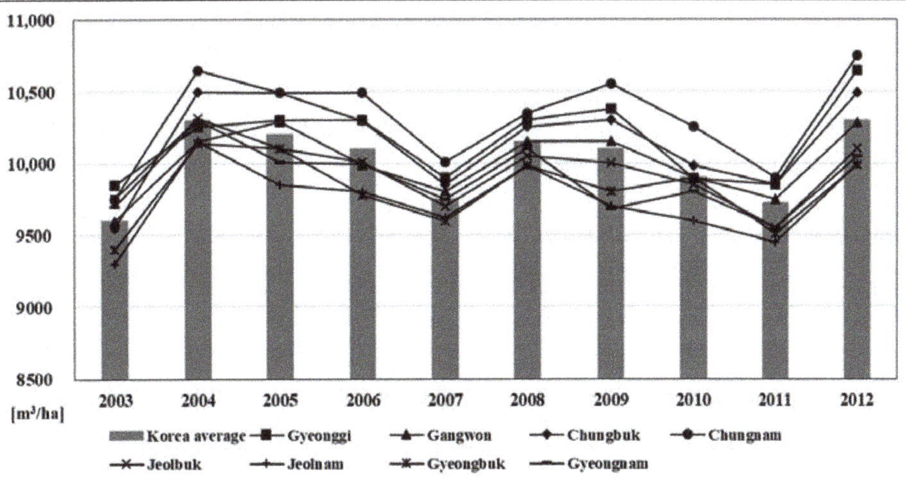

Figure 9. Comparison of direct water consumption per ha by region and year.

Figure 10 shows a graph comparing green water consumption by region and year. In 2003, the consumption of green water (6000 m^3) was the largest to produce rice at 1 ha. However, green water consumption was lowest in 2004, 2008, and 2009. Whereas there was a difference in precipitation by region per year, Jeollabuk Province had a relatively large consumption of green water. Gyeongsangbuk Province, however, had relatively less. In fact, no consistent trends were seen in the use of green water per year.

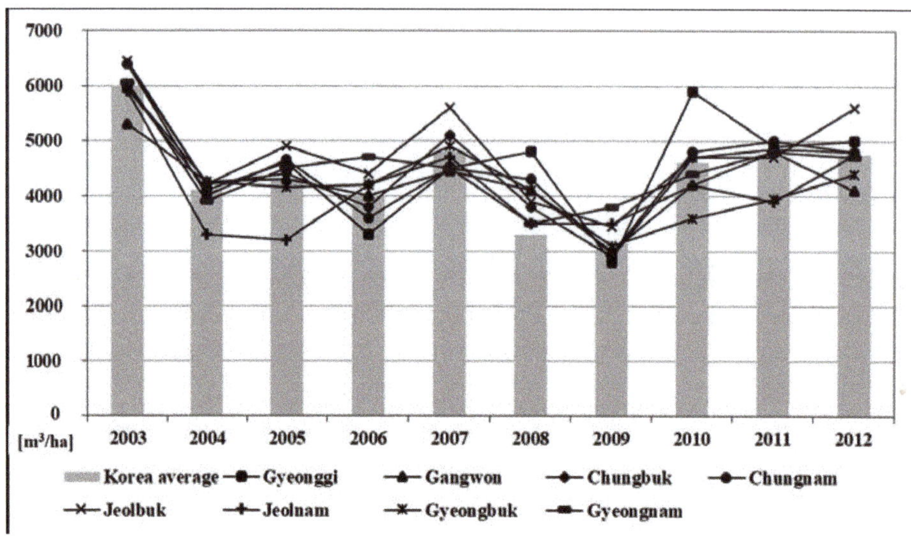

Figure 10. Comparison of green water consumption per ha by region and year.

Figure 11 illustrates the amount of blue water consumption by region and year. It shows that the tendency toward blue water consumption by region and year is exactly opposite of the consumption trend of green water, because the consumptive water use per crop, which is same as evapotranspiration, is constant. In fact, Jeollabuk Province had the least amount of blue water usage, whereas Gyeongsangbuk Province had the highest amount.

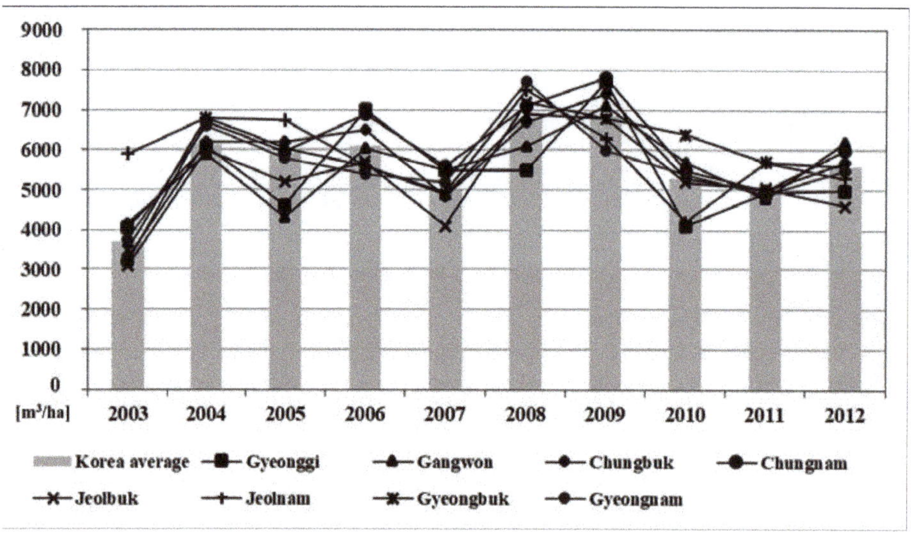

Figure 11. Comparison of blue water consumption per ha by region and year.

Figure 12 shows the region- and year-wise distribution of direct water consumption per ton. Comparing the distribution shown in Figure 12 with that of Figure 9, direct water consumption per ton was relatively constant compared with consumption per ha, depending on the year and region.

Consequently, direct water consumption is more closely related to yield than to land area. The average consumption of direct water per ton was 2200 m^3, but in 2009, it was 1900 m^3 lower than the average consumption. This study investigated why the direct consumption in 2009 was less than other years using Figure 5. In 2009, the sunshine duration was longer, and the average wind speed and relative humidity were the lowest compared to other years. Precipitation was also relatively less than other years. Therefore, the amount of direct water consumption per ton in 2009 decreased, because of increased yield depending on weather conditions.

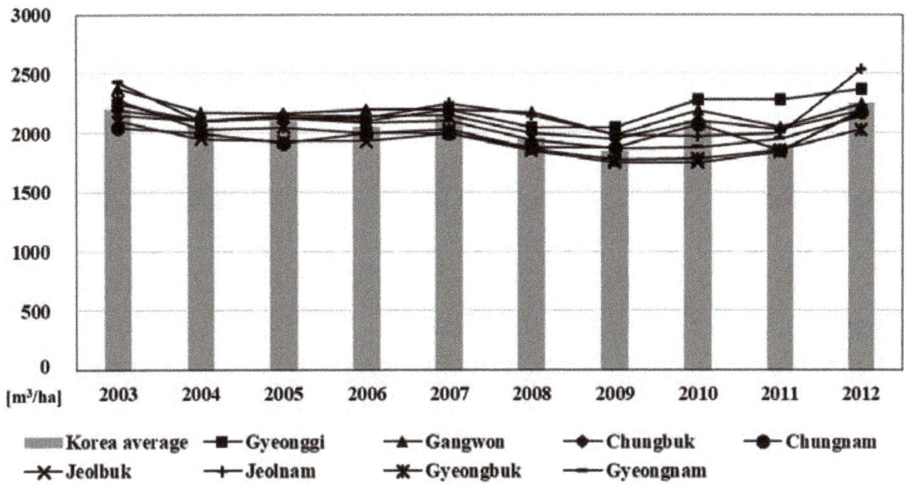

Figure 12. Comparison of direct water consumption per ton by region and year.

3.4. Estimation of Direct Water Footprint of a National Scale in Agricultural Sector

On the basis of the direct water footprints for 42 agricultural and three livestock products, this study estimated the total water footprint for all agricultural and livestock products consumed in the Republic of Korea in 2014. For the purpose of the study, it was assumed that all agricultural and livestock products consumed were produced domestically. In 2014, the total consumptive amounts of each agricultural and livestock product group were based on data provided by the National Statistical Office.

Figure 13 shows a bar graph representing the direct water footprint required for the production of agricultural and livestock products at the national level. The extrapolation method was used to extend the water footprints per ton of 45 agricultural and livestock products to the water footprint from the total consumption of 45 products. Accordingly, the total water footprint was estimated at 37.6 billion m^3, which is equivalent to 27.9% of the total available water resources of Korea, with an average 134.9 billion m^3 per year [26]. Next, the total blue water consumption was analyzed to be 69.3% of the national total water footprint. This is equivalent to 19.3% of the total available water resources in Korea. Of the total water footprints, meats accounted for 43.0%, followed by crops and vegetables by 34.1% and 13.3%, respectively. In particular, it was analyzed that the indirect water of vegetables cultivated mainly in facility was close to 90% of the total water footprint of vegetables, and the consumption of blue water by meat accounted for 66.2% of the total water footprint of meat, because the meat consumed mainly processed feed.

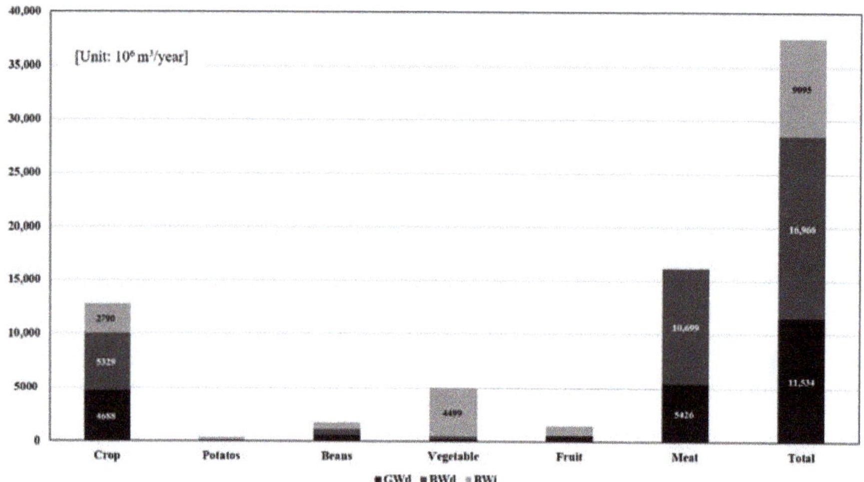

Figure 13. Direct water footprint by agricultural and livestock products at a national level.

3.5. Comparison of Water Footprints between Different Studies

To verify the reliability of this study, we compared the results of water footprint studies of different rice sources. The comparative sources are Korea Rural Community Corporation and WFN, representative water footprint research institutes and Figure 14 shows a bar graph comparing the water footprint of rice among three different sources [27,28]. According to the results, the direct water footprint as the sum of green, blue, and gray water, excluding indirect water, was almost the same as the three sources. In particular, the results of green water calculated on the basis of weather conditions being in close agreement with the results of WFN. However, the results of blue water were somewhat different from the three studies. This could be caused by differences in the method and timing of collecting statistical data at the farm and open field. However, compared to other studies, the characteristic of this study is that it considers indirect water consumption by agricultural materials according to the requirements of ISO 14044 and ISO 14046. Thus, it concluded that the inclusion of indirect water in the water footprint resulted in about 17% increase.

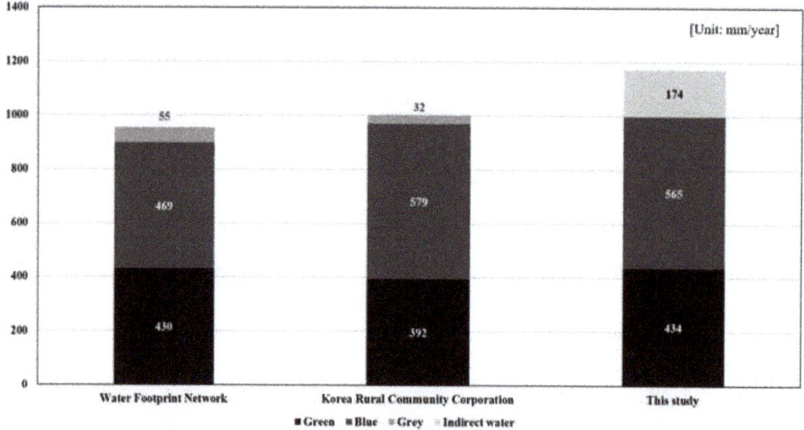

Figure 14. Comparison of water footprints on rice at different studies.

4. Conclusions

This study measured the water footprints of 42 agricultural products and three livestock products using a method developed by integrating the characteristics of two methods of measuring internationally accepted water footprints. To estimate accurate and representative water footprints, this study collected massive weather data on six indicators from 79 weather stations in Korea from 2003 to 2012. From the results of the water footprint, we confirmed that the average water footprint of the vegetables grown at facilities was 7.9 times larger per ton than those of the vegetables cultivated in an open field. Moreover, we found that the water footprint per ton of beef was about 4.2 times the average water footprint per ton of vegetables grown in facility. Assuming that all agricultural and livestock products consumed in Korea were produced domestically in 2014, the water footprint estimates accounted for 27.9% of the total domestic water resources. This study is meaningful in that it estimated the water footprint of agricultural and livestock products using massive meteorological data measured in Korea. In a future study, we will analyze the amount of the virtual water trade of Korea via the import of food. The results of that study should be effectively applied to overall national agricultural water management, including virtual water trade.

Author Contributions: Author 1 (I.K.) wrote the directions and main contents of this paper, and author 2 (K.-s.K.) supported the direction of this paper.

Funding: This research was funded by the Institute of Planning and Evaluation for Technology in Food, Agriculture and Forest (IPET) and the Rural Development Administration (RDA).

Acknowledgments: This study was carried out with support from the Institute of Planning and Evaluation for Technology in Food, Agriculture and Forest (IPET) and the Rural Development Administration (RDA). We sincerely appreciate the work and support of the IPET and RDA.

Conflicts of Interest: The authors declare no conflicts of interest.

References

1. United Nations Department of Economic and Social Affairs/Population Division. *World Population Prospects: The 2017 Revision, Key Findings and Advance Tables*; United Nations Department of Economic and Social Affairs/Population Division: New York, NY, USA, 2017.
2. Kim, S.J. The value of agricultural water and the right direction of demand management in the era of water crisis. In Proceedings of the Symposium for the Policy on Water Saving Movement, Seoul, Korea, 4 April 2003.
3. Kim, S.S.; Jang, U.; Yoon, I.J.; Kwon, K.H.; Kim, W.J.; Lee, Y.J. *A Study on the Sustainable Water Management Legislation for Climate Change Response*; Yonsei University: Seoul, Korea, 2012.
4. OECD. *Environmental Outlook to 2050: The Consequences of Inaction*; OECD: Paris, France, 2012.
5. Ahn, H.J.; Sung, H.S.; Forman, A. *Korea Ecological Footprint Report 2016: Measuring Korea's Impact on Nature*; World Wide Fund for Nature-Korea: Seoul, Korea, 2016.
6. Lee, C.; Moon, H.; Choi, Y.; Kim, Y.; Yoo, M.; Son, H. *The Problems and Improvements of Food Security in Korea*; The Korea Academy of Science and Technology: Seongnam-si, Korea, 2009.
7. Myung, H.; Choi, J.; Kim, T.; Jung, H.; Park, J. *Major Land Crop Industry Revitalization Strategy*; Korea Rural Economic Institute: Naju-si, Korea, 2009.
8. Mubako, S.T.; Lant, C.L. Agricultural virtual water trade and water footprint of U.S. states. *J. Ann. Assoc. Am. Geogr.* **2013**, *103*, 385–396. [CrossRef]
9. Lamastra, L.; Miglietta, P.P.; Toma, P.; De Leo, F.; Massari, S. Virtual water trade of agri-food products: Evidence from Italian-Chinese relations. *Sci. Total Environ.* **2017**, *599–600*, 474–482. [CrossRef] [PubMed]
10. Hoekstra, A.Y.; Chapagain, A.K.; Aldaya, M.M.; Mekonnen, M.M. *The Water Footprint Assessment Manual: Setting the Global Standard*; Water Footprint Network: London, UK, 2011.
11. Weather Data Release Portal. Available online: https://data.kma.go.kr/cmmn/main.do (accessed on 7 May 2019).
12. Lee, Y.H. *Agricultural and Livestock Income Collection*; Rural Development Administration: Jeonju-si, Korea, 2014.

13. Korea Statistical Information Service. Available online: https://kosis.kr/index/index.do (accessed on 7 May 2019).

14. ISO. *ISO 14046:2014: Life Cycle Assessment—Water Footprint—Principles, Requirements and Guidelines*; ISO: Geneva, Switzerland, 2014.

15. ISO. *ISO 14044:2006: Environmental Management—Life Cycle Assessment—Requirements and Guidelines*; ISO: Geneva, Switzerland, 2006.

16. Jefferie, D.; Muñoz, I.; Hodges, J.; King, V.J.; Aldaya, M.; Ercin, A.E.; Canals, L.M.; Hoekstra, A.Y. Water footprint and life cycle assessment as approaches to assess potential impacts of products on water consumption. Key learning points from pilot studies on tea and margarine. *J. Clean. Prod.* **2010**, *33*, 155–166. [CrossRef]

17. Zhuo, L.; Mekonnen, M.M.; Hoekstra, A.Y. *Seisitivity and Uncertainty in Crop Water Footprint Accounting: A Case Study for the Yellow River Basin*; UNESCO-IHE: Delft, The Netherlands, 2013.

18. Serio, F.; Miglietta, P.P.; Lamastra, L.; Ficocelli, S.; Intini, F.; De Deo, F.; De Donno, A. Groundwater nitrate contamination and agricultural land use: A grey water footprint perspective in Southern Apulia Region (Italy). *Sci. Total Environ.* **2018**, *645*, 1425–1431. [CrossRef] [PubMed]

19. Allen, R.G.; Pereira, L.S.; Raes, D.; Smith, M. *Crop Evapotranspiration: Guidelines for Computing Crop Water Requirements*; Food and Agriculture Organization of the United Nations: Roma, Italy, 1998.

20. Ghufran, M.A.; Batool, A.; Irfan, M.F.; Butt, M.A.; Farooqi, A. Water footprint of major cereals and some selected minor crops of Pakistan. *J. Water Resour. Hydraul. Eng.* **2015**, *4*, 358–366. [CrossRef]

21. Zotarelli, L.; Dukes, M.D.; Romero, C.C.; Migliaccio, K.W.; Morgan, K.T. *Agricultural and Biological Engineering Department*; AE459; UF/IFAS Extension: Gainesville, FL, USA, 2010.

22. Richard, G.A.; Luis, S.P.; Dirk, R.; Martin, S. *Crop evapotranspiration: Guidelines for computing crop water requirements*; FAO: Rome, Italy, 1998.

23. Han, K. *Agricultural Technology: Method of Estimating the Crop Coefficient and Crop Evapotranspiration*; Rural Development Administration: Jeonju-si, Korea, 2017.

24. Yano, S.; Hanasaki, N.; Itsubo, N.; Oki, T. Water scarcity footprints by considering the differences in water sources. *Sustainability* **2015**, *7*, 9753–9772. [CrossRef]

25. Lovarelli, D.; Bacenetti, J.; Fiala, M. Water footprint of crop productions: A review. *Sci. Total Environ.* **2016**, *548–549*, 236–251. [CrossRef] [PubMed]

26. My Water Portal. Available online: https://www.water.or.kr/knowledge/educate/educate_05.do (accessed on 7 May 2019).

27. Shin, A.; Kim, Y.; Kim, J.; Kim, K.; Lee, S.; Choi, W.; Mun, S.; Choi, J.; Lee, S.; Hong, E.; et al. *Water Footprint Estimation and Application for Sustainable Water Resources Use*; Korea Rural Community Corporation: Naju-si, Korea, 2014.

28. Chapagain, A.K.; Hoekstra, A.Y. The blue, green and grey water footprint of rice from production and consumption perspectives. *Ecol. Econ.* **2011**, *70*, 749–758. [CrossRef]

 *sustainability* 

Article

Nitrate Vulnerable Zones Revision in Poland—Assessment of Environmental Impact and Land Use Conflicts

Ewa Szalińska [1],*, Paulina Orlińska-Woźniak [2] and Paweł Wilk [2]

[1] Faculty of Geology, Geophysics and Environmental Protection, AGH University of Science and Technology, Krakow 30-059, Poland
[2] Section of Modeling Surface Water Quality, Institute of Meteorology and Water Management, 01-673 Warsaw, Poland; paulina.wozniak@imgw.pl (P.O.-W.); pawel.wilk@imgw.pl (P.W.)
* Correspondence: eszalinska@agh.edu.pl; Tel.: +48-12-627-4950

Received: 1 August 2018; Accepted: 10 September 2018; Published: 14 September 2018

Abstract: Despite concerted efforts through the European territory, the problems of nitrogen pollution released from agricultural sources have not been resolved yet. Therefore, infringement cases are still open against a few Member States, including Poland, based on fulfilment problems of commitments regarding the Nitrate Directive. As a result of the litigation process, Poland has completely changed its approach to nitrate vulnerable zones. Instead of just selected areas, the measured actions will be implemented throughout the whole Polish territory. Additionally, further restrictions concerning the fertilizer use calendar will be introduced in areas indicated as extremely cold or hot, based on the average temperature distribution (poles of cold, and heat). Such a change will be of key importance to farmers, whose protests are already audible throughout the country, and can be expected to intensify. To assess the impact of the introduced modifications a modelling approach has been adopted. The use of the Macromodel DNS/SWAT allowed for the development of baseline and variant scenarios incorporating details of stipulated changes in the fertilizer use for a pilot catchment (Słupia River). The results clearly indicate that the new restriction will have a substantial effect on the aquatic environment by altering the amount of released total nitrogen.

Keywords: nitrogen; agriculture; Nitrate Vulnerable Zones; macromodel DNS/SWAT

1. Introduction

Nitrogen (N) is the most important nutrient controlling agricultural primary production. To support a rapidly growing population and its food demand this production requires the external application of N in form of fertilizers. Despite introducing new N use management techniques, e.g., precision agriculture, or synchronizing of N application with crop demand [1,2], considerable amounts of N are still being lost to the environment, contributing to surface and groundwater pollution, eutrophication, and their subsequent effects on aquatic ecosystems. The anthropogenic nutrient over-enrichment is considered a main problem in many regions of the world [3]. In Europe the average use of N fertilizers has substantially grown in recent times, from 44.9 kg N/ha in 2004, up to 49.8 kg N/ha in 2015 [4].

To prevent and reduce water pollution from nutrients arising from agricultural sources, the European Union (EU) introduced a range of control measures, with general rules set by the Nitrate Directive (ND) (91/676/EEC). This document obliges Member States to designate areas vulnerable to nitrate pollution and to concert the efforts to reduce this pollution through national action programmes. These programmes require land managers to follow a range of measures, such as, among others, controlling the timing and quantities of fertilizers applied to the land, and ensuring proper storage capacity for livestock manure. It has been already concluded that implementation of the ND decreased both N leaching losses to ground and surface waters, and gaseous emissions to the atmosphere in EU-27 [5]. However, the problem of nutrient pollution has not been completely solved in the EU so far [6].

Poland is one of the countries located in the Baltic Sea (BS) catchment. Due to its proportion of the catchment area (ca. 18%), and very large share in agricultural land area (ca. 50%) [7], it is considered as one of the main contributors to the excessive nutrient loads into this sea. Indeed, despite the concerted actions in all the Baltic countries to focus on reduction of nutrient loadings, 96% of the total surface area in the BS remains affected by eutrophication [8]. Nutrient loads from Poland are being discharged mostly from riverine sources, via two main rivers (Vistula and Oder), draining ca. 88% of the country's territory, and estimated under the National Environment Monitoring Program [9]. A substantial downward trend has been observed in the total nitrogen (TN) load from the Polish territory, from 262 thousand tons in 1995 to 170 thousand tons in 2014 [10]. However, it should also be noted that 9 small rivers from the Pomerania region, carry their waters directly into the BS. These rivers, including the Słupia River—which is the one used as a pilot catchment in this study, are not monitored for nutrient loads [11].

Of the total area of Poland, agricultural land occupies ca. 60% of the surface, with the industry profile dominated by private farms [12]. The restructuring trend observed in Polish agriculture after 1990, also enforced by adoption of EU laws, resulted in a general decrease of pressures on aquatic environment quality (e.g., improvement of sewage system management) [9,13]. However, this resulted in Polish farmers striving to achieve competitiveness in the EU agricultural market and, thus, increased the use of fertilizers (from 47 kg N/ha in 1995 to 72 kg N/ha in 2016) [14,15]. Since agriculture production, as well as fertilizer use, shows strong regional differentiation in Poland, based on the variability of natural and economic factors [12], the measures to approach nutrient discharge problems should be different. The paper aims to present possible changes in the N load discharge from Polish catchments under the new measures imposed by the EU. To assess possible changes for the aquatic environment a modelling approach was adopted, allowing analysis of different variant scenarios of fertilizer use. Moreover, the 15-year period of meteorological data used in this study also allowed to draw conclusions on possible impacts of climate changes on the N yield from the catchments.

2. Nitrate Vulnerable Zone (NVZ) Approach

2.1. NVZ Status Quo in Europe

Agriculture (i.e., arable land, permanent crops, pastures, and mosaics) occupies nearly 42% (2012) of the EU territory. However, a visible trend of agricultural land loss through the abandonment or withdrawal from farming activities has been observed recently (—0.5% of the agricultural area, 2006–2012) [16]. It provides indisputable benefits to society, although some farming activities cause substantial impacts on water bodies [6]. Although an improvement of point-source management has been observed in recent decades in Europe [17], most studies point out farming practices as a major source of nutrients in the aquatic environment [18,19]. Crop and livestock production, as well as fertilizer use, are considered as main accelerators of the nutrient cycle, as global surpluses of N continue to increase (+23% N) [20].

In general, the nutrient balance (defined as difference between the nutrient inputs entering and leaving farming systems) is being altered in many areas of Europe. In the period 2012–2014, all Member States, except Romania, had a surplus of N, with values exceeding 50 kg/ha in many countries [6]. The nutrient agricultural discharges alter quality of aquatic resources and result in an increased eutrophication risk [21], and over-fertilization pressure exerts more awareness as climate change and deterioration of fresh water become more critical. According to the Fifth IPCC Assessment Report (AR5) conclusions, and the National Oceanic and Atmospheric Administration (NOAA) findings, climate change will affect (with a different probability) the global water cycle, causing changes in precipitation distribution, and its intensification over land areas, as well as the frequency of extreme weather events [22,23]. These changes will likely affect the crop cycle and, therefore, its N uptake capacity. Moreover, results of modelling studies [24] show that, under specific climate change scenarios, an increasing trend of the nitrate leaching pattern is possible.

To tackle the issue of water pollution caused by nitrates from agricultural sources at a catchment level, nitrate vulnerable zones (NVZ) have been introduced under the ND throughout the EU. The NVZs are defined as those waters, surface and/or groundwater in which the nitrate levels exceed or are likely to exceed 50 mg/L from agricultural sources. The extent of the ND appliance depends upon the interpretation of the requirements by Member States. Particularly, the interpretation of "vulnerable", since this affects the range of the area subjected to mandatory requirements. In accordance with the provisions of the ND, Member States had two options to implement its stipulations. The first method relied on identification of the waters affected by nitrate pollution and waters which could be affected in case of a lack of any action pursued, and designating NVZs in their territories. The second method allowed the Member State to be exempted from the obligation of NVZs identification through implementation of action programmes throughout their national territory. Those countries opting for the second method were: Austria, Denmark, Finland, Germany, Ireland, Lithuania, Luxembourg, Malta, the Netherlands, Romania, Slovenia, the Region of Flanders, and Northern Ireland.

In the remaining countries the criteria for NVZs designation had to be developed, and most of them depended on the results of various computer simulations [25,26]. The total area of NVZ's, including countries that apply for a whole-territory approach, represented approximately 61% of EU agricultural area in 2015. Therefore, there are still areas in Europe with potential water pollution that are not included in any action programme. Moreover, in some countries, the designed territories are limited to reduced areas putting in question the potential effectiveness of the action programmes [6].

2.2. NVZ Revision in Poland

The provisions of the ND have been officially introduced in Poland in 2004. Delimitation of the initial range of NVZs was performed by the Regional Water Management Authorities in each of the seven surface water districts and based on: (i) content of N compounds in surface and underground waters; (ii) eutrophication of surface waters (including internal and coastal sea waters); (iii) agricultural land use structure and soil typology; (iv) type of agricultural activity, and concentration of animal production; and (v) prevailing meteorological-, hydrological-, and hydrogeological conditions [27]. Originally, in the first period of the ND being in force in Poland (2004–2008), 21 NVZs had been delimited, which accounted for 2% of the total area of the Polish territory. Moreover, the Action Programmes for Reduction of Outflow of Nitrogen from Agricultural Sources, and the Code of Good Agricultural Practices had been prepared and implemented [28]. Standards included in these documents related primarily to the requirements of fertilizers and plant protection products management, water and soil conservation, rational use of wastewater and sewage sludge, conservation of valuable habitats and species found in agricultural areas, as well as maintaining cleanliness and order on farms [29].

Following the ND requirements the eutrophic state of the waters, action programmes, and extension of NVZs should be reviewed every four years. Hence, such actions have been performed in Poland in 2008, and 2012 based on the results of monitoring tests, expertise [30], as well as modelling results [31]. In the second cycle (2008–2012) 19 NVZs were designated, covering approximately 1.5% of the country's area. Then, in the third cycle (2012–2016) 48 NVZs constituted 4.46% of the Polish territory. Process of the ND implementation in Poland has been followed and assessed by the European Commission (EC) [32], and insufficient surface areas designated as NVZs had been indicated. Due to the fact that Poland had not complied with the EC recommendations, a litigation process was started in 2013 (Case C 356/13). As Poland has failed to fulfil the ND obligations by inadequate identification and classification of nitrate vulnerable waters, and adopting incompatible measures in action programmes, a change of the approach to the ND has been imposed. This change resulted in switching into the second option of the ND provision implementation, i.e., an indication of the whole country's territory as a nitrate vulnerable zone, and development of a new approach to tackle the nitrogen pollution issue.

2.3. Consequences of the NVZ Approach Change

Apart from extending the NVZs range, the new approach to the implementation of the ND requirements will be based on the on the new Action Programme adopted in June 2018 [33]. This programme specifies, among others: periods of fertilizer application, storage conditions for natural fertilizers, doses, and methods of nitrogen fertilization, and the way in which farmers are to document the implementation of these requirements. According to this document, mineral nitrogen fertilizers on arable land can be applied from 1 March to the end of October, other fertilizers (solid) from 1 March to 30 November, and liquid at such dates as nitrogen ones (Table 1). Selected fertilizers (mineral nitrogen and other liquid fertilizers) can be applied from 15 February on grounds that are not frozen, covered with snow or water, or saturated with water. Additionally, in accordance with this regulation liquid and solid manure fertilizers should be stored in a way that prevents leachates from entering the ground and water during the period when they are not used. Moreover, an increase of natural fertilizer tank capacity should be ensured, to enable their storage for a period of six months, not four as before. The document also indicates that farmers are obliged to keep records of nitrogen fertilization, and agree them with the regional agricultural stations.

Table 1. Comparison of the fertilization periods in the pole areas and the rest of the country.

Type of Land	Fertilizers, Excluding Mineral Nitrogen Fertilizers		Mineral Nitrogen Fertilizers
	Solid	Liquid	
Remaining area	01.03–30.11		01.03–31.10
Arable land in the "pole of cold" area	05.03–15.11		05.03–20.10
Arable land in the "pole of heat" area	15.02–30.11		15.02–15.11

One of the most important elements of the ND is the control of doses and periods of fertilization on selected types of land. Due to the fact that Poland extends in the north–south dimension over 649 km, and 689 km in the east–west dimension, there are significant differences in the average annual temperature of air between the extreme regions of the country. This fact has a direct influence on plant vegetation periods, and subsequently on the periods of fertilizer application. Therefore, it was considered necessary to designate areas of the country where the dates of fertilizer application will be shortened or extended. This task performed by the Institute of Meteorology and Water Management (IMGW), resulted in delimitation of so-called "pole of cold" and "pole of heat" (Figure 1). The confines of the poles are based on the administrative borders of municipalities (communes), and the delimitation process was based on:

- Data from the 1750 stations belonging to the State Hydrological and Meteorological Service (PSHM) carried out by the IMGW, used for preparation of the yearly average temperature distribution maps for the period of 1981–2014—to select commune with the lowest and highest average temperatures;
- Data from the Institute of Soils Science and Plant Cultivation (IUNG) describing estimated length of the growing seasons in the period of 2011–2020 [34]—to select commune with the shortest and longest growing seasons; and
- Data from the Plan of the Rural Areas Development [35] delimiting areas with unsuitable agricultural conditions (in the mountain areas).

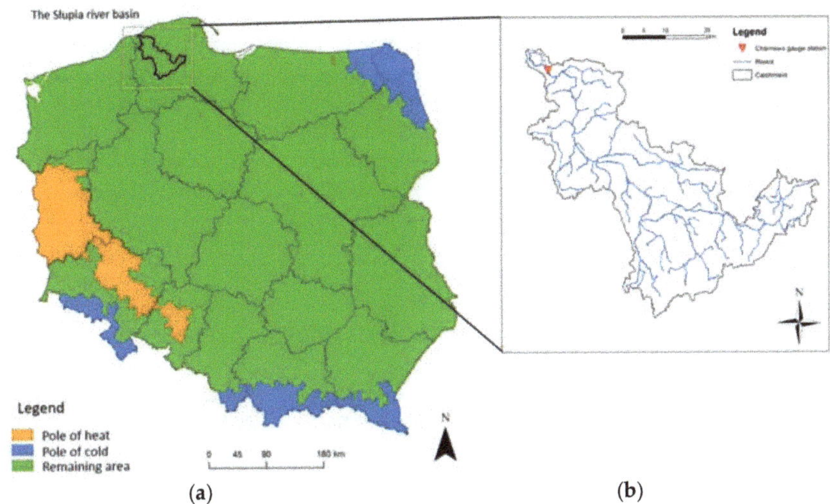

Figure 1. (a) Areas delimited as "pole of cold" and "pole of heat"; (b) modelling area—the Słupia River catchment.

Eventually, the "pole of cold" and "pole of heat" areas covered the range of 5.8% and 5.7% of the country's territory, respectively. The "pole of hot" covers three provinces (Opolskie, Dolnośląskie, and Lubuskie) and is a single area. The "field of cold", on the other hand, consists of three areas covering mountainous areas in the south of the country in the Dolnośląskie, Śląskie, Małopolskie, and Podkarpackie voivodships as well as the northeastern Polish area in the Warmian-Masurian and Podlasie voivodships.

According to the new regulations, the available time for fertilization in the "pole of cold" area has been shortened by a total of 20 days for solid organic fertilizers, and 16 days for solid nitrogen and nitrogenous mineral fertilizers. For the "pole of heat" area these periods have been extended by 15 and 30 days, respectively (Table 1).

3. Modeling of NVZ Revision' Impact—Variant Scenarios

To estimate the impact of introducing additional restrictions on fertilization periods in the "pole" areas, model calculations have been performed using the macromodel DNS/SWAT. Its construction and applicability have been described elsewhere [36]. Briefly, the macromodel combines existing mathematical models and equations of hydrological transport in a catchment with the benefits of the SWAT module, which is generally used to model continuous long-term yields within hydrologic response units. The numerical model of a catchment created with the use of the macromodel DNS/SWAT enables to analyse different scenarios of the catchment exploitation in different meteorological and hydrologic conditions. This tool is also used to analyse the yield of nutrients at any selected control point of the river [11,37].

In the current study, the Słupia River catchment (Figure 1) has been selected as a pilot area for the modelling purposes. Although, this catchment is not located in any of the designated "pole" areas, covered by additional restrictions, its choice was promoted by many factors. Above all, the Słupia River discharges directly into the Baltic Sea. Moreover, this catchment has been used for total nitrogen (TN) field research by the IMGW since 2013. These results are beneficial in verification of the TN calculations with use of the macromodel DNS/SWAT. The aforementioned research conducted also at the neighbouring catchments confirmed that the Słupia River may be considered a representative catchment for the whole Pomeranian region. Therefore, it plays an important role in the estimation of TN loads from the territory of Poland into the Baltic Sea [38–40]. The Słupia River has a length of 139 km, and its catchment covers an area of 1623 km^2. The entire catchment is located in the northwestern part of the Pomeranian Voivodeship (Figure 1), and its area is dominated by agricultural use (almost 50% of the total area). Forest area covers 44% and occurs mainly in its central part. Buildings and anthropogenic areas constitute about 4% of the catchment area, with the main urban centres in Słupsk, Ustka, and Bytów. Floods in this catchment occur in the snow melting period (spring) and are inconsequential. However, the outlet area is frequently affected by sea water storm surges. As a main calculation profile for the macromodel DNS/SWAT calculation, the Charnowo profile was selected. This cross-section is located close to the mouth of the river (Figure 1), but at the same time far enough to be insusceptible to the impact of backwater from the sea. The Charnowo profile, a semi-automatic device (autosampler) has been also installed to collect water samples for the total nitrogen (TN) analyses. The macromodel DNS/SWAT generates datasets for the flow rate and loads of selected pollutants such as TN. Data is generated with a daily time step for any selected profile on the river.

The baseline scenario of the Słupia River model (VS0), also referred in the current study as a reference or control one, has been created with the goal of the best accurate representation of the real conditions in the catchment. To create VS0, the input data have been used, namely:

- Digital elevation model (DEM) on a scale of 1:50,000;
- Map of hydrographical divisions of Poland;
- Data on discharges of pollutants from sewage treatment plants, containing geographical coordinates of discharge points, amount of municipal waste water treated (m^3/year), total suspended soil (mg/dm^3), total nitrogen and total phosphorus (mg/dm^3)—this data came from the National Sewage Treatment Program and did not require processing;
- Meteorological data on precipitation, temperature, humidity, sunlight, and wind speed and direction derived from historical database IMGW-PIB;
- Digital maps for soil classes, at a scale of 1:100,000 divided into 23 classes of soils. These classes were created by the generalization of all types of soil occurring in this area;Land use maps developed on the basis of Coordination of Information on the Environment programme (CORINE Land Cover). Maps that divide the catchment area into urbanized, agricultural, forested, wetlands, and water bodies. The agricultural land was further divided into specific crops; and
- Fertilizer data from the Local Data Bank and research carried out by the Warsaw University of Life Sciences.

Since the processes affecting the outflow of N and P loads from the catchment may be completely different, depending on the hydrological characteristics of a given year, the VS0 was based on an uninterrupted period of 15 years, including years with dry, average, and wet hydrological conditions. Then, the SV0 was calibrated and verified with use of the TN data from the IMGW monitoring station.

The use of the macromodel DNS/SWAT allows also for incorporation of the data on agrotechnical treatments, and so-called fertilization calendar. The fertilization calendar is based on data on the maximum permissible fertilizer dose that can be applied, taking into account the requirements of cultivated plants, soil types, and slope. The calendar divides fertilizers into mineral and organic and distributes them for specific months and days. All three prepared variant scenarios were based on modification of the fertilization calendar. For the scenarios representing the "field of cold" conditions, the fertilization periods were shortened, whereas for the scenario representing the "field of heat" conditions, they were extended. To assess the effects of the fertilization restrictions in the "pole of cold" and "pole of heat" areas in the Słupia River three variant scenarios (VS1-3) have been created:

- VS1—"pole of cold" variant scenario 1—assumed shortening of the fertilization period by 20 days, and thus reducing of the amount of both mineral and organic fertilizers used during the allowed period;

- VS2—"pole of cold" variant scenario 2—assumed shortening of the fertilization period by 20 days, and maintaining the VS0 amount of fertilizer applied in the catchment through increase of the fertilizer dose in the remaining allowed period; and
- VS3—"pole of heat" variant scenario—assumed prolongation of the fertilization period by 30 days in total, while maintaining the VS0 amount of fertilizer applied in the catchment.

4. Results

All scenario simulations were performed on the monthly data for the period of 2002–2016. In the current study, to underline the seasonal differences, the yearly data and subsequent calculations were divided into two periods: summer (April–September), and winter (October–March), for which the monthly average TN load values for all the prepared scenarios have been calculated (Table 2). The differences in the TN loads between summer and winter months are clearly visible already for the VS0, with the average difference between the TN loads for the summer and winter months of 18,494 kg/month. Even larger differences between the average seasonal TN loads at the Charnowo calculation profile were detected during the variant scenario simulations, with the highest value for the VS3, reaching 42,248 kg/month (VS1—36,218 kg/month, and VS2—38,295 kg/month, respectively). For the VS0, and all three variant scenarios, the percentage of the TN load reduction (assuming the TN value for VS0 as 100%, Table 2), and also the resultant dispersion, have also been calculated. The dispersion was based on Equation (1):

$$R = \frac{X_{max} - X_{min}}{\overline{X}},\tag{1}$$

where R—data dispersion within the month; X_{max}—the maximum value of the measurement during the month; X_{min}—the minimum value of the measurement during the month; $\overline{X}$—mean value of the measurement during the month.

Simulation results showed that the reduction of the fertilization period by 20 days and, thus, reduction of the fertilizers amount by 55 kg/ha TN for each of the five crop types included in the fertilizer calendar, (VS1) could bring 8.61% of the TN load reduction (average for the period of 2002–2016) in the Słupia River catchment. Dispersion of the yearly results was very high for this scenario (775%), and clear differences were visible between the seasons, and among particular years (Table 2, Figure 2). The reduction of the TN load varied from 0.86% to 6.85% for the summer period, and from 9.87% to 19.16% for the winter period. The average TN load difference at the selected calculation profile between the baseline and "pole of cold 1" scenarios was 762 kg/month (coefficient of variation, cv = 60%) during the summer period, and 7171 kg/month (cv = 21%) during the winter period on average. As for the particular years, the lowest values of the VS0 and VS1 difference were observed in 2002–2003 and 2011, and the highest for 2006 and 2015 (for the summer and winter period, respectively). In general, in the all analysed years, the results of the variant VS1 scenario maintained a constant trend observed in the catchment, i.e., a high TN load values in winter periods (From October to March) and low TN load values in summer periods (from April to September, but the months in which the TN load was particularly low were July, August, and September). For the last two years (2014–2016) specified in Figure 2, the lowest TN load was slightly over 4700 kg/month (September 2014), while the highest TN load reached almost was 62,000 kg/month (January 2015). In practice, this means that the TN load values for the winter period are over 15-times higher than in the summer period. Figure 2 shows that in some earlier years the differences described were even greater.

Table 2. Monthly average values of TN loads in the Charnowo profile for the summer and winter periods for the baseline and variant scenarios.

| Years | Summer Season (IV-IX) | | | | | | | | | |
| | Average Value kg/month | | | | | | % | | |
	VS0	VS1	VS2	VS3	Difference (VS0-VS1)	Difference (VS0-VS2)	Difference (VS0-VS3)	Reduction (VS1)	Reduction (VS2)	Reduction (VS3)
2002	21,470	21,231	21,331	42176	239	139	−20,705	0.86	0.47	−145.09
2003	10,005	9850	9865	22,423	155	140	−12,418	0.94	0.56	−138.23
2004	31,037	29,796	30,089	10,239	1242	948	20,798	3.12	2.25	67.77
2005	39,360	38,946	38,604	31,120	414	756	8240	1.15	1.25	−42.93
2006	21,776	19,732	19,439	41,330	2044	2337	−19,554	6.85	7.55	−76.27
2007	25,298	24,827	24,893	21,996	470	405	3301	2.08	1.56	20.15
2008	18,274	17,328	17,217	26,440	946	1057	−8167	3.67	3.30	−59.12
2009	14,380	13,881	13,935	19,465	499	444	−5086	3.27	2.70	−26.86
2010	32,196	30,968	31,586	14,535	1228	610	17,662	3.27	1.46	53.93
2011	23,057	22,279	22,338	33,684	778	718	−10,627	2.83	2.38	−64.54
2012	23,747	22,891	22,912	22,800	856	835	947	3.25	2.99	6.57
2013	19,146	18,345	18,545	23,849	801	602	−4703	3.29	2.36	−29.29
2014	12,714	11,960	11,999	19,426	753	714	−6713	3.57	3.23	−96.36
2015	13,118	12,554	12,284	12,784	564	833	334	2.16	3.23	0.68
2016	16,724	16,288	16,523	13,340	436	201	3385	2.77	1.35	14.32
average	21,487	20,725	20,771	23,707	762	716	−2220	2.87	2.44	−34.35
cv	36	37	37	40	61	72	−520	49	67	−179

Table 2. *Cont.*

Years	Winter Season (X-III)									
	Average Value kg/month							%		
	VS0	VS1	VS2	VS3	Difference (VS0-VS1)	Difference (VS0-VS2)	Difference (VS0-VS3)	Reduction (VS1)	Reduction (VS2)	Reduction (VS3)
2002	74,355	68,198	70,693	60,496	6157	3662	13,859	11.08	7.58	−16.87
2003	46,297	40,351	41,680	73,747	5946	4617	−27,450	14.99	10.72	−83.29
2004	104,786	95,948	98,720	48,993	8838	6066	55,793	13.24	6.99	10.00
2005	82,055	75,860	78,035	102,288	6195	4019	−20,233	11.98	5.94	−176.14
2006	39,890	34,333	36,216	83,804	5557	3674	−43,914	13.01	8.51	−125.33
2007	99,298	92,311	94,022	38,873	6988	5276	60,425	12.65	9.67	30.21
2008	52,587	43,205	46,477	103,619	9382	6110	−51,032	18.74	12.05	−138.09
2009	54,470	46,677	49,302	54,153	7793	5168	317	16.04	10.46	−6.29
2010	59,727	52,528	54,785	54,281	7198	4942	5445	12.78	8.18	−30.88
2011	90,962	85,965	85,549	62,119	4997	5413	28,843	9.87	8.10	−67.05
2012	82,189	73,941	75,514	92,209	8248	6675	−10,020	14.97	10.74	−8.35
2013	53,105	46,690	48,578	83,220	6415	4527	−30,115	15.20	10.32	−75.85
2014	29,563	24,486	25,825	52,608	5077	3738	−23,046	14.14	9.94	−129.50
2015	49,737	39,927	43,116	29,844	9809	6621	19,892	17.36	12.03	37.71
2016	42,689	33,720	37,468	49,072	8969	5221	−6382	19.16	11.07	−15.76
average	64,114	56,943	59,065	65,955	7171	5049	−1841	14.35	9.49	−53.03
cv	35	39	37	33	21	20	−1742	18	19	−121
Yearly Data (I-XII)										
average all	42,800	38,834	39,918	44,831	3966	2882	−2031	9	6	−44
cv all	106	113	108	103	126	116	−2686	105	97	−290

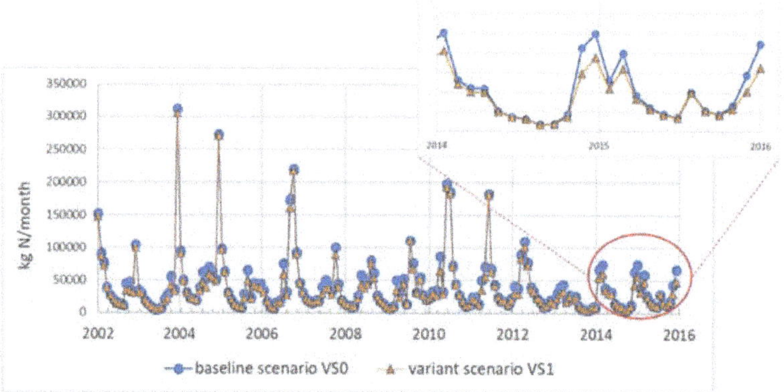

Figure 2. Comparison of the TN loads (kg N/month) for the baseline scenario (VS0) and pole of cold variant scenario 1 (VS1) at the Charnowo profile.

For the next "pole of cold" scenario (VS2), with shortening of the fertilization calendar, but without reduction of the fertilizer amount, the percentage of the TN load reduction was slightly lower compared to the VS1 scenario (5.96%). Dispersion for the VS2 results was at the similar level (737%), and the similar pattern was maintained for the seasonal changes. However, slightly lower ranges of the reduction for the summer (0.47–7.55%), and winter (5.94–12.05%) periods were detected (Table 2). The average TN loads both for summer and winter were also lower than those calculated for the VS1 scenario, by 0.2% and 3.59%, respectively. Comparing results of the VS2 with the baseline scenario the TN load average difference by 716 kg/month (cv = 72%) for the summer period, and 5049 kg/month (cv = 20%) for the winter period was concluded. As for the yearly pattern of the TN load distribution, the extreme values were detected in the same years as for the VS1 (2002–2003, and 2006) for the summer period. During the winter period the extreme values were observed in 2002 and 2012. Similarly, to the variant scenario VS1, also for VS2, the dependence of the TN load value on time remained the same (high values in the winter months and low in the summer months) (Figure 3). For the years 20014–2016, the lowest TN load was 4,810 kg/month (September 2014), while the highest TN load was 64,666 kg/month (December 2015), which again highlights the 15-times difference between winter and summer periods.

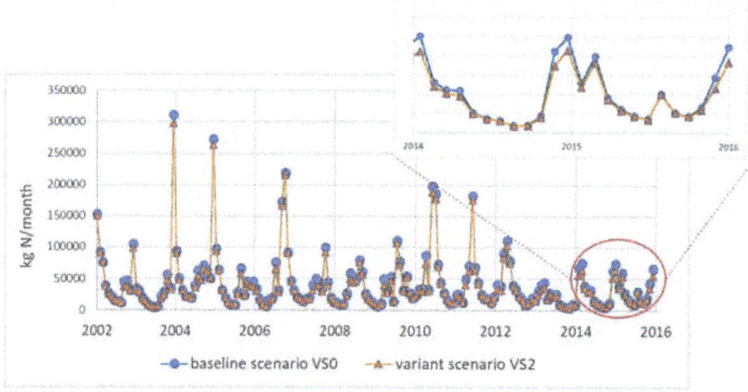

Figure 3. Comparison of the TN loads (kg N/month) for the baseline scenario (VS0) and pole of cold variant scenario 2 (VS2) at the Charnowo profile.

Extension of the fertilization period by 30 days (with maintaining the yearly fertilizer consumption) assumed in the "pole of heat" scenario (VS3) showed opposite trends. For the discussed pilot catchment such a change could result in a 43.69% increase of the TN load comparing to the baseline scenario. The reduction percentage for the VS3 compared to the VS0 scenario ranged between 67.77% and −145.09% (increase) for the summer period, and from 37.71% to −176.14% (increase) for the winter period (Table 2). Dispersion of the results was smaller than in the case of the VS1 and VS2 scenarios and amounted to 677%, and dramatic changes could be observed for the particular years (Figure 4). For the summer period the extreme values were observed for 2002, and 2004, while the contrasting (lowest-highest) values were detected for 2007, and 2008. The average TN values showed an increase by 2220 kg/month (cv = 520%) for the summer period, and by 1841 kg/month (cv = 1742%) for the winter period. Additionally, in the case of the variant VS3 scenario, the trend has been preserved for most years. In the case of this scenario, however, there is a clear difference in the TN load values obtained for selected months compared to the VS0 scenario. For the 2014–2016 period highlighted in Figure 4, the lowest TN load was about 4800 kg/month (September 2015), while the highest value (~107,000 kg/month) reached in January 2014. In comparison to the VS1 and VS2 scenarios, the difference between TN loads for summer and winter in the period 2014–2016 has even increased, being 26-times higher during the winter period.

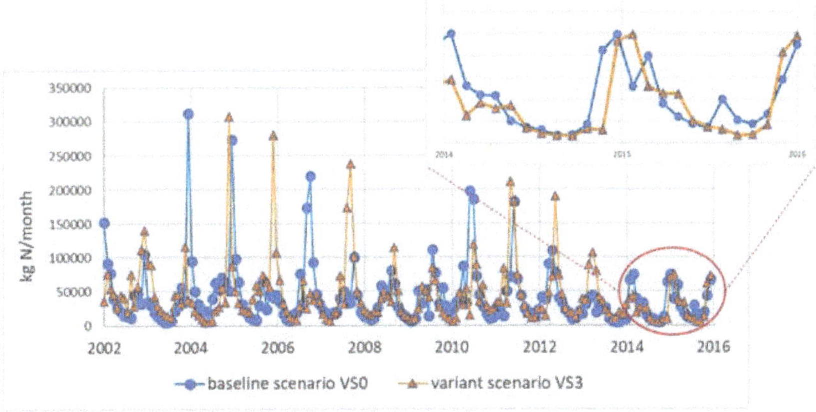

Figure 4. Comparison of the TN loads (kg N/month) for the baseline scenario (VS0) and pole of heat variant scenario (VS3) at the Charnowo profile.

5. Discussion

Introduction of the new NVZ approach in Poland is supposed to bring vast changes for the environment, but also for communities living from agriculture, which create an important part of the Polish society. After adoption of the new legislation Poland will become one large area of NVZ, which will result in additional restrictions that will apply to farmers in all regions of the country. These restrictions are meant to bring environmental benefits, however, conflict between this large and influential social group and the legislator are expected. Fear of additional costs related to the required farm investments, and an extra effort to put into fertilizer use, reporting, and management, has been already observed through the press and social media activity focused on agribusiness issues. Even a more explicit social response is expected from the areas assigned to the "pole of cold", and "pole of heat". It should be noted that restrictions related to the fertilizer use calendar will be imposed on farms based on administrative borders of communes, without recognition of the actual range of corresponding crops. Areas defined as the "cold pole", especially in mountain areas (Figure 1), are already less competitive compared to the rest of the country (poorer soil quality), therefore, introducing new regulations may further aggravate the current situation of farmers. Moreover, one should expect particularly strong social resistance related to the introduction of different regulations than in other parts of the country. The "pole of heat" areas will likely experience fewer conflicts, as farms located in this area will benefit from an extended period of fertilization. However, the overuse of fertilizers, exceeding the available dose, could be expected through the application time extension.

Despite, the possible social conflicts, the results obtained from the macromodel DNS/SWAT clearly show that restrictions imposed on farmers in the "pole of cold" areas will bring sound ecological effects. The simulations for the variant scenarios including conditions to be met in these regions show a noticeable reduction of the yearly TN load (8.61%, and 5.96% for the VS1, and VS2, respectively) in the pilot catchment. Since, the requirement of the fertilizer calendar shortening does not necessarily have to guarantee reduction of the total dose of the chemicals used in the catchments, then the second "pole of cold" scenario (VS2) should be considered as more probable. In the case of the pilot catchment, this scenario would result in an average decrease of the TN load by ca. 3000 kg/month. On the contrary, calculations for the "pole of heat" scenario (VS3) resulted in a distinct increase of the yearly TN load (43.69%) due to the extension of the fertilizer use period. Such an increase has to be considered as significant, especially taking into consideration that the size of the pilot catchment does not exceed 1620 km^2. For the catchment of the Warta River, located in the central part of Poland, with a size of 54,529 km^2 and where intensive agricultural activity is conducted [36] (with ca. 12% of the total area belonging to the "pole of heat" area), such a level of increase would likely result in hundreds of thousands of TN (kg) surplus per month. However, it should be remembered that the studied Słupia River catchment does not belong to the pole areas, therefore, only general restrictions imposed by the NVZ's new regulations will be introduced. Even in that case, the average TN load discharged into the Baltic Sea by this river (ca. 43,000 kg/month) should be reduced. Additionally, taking into consideration the presence of eight other small rivers in the Pomerania region with comparable catchments would bring a highly expected reduction of TN load discharged from Poland, from the point of view of pollution to the Baltic Sea. At this stage it is difficult to say whether the remaining catchments in this region will be comparable to the implementation of the new regulations, but it is nevertheless correct to assume that the TN load will be reduced throughout the area.

The justification of the obtained ecological results should be, however, discussed taking into consideration seasonal and yearly pattern of TN load changes. The significant differences between the winter and summer periods in the TN loads have been observed for the modelling results (VS0-3) at the study area. The summer TN loads in the Słupia River at the calculation profile of Charnowo were noticeably smaller (by average ca. 64% for all discussed scenarios) then the winter loads. This phenomenon, called a flattening phenomenon, has been observed in different catchments in Poland. It consists in periodic reduction of nutrient compounds released to surface waters, due to plant cover influence on retention of water and nutrients [41]. Although, in the case of the performed simulations, this phenomenon was additionally altered through the changes in the fertilization period. Both circumstances contributed to the high values of result dispersion. The obtained values were at the level of 737–775% for the "pole of cold" scenarios, and slightly lower (677%) for the "pole of heat" calculations. With the latter value related most likely to the extension of the fertilization period in the VS3 scenario. These results confirm natural variability of the TN loads depending on the season of the year, which is characteristic for the majority of the catchments at this latitude. For example, the dispersion for the TN load for the already mentioned Warta River was at the similar level (ca. 674%).

As for the yearly pattern of differences, a large variability of the average TN loads has been observed even for the reference scenario (VS0). For this variant, the extreme TN load values were recorded in 2003–2005, and 2014 (Table 2). To investigate this pattern, 1-min data from the Ustka meteorological station, located directly at the Słupia River catchment, were used. These data were calculated at the IMGW-PIB for the needs of the Polish Atlas of Rainfall Intensity (PANDa) currently being prepared by the RETENCJAPL [42]. This information allowed identifying sudden short-term precipitation events and a frequency of their occurrence on particular days and months of the year. Detailed information was retrieved for the years where the extremely high or low values of the TN load reduction were observed in each season. Thus, the years of 2003 and 2005 were selected as charged with extreme values during the summer season (9850, and 38,604 kg/month, respectively, for the both "pole of cold" scenarios). While, 2014 and 2004 were bearing the extreme values for the winter period of VS1 and VS2 (24,486, and 98,720 kg/month, respectively). For these years the seasonal precipitation was examined at the Ustka station. The average annual rainfall for this station is quite high and fluctuates from 497 mm during the winter period to over 873 mm in the summer one, with the highest sums recorded from July to October (over 60 mm), and the smallest (below 40 mm) in January, and April. For the Ustka station, both 2003 and 2005 were classified as dry years. In 2003, heavy rain in the summer season was low, while the summer period of 2005 was extremely dry. On the other hand, from October to December numerous short-lived but intense rainfalls were noted. The years selected as extreme from the point of view of the TN load were considered as wet. The recorded rainfall data for 2004 clearly justifies the maximum TN load in the winter for this year. Heavy rainfalls had been occurring since September of that year. Conversely, the situation looked like in 2014 where, after intensive summer rainfall, there was a dry autumn and winter explaining the low TN load values for both VS1 and VS2. The similar analysis performed for the "pole of heat" scenario VS3 indicated years 2004 and 2002 as bearing the minimal and maximal TN loads (10,239 and 42,176 kg/month, respectively). The year of 2004, as already mentioned, was characterized by exceptionally low rainfall values during summer, with only precipitation that could have a significant impact on the surface runoff volume observed from September of this specific year. In turn, 2002 was full of numerous short-lived, but intense, precipitation during the summer months. For the winter period, the values of the minimal and maximal TN loads were observed in 2015 and 2008 (29,844 and 103,619 kg/month, respectively). Unfortunately, the data for 2015 is difficult to analyse, since they are very limited and do not allow establishing the relationship between the precipitation and TN load. As for 2008, numerous intense rainfalls from August to December were detected. There is no doubt that atmospheric precipitation, its frequency and intensity, are of great importance for the release of nutrients into surface waters. Throughout Poland signs of climate change and its impact on surface waters have been observed, and is increasing year by year. Increased frequency of extreme events, such as intense rainfalls, in turn, will lead to an increase in the amount of nutrients entering surface waters from areas used for agriculture [43] and clearly should be taken into consideration while estimating the impact of the NVZ regulations.

The results obtained, although they should be treated as preliminary, indicate the positive environmental aspects of the implementation of the new NVZ action program. The introduction of additional restrictions and requirements to limit the release of nitrogen from agricultural sources to surface waters and further into the Baltic Sea will undoubtedly reduce excessive amounts of nutrients in the aquatic environment and dependent water by improving the functioning of ecosystems in the long term. While there is no doubt that any initiative aimed at improving the quality of the environment is extremely important, it is also necessary to remember about the social aspect which, if not well thought out already at the planning stage, can effectively hamper and sometimes even prevent changes. Already at the stage of strongly truncated social consultations and reactions of institutions representing the interests of farmers, it could be concluded that the new regulations will generate a very large number of conflicts between the legislator and farmers.

6. Conclusions

The current study made the first attempt to analyse an impact of the new action program aimed to reduce nitrogen pollution caused by agricultural sources in Poland. Implementation of the general requirements imposed through this program on the whole country will, unfailingly, bring benefits to the environment. However, their financial and organizational costs will likely meet discontentment from farmers, which is inevitable when one has to choose between the competitiveness of farms and the improvement of the state of the environment. In addition to these changes, more specific restrictions will be imposed on the territories designated for reduction or extension of the period when the use of nitrogen fertilizers is allowed.

To incorporate details of the stipulated changes in the fertilizer use calendar, the modelling approach was adopted. With the use of the macromodel DNS/SWAT the baseline model, and three variant scenarios, were created for the pilot catchment (Słupia River). The obtained results have shown that introduction of more rigorous restrictions on nitrogen fertilizer use would have a considerable impact on TN load reduction in the areas subjected to such changes ("pole of cold"). Therefore, the total load of nitrogen discharged from the Polish territory could be also reduced. However, the extension of the fertilizer use period will likely result in an increase of total nitrogen load released from the catchments located at the "pole of heat" region. The analyses described in the article have also confirmed the strong relationship of atmospheric precipitation with the amount of nutrients in surface waters. Climate change is becoming more and more visible throughout Europe, which will, inter alia, cause intensification of violent meteorological phenomena such as rapid rainfall. As a result, more and more nitrogen could be released from cultivated fields, through run-off to surface waters, causing algal blooms in water bodies, including the Baltic Sea. Therefore, new programs of measures limiting the use of nutrients are necessary.

The general questions, i.e., will the described changes in legislation help to improve the general quality of surface waters, and/or will the costs incurred as a result of these changes and conflicts between the legislator and the farmers be consequently considered to be profitable, are still premature and need to be answered. However, it must not be forgotten that any introduction of new legislations, especially those that cover the whole area of the country, require long and well-prepared social consultations and information programs, as well as transitional periods. Only the combination of these three elements could limit the aforementioned conflicts. At the same time, it must never be forgotten that farmers will face the main burden of these provisional implementations, and the role of the state, in addition to caring for the environment, should also be concerned for the competitiveness and good conditions of farms.

Author Contributions: Conceptualization, E.S. and P.W.; methodology: P.W.; software: P.W. and P.O.-W.; validation: P.W.; formal analysis: P.W.; investigation: E.S. and P.W.; resources: P.W. and P.O.-W.; data curation: P.W. and P.O.-W.; writing—original draft preparation: E.S. and P.W.; writing—review and editing: E.S.; visualization: P.W.; supervision: E.S.

Funding: This research received no external funding.

Conflicts of Interest: The authors declare no conflict of interest.

References

1. Sharma, L.K.; Bali, S.K. A review of methods to improve nitrogen use efficiency in agriculture. *Sustainability* **2018**, *10*, 51. [CrossRef]
2. Rütting, T.; Aronsson, H.; Delin, S. Efficient use of nitrogen in agriculture. *Nutr. Cycl. Agroecosyst.* **2018**, *110*, 1–5. [CrossRef]
3. Sutton, M.A.; Oenema, O.; Erisman, J.W.; Leip, A.; van Grinsven, H.; Winiwarter, W. Too much of a good thing. *Nature* **2011**, *472*, 159–161. [CrossRef] [PubMed]
4. FAOSTAT. Food and Agriculture Organization of the United Nations: Rome, Italy. Available online: http://faostat.fao.org/ (accessed on 16 July 2018).
5. Velthof, G.L.; Lesschen, J.P.; Webb, J.; Pietrzak, S.; Miatkowski, Z.; Pinto, M.; Kros, J.; Oenema, O. The impact of the Nitrates Directive on nitrogen emissions from agriculture in the EU-27 during 2000–2008. *Sci. Total Environ.* **2014**, *468–469*, 1225–1233. [CrossRef] [PubMed]
6. European Commission (EC). *Report from the Commission to the Council and the European Parliament on the Implementation of Council Directive 91/676/EEC Concerning the Protection of Waters against Pollution Caused by Nitrates from Agricultural Sources based on Member State Reports for the Period 2012–2015*; EC: Brussels, Belgium, 2018.

7. Pastuszak, M. Description of the Baltic Sea catchment area—Focus on the Polish sub-catchment. In *Temporal and Spatial Differences in Emission of Nitrogen and Phosphorus from Polish Territory to the Baltic Sea*; Pastuszak, M., Igras, J., Eds.; National Marine Fisheries Research Institute: Gdynia, Poland, 2012; pp. 13–44, ISBN 978-83-61650-08-9.

8. Baltic Marine Environment Protection Commission (HELCOM). *State of the Baltic Sea—Second HELCOM Holistic Assessment 2011–2016*; Baltic Sea Environment Proceedings 155; HELCOM: Helsinki, Finland, 2018.

9. Pastuszak, M.; Witek, Z. Discharges of water and nutrients by the Vistula and Oder rivers draining Polish territory. In *Temporal and Spatial Differences in Emission of Nitrogen and Phosphorus from Polish Territory to the Baltic Sea*; Pastuszak, M., Igras, J., Eds.; National Marine Fisheries Research Institute: Gdynia, Poland, 2012; pp. 309–354, ISBN 978-83-61650-08-9.

10. Baltic Marine Environment Protection Commission (HELCOM). *Sources and Pathways of Nutrients to the Baltic Sea*; Baltic Sea Environment Proceedings No. 153; HELCOM: Helsinki, Finland, 2018.

11. Ostojski, M.S. *Modeling of Drainage Processes to the Baltic Sea of Biogenic Compounds: On the Example of Nitrogen and Total Phosphorus*; Wydawnictwo Naukowe PWN: Warsaw, Poland, 2012; p. 255. (In Polish)

12. Krasowicz, S.; Górski, T.; Budzyńska, K.; Kopiński, J. Agricultural characteristics of the territory of Poland. In *Temporal and Spatial Differences in Emission of Nitrogen and Phosphorus from Polish Territory to the Baltic Sea*; Pastuszak, M., Igras, J., Eds.; National Marine Fisheries Research Institute: Gdynia, Poland, 2012; pp. 45–104, ISBN 978-83-61650-08-9.

13. Szalinska, E. Water Quality and Management Changes Over the History of Poland. *Bull. Environ. Contam. Toxicol.* **2018**, *100*, 26–31. [CrossRef] [PubMed]

14. GUS. Statistical Yearbook of the Republic of Poland, 2001. Statistics Poland. Available online: https://stat.gov.pl/ (accessed on 20 July 2018).

15. GUS. Statistical Yearbook of the Republic of Poland, 2017. Statistics Poland. Available online: https://stat.gov.pl/ (accessed on 20 July 2018).

16. European Environment Agency (EEA). *Landscapes in Transition: An Account of 25 Years of Land Cover Change in Europe*; European Environment Agency Report No. 10/2017; EEA: Copenhagen, Denmark, 2017.

17. Van Drecht, G.; Bouwman, A.F.; Harrison, J.; Knoop, J.M. Global nitrogen and phosphate in urban wastewater for the period 1970 to 2050. *Glob. Biogeochem. Cycles* **2009**, *23*, GB0A03. [CrossRef]

18. Bouraoui, F.; Grizzetti, B. Long term change of nutrient concentrations of rivers discharging in European seas. *Sci. Total Environ.* **2011**, *409*, 4899–4916. [CrossRef] [PubMed]

19. Withers, P.J.A.; Neal, C.; Jarvie, H.P.; Doody, D.G. Agriculture and eutrophication: Where do we go from here? *Sustainability* **2014**, *6*, 5853–5875. [CrossRef]

20. Bouwman, L.; Goldewijk, K.K.; Van Der Hoek, K.W.; Beusen, A.H.W.; Van Vuuren, D.P.; Willems, J.; Rufino, M.C.; Stehfest, E. Exploring global changes in nitrogen and phosphorus cycles in agriculture induced by livestock production over the 1900–2050 period. *Proc. Natl. Acad. Sci. USA* **2013**, *110*, 21195. [CrossRef] [PubMed]

21. Dupas, R.; Delmas, M.; Dorioz, J.-M.; Garnier, J.; Moatar, F.; Gascuel-Odoux, C. Assessing the impact of agricultural pressures on N and P loads and eutrophication risk. *Ecol. Indic.* **2015**, *48*, 396–407. [CrossRef]

22. Intergovernmental Panel on Climate Change (IPCC). *Climate Change 2014: Synthesis Report*; Core Writing Team, Pachauri, R.K., Meyer, L.A., Eds.; IPCC: Geneva, Switzerland, 2015; p. 151.

23. National Oceanic and Atmospheric Administration (NOAA). National Centers for Environmental Information, State of the Climate: Global Climate Report for December 2016. Available online: https://www.ncdc.noaa.gov/sotc/global/201612 (accessed on 17 January 2017).

24. Basso, B.; Giola, P.; Dumont, B.; Migliorati, M.D.A.; Cammarano, D.; Pruneddu, G.; De Antoni Migliorati, M.; Cammarano, D.; Pruneddu, G.; Giunta, F. Tradeoffs between Maize Silage Yield and Nitrate Leaching in a Mediterranean Nitrate-Vulnerable Zone under Current and Projected Climate Scenarios. *PLoS ONE* **2016**, *11*, e0146360. [CrossRef] [PubMed]

25. Worrall, F.; Spencer, E.; Burt, T.P. The effectiveness of nitrate vulnerable zones for limiting surface water nitrate concentrations. *J. Hydrol.* **2009**, *370*, 21–28. [CrossRef]

26. Wall, D.; Jordan, P.; Melland, A.R.; Mellander, P.E.; Buckley, C.; Reaney, S.M.; Shortle, G. Using the nutrient transfer continuum concept to evaluate the European Union Nitrates Directive National Action Programme. *Environ. Sci. Policy* **2011**, *14*, 664–674. [CrossRef]

27. Banaszak, K. Verification of areas particularly exposed to pollution with nitrogen compounds from agricultural sources on the territory of the Regional Management of Water Management in Gliwice. *Woda-Środowisko-Obszary Wiejskie* **2009**, *9*, 9–18. (In Polish)

28. Nagrabska, J.; Pastuszczak, K. Problems associated with with implementation of action programmes in areas vulnerable to pollution by nitrogen compounds from agricultural sources in Wielkopolska. *Woda-Środowisko-Obszary Wiejskie* **2009**, *9*, 49–61. (In Polish)

29. Król, M.A. Legal framework of environmental law for agricultural production in Poland. *Polityki Europejskie Finanse Mark.* **2015**, *13*, 86–106.

30. Woźniak, E.; Nasiłowska, S.; Jarocińska, A.; Igras, J.; Stolarska, M.; Bernoussi, A.S.; Karaczun, Z. *Designation of Areas under Real Pressure of Agricultural Activity due to Water Pollution Caused by Nitrogen Compounds*; Prepared on commission of President of National Water Management Authority by University of Warsaw; University of Warsaw: Warsaw, Poland, 2011; p. 30. (In Polish)

31. Instytut Uprawy Nawożenia i Gleboznawstwa-Państwowy Instytut Badawczy (IUNG-BIP). *Assessment of Agricultural Pressure on the Condition of Surface and Underground Waters and the Indication of Areas Particularly Exposed to Nitrates from Agricultural Sources*; Prepared on Commission of Ministry of Agriculture and Rural Development (MRiRW) by Institute of Soils Science and Plant Cultivation (IUNG-BIP); IUNG-BIP: Pulawy, Poland, 2011; p. 38. (In Polish)

32. Alterra Research Institute, Wageningen UR. Assessment of the designation of nitrate vulnerable zones in Poland. In *Environmental Sciences*; Report; Wageningen University & Research Centre: Wageningen, The Netherlands, 2007; p. 126.

33. Dziennik Ustaw Rzeczpospolitej Polskiej (Dz.U.2018.1339). Action Programme Aimed at Reducing Pollution Water from Nitrates from Agricultural Sources and Prevention Further Pollution. 2018. Available online: http://www.dziennikustaw.gov.pl/du/2018/1339/1 (accessed on 14 September 2018). (In Polish)

34. Nieróbca, A.; Kozyra, J.; Mizak, K.; Wróblewska, E. Change in the length of the growing season in Poland. *Woda-Środowisko-Obszary Wiejskie* **2013**, *13*, 81–94.

35. Ministry of Agriculture and Rural Development (MRiRW). *Plan of the Rural Areas Development*; MRiRW: Warsaw, Poland, 2004; p. 186. (In Polish)

36. Wilk, P.; Orlińska-Woźniak, P.; Gębala, J. The river absorption capacity determination as a tool to evaluate state of surface water. *Hydrol. Earth Syst. Sci.* **2018**, *22*, 1033–1350. [CrossRef]

37. Gebala, J.; Orlinska-Wozniak, P.; Wilk, P. Application of the SWAT model in the methodologies for identifying pollution sources of biogenic compounds in catchments. In *Selected Examples of Systems Decision Support and Modeling in Water Management*; IMGW: Warsaw, Poland, 2014; pp. 87–102. (In Polish)

38. Wilk, P.; Gębala, J.; Orlińska-Woźniak, P.; Ostojski, M. The importance of hourly nutrient concentration variability in terms of assessment of the surface water state in Słupia Pilot River. *Meteorol. Hydrol. Water Manag.* **2014**, *4*, 13–24. [CrossRef]

39. Wilk, P.; Orlinska-Wozniak, P.; Gebala, J. Variability of nitrogen and phosphorus concentration ratio for selected catchments of the Pomeranian rivers. *Sci. Rev. Eng. Environ. Sci.* **2017**, *26*, 55–65.

40. Wilk, P.; Orlińska-Woźniak, P.; Grabarczyk, A.; Gębala, J. Rozkład stężenia zanieczyszczeń azotu ogólnego i fosforu ogólnego w profilu poprzecznym rzeki Słupi. *Gospodarka Wodna* **2018**, *3*, 87–93. (In Polish)

41. Wilk, P.; Orlińska-Woźniak, P.; Gębala, J.; Ostojski, M. The flattening phenomenon in a seasonal variability analysis of the total nitrogen loads in river waters. *Czasopismo Techniczne* **2017**, *11*, 137–159.

42. Burszta-Adamiak, E.; Licznar, P. An analysis of the time structure of maximum precipitation of the Polish Atlas of Rainfall Intensity (PANDa). *Instal* **2018**, *3*, 49–53.

43. Szalińska, E.; Wilk, P.; Grabarczyk, A. Tools for evaluation of the quantity and quality of the floating river load in the context of climate change. *Gospodarka Wodna* **2018**, *4*, 101–106. (In Polish)

 *sustainability* 

Article

Analysis of the Impact of Rural Households' Behaviors on Heavy Metal Pollution of Arable Soil: Taking Lankao County as an Example

Shixin Ren [1], Erling Li [1,2,*], Qingqing Deng [1], Haishan He [1] and Sijie Li [1]

[1]   Key Research Institute of Yellow River Civilization and Sustainable Development & Henan Collaborative Innovation Center for the Yellow River Civilization Heritage and Modern Civilization Construction, College of Environment and Planning/Agricultural and Rural Sustainable Development Research Institute, Henan University, Kaifeng 475004, China; sxren@vip.henu.edu.cn (S.R.); dqq0801@126.com (Q.D.); seamountainriver@163.com (H.H.); 18739992308@163.com (S.L.)
[2]   Collaborative Innovation Center of Urban-Rural Coordinated Development, Henan Province, Academician Laboratory for Urban and Rural Spatial Data Mining, Zhengzhou 450000, China
*   Correspondence: erlingli@henu.edu.cn

Received: 28 October 2018; Accepted: 21 November 2018; Published: 23 November 2018

Abstract: As heavy metal pollution of arable soil is a significant issue concerning the quality of agricultural products and human health, the rural households' behaviors have a direct impact on heavy metal content in arable soil and its pollution level, but only a few researches have been done at such microscopic scale. Based on 101 field questionnaires of rural households in Lankao County and the monitoring data on heavy metal of arable soil of each rural household, the kind of rural households' behaviors which impose obvious influence on heavy metal content of arable soil are investigated via single-factor pollution index, Nemerow pollution index and econometric model in this study. The results show that, rural households' land utilization mode affects heavy metal content in soil, e.g., the degree of heavy metal pollution of soil for intensive planting is higher than that of traditional planting, viz. vegetable greenhouse > garlic land > traditional crop farmland. The management of cultivated land with due scale is beneficial to reducing heavy metal content in soil, that is, the land fragmentation degree is in direct proportion to heavy metal content in soil, so rural households are encouraged to carry out land circulation and combine the patch into a large one. Excess application of fertilizer, pesticide and organic fertilizer will lead to heavy metal pollution of soil, while agricultural technical training organized by government department and the foundation of agricultural cooperative can promote the technical level and degree of organization of rural households and enable them to be more scientific and rational in agrochemicals selection and application, hence reducing or avoiding heavy metal pollution of soil. Single factor pollution level of heavy metal in the soil for planting various crops is different, so it is recommended to prepare various pollution reduction programs for different land types and pollution levels for the harmony and unity of human-nature system.

Keywords: rural households' behaviors; arable soil; heavy metal pollution assessment; Lankao county

---

1. Introduction

Under the background of acceleration of regional industrialization, urbanization, and agricultural modernization, environmental issues attributed to human activities become increasingly obvious, and heavy metal pollution also becomes one of the major factors affecting the quality of arable soil [1]. According to Analysis of the Report on the National General Survey of Soil Contamination released by China in 2014 for the first time, the over-limit ratio of arable soil in China reaches 19.4% mainly due to such pollutants as heavy metal elements including Ni, Cu, Cd, Pb, As, and Hg, as well as DDT and PAHs [2], so heavy metal pollution of arable soil is an extremely urgent problem. Heavy metal content of arable soil is mainly affected by such natural factors as soil parent material (internal factor) and human activities (external factor), and the latter has been the major factor affecting heavy metal pollution of arable soil along with social and economic development. As the subject of agricultural production and management, rural households impose a direct impact on heavy metal content of soil by applying fertilizer and pesticide or other production activities [3,4]. Therefore, in the context of serious heavy metal pollution of arable soil and the increasing importance of the safety of agricultural products and human health, exploring the influencing mechanism of rural households' behaviors on heavy metal pollution of arable soil is of great significance to the government's preparation of corresponding policies for standardizing rural households' behaviors, improving scientific cognition and technical level of rural households, and pushing forward agro-ecological civilization construction and environment-friendly development of China.

Heavy metal pollution issue of arable soil receives wide concern from scholars at home and abroad due to its universality, management difficulty, and harmfulness, therefore a number of researches focus on the heavy metal content of arable soil [5,6], source analysis [7], pollution assessment [8–10], ecology [1,11], and health risk [12,13] evaluation, etc., mainly by applying the following approaches: single factor pollution index [4], geo-accumulation index [14], Nemerow pollution index [15], pollution load index [16], potential ecological risk index [17], and health risk assessment model [18], etc. In terms of selecting studied area, some scholars also carried out studies on the heavy metal content of arable soil around the functional area in addition to farmland of common areas [19]. For example, cultivated land around the mining area is susceptible to mining, smelting, and dump slag, which will aggravate heavy metal pollution of soil [20]; heavy metal content of arable soil in suburban areas increases obviously due to influences of urban construction, industrial and domestic pollution discharge [21]. As a key factor affecting the heavy metal content in arable soil, it is widely proved that agricultural production promotes the heavy metal accumulation in arable land [22,23]. Relevant researches on heavy metal pollution remediation of arable soil indicates that the adoption of appropriate treatment measures can reduce the heavy metal content of arable soil to some extent, such as the adjustment of planting patterns, deep ploughing for soil amelioration, formula fertilization, and adoption of phytoremediation, etc. [24].

With the transformation of traditional agriculture to modern intensive agriculture speeding up, agricultural production has been one of the major factors affecting heavy metal pollution of arable soil, and it is widely proved that agricultural production promotes the heavy metal accumulation in arable land [22,23]. As the basic economic unit of agricultural production [25], rural households play a crucial role in the process. In earlier research, many scholars incorporate rural households' behaviors into their research system from various aspects [26–29]. Along with increasingly outstanding agricultural pollution issues, the impact of rural households' behaviors on the agricultural environment also becomes a focus of this academic circle [30]. Related researches show that rural households' behaviors impose important impacts on the heavy metal content of arable soil. Such behaviors include rural households' resource utilization, operation, technology application, planting selection, and cognition. Among them, the number of rural households' available resources and the utilization pattern affect heavy metal content in soil. Rural households' operation behaviors such as input into fertilizer, pesticide, and organic fertilizer directly affect the heavy metal content of arable soil. Different scales of cultivated land operation may lead to the differentiation of agricultural environment, for example, small-scale cultivated land operation and high fragmentation will go against the adoption of new farming technology and may lead to increasing dose of agricultural chemicals in unit area [31]. The promotion of agricultural technology level can assist rural households to master or adopt environment-friendly field management technology and improve the utilization rate of fertilizer and pesticide [32]. Rural households' environmental cognitive level is an important factor affecting their behavior [33,34], so the emergence and aggravating of heavy metal pollution of arable soil is largely related to their low environmental cognitive level and weak environmental awareness. Rural households' planting [35] selection may change agricultural landscape structure and production element input, thus affecting the heavy metal content of arable soil to varying degrees, so comparative study by classification will be done in this paper. Besides, the basic attribute of household head and livelihood features of the family are the main motivation of different rural households' behaviors. For example, the higher the education level of the household head is, the higher the environmental awareness will be. Age will affect the rural households' receptivity toward new technology. The household income will affect the input of agricultural means of production, etc. [31,36]. As a result, the attributes of rural households are also included into the analysis variables.

To sum up, existing literatures show in-depth research on heavy metal pollution of arable soil, but most focus on macroscopic scale and a few combine rural households' behaviors with heavy metal content of soil at microscopic scale [37], and even fewer researches combine outdoor questionnaire data of rural households with indoor soil sample monitoring data for quantitative analysis. Therefore, on account of man–land coupling idea and following the research direction of increasing integration of physical geography and human geography, this paper takes rural households' behaviors theory as the basis and Lankao County of Henan as the example based on field investigation and laboratory experiment via such approaches as single factor pollution index, Nemerow Pollution Index and stepwise regression analysis. In this paper, the scientific issue "what kind and how much of the impact imposed by rural households' behaviors on heavy metal content of arable soil" is discussed, so as to provide reference for different regions to prepare scientific pollution reduction programs.

2. Materials and Methods

2.1. Study Area and Sample Selection

Lying between $114°40'\sim115°16'$ E and $34°44'\sim35°01'$ N and northeast of Henan Province at the middle and lower reaches of the Yellow River, Lankao County is a typical plain rural area and its present agricultural development has certain representativeness for it covers the common features of agricultural development in the middle area of China. In recent years, under the support of industrial poverty shake-off policy, featured agriculture like greenhouse vegetable planting is developed in Lankao County and the planting mode becomes more professional and distinctive in a short period of time, which provides comparable sample for accumulation of heavy metal content in soil, showing certain representativeness.

Sample villages and sample rural households are randomly selected in Lankao County. In order to focus on the impact of rural households' behaviors on heavy metal content of soil, land parcels far from factory, company, highway and construction land are chosen during sampling. Moreover, Lankao County is dominated by subregion climate which generates minor difference in natural factors like climate and parent material, so this can maximally lower the impact of surroundings and natural factors on heavy metal content of soil. By following these principles, we selected 20 sample villages (Figure 1) and 105 sample rural households in total. We handed out designed semi-structured questionnaire to each rural household and carried out "face-to-face" in-depth interviews and collected soil sample from the parcel of each rural household to monitor the heavy metal content of soil. Finally, we acquired 101 valid questionnaires and 101 soil samples accordingly, with effective rate of 96.19%. According to investigation on the types of crops planted in the parcels, farmland of the studied area was classified into traditional crop farmland (for traditional crops: wheat, corn, peanut and cotton), garlic land (for garlic and traditional crops) and vegetable greenhouse (parcels with greenhouse or arched shed for vegetable planting) in a sample size of 64:11:26 in this paper, so as to analyze the influence of rural households' behaviors on heavy metal content of soil under various land utilization modes. This investigation was done by the author and 14 members in July 2017. Questionnaire survey covers the attribute features of household head, status of land resource, agricultural production and operation, rural households' technological utilization and environmental cognitive level.

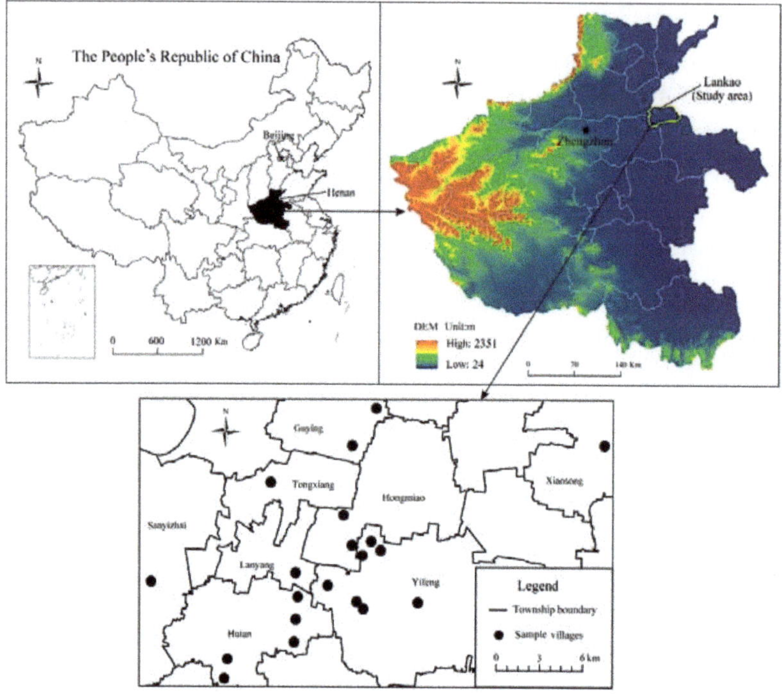

Figure 1. Diagram of study area and sample villages.

2.2. Soil Sample Collection and Sample Analysis

Taking and processing of 101 soil samples: according to "S" distribution and sampling depth of 0–20 cm, the samples were processed by quartering after removal of impurities like burr, plant & vegetable body and uniform mixing, with the rest 500 g soil analysis sample stored in a clean ziplock bag marked with corresponding No. of rural households' questionnaire and then carried it back to the laboratory. After natural air-drying, the collected samples were crushed and screened by a 2 mm nylon sieve, and then distributed evenly on plastic cloth after thorough mixing. After that, about 50 g was taken by multi-point sampling and further ground by agate mortar after remixing to make the samples all pass 0.15 mm nylon sieve, which was stored in ziplock bag for use.

"HNO_3-HF-$HClO_4$" digestion system was adopted for samples, with X-Series inductively coupled plasma source mass spectrometer (ICP-MS, Thermofisher) applied for measuring the content of Cr, Ni, Cu, Zn, Cd and Pb in soil. During the experiment, parallel test and national standard soil sample (GSS-5) recovery test are adopted for quality control. Relative deviation of secondary parallel test is within 5% and recovery rate of standard sample is 92.1%~106.3%.

2.3. Approaches for Evaluating Heavy Metal Pollution of Soil

In this paper, single factor pollution index and Nemerow Pollution Index are adopted to evaluate the heavy metal pollution of soil in the studied area.

2.3.1. Single Factor Pollution Index Approach

Single factor pollution index is one of the common approaches to evaluate the pollution level of a certain pollutant in soil [7]. The calculation equation (Equation (1)) is:

$$P_i = C_i / S_i \tag{1}$$

where P_i refers to the single factor pollution index of heavy metal i in soil; Ci refers to the measured value of i in soil (mg·kg^{-1}); Si refers to the evaluation criterion of i (mg·kg^{-1}), with class II standard in Environmental Quality Standard for Soils (GB 15618—1995) [38] adopted. This standard is the soil limit for guaranteeing agricultural production and maintaining human health [39]. Pollution classification standard of P_i is: soil is clean when $P_i \leq 0.7$, relatively clean but reaches safety warning state when $0.7 < P_i \leq 1$; slight pollution when $1 < P_i \leq 2$, moderate pollution when $2 < P_i \leq 3$ and severe pollution when $P_i > 3$. In the last case, heavy metal pollution of soil is quite serious [15,39].

2.3.2. Nemerow Pollution Index Approach

Heavy metal pollution is usually a kind of combined pollution in soil environment. Therefore, it is necessary to synthesize the pollution index of different pollutants in the same sampling point and different samples of the same pollutant on the basis of single factor pollution index, so as to make a comprehensive evaluation of the result. Nemerow Pollution Index [14,15] approach is widely applied in evaluating soil pollution level, and can comprehensively reflect the level of various pollutants in regional soil [40] and highlight the action of pollutants with heavy pollution [41]. The calculation equation (Equation (2)) is given below:

$$P_N = \sqrt{\frac{(C_i/S_i)^2{}_{max} + (C_i/S_i)^2{}_{ave}}{2}} \tag{2}$$

where P_N is Nemerow Pollution Index of heavy metal element in soil; $(C_i/S_i)_{max}$ is the maximum value of single factor pollution index in heavy metal element participating in soil evaluated; $(C_i/S_i)_{ave}$ is the average of single factor pollution index of various heavy metal elements. Pollution classification standard of P_N is: soil is clean when $P_N \leq 0.7$, reaches safety warning state when $0.7 < P_N \leq 1$; slightly polluted when $1 < P_N \leq 2$, moderately polluted when $2 < P_N \leq 3$ and severely polluted when $P_N > 3$ [15,39].

2.4. Setup of Econometric Model

2.4.1. Model Construction

The following econometric model is set up to analyze the impact of rural households' behaviors on heavy metal pollution of arable soil (Equation (3)):

$$y = \alpha + \sum_i \gamma_i X_i + \varepsilon \tag{3}$$

where y is the heavy metal pollution level of arable soil; α is a constant; γ_i refers to regression coefficient, representing the contribution rate of various factors to y; X_i refers to factors affecting heavy metal pollution level of soil; ε is a random disturbance term.

2.4.2. Variable Selection and Assignment

Based on the above literature analysis, the analysis framework for influence of rural households' behaviors on heavy metal content of arable soil (Figure 2) is established in this paper with the following main idea: under the influence of attributes of different individuals and families, rural households directly or indirectly affect the heavy metal content of arable soil through resource utilization, operation input, technology application, planting selection and cognition, and certain accumulation of heavy metal content will give rise to heavy metal pollution. It is necessary to divide the above behaviors into specific influencing factors, so as to select the corresponding analysis variables based on detailed factors and design the questionnaire for final field investigation.

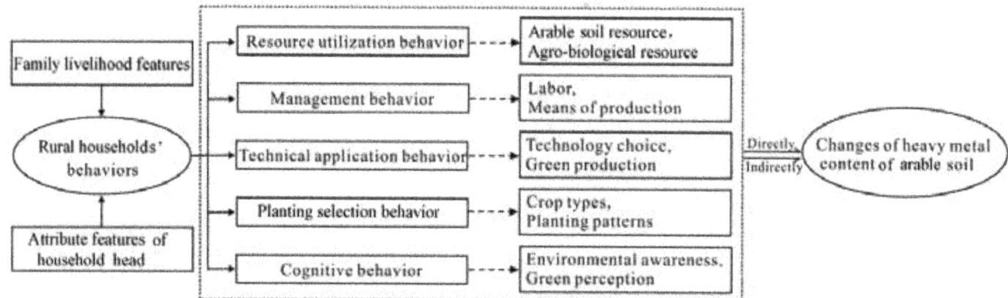

Figure 2. Framework for analysis on the impact of rural households' behaviors on heavy metal content of arable soil.

Dependent variable(y) stands for the heavy metal pollution level of arable soil and is represented by Nemerow Pollution Index (P_N). Based on theoretical analysis mentioned above, such indicators as attribute features of household head (X_1~X_4), family livelihood features (X_5~X_8), land resource endowment (X_9, X_{10}), input of agricultural means of production (X_{11}~X_{13}), environment cognition (X_{14}, X_{15}) and agricultural technology level (X_{16}) are selected as the independent variables (Table 1) in this paper. According to investigation result, planting selection of rural household is embodied in classification of vegetable greenhouse, garlic land and traditional crop farmland. Econometric analysis is shown below for the above three types of land.

Table 1. Variable selection and assignment of rural households' behaviors.

Variable	Name	Assignment
Attribute features of household head	Age (X_1)	1 stands for household head bellow 30 years old, 2 stands for those aged from 30 to 40, 3 stands for those aged from 40 to 50, 4 stands for those aged from 50 to 60, 5 stands for those above 60 years old
	Village cadres or not (X_2)	1—Yes, 0—No
	Education level (X_3)	1—Illiteracy or semiliterate, 2—Primary school level, 3—Junior high school, 4—Senior high school, 5—Junior college and above
	Years of agricultural production (X_4)	1—Under 10 years, 2—10–20 years, 3—20–30 years, 4—Above 30 years.
Family livelihood features	Member of agricultural cooperative or not (X_5)	1—Yes, 0—No
	Number of family members engaged in agriculture (X_6)	Actual labor force
	Proportion of agricultural income in family income (X_7)	Actual proportion
	Annual household income per capita (X_8)	Annual household income per capita
Land resource endowment	Arable area of the family (X_9)	Actual arable area
	Land fragmentation degree (X_{10})	Land fragmentation distribution
Input of agricultural means of production	Fertilizer input intensity (X_{11})	Fertilizer input of unit area
	Intensity of pesticide application (X_{12})	Amount of pesticide input in unit area
	Applying organic fertilizer or not (X_{13})	1—Yes, 0—No
Environment cognition	Impact of pesticide and fertilizer on environment (X_{14})	1—Negative impact, 0—No impact
	Environmental awareness (X_{15})	1—Never concern, 2—Occasional concern, 3—Frequently concern
Agricultural technology level	Attending agricultural technical training or not (X_{16})	1—Yes, 0—No

2.4.3. Determination of Data Processing and Regression Approach

It is necessary to first take the logarithm of continuous variable to eliminate the impact of variable heteroscedasticity [42]. In order to eliminate the multicollinearity among variables [43], stepwise regression approach is adopted for model estimation.

3. Assessment on Heavy Metal Pollution of Arable Soil

3.1. Characteristic Analysis on Heavy Metal Content of Arable Soil

Descriptive statistical analysis is performed on heavy metal content of 101 soil samples taken from the studied area, and the results are given in Table 2. The average element content of heavy metal Cr, Ni, Cu, Zn, Cd, and Pb is 53.802 mg·kg^{-1}, 28.560 mg·kg^{-1}, 44.376 mg·kg^{-1}, 125.395 mg·kg^{-1}, 0.350 mg·kg^{-1} and 50.360 mg·kg^{-1} respectively. Compared with class II standard (pH > 7.5) in Environmental Quality Standard for Soils (GB15618 -1995), in the studied area, the content of Ni, Cu, Zn, Cd, and Pb, except Cr, in agricultural soil samples exceeds the standard by 1.98%, 1.98%, 4.95%, 6.93% and 2.97% respectively, indicating varying degrees of Ni, Cu, Zn, Cd, and Pb pollution in the studied area. The average content of Ni, Cu, Zn, Cd, and Pb, except Cr, is higher than background value of soil in Henan, showing that heavy metal in arable soil of the studied area presents obvious accumulation trend due to great influence of agricultural production and other human activities. The standard deviation of Zn and Pb are significantly higher than other heavy metal elements, which shows that the sample data are discrete and have major variability. Variable coefficient can reflect the average variation degree of heavy metal element content and that of six heavy metals is: Pb > Zn > Cd > Cu > Ni > Cr. According to classification of variation degree [44], Cr, Ni, and Cu (0.243, 0.268 and 0.345) show moderate variation (0.15 < Cv < 0.36), while Zn, Cd, and Pb (1.034, 0.783 and 1.416) show high variation (Cv > 0.36), especially the variable coefficient of Zn and Pb are all above 1. This indicates that allogenic material enters into some sampling points, which is strongly influenced by human activities.

Table 2. Statistics on Heavy Metal Content of Arable Soil in the Studied Area (*n* = 101).

Element	Content/ mg·kg^{-1}	Mean/ mg·kg^{-1}	Standard Deviation/ mg·kg^{-1}	Variable Coefficient	Background Value of Soil in Henan/ mg·kg^{-1}	National Soil Environment Quality Standard (Class II)/ mg·kg^{-1}
Cr	17.598~123.977	53.802	13.077	0.243	63.800	250
Ni	17.462~67.362	28.560	7.640	0.268	26.700	60
Cu	23.795~118.839	44.376	15.300	0.345	19.700	100
Zn	55.142~795.170	125.395	129.677	1.034	60.100	300
Cd	0.133~2.321	0.350	0.274	0.783	0.074	0.6
Pb	18.338~386.778	50.360	71.331	1.416	19.600	350

By comparing the average heavy metal content in soil in various types of farmland (Figure 3), we can see that the average content of Cr, Ni, Cu, and Cd is the highest in vegetable greenhouse. The sequence of Cr content from high to low is vegetable greenhouse > traditional crop farmland > garlic land, while the average content of Ni, Cu, and Cd is vegetable greenhouse > garlic land > traditional crop farmland. In garlic land, the content of Zn and Pb is higher than that in vegetable greenhouse and traditional crop farmland, in the following sequence: garlic land > vegetable greenhouse > traditional crop farmland (in terms of Zn); garlic land > traditional crop farmland > vegetable greenhouse (in terms of Pb). From a whole view, the content of heavy metal in vegetable greenhouse and garlic land is higher than that in traditional crop farmland.

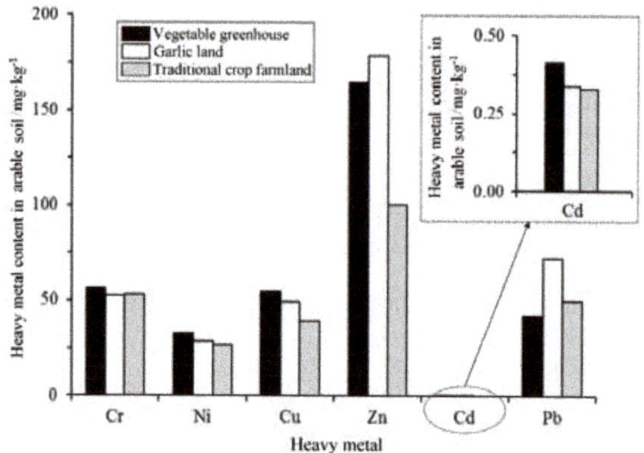

Figure 3. Average content of soil heavy metal in different types of farmland.

3.2. Analysis on Characteristics of Heavy Metal Pollution of Arable Soil

Single factor pollution index (P_i) and Nemerow Pollution Index (P_N) of heavy metal elements are calculated by formulas (1) and (2). Range of single factor pollution index of Cr, Ni, Cu, Zn, Cd, and Pb is 0.070~0.496, 0.291~1.123, 0.238~1.188, 0.184~2.651, 0.222~3.869 and 0.052~1.105, with the average sequence of Cd (0.584) > Ni (0.476) > Cu (0.444) > Zn (0.418) > Cr (0.215) > Pb (0.144). It can thus be seen that the average pollution index of all six heavy metal elements is smaller than 0.7, i.e. within the warning limit. Nemerow Pollution Index is 0.256~2.829 with an averaging of 0.578, which shows heavy metal pollution of some sampling points in the studied area, but unpolluted state on the whole.

Frequency distribution of pollution index can further show the pollution of heavy metal elements in the studied area (Table 3). Seen from frequency distribution of single factor pollution index (P_i), the frequency of Cr in unpolluted state among the six heavy metal elements is 100%. But Ni, Cu, and Pb show a certain degree of slight pollution and the frequency is 1.98%, 1.98% and 2.97% respectively, with 1.98%, 2.97%, and 1.98% sampling points reaching the warning state; Zn shows moderate and slight pollution in 2.97% and 1.98% sampling points with 3.96% reaching the warning state. Through comparison, we can learn that the pollution frequency and degree of Cd are the highest, that is, severe pollution occurs in 0.99% arable soil, moderate and slight pollution occurs in 0.99% and 4.95% soil, and 13.86% sampling points reach the warning state. Seen from frequency distribution of Nemerow Pollution Index (P_N), the frequency of moderate and slight pollution in sampling points is 0.99% and 5.94%, with 7.92% reaching the warning state, while other sampling points are in unpolluted state.

Table 3. Frequency distribution of soil heavy metal element pollution index.

Pollution Degree	Frequency Distribution of Single Factor Pollution Index (P_i)/%						Frequency Distribution of Nemerow Pollution Index (P_N)/%
	Cr	Ni	Cu	Pb	Zn	Cd	
Unpolluted ($P \leq 0.7$)	100.00	96.04	95.05	95.05	91.09	79.21	85.15
Warning state ($0.7 < P \leq 1$)	0.00	1.98	2.97	1.98	3.96	13.86	7.92
Slight pollution ($1 < P \leq 2$)	0.00	1.98	1.98	2.97	1.98	4.95	5.94
Moderate pollution ($2 < P \leq 3$)	0.00	0.00	0.00	0.00	2.97	0.99	0.99
Heavy pollution ($P > 3$)	0.00	0.00	0.00	0.00	0.00	0.99	0.00

Note: P stands for single factor pollution index (P_i) or Nemerow Pollution Index (P_N).

In order to analyze the average pollution level of various arable soils, the average values of single factor pollution index and Nemerow pollution index are calculated. According to the calculation results of the average values of single factor pollution index (Figure 4), we can learn that the average value of single factor pollution index of heavy metals in three types of arable soil is smaller than 1, indicating no heavy metal pollution of each type on the whole. Pollution index of Pb is the lowest, and that in vegetable greenhouse, garlic and traditional crop farmland is 0.120, 0.207, and 0.143 respectively, while that of Cd, Cu, Ni, and Zn is relatively high. The average value of Nemerow pollution index of heavy metal in three types of arable soil is: vegetable greenhouse (0.703) > garlic land (0.624) > traditional crop farmland (0.519), with no heavy metal pollution on the whole. However, it is worth noting that, the heavy metal pollution index of vegetable greenhouse with high land use intensity has reached the warning state and is higher than that of garlic land and traditional crop farmland, so it is necessary to properly adjust the management and planting strategy of agricultural production in future. It is discovered in investigation that organic fertilizer such as chicken manure, pig manure and excrements of other livestock is widely applied as base fertilizer in vegetable cultivation and some rural households also apply it in garlic land and traditional crop farmland, but excrements of livestock generally contain excess heavy metals like Cd, Cu, Zn, Pb, Cr, and Ni [45,46]. In addition, it is found from questionnaire that the application of fertilizers and pesticides in vegetable greenhouse is also higher than that in garlic land and traditional crop farmland, thus the heavy metal content of various arable soils is closely related to the input of fertilizer, pesticide, farmyard manure, etc.

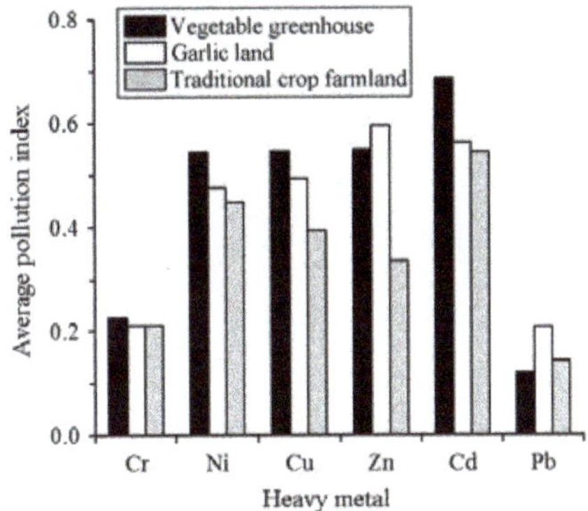

Figure 4. Average pollution index of soil heavy metal in different types of farmland.

4. Impact Degree of Rural Households' Behaviors on Heavy Metal Pollution of Arable Soil

The stepwise regression model is set up by taking Nemerow Pollution Index as the dependent variable and the rural households' behaviors indicator as the independent variable. In order to better distinguish the impact of rural households' behaviors on heavy metal pollution of soil in different types of arable land, model estimation is first conducted on all samples in this study, and then regression is performed on samples of vegetable greenhouse, garlic land and traditional crop farmland respectively, with the results given in Table 4 (in the table, models I, II, III, and IV respectively show the regression results of all samples from vegetable greenhouse, garlic land and traditional crop farmland).

(1) Overall sample (model I). The stepwise regression result shows that variables included into the model are: applying organic fertilizer or not (X_{13}), agricultural technology level (X_{16}), fertilizer input intensity (X_{11}) and member of agricultural cooperative or not (X_5). The following equation (Equation (4)) is established:

$$y = -1.781 + 0.21X_{13} - 0.163X_{16} + 0.241X_{11} - 0.179X_5 \tag{4}$$

Among them, X_{13} and X_{11} impose obvious positive effect on heavy metal pollution level of arable soil, while X_{16} and X_5 impose negative effect on it. However, the overall goodness of fit of the equation is relatively low, and the regression results of different types of land are analyzed below.

Table 4. Results of the regression analysis on the impact of rural households' behaviors on heavy metal pollution of arable soil.

Model	Independent Variable	Regression Coefficient	Std. Error	t-Statistic	Adjusted R-squared	Prob. (F-statistic)
I	Constant	−1.781	0.393	−4.532 ***	0.159	0.000
	X_{13}	0.210	0.087	2.406 **		
	X_{16}	−0.163	0.080	−2.027 **		
	X_{11}	0.241	0.092	2.628 **		
	X_5	−0.179	0.099	−1.808 *		
II	Constant	−2.083	1.043	−1.998 *	0.339	0.007
	X_{16}	−0.341	0.171	−1.988 *		
	X_5	−0.469	0.184	−2.552 **		
	X_{12}	0.315	0.163	1.927 *		
III	Constant	−0.258	0.403	−0.640	0.396	0.054
	X_{13}	0.680	0.283	2.403 **		
	X_{15}	−0.348	0.178	−1.958 *		
IV	Constant	−1.673	0.517	−3.236 ***	0.231	0.000
	X_8	0.239	0.061	3.898 ***		
	X_{10}	0.166	0.071	2.330 **		
	X_{11}	0.275	0.122	2.265 **		

Note: ***, ** and * respectively represent that it is significant at 1%, 5% and 10% level.

(2) Vegetable greenhouse (model II). The regression result shows that the key factors affecting heavy metal pollution level of vegetable greenhouse are: agricultural technology level, member of agricultural cooperative or not, and pesticide input intensity. The following equation (Equation (5)) is established:

$$y = -2.083 - 0.341X_{16} - 0.469X_5 + 0.315X_{12} \tag{5}$$

Agricultural technology level (X_{16}) of rural households imposes negative effect on heavy metal pollution level of vegetable greenhouse, that is, rural households' participation in agricultural technical training is conductive to lowering the pollution level. Rural households having participated in agricultural technical training will acquire more knowledge and information on agricultural inputs and agricultural production, so that they can "just shoot the problem" in deciding the type and dosage of agricultural inputs [32], which can help to avoid excess input of agricultural chemicals containing heavy metal element. Among the investigated samples, some rural households have rich experiences in planting with arched shed or other simple shed but lack knowledge in planting and management of greenhouse vegetable emerging in recent years, so the training and guidance of related technologies (e.g.: control of greenhouse temperature and humidity, base fertilizer, top application, pesticide application rate and cycle, irrigation cycle, etc.) is of vital importance to guarantee the sustainability of vegetable cultivation.

Member of agricultural cooperative or not (X_5) imposes obvious negative effect on heavy metal pollution level of farmland soil, indicating that agricultural cooperative and other effective agricultural organization mode can lower the heavy metal pollution level of soil in greenhouse agriculture. The emergence of agricultural cooperative relieves or eliminates the difficulty of rural households participating in large market [47] to a certain extent, hence beneficial for rural households' acquiring large-scale returns, enhancing market competitiveness and avoiding risks. In the investigated area, some vegetable cooperatives play an important role in vegetable cultivation and selling, which improves the technological level and quality of rural households by providing technical support and consultation services to enable more scientific fertilizer (pesticide) selection and application, thus reducing or avoiding irrational agricultural input and lowering the risk of heavy metal pollution in soil.

Heavy metal pollution level of vegetable greenhouse is in direct proportion to pesticide input intensity (X_{12}). Repeated application of pesticide for insect and disease prevention is required in the growth cycle of vegetable, but the catalyst used in synthetic raw material of pesticide contains heavy metals like Cu, Pb, and Cr [4,48]. During pesticide spraying, there will be a part left on the ground or entering into the soil together with crop leaves, resulting in heavy metal accumulation in soil. Therefore, pesticide application rate bears a positive correlation with heavy metal pollution level of soil.

(3) Garlic land (model III). The regression result shows that application of organic fertilizer and rural households' environmental awareness are the key factors affecting heavy metal pollution level of soil in garlic land and the two factors impose positive and negative effect on the variable respectively. However, seen from the overall fitting effect of the model, observed value of F-statistics fails to pass the significance test ($p = 0.054 > 0.05$), which presents low reliability of statistical results. This may be attributed to insufficient samples of garlic land, so this model will not be analyzed.

(4) Traditional crop farmland (model IV). The key factors affecting the heavy metal pollution level of traditional crop farmland are: annual per capital income of family, land fragmentation degree and fertilizer input intensity successively. Its regression equation (Equation (6)) is:

$$y = -1.673 + 0.239X_8 + 0.166X_{10} + 0.275X_{11} \tag{6}$$

Annual household income per capita (X_8) imposes obvious positive effect on heavy metal pollution level of traditional crop farmland, viz. the higher the annual per capital income of family is, the higher the pollution level will be. Families with high-income level mostly acquire income by being engaged in industry or project or other part-time jobs, and relatively pay less attention to agricultural production, thus leading to irrational selection and input of agricultural means of production.

Land fragmentation degree (X_{10}) bears significant positive correlation with heavy metal pollution level of farmland soil, viz. in direct proportion to pollution level. As a prominent feature in traditional agricultural production of China [49], the land fragmentation may increase the production cost (labor cost, fertilizer cost, etc.) of farmers [50], which increases the dosage of heavy metal source to a certain degree and highlights the necessity of large-scale operation of land in a sense.

Fertilizer input intensity (X_{11}) imposes positive effect on heavy metal pollution level of traditional crop farmland, that is, the higher the fertilizer input intensity is, the higher the heavy metal pollution level will be. The commonly sold fertilizers in market of China, such as urea, superphosphate and compound fertilizer contain different levels of heavy metal element [51], and these elements may accumulate in soil along with long-term application of fertilizers. Moreover, some studies point out that fertilizer application rate in China at present is over the optimal rate in economic sense [52], and excess fertilizer application leads to heavy metal element over the standard, which aggravates the pollution.

5. Discussion

The existing related studies on rural households' behavior mainly analyzed the laws and problems in social and economic development of rural area at a microscopic scale, for example, rational peasant school [27] regarded peasant as the "economic man" and stressed discussing peasants' economic behavior; Shi Qinghua [28] concerned peasants' sociality and emphasizes that peasant would give equal consideration to social or collective interests while pursuing personal interests; Li Xiaojian [29] thought that peasants' behaviors were in close relation to the environment from a geographic view, etc. It can thus be seen that the existing peasants' behavior theory lacks organic connection of the production, consumption, technical application, cognition and labor supply of peasants with the environmental effect [26]. However, this study found that different planting types and planting modes had different effects on soil heavy metal pollution.

In terms of the heavy metal pollution degree of various arable soils, the pollution degree of vegetable greenhouse under intensive planting mode is relatively high due to more application of agro-chemicals and organic fertilizer—one of the reasons. Related researches indicate that some commonly used chemical fertilizer, pesticide and organic fertilizer contain different amounts of heavy metal element, and the unreasonable application of them is one of the main reasons for the increase of heavy metal content in arable soil. As a result, implementing strict control over the production standards of associated industries including livestock feed and fertilizer and promoting environment-friendly transformation of all links in the whole industry is an effective means to lower heavy metal pollution of soil.

Moreover, the regression results show that land fragmentation degree bears positive correlation with heavy metal pollution level of arable soil, which highlights the necessity of moderate scale management of arable soil. Thus, rural households are encouraged to carry out various land circulations and combine the patch into a large one. In "three powers separation" (separate setting of property, contracting right, and management right of agricultural land) proposed in the new-round land reform at the end of 2014, and "land contracting period extended for 30 years more" proposed in "The 19th National Congress of the Communist Party of China", land circulation of rural households is encouraged and important measures for moderate scale management of soil are promoted, which will further improve the environmental effect of agricultural planting. At the same time, we also find that agricultural technical training and the guidance provided by the agricultural cooperative can help rural households acquire knowledge about agricultural production and planting management. This will help rural households avoid excessive use of agro-chemicals, thus reducing the probability of heavy metal pollution in arable soil. By providing agricultural means of production, agricultural products selling, processing, transportation and storage as well as technology and information relating to agricultural production for its members, agricultural cooperative effectively realizes the merging of dispersed rural households and large market, thus improving the overall large-scale agricultural and environmental benefits. Therefore, agricultural departments at county and town levels shall strengthen the technical training on rural households, perfect the operation and management mechanism of agricultural cooperative and encourage the rural households to join in the cooperative, so as to enhance their scientific cognition and realize the modern agricultural development with equal consideration to environmental protection and quality of agricultural products.

6. Conclusions

Empirical analysis is performed on the impact of rural households' behaviors on heavy metal pollution of arable soil based on investigation and experiment data to obtain the following conclusions.

Rural households' land utilization mode affects soil heavy metal content, for example, heavy metal pollution degree of soil for intensive planting is higher than that of traditional planting, viz. vegetable greenhouse > garlic land > traditional crop farmland. This means that the risk of heavy metal pollution of soil is also enlarged during transformation from traditional agriculture to modern agriculture.

The higher the land fragmentation degree is, the higher the heavy metal content of soil will be. This indicates that moderate scale management of land is conductive to lowering the heavy metal content of arable soil. Under proper environmental and economic conditions, moderate scale management of land can lower the land dispersity and enable the optimal combination and effective operation of land, fund, equipment, operation management and information, hence achieving maximum economic benefit and environmental effect.

Excess application of fertilizer, pesticide and organic fertilizer by rural households may result in heavy metal pollution of soil, while agricultural technical training organized by government department can promote the technical level of rural households and enable them to be more scientific and rational in fertilizer (pesticide) selection and application, thus reducing or avoiding heavy metal pollution of soil. Meanwhile, the foundation of agricultural cooperative changes the organization mode of agricultural production.

For soils under different cropping patterns, the pollution level of heavy metal varies. Under the influences of agricultural production and other human activities, heavy metal elements Cr, Ni, Cu, Zn, Cd, and Pb in soil of the studied area accumulate to a certain extent, causing different pollution. The average P_i of six heavy metals is Cd > Ni > Cu > Zn > Cr > Pb. Among them, Cd pollution frequently appears and even reaches severe level in some sampling points. As a result, various pollution reduction programs shall be developed for different crop types and heavy metal pollution levels, and measures such as formulated fertilization and adjustment to local conditions shall be adopted to achieve harmony and sustainable development of human-nature system.

Author Contributions: E.L. conceived the research, designed the analytical framework and revised the manuscript of this study; S.R. performed the statistical analysis and wrote the manuscript; Q.D. participated in the process of questionnaire design and survey; H.H. and S.L. carried out the determination of heavy metal content. All authors have read and approved the final manuscript.

Funding: This study was supported by National Natural Science Foundation of China (41471105, 41430637), Natural Science Foundation of Henan Province (182300410144), Program for Innovative Research Team (in Science and Technology) in University of Henan Province (16IRTSTHN012) and Key Project of the Humanities and Social Sciences Research Base in Ministry of Education (15JJDZONGHE008).

Conflicts of Interest: The authors declare no conflict of interest.

References

1. Eziz, M.; Mamut, A.; Mohammad, A.; Ma, G.F. Assessment of heavy metal pollution and its potential ecological risks of farmland soils of oasis in Bosten Lake Basin. *Acta Geogr. Sin.* **2017**, *72*, 1680–1694.

2. Ministry of Environmental Protection; Ministry of Land and Resources. Report on the National Soil Pollution Survey of China. Available online: http://www.zhb.gov.cn/gkml/hbb/qt/201404/t20140417_270670.htm (accessed on 20 December 2017).

3. Ma, J.H.; Ma, S.Y.; Chen, Y.Z. Migration and accumulation of heavy metals in soil-crop-hair system in a sewage irrigation area, Henan, China. *Acta Sci. Circum.* **2014**, *34*, 1517–1526.

4. Zhu, C.Y.; Wang, T.Y.; Xu, L.; Pang, B.; Li, Q.F.; Lv, Y.L. Contamination and risk assessment of heavy metals in soils from pesticide factory. *China Pop. Resour. Environ.* **2013**, *23*, 67–72.

5. Dudas, M.J.; Pawluk, S. Heavy metals in cultivated soils and in cereal crops in Alberta. *Can. J. Soil Sci.* **1977**, *57*, 329–339. [CrossRef]

6. Fan, Y.; Li, H.; Xue, Z.; Zhang, Q.; Cheng, F. Accumulation characteristics and potential risk of heavy metals in soil-vegetable system under greenhouse cultivation condition in Northern China. *Ecol. Eng.* **2017**, *102*, 367–373. [CrossRef]

7. Nicholson, F.A.; Smith, S.R.; Alloway, B.J.; Carlton-Smith, C.; Chambers, B.J. An inventory of heavy metals inputs to agricultural soils in England and Wales. *Sci. Total Environ.* **2003**, *311*, 205–219. [CrossRef]

8. Ustyak, S.; Petrikova, V. Heavy metal pollution of soils and crops in Northern Bohemia. *Appl. Geochem.* **1996**, *11*, 77–80. [CrossRef]

9. Blanco, G.; Sergio, F.; Frías, Ó.; Salinas, P.; Tanferna, A.; Hiraldo, F. Integrating population connectivity into pollution assessment: overwintering mixing reveals flame retardant contamination in breeding areas in a migratory raptor. *Environ. Res.* **2018**, *166*, 553–561. [CrossRef] [PubMed]

10. Yang, Q.; Li, Z.; Lu, X.; Duan, Q.; Huang, L.; Bi, J. A review of soil heavy metal pollution from industrial and agricultural regions in China: Pollution and risk assessment. *Sci. Total Environ.* **2018**, *642*, 690–700. [CrossRef] [PubMed]

11. Baran, A.; Wieczorek, J.; Mazurek, R.; Krzysztof, U.; Klimkowiczpawlas, A. Potential ecological risk assessment and predicting zinc accumulation in soils. *Environ. Geochem. Health* **2018**, *40*, 435–450. [CrossRef] [PubMed]

12. Briki, M.; Ji, H.B.; Li, C.; Ding, H.J.; Gao, Y. Characterization, distribution, and risk assessment of heavy metals in agricultural soil and products around mining and smelting areas of Hezhang, China. *Environ. Monit. Assess.* **2015**, *187*, 1–21. [CrossRef] [PubMed]

13. Keshavarzi, B.; Hassanaghaei, M.; Moore, F.; Rastegari Mehr, M.; Soltanian, S.; Lahijanzadeh, A.R.; Sorooshian, A. Heavy metal contamination and health risk assessment in three commercial fish species in the persian gulf. *Mar. Pollut. Bull.* **2018**, *129*, 245–252. [CrossRef] [PubMed]

14. Song, J.X.; Zhu, Q.; Jiang, X.S.; Zhao, H.Y.; Liang, Y.H.; Luo, Y.X.; Wang, Q.; Zhao, L.L. GIS-based heavy metals risk assessment of agricultural soils—A case study of Baguazhou, Nanjing. *Acta Pedol. Sin.* **2017**, *54*, 81–91.

15. Guo, W.; Sun, W.H.; Zhao, R.X.; Zhao, W.J.; Fu, R.Y.; Zhang, J. Characteristic and evaluation of soil pollution by heavy metal in different functional zones of Hohhot. *Environ. Sci.* **2013**, *34*, 1561–1567.

16. Tomlinson, D.L.; Wilson, J.G.; Harris, C.R.; Jeffrey, D.W. Problems in the assessment of heavy metals levels in estuaries and the formation of pollution index. *Helgol. Meeresunters.* **1980**, *33*, 566–575. [CrossRef]

17. Hakanson, L. An ecological risk index for aquatic pollution control: A sedimentological approach. *Water Res.* **1980**, *14*, 975–1001. [CrossRef]

18. Zeng, F.F.; Wei, W.; Li, M.S.; Huang, R.X.; Yang, F.; Duang, Y.Y. Heavy metal contamination in rice-producing soils of Hunan Province, China and potential health risks. *Int. J. Environ. Res. Public Health* **2015**, *12*, 15584–15593. [CrossRef] [PubMed]

19. Antibachi, D.; Kelepertzis, E.; Kelepertsis, A. Heavy metals in agricultural soils of the Mouriki-Thiva Area (Central Greece) and environmental impact implications. *Soil Sediment Contam.* **2012**, *21*, 434–450. [CrossRef]

20. Bhuiyan, M.A.H.; Parvez, L.; Islam, M.A.; Dampare, S.B.; Suzuki, S. Heavy metal pollution of coal mine-affected agricultural soils in the northern part of Bangladesh. *J. Hazard. Mater.* **2010**, *173*, 384–392. [CrossRef] [PubMed]

21. Huang, S.S.; Liao, Q.L.; Hua, M.; Wu, X.M.; Bi, K.S.; Yan, C.Y.; Chen, B.; Zhang, X.Y. Survey of heavy metal pollution and assessment of agricultural soil in Yangzhong district, Jiangsu Province, China. *Chemosphere* **2007**, *67*, 2148–2155. [CrossRef] [PubMed]

22. Bai, L.Y.; Zeng, X.B.; Li, L.F.; Peng, C.; Li, S.H. Effects of land use on heavy metal accumulation in soils and source analysis. *Sci. Agric. Sin.* **2010**, *43*, 96–104. [CrossRef]

23. Li, L.F.; Zeng, X.B.; Bai, L.Y. Accumulation of copper and zinc in soils under different agricultural and natural field. *Acta Ecolo. Sin.* **2008**, *28*, 4372–4380.

24. Hammer, D.; Kayser, A.; Keller, C. Phytoextraction of Cd and Zn with Salix viminalis in field trials. *Soil Use Manage.* **2003**, *19*, 187–192. [CrossRef]

25. Li, E.L.; Liu, P. The response capability of households to urbanization in undeveloped rural areas. *China Rural Surv.* **2010**, *31*, 34–44.

26. Wu, X.L. *The Study of Farm Household Behavior and Agricultural Non-Point Source Pollution: Based on the Survey of 189 Households in the Eleven Counties in Jiangxi Province*; Jiangxi Agricultural University: Nanchang, China, 2011.

27. Schultz, T.W. *Transforming Traditional Agriculture*; Yale University Press: New Haven, CT, USA, 1964.

28. Shi, Q.H. *Study on Household Economic Growth and Development*; China Agricultural Press: Beijing, China, 1999.

29. Li, X.J. Reductionism and Geography of Rural Households. *Geogr. Res.* **2010**, *29*, 767–777.

30. Wang, Y.; Yang, J.; Liang, J.; Qiang, Y.; Fang, S.; Gao, M.; Fan, X.; Yang, G.; Zhang, B.; Feng, Y. Analysis of the environmental behavior of farmers for non-point source pollution control and management in a water source protection area in china. *Sci. Total. Environ.* **2018**, *227*, 1126–1135. [CrossRef] [PubMed]

31. Liang, L.T.; Zhai, B.; Fan, P.F. Environmental influence mechanism of household production behavior: A case of traditional agricultural areas in Henan Province. *J. Nanjing Agric. Univ. (Soc. Sci.)* **2016**, *16*, 145–153.

32. Ying, R.Y.; Zhu, Y. The impact of agricultural technical training on farmers' agrochemical use behavior: evidence from experimental economics. *China Rural Survey* **2015**, *36*, 50–59.

33. Aregay, F.A.; Minjuan, Z.; Tao, X. Knowledge, attitude and behavior of farmers in farmland conservation in china: an application of the structural equation model. *J. Environ. Plan. Man.* **2018**, *61*, 1–23. [CrossRef]

34. Zhang, L.; Li, X.; Yu, J.; Yao, X. Toward cleaner production: what drives farmers to adopt eco-friendly agricultural production? *J. Clean Prod* **2018**, *184*, 550–558. [CrossRef]

35. Cao, Yu.; Du, Z.; Wan, G. Family farms' choice of cropping behavior under different pesticide residue standards. *Systems Eng. Theor. Practice* **2018**, *38*, 1492–1501.

36. Hou, J.D.; Lv, J.; Yin, W.F. Effects of farmer households' production and operation behaviors on rural eco-environment. *China Pop. Resour. Environ.* **2012**, *22*, 26–31.

37. Wong, Z.L. Research progess and reviews of households, theory and application. *Issues Agric. Econ.* **2008**, *29*, 29–93.

38. Ministry of Environmental Protection. *Soil Environmental Quality Standard (GB15618-1995)*; Standards Press of China: Beijing, China, 1997.

39. Zheng, G.Z. Investigation and assessment on heavy metal pollution of farming soil in the Jinghe River Basin. *Arid Zone Res.* **2008**, *25*, 626–630.
40. Gong, M.D.; Zhu, Y.Q.; Gu, Y.Q.; Li, S.Y.; Jianati, T. Evaluation on heavy metal pollution and its risk in soils from vegetable bases of Hangzhou. *Environ. Sci.* **2016**, *37*, 2329–2337.
41. Wang, Y.Q.; Bai, Y.R.; Wang, J.Y. Distribution of soil heavy metal and pollution evaluation on the different sampling scales in farmland on Yellow River Irrigation Area of Ningxia: A case study in Xingqing County of Yinchuan City. *Environ. Sci.* **2014**, *35*, 271–2720.
42. Hao, F.Y.; Sun, X.G.; Du, F.L. Empirical analysis on the influencing factors of householders' willingness of participating in arable land circulation. *J. Inner Mongolia Agric. Univ. (Soc. Sci.)* **2016**, *18*, 31–36.
43. Zhong, X.L.; Li, J.T.; Feng, Y.F.; Li, J.G.; Liu, H.H. Farmland transfer willingness and behavior in the perspective of farm household cognition in Guangdong Province. *Resour. Sci.* **2013**, *35*, 2082–2093.
44. Wang, Y.Q.; Bai, Y.R.; Wang, J.Y. Distribution of urban soil heavy metal and pollution evaluation in different functional zones of Yinchuan City. *Environ. Sci.* **2016**, *37*, 710–716.
45. Liu, Q.D.; Jiang, D.H.; Gao, L.J.; Li, J.J.; Sun, Q.P.; Liu, B.S. Research progress on heavy metal accumulation and migration of livestock dung organic fertilizer in soil-vegetable system. *Chin. J. Soil Sci.* **2014**, *45*, 252–256.
46. Wang, M.; Li, S.T. Heavy metals in fertilizers and effect of the fertilization on heavy metal accumulation in soils and crops. *J. Plant Nutr. Fert.* **2014**, *20*, 466–480.
47. Cai, R. Contractual arrangements for agricultural cooperatives: The allocation of the decision-making power in production-an empirical analysis from perspective of householders. *Chin. Rural Econ.* **2013**, *29*, 60–70.
48. Shi, N.N.; Ding, Y.F.; Zhao, X.F.; Wang, S.Q. Heavy metals content and pollution risk assessment of cropland soils around a pesticide industrial park. *Chin. J. Appl. Ecol.* **2010**, *21*, 1835–1843.
49. Xu, Y.G.; Yang, G.Q.; Wen, G.H. Impacts of arable land fragmentation on land use efficiency: An empirical analysis based on farms of different scales. *Res. Agric. Modern.* **2017**, *38*, 688–695.
50. Lu, H.; Hu, H. Does land fragmentation increase agricultural production costs? A microscopic investigation from Jiangsu Province. *Econ. Rev.* **2015**, *36*, 129–140.
51. Chen, L.H.; Ni, W.Z.; Li, X.L.; Sun, J.B. Investigation of heavy metal concentrations in commercial fertilizers commonly-used. *J. Zhejiang Sci-Tech Univ.* **2009**, *26*, 223–227.
52. Chou, H.G.; Luan, H.; Li, J.; Wang, Y.J. The effect of risk aversion on the behavior of farmers' fertilizer application. *Chin. Rural Econ.* **2014**, *30*, 85–96.

 *sustainability*

Article

Land Use/Cover Change Effects on River Basin Hydrological Processes Based on a Modified Soil and Water Assessment Tool: A Case Study of the Heihe River Basin in Northwest China's Arid Region

Xin Jin, Yanxiang Jin * and Xufeng Mao

Key Laboratory of Physical Geography and Environmental Processes, School of Geographical Science, Qinghai Normal University, Xining 810016, China; jinx13@lzu.edu.cn (X.J.); maoxufeng@yeah.net (X.M.)
* Correspondence: jinyx13@lzu.edu.cn

Received: 4 January 2019; Accepted: 14 February 2019; Published: 19 February 2019

Abstract: Land use/cover change (LUCC) affects canopy interception, soil infiltration, land-surface evapotranspiration (ET), and other hydrological parameters during rainfall, which in turn affects the hydrological regimes and runoff mechanisms of river basins. Physically based distributed (or semi-distributed) models play an important role in interpreting and predicting the effects of LUCC on the hydrological processes of river basins. However, conventional distributed (or semi-distributed) models, such as the soil and water assessment tool (SWAT), generally assume that no LUCC takes place during the simulation period to simplify the computation process. When applying the SWAT, the subject river basin is subdivided into multiple hydrologic response units (HRUs) based on the land use/cover type, soil type, and surface slope. The land use/cover type is assumed to remain constant throughout the simulation period, which limits the ability to interpret and predict the effects of LUCC on hydrological processes in the subject river basin. To overcome this limitation, a modified SWAT (LU-SWAT) was developed that incorporates annual land use/cover data to simulate LUCC effects on hydrological processes under different climatic conditions. To validate this approach, this modified model and two other models (one model based on the 2000 land use map, called SWAT 1; one model based on the 2009 land use map, called SWAT 2) were applied to the middle reaches of the Heihe River in northwest China; this region is most affected by human activity. Study results indicated that from 1990 to 2009, farmland, forest, and urban areas all showed increasing trends, while grassland and bare land areas showed decreasing trends. Primary land use changes in the study area were from grassland to farmland and from bare land to forest. During this same period, surface runoff, groundwater runoff, and total water yield showed decreasing trends, while lateral flow and ET volume showed increasing trends under dry, wet, and normal conditions. Changes in the various hydrological parameters were most evident under dry and normal climatic conditions. Based on the existing research of the middle reaches of the Heihe River, and a comparison of the other two models from this study, the modified LU-SWAT developed in this study outperformed the conventional SWAT when predicting the effects of LUCC on the hydrological processes of river basins.

Keywords: land use/cover change; SWAT; hydrological processes

---

1. Introduction

Resulting from the long-term interaction between human needs and natural processes [1–3], land use/cover change (LUCC) affects canopy interception, soil infiltration, land-surface evapotranspiration (ET), and other hydrological parameters during rainfall, which in turn affects the hydrological regimes and runoff mechanisms of river basins [4–7]. The effects of LUCC on hydrological processes vary based on unique site characteristics. For example, the presence of forest is related to the occurrence of different hydrological functions under a region's unique climate, soil type, geomorphology, and topography [8–11].

The U.S. Geological Survey (USGS) identified the study of 'land use and land cover change rates, causes, and consequences' (including their effect on hydrological processes) as one of its seven major goals in its 2013 *Climate and Land Use Change Science Strategy*, which is to be implemented over a period of 10 years. Moreover, Future Earth—a global platform sponsored by the International Council for Science (ICSU), the International Conference on Sustainability Science (ICSS), and other international organizations—identified a 'dynamic planet' theme as one of its three major research areas, which aims to understand the interactions between natural and social components and their effect on the Earth's systems. As a result of these targeted areas of focus, LUCC effects on hydrological processes in river basins has garnered recent attention and emerged as a critical frontier of international geo-scientific research [1–4].

Several methods exist for the determination of LUCC effects on hydrological processes in river basins, including (1) experimental paired-watershed methods, (2) lumped hydrological models, and (3) distributed hydrological models [12–14]. The experimental paired-watershed method is typically applied only to small river basins, requires a long-term study period, and has limited comparability [15]. Lumped hydrological models, which treat the entire river basin as a single unit, often fail to reflect the variability of river basin parameters and the associated regional differences in LUCC effects on hydrological processes [16–18]. Physically based distributed (or semi-distributed) models more accurately reflect the spatial variability of hydrological processes than lumped models and thus play an important role in interpreting and predicting LUCC effects on hydrological processes in river basins [19–23]. However, distributed models generally assume that no LUCC occurs during the simulation period to simplify the computation process.

The soil and water assessment tool (SWAT), supported by the U.S. Department of Agriculture, is an example of a semi-distributed hydrological model. Because of its open-source code, strong functionality, and excellent simulation results in multiple watersheds, the SWAT has been applied worldwide [24–26]. When applying the SWAT, the subject river basin is subdivided into multiple hydrologic response units (HRUs) based on the land use/cover type, soil type, and surface slope. The land use/cover type is assumed to remain constant throughout the simulation period [27], which limits the ability to interpret and predict the effects of LUCC on hydrological processes in the subject river basin.

To overcome this limitation, select researchers have divided the entire simulation period into uniform time intervals (e.g., 5 years intervals) and performed interval simulations using land use/cover data from a single year within each interval [21–23]. Although this method provides some of the necessary variability in LUCC, it also complicates the model development and simulation processes and may fail to reflect year-to-year changes in land use/cover.

This study sought to improve upon these prior efforts. In this study, a modified SWAT (LU-SWAT) was developed that incorporates annual land use/cover data to simulate LUCC effects on hydrological processes under different climatic conditions. To validate this approach, this modified model together with two other conventional SWAT models (SWAT 1 and SWAT 2) based on different land use maps in different years (2000, the middle year of the study period, and 2009, the last year of the study period) was applied to the middle reaches of the Heihe River in northwest China.

In northwest China's arid region, inland river basins form the main hydrological system and occupy 35% of the total land area of the country. Most of the runoff in this region comes from the mountains and dissipates in the piedmont basin [24,28]. The Heihe River Basin is a typical inland river basin and is the second largest river basin in northwest China. The entire runoff of the Heihe River Basin dissipates in its middle reaches, which is also the region most affected by human activity. The ecosystem of the Heihe River Basin is presently at risk [29]. Before 2000, a number of human disturbances, including deforestation, overgrazing, and urbanization, caused drastic changes in the Heihe River Basin's land use/cover, destabilizing its entire ecosystem. For example, some downstream terminal lakes disappeared as the groundwater table in the basin's middle reaches dropped [28]. Since 2000, China has implemented a series of environmental protection measures, such as reforestation and regrassing of farmland, causing further spatial and temporal LUCC in the Heihe River Basin. These changes have affected the basin's hydrological cycle in a very complex and multifaceted way [9].

Finding an effective means for studying LUCC effects on the hydrological processes in the middle reaches of the Heihe River is particularly crucial for this region. However, the results of this study will more generally reveal how LUCC affects hydrological processes in the water consumption areas of inland river basins in arid regions, providing a scientific basis for the effective management and sustainable use of all inland river basin water resources.

2. Materials and Methods

2.1. Study Area

The middle reaches of the Heihe River are located between the Qilian Mountains and the Beishan Mountains. After passing through the Yingluoxia hydrometric station, the Heihe River flows through the plains of the Hexi Corridor, passing through Zhangye City to the Zhengyixia hydrometric station. Figure 1 shows a map of the Heihe River Basin study area.

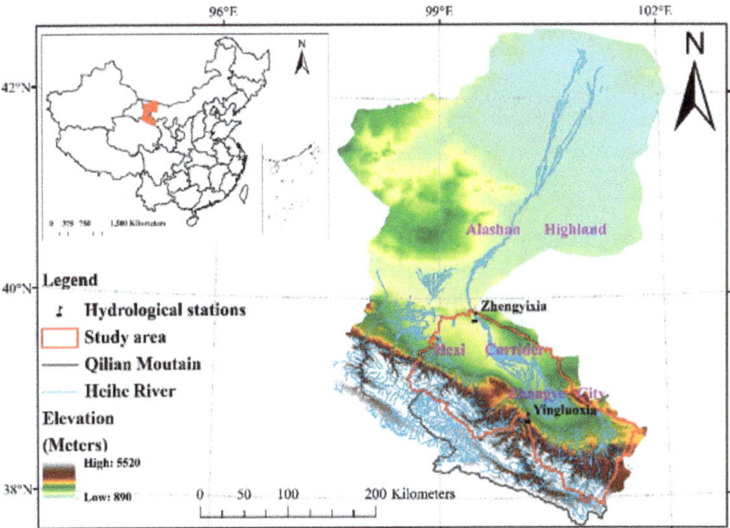

Figure 1. Map of the Heihe River Basin study area.

The topography of the middle reaches of the Heihe River slopes from south to north. The terrain is higher in the south and west and lower in the north and east, with an average altitude of 1400–1700 m. This 185-km section of the Heihe River is a primary area for runoff utilization and water consumption [24].

The middle reaches of the Heihe River experiences abundant heat and sunlight, making it suitable for crop growth and agricultural development. This region is designated as an irrigated agro-economic zone. Approximately 61% of the soil in this region is grey-brown desert soil. Other soil types in this region include chestnut soil, light chestnut soil, brown desert soil, desert sandy soil, and a small number of azonal soils, such as anthropogenic-alluvial soil, meadow soil, and marsh soil [29].

Due to the impact of human activity, many irrigation oases are distributed throughout the piedmont alluvial fan and alluvial plain in the lower middle reaches and upper river basin, respectively, forming a landscape dominated by artificially grown vegetation [9]. Precipitation in the plains is high in the east (~250 mm) and low in the west (≤50 mm) [29]. According to the Köppen Geiger classification, the climate in the study area is BSK (cold semi-arid).

2.2. Land Use/Cover Data Acquisition

To support this study, annual land use/cover data—in the form of Landsat Thematic Mapper (TM) images—were obtained for the middle reaches of the Heihe River from 1990 to 2009.

Using a 1:50,000 topographic map as the datum and the Albers projection, the remote sensing images were geometrically corrected using a quadratic polynomial model. The interpretation keys of the remote sensing images were established using land use maps and observed data (we have 30 ground control points to check the accuracy) corresponding to the same period. Next, supervised human-machine classifications and image interpretations were performed using ArcGIS 10.0 (ESRI, Redlands, CA, USA) and ENVI 5.1 (Harris Geospatial Solutions, Inc., Broomfield, CO, USA) image processing software, and the results were compared with land use maps of the study area for the corresponding period. On-site verification revealed that the qualitative accuracy of the data classification exceeded 95%. Compared with the existing land use maps of the study area, the kappa coefficients of the interpreted land use maps in this study are all over 0.93.

Finally, based on *China's Land Use Classification System* and the land use classification system used by the SWAT, the land use categories selected for use in this study included farmland, forest, grassland, water, residential, and bare land.

2.3. Conventional SWAT Assessment

The conventional SWAT is a semi-distributed hydrological model that first subdivides the entire river basin into a number of sub-basins based on factors, such as topography and river-network distribution [27]. Next, the SWAT further subdivides the sub-basins into HRUs based on the land use classifications, soil classifications, and terrain slopes in the river basin. The land use/cover in the study area is assumed to remain constant throughout the simulation period. For each individual HRU, a conceptual model is used to estimate its precipitation, runoff, sediment, and other factors. After completing these calculations, river confluence calculations are made [27].

Based on the water balance principle, the SWAT calculates the water volume as follows [27]:

$$S_t = S_0 + \sum_{i=1}^{t} \left(R_{day} - Q_{surf} - E_t - S_{seep} - Q_{gw} \right),$$

where S_t is the soil water content (mm), S_0 is the initial soil water content (mm), t is the total simulation time (days), R_{day} is the precipitation on day i (mm), Q_{surf} is the surface runoff on day i (mm), E_t is the actual ET rate on day i (mm), S_{seep} is the soil permeability on day i (mm), and Q_{gw} is the baseflow (mm).

2.4. Conventional SWAT Modification

In the conventional SWAT, HRUs are the basic computation elements, each with a commonly defined land use, soil type, and slope. An individual HRU consists of multiple grid units that can be spatially adjacent or apart from one another. The number, surface areas, and spatial locations of HRUs are determined based on the combined number, surface areas, and spatial locations of patches on the land use, soil type, and slope maps.

In the present study, the soil type and slope data used to generate HRUs remained constant, whereas the land use data changed from year to year. To incorporate annual land use/cover data in the conventional SWAT (via a modified SWAT or LU-SWAT), the number, surface areas, and spatial locations of the HRUs generated from the multi-year land use/cover data must remain unchanged. To ensure that this condition is met in the LU-SWAT, the land use datasets for each year in the study period are spatially superimposed (based on successive years) to generate a land use overlay map and obtain its corresponding attribute table showing each patch number and its corresponding attribute (e.g., land use type) for each year prior to generating the HRUs.

The spatially superimposed land use map is next superimposed with the soil type and terrain slope maps. During this superimposition process, the annual land use types corresponding to each patch listed in the attribute table are invoked by year to generate annual HURs. By matching the original land use maps for each year to their corresponding superimposed maps, the number, surface areas, and locations of patches for each year remain identical, but the patch attributes (e.g., land use types) may vary from year to year. Concurrently, the number, surface areas, and spatial distributions of the previously generated HRUs for each year remain constant, enabling the SWAT to carry out subsequent calculations based on land use changes. Figure 2 shows this modified HRU generation process based on the annual land use/land cover (LULC) maps.

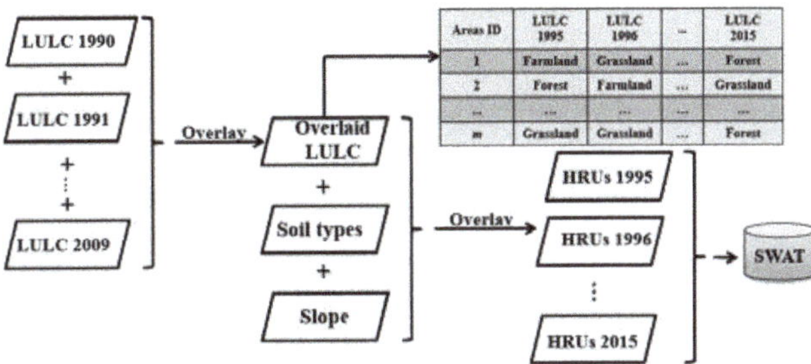

Figure 2. Modified hydrological response unit (HRU) generation process based on annual land use/land cover (LULC) maps.

In addition to modifying the HRU generation process in the conventional SWAT, its computation flow also required modification. The conventional SWAT performs parameter initialization on a daily cycle, which prevents the use of annual land use data. Unlike the conventional SWAT model, the LU-SWAT runs the yearly loop subroutine prior to parameter initialization, allowing the current year's data (HRUs) to be input prior to initializing the parameters and running the daily loop subroutine. After the last day in a year has been simulated, results are saved in a file and used as input data for the subsequent year. Specifically, the data are reloaded and initialized with the corresponding initialization parameters, and the daily loop subroutine is run again. If the preceding year of the current input year is the last year of the simulation period, the simulation is finished. Figure 3 compares the computation flows of the conventional SWAT and the LU-SWAT developed in this study.

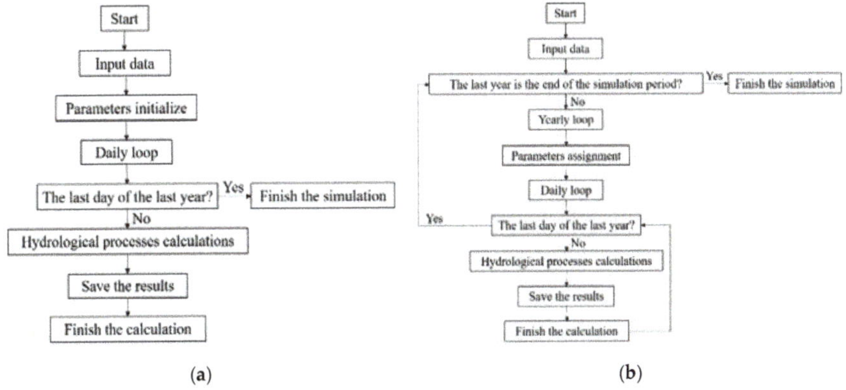

Figure 3. Comparative computation flows: (**a**) Conventional SWAT and (**b**) proposed LU-SWAT.

To implement this modified computation flow, the conventional SWAT code was rewritten using Fortran language in the Microsoft Visual Studio 2010 programming environment (Rev. 635). After successful code modification and compilation, the original SWAT.exe file was replaced with a new executable file.

2.5. Conventional and Modified SWAT Application

To validate this approach, the modified SWAT or LU-SWAT together with two other conventional SWAT models based on different land use maps in different years (2000 and 2009) were applied to the middle reaches of the Heihe River in northwest China. To develop the hydrological model for this region, soil type data was obtained from a 1:1,000,000 soil map of Gansu Province. A digital elevation model (DEM) with a spatial resolution of 30 × 30 m provided topographic data. To account for the impact of human activity on the river network, supplemental topographic survey data digitized from a 1:100,000 topographic map was used to adjust the DEM's river channel data.

Meteorological data, including the precipitation, temperature, wind speed, relative humidity, and sunshine duration, were obtained from the Cold and Arid Regions Science Data Centre (http://westdc.westgis.ac.cn). These data were measured at 12 meteorological stations in the Heihe River Basin, including the Gaotai, Jinta, Jiuquan, Linze, Minyue, Shandan, Sunan, Zhangye, Qilian, Tuolei, Yeniugou, and Yongchang stations.

Because the middle reaches of the Heihe River offer abundant sunlight, rich natural resources, and a flat topography, approximately 97.6% of the entire Heihe River Basin's population and 98.5% of the cultivated land in the upper and middle reaches are concentrated here. As such, agricultural (farmland) management measures and domestic water consumption in the study area were deemed important to this study. Farmland management data included irrigation measures and cultivation/harvesting times in the river basin. In the middle reaches of the Heihe River, the amount of water used for irrigation and the corresponding water sources vary among the different irrigational districts. The proposed LU-SWAT accounts for this time and spatial heterogeneity when defining the irrigation measures because of the dynamic HRUs. The two other conventional SWAT models only account for spatial heterogeneity in the irrigation.

2.6. Simulation Evaluation

The validity of the proposed LU-SWAT in this application was evaluated using the Nash-Sutcliffe efficiency (NSE) parameter, percent bias (PBIAS), and the ratio of the root mean square error (RSME) to the standard deviation of observations (RSR) [30]. The NSE parameter ranges from $-\infty$ to 1. An optimal NSE value of 1 indicates good model performance and high model credibility. As the NSE value approaches 0.5, the simulation results approach the average observed values, indicating satisfactory model performance. Similarly, PBIAS values ranging from -10% to 10% indicate good model performance. Finally, smaller RSR values indicate better model performance. An existing research [30] details the calculation processes and significance of these three simulation evaluation parameters.

3. Results and Discussion

3.1. Historic Land Use/Cover Changes in the Heihe River Basin

Table 1 shows the land areas by use type in the middle reaches of the Heihe River measured annually from 1990 to 2009. Farmland, forest, grassland, and bare land consistently accounted for most of the land area in this region. From 1990 to 2009, farmland, forest, and urban areas all showed increasing trends. Urban areas developed most rapidly and extensively, doubling in surface area over this period. However, urban areas accounted for only a small portion of the total study area. Forest areas also grew steadily, increasing in surface area by 55.70% over this period. Figure 4 shows the land use changed and no changed area of the study region from 1990 to 2009. It is obvious that most of the study area experienced land use changes in this 20-year period.

Table 1. Land areas by use type in the middle reaches of the Heihe River from 1990 to 2009 (units: km^2).

Year	Farmland	Forest	Grassland	Water	Urban	Bare land
1990	3812.58	2766.74	7590.15	585.397	325.22	9936.87
1991	3847.61	2791.93	7492.58	570.387	375.295	9939.16
1992	3887.59	2859.44	7385.07	725.492	415.322	9744.04
1993	3932.6	2911.93	7292.58	650.441	442.807	9786.60
1994	3975.11	2981.93	7277.57	612.916	495.302	9674.14
1995	4015.22	3049.43	7170.06	567.885	525.309	9689.05
1996	4057.32	3110.02	7094.67	647.96	578.102	9528.89
1997	4105.41	3200.05	6999.68	685.465	620.604	9405.75
1998	4145.18	3285.07	6907.18	692.97	688.109	9298.45
1999	4187.21	3355.11	6889.67	682.963	728.116	9173.89
2000	4218.52	3410.63	6897.18	680.461	758.122	9052.05
2001	4230.22	3530.16	6881.83	670.455	783.133	8921.16
2002	4245.24	3662.69	6882.17	667.953	838.136	8720.78
2003	4268.63	3795.21	6867.16	682.963	843.139	8559.86
2004	4283.55	3835.24	6864.65	685.465	868.156	8479.90
2005	4305.31	3972.76	6842.14	682.963	888.17	8325.62
2006	4319.42	4015.29	6837.14	672.956	928.177	8243.98
2007	4335.11	4147.81	6824.63	670.455	955.682	8083.28
2008	4358.22	4277.83	6827.13	662.949	983.187	7907.64
2009	4376.22	4307.85	6819.62	667.953	1005.69	7839.62

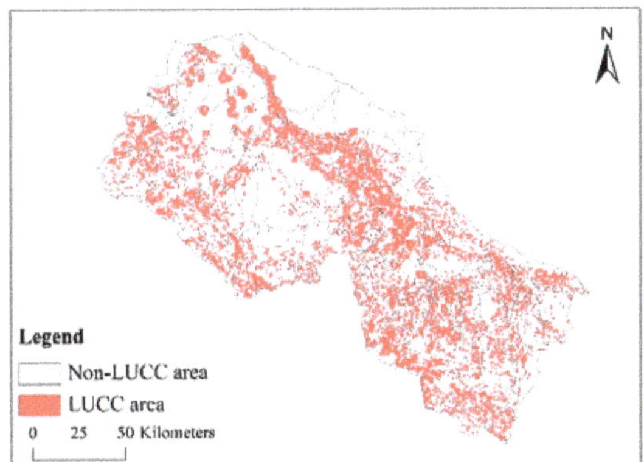

Figure 4. Land use changed and unchanged area in the study region.

Implementation of the Heihe Water Diversion Project (HWDP) and the farmland reforestation and regrassing measures in 2000 directly affected land use and cover in this region. From 1990 to 2000, farmland and water areas increased by 10.65% and 16.24%, respectively. Grassland and bare land areas decreased over this same period; grassland areas decreased by 9.13%. From 2000 to 2009, farmland areas continued to increase, but at a much slower rate of 3.45% and water areas gradually decreased by 0.37%. In response to the reforestation and regrassing measures implemented in 2000, grassland areas continued to decrease, but at a much slower rate of 0.90%.

Spatial overlay analysis of the land use data in the middle reaches of Heihe River revealed that the significant land use conversion trends from 1990 to 2009 were from grassland to farmland and from bare land to forest. From 2001 to 2009, a single significant land use conversion trend from bare land to forest was observed in the study area.

3.2. Calibration and Validation of the Models

In this study, the HRU area ratio (land use percentage) was set to 2%. For the LU-SWAT model, the number of HRUs is 100,168. For SWAT 1 and SWAT 2, the number is 2314 and 2540, respectively.

The LU-SWAT has more HRUs than the conventional SWAT model, which is due to the use of the overlaid land use map in LU-SWAT, which has more patches than the single year land use maps. The large numbers of HRUs may lead to model complexity.

The conventional SWAT and proposed LU-SWAT are based on the same physical processes. As such, the sensitivities of their respective model parameters were assumed as consistent. This assumption enabled the use of the conventional SWAT to support the calibration of the proposed LU-SWAT. We set 1988–1989 as the initial period for model initialization, 1990–2000 as the calibration period, and 2000–2009 as the validation period. Referring to the existing study [31], a sensitivity analysis was performed using the conventional SWAT and was based on 22 parameters related to the water cycle process. Table 2 shows the results of the model parameter sensitivity analysis, where t is the sensitivity of each parameter (as $|t|$ increases, the parameter sensitivity increases), and p is the statistical significance of the parameter sensitivity (as p approaches 0, the parameter sensitivity increases.

The 10 most sensitive model parameters from Table 2 were selected for use in the initial calibration of the proposed LU-SWAT. These include the effective hydraulic conductivity of the main channel alluvium (CH_K2), initial Soil Conservation Service (SCS) runoff curve number for moisture condition II (CN2), baseflow recession constant (ALPHA_BF), Manning's n value for the main channel (CH_N2), threshold water level in the shallow aquifer for the base flow (GWQMN), melt factor on 21 December (SMFMN), groundwater revaporization coefficient (GW_REVAP), delay time for the aquifer recharge (GW_DELAY), snowfall temperature (SFTMP), and snow temperature lag factor (TIMP).

Table 2. Initial model parameter sensitivity analysis results using the conventional SWAT.

	Hydrological Parameter	t	p
ESCO	Soil evaporation compensation coefficient	0.39	0.86
CANMX	Maximum canopy storage	−0.41	0.85
HRU_SLP	Average slope of the sub-basin	0.44	0.81
RCHRG_DP	Aquifer percolation coefficient	0.49	0.78
SURLAG	Surface runoff lag coefficient	0.52	0.76
OV_N	Manning's n value for overland flow	−0.55	0.68
EPCO	Plant uptake compensation factor	0.6	0.67
BIOMIX	Biological mixing efficiency	−0.63	0.63
SLSUBBSN	Average slope length	0.64	0.6
SMTMP	Snow melting accumulated temperature	1.86	0.06
SMFMX	Melt factor on 21 December	−0.74	0.41
REVAPMN	Threshold water level in shallow aquifer for revaporization	0.92	0.37
TIMP	Snow temperature lag factor	0.97	0.34
SFTMP	Snowfall temperature	−0.99	0.32
GW_DELAY	Delay time for aquifer recharge	1.02	0.31
GW_REVAP	Groundwater revaporization coefficient	−1.15	0.25
SMFMN	Melt factor on 21 December	−1.37	0.17
GWQMN	Threshold water level in shallow aquifer for base flow	0.7	0.45
CH_N2	Manning's n value for the main channel	−2.49	0.01
ALPHA_BF	Baseflow recession constant	6.11	0
CN2	Initial Soil Conservation Service (SCS) runoff curve number for moisture condition II	8.59	0
CH_K2	Effective hydraulic conductivity of main channel alluvium	−14.09	0

Note: For each parameter, t is the sensitivity (as $|t|$ increases, parameter sensitivity increases), and p is the statistical significance of the sensitivity (as p approaches 0, parameter sensitivity increases).

The calibration process for a hydrological model is not as simple as fitting the selected parameters to observed data. Rather, based on a comprehensive consideration of the river basin characteristics, the simulated data is closely calibrated to fit the observed data, without exceeding a reasonable range of parameter values. In this study, a two-step process was followed: (1) A reasonable range of parameter values was defined based on the existing research of the Heihe River Basin and (2) a subsequent multi-step manual calibration method [27] was followed. The ranges of the 10 parameters are listed in Table 3.

Table 3. Ranges of the calibrated parameters.

Parameters	Max Value	Min Value
TIMP [v]	1	0.01
SFTMP [v]	0.5	1.5
GW_DELAY [v]	0	300
GW_REVAP [v]	0.2	0.02
SMFMN [v]	10	0
GWQMN [v]	150	350
CH_N2 [v]	0.2	0.01
ALPHA_BF [v]	0.30	0.00
CN2 [r]	0.50	−0.50
CH_K2 [v]	300	0

[v] Parameter value is replaced by a given value; [r] Parameter value is multiplied by (1 + a given value).

The SWAT 1 and SWAT 2 models were calibrated using observed runoff data at the Zhengyixia station based on the parameters listed in Table 3 according to the existing study [27]. The proposed LU-SWAT uses annual land use/cover data to reflect the LUCC effects on model parameters, this study used a subsequent dynamic parameter calibration method following initial parameter calibration to match the various land use data with the optimal parameter combinations. Of the 10 most sensitive model parameters identified in the initial calibration of the LU-SWAT, only four of these parameters (CH_K2, CN2, ALPHA_BF, and GW_REVAP) were potentially affected by LUCC. These four parameters were subsequently selected for calibration of the LU-SWAT, while all other parameter values remained unchanged. Ultimately, 20 sets of optimal parameters were identified based on LU-SWAT simulations that considered annual land use/cover data from 1990 to 2009 and that were corrected using runoff data measured at the Zhengyixia station. The calculation time of the SWAT 1 and SWAT 2 were about 20–25 s (10 years) and for the LU-SWAT model, the time was 2.5–3 min.

Figure 5 compares the estimated and observed monthly runoff in the middle reaches of the Heihe River based on simulations from the proposed LU-SWAT, SWAT 1, and SWAT 2 following calibration and measurements from the Zhengyixia station. Figure 6 presents the same comparisons for the annual runoff in the region. Simulated results from the proposed LU-SWAT were generally consistent in both magnitude and direction when compared with the measured data, demonstrating its validity for broader applications. The performance of the SWAT 1 and SWAT 2 models was sufficient, but their NSE values are lower and RSR values are higher than the LU-SWAT model.

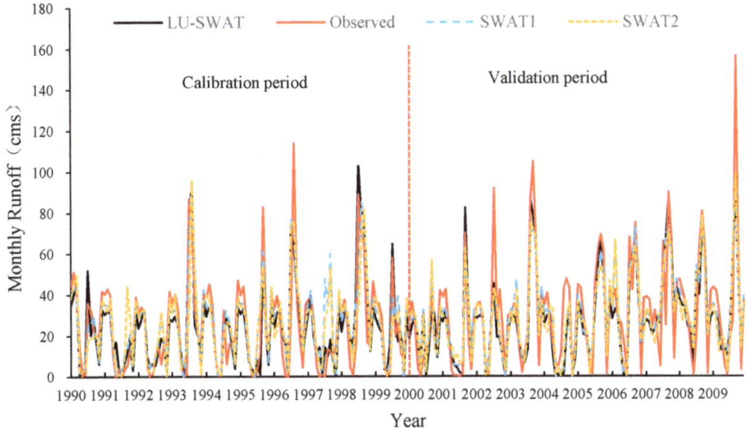

Figure 5. Estimated and observed monthly runoff in the middle reaches of the Heihe River based on LU-SWAT simulations and Zhengyixia hydrometric station measurements.

For the LU-SWAT, the NSE are higher and RSR are lower than SWAT 1 and SWAT 2, and the PBIAS are all between −10% and 10%. That means the performance of the LU-SWAT was the best. Relative to the SWAT 1 and SWAT 2, the proposed LU-SWAT achieved NSE values of 0.75 and 0.82, PBIAS values of 4.43% and 4.43%, and RSR values of 0.50 and 0.42 in the calibration period when simulating the monthly and annual runoff in the middle reaches of the Heihe River, respectively. Additionally, NSE values of 0.72 and 0.80, PBIAS values of 7.97% and 7.97%, and RSR values of 0.53 and 0.45 in the calibration period when simulating the monthly and annual runoff were found. For the LU-SWAT model, NSE, RSR, and PBIAS values were better for the calibration periods than the validation periods. Table 4 summarizes these results, which are possibly due to the fact that the LU-SWAT accounted for the time and spatial heterogeneity in LUCC and irrigation, whereas SWAT 1 and SWAT 2 only used one-year land use data. In addition, the performance of SWAT 1 and SWAT 2 were similar.

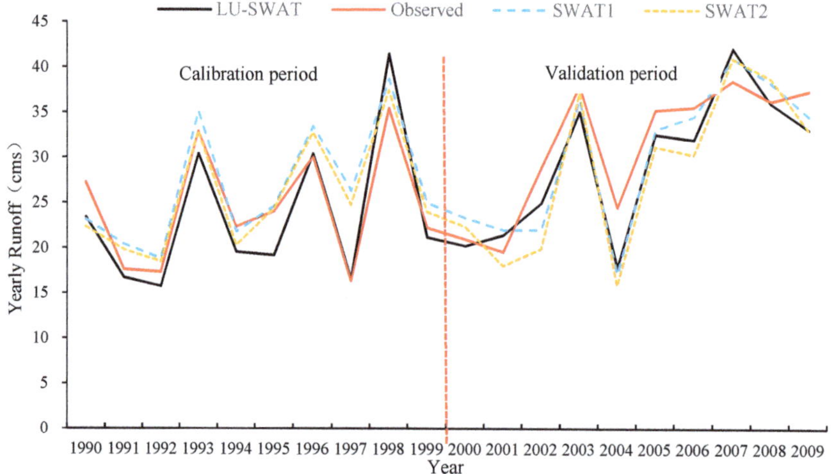

Figure 6. Estimated and observed annual runoff in the middle reaches of the Heihe River.

Table 4. Evaluation parameters for SWAT 1, SWAT 2, and LU-SWAT when simulating monthly and annual runoff in the middle reaches of the Heihe River.

			SWAT 1	SWAT 2	LU-SWAT	[32]	[33]
Monthly	NSE	Calibration	0.63	0.58	0.75	0.63	0.75
		Validation	0.69	0.68	0.72	0.60	0.70
	PBIAS	Calibration	−8.97	−4.71	4.43	Na	Na
		Validation	5.05	10.70	7.97	Na	Na
	RSR	Calibration	0.61	0.65	0.50	0.61	0.50
		Validation	0.56	0.57	0.53	0.77	0.55
Yearly	NSE	Calibration	0.70	0.77	0.82	Na	Na
		Validation	0.77	0.75	0.80	Na	Na
	PBIAS	Calibration	−8.97	−4.71	4.43	Na	Na
		Validation	5.05	10.70	7.97	Na	Na
	RSR	Calibration	0.55	0.48	0.42	Na	Na
		Validation	0.48	0.59	0.45	Na	Na

When compared with the existing research of the middle reaches of the Heihe River [32,33], the proposed LU-SWAT also outperformed the conventional SWAT model (Table 4): The NSE value increased 0%–20% and the RSR value decreased by about 0%–31%. In addition, the calibration and validation period (5 years) in the previous research were shorter than that in this study. The runoff measured at the Zhengyixia station is mainly affected by agricultural irrigation in the middle reaches of the Heihe River. The annual water volume used for agricultural irrigation is in turn affected by the farmland surface area and annual crop varieties. Unlike the conventional SWAT, the LU-SWAT proposed in this study incorporates both annual land use/cover data and detailed annual agricultural irrigation data for different irrigation districts and crops in the HRUs. As noted previously, agricultural irrigation is an important factor affecting this region's water cycle. By incorporating detailed agricultural irrigation data and accounting for inherent spatial variation, the proposed LU-SWAT provided better simulation of the runoff in the middle reaches of the Heihe River.

The proposed LU-SWAT was next applied more broadly to estimate various hydrological parameters over time (from 1990 to 2009). Table 5 summarizes these simulation results. The variations in the surface runoff, groundwater runoff, lateral flow, infiltration, and ET differed from that of the precipitation. The hydrological processes in the middle reaches of the Heihe River are primarily affected by human activity (e.g., agricultural irrigation) rather than by precipitation.

The recharge of river water in this region occurs primarily through groundwater runoff and lateral flow; surface runoff recharge is lower, especially during dry years (years with low precipitation). During wet years, recharge through surface runoff increases. The conventional SWAT model reflects this by using the SCS runoff curve number method to compute surface runoff. In addition, the total water yield in the middle reaches of the Heihe River accounts for only a small proportion of the total rainfall. These water consumption characteristics are consistent with the hydrological characteristics of inland river basins in northwest China's arid region [28,34].

Table 5. Hydrological parameters in the middle reaches of the Heihe River from 1990 to 2009 estimated using the proposed LU-SWAT (units: mm).

Year	Precipitation	Surface Runoff	Lateral Flow	Groundwater Runoff	ET	Total Water Yield
1990	206.48	6.23	9.35	17.78	231.60	33.36
1991	149.23	1.94	3.60	15.62	182.41	21.16
1992	207.28	2.67	3.68	12.71	209.88	19.06
1993	249.87	9.30	15.17	15.40	233.35	39.87
1994	181.93	2.89	4.16	17.34	201.64	24.39
1995	202.70	3.71	6.05	16.97	213.05	26.73
1996	204.41	6.39	11.87	14.82	201.89	33.09
1997	133.55	2.23	2.84	14.21	180.64	19.28
1998	250.29	15.75	14.54	13.71	233.54	44.00
1999	172.33	5.16	6.84	13.91	198.09	25.91
2000	206.68	4.11	6.71	13.03	221.00	23.85
2001	168.84	3.19	4.59	12.62	186.43	20.40
2002	212.57	7.46	12.17	14.21	211.77	33.85
2003	234.08	10.06	18.68	15.19	206.02	43.93
2004	161.14	7.65	7.06	11.75	206.73	26.47
2005	229.29	11.22	14.87	14.52	226.26	40.61
2006	137.35	11.27	18.39	16.06	166.46	45.72
2007	247.77	13.39	20.08	14.77	209.73	48.24
2008	178.13	9.75	18.11	17.14	203.00	45.00
2009	143.54	11.19	15.46	15.04	170.00	41.69

3.3. Land Use/Cover Change Effects on the Hydrological Processes in the Heihe River Basin

Following the initial calibration and validation of the proposed LU-SWAT, the model was used to simulate the effects of LUCC on hydrological processes in the middle reaches of the Heihe River. The control variable method was used to eliminate any confounding climatic factor effects on the hydrological processes.

Prior to simulation, climatic conditions in this region were classified as dry, wet, and normal based on annual precipitation volumes (listed previously in Table 5). From 1990 to 2009, 1997, 1998, and 1995 were identified as representative dry, wet, and normal years, respectively, with corresponding precipitations of 133.55, 250.29, and 202.70 mm. Climatic data (precipitation, temperature, relative humidity, wind speed, and solar radiation in this model) was combined with irrigation data for the middle reaches of the Heihe River, and inflow runoff data from the Yingluoxia station in the upper reaches of the Heihe River for these three years to simulate the runoff at the Zhengyixia station in the middle reaches of the Heihe River from 1990 to 2009. Based on these simulations, the effects of LUCC on the river basin hydrological processes under dry, wet, and normal climatic conditions were determined.

Figures 7–9 show the variations over time (from 1990 to 2009) in the annual surface runoff, lateral flow, groundwater runoff, ET volume, and total water yield in the middle reaches of the Heihe River under dry, wet, and normal climatic conditions, respectively.

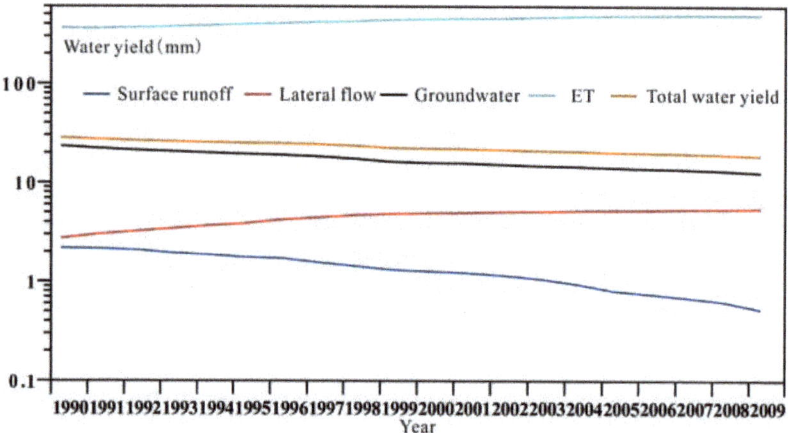

Figure 7. Hydrological parameter variation in the middle reaches of the Heihe River under dry conditions from 1990 to 2009.

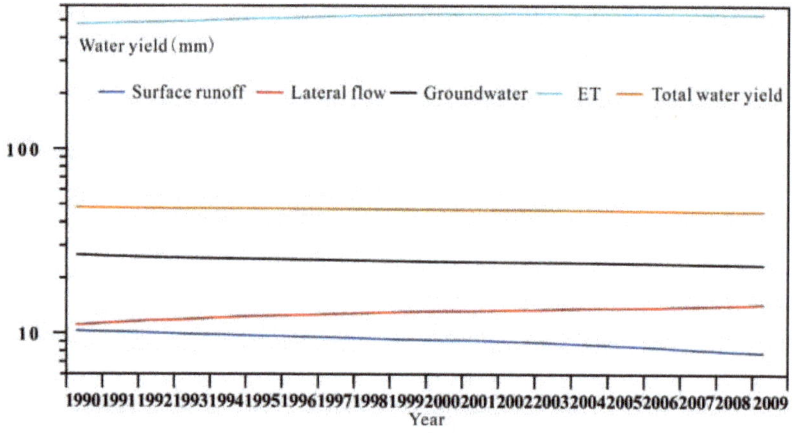

Figure 8. Hydrological parameter variation in the middle reaches of the Heihe River under wet conditions from 1990 to 2009.

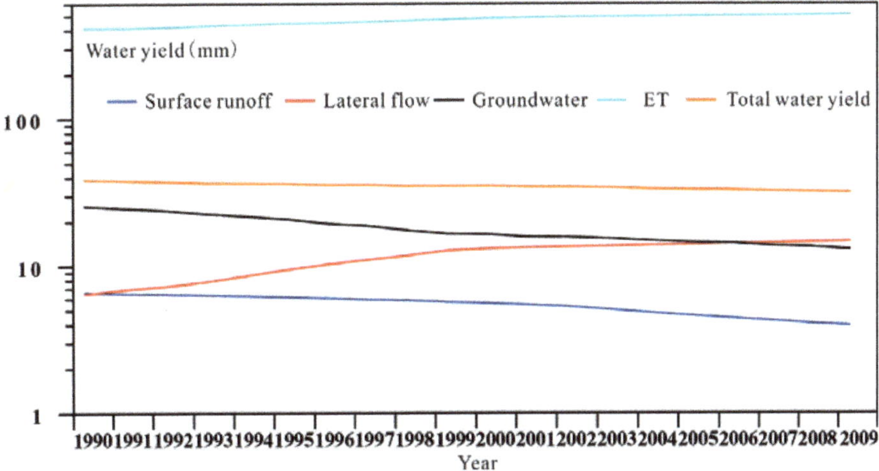

Figure 9. Hydrological parameter variation in the middle reaches of the Heihe River under normal conditions from 1990 to 2009.

As described previously in Section 3.1, the extent of land use conversion was lower from 2001 to 2009 than from 1990 to 2000 because of the HWDP and farmland reforestation and regrassing measures. Implementation of the HWDP and farmland reforestation and regrassing measures in 2000 also directly affected the hydrological processes in the region. Similar to the land use conversion trends, the magnitudes of changes in the estimated hydrological parameters were lower from 2001 to 2009 than from 1990 to 2000. Despite this variation before and after 2000, the overall trends for the hydrological parameters were consistent over time (from 1990 to 2009). As such, this study considered the overall changes in hydrological parameters from 1990 to 2009.

Under dry conditions, surface runoff, groundwater runoff, and total water yield showed decreasing trends over time, while lateral flow and ET volume showed increasing trends. Surface runoff, groundwater runoff, and total water yield decreased by 75.93%, 45.73%, and 33.74%, respectively, from 1990 to 2009. Lateral flow and ET volume increased by 99.93% and 35.11%, respectively, during this same period.

Similar trends over time were observed under wet conditions, but smaller changes in the hydrological parameters were estimated. Under wet conditions, surface runoff, groundwater runoff, and total water yield decreased by 23.61%, 11.42%, and 4.23%, respectively, from 1990 to 2009. Lateral flow and ET volume increased by 30.22% and 13.10%, respectively, during this same period.

Under normal conditions, similar trends over time were again observed, but the magnitudes of changes in the hydrological parameters differed from those under dry and wet conditions. Surface runoff, groundwater runoff, and total water yield decreased by 40.44%, 49.55%, and 18.47%, respectively, from 1990 to 2009. Lateral flow and ET volume increased by 120.21% and 23.60%, respectively, during this same period.

The primary water consumption zone of the Heihe River Basin lies in the middle reaches of the Heihe River. The water consumption of this area mainly occurs through agricultural irrigation, and flooding irrigation, which requires the extraction of large amounts of water from the river and underground runoff, is the most commonly used method of irrigation in this area. The agricultural irrigation module in the SWAT model provides a realistic model of flooding irrigation: Water losses from the water source to the soil (including transfer losses and evaporation losses) are indicated by the irrigation efficiency. The surface runoff flowing away from the farmlands as a percentage of the irrigation water is indicated by the "surface runoff ratio". The remaining water either seeps into the soil or is evaporated into the air, and the soil water module in the SWAT model is used to model the movement of water in the soils. During the irrigation season, large amounts of river water and underground water are extracted to irrigate the farmlands. A considerable amount of water is not used by the crops; this water either evaporates into the air, seeps into the soil to form lateral flows, or seeps through the ground to supplement underground water sources. Therefore, the increase in evaporation and lateral flows in the basin is mainly caused by the continuous increase of the farmland area. The effects of this change in farmland area on the surface runoff are relatively small. This is because the "surface runoff ratio" for agricultural irrigation was set to 0 according to the information queries that were made during this study; in other words, irrigation water does not flow from the farmlands into the river via surface runoff. Furthermore, the interception of rainfall by the canopy of the woodlands will reduce surface runoff. Therefore, the surface runoff of the basin decreases due to the continuous increase in farmland (in the middle reaches of the Heihe River) and woodland areas, and the continuous decrease in barren land and grassland areas. Underground runoff in the middle reaches of the Heihe River is decreasing over time, because river water and varying amounts of groundwater are being extracted for agricultural irrigation in the irrigation zones of this area; this is compounded by the increase in farmland area over time.

Therefore, the increase in farmland area is the main cause of the increase in lateral flows and evaporation in the middle reaches of the Heihe River. The increase in woodland area and the continuous decrease in barren land and grassland areas are the main reasons for the decrease in surface runoff. The increase in farmland area and the corresponding increase in irrigation water are the main causes of the decrease in underground runoff. Furthermore, it may be observed that the decrease in underground runoff is the greatest in drought and normal rainfall conditions. This is because very large amounts of water are still being extracted from groundwater sources and the river during drought and normal rainfall conditions. Since the amount of rainfall in these conditions is less than that in wet conditions, the groundwater sources do not receive a large amount of replenishment. Therefore, the significant decrease in underground runoff during drought and normal rainfall conditions may be attributed to the aforementioned causes.

As shown in Figures 7–9, changes in the hydrological processes in the middle reaches of the Heihe River were most evident under dry and normal climatic conditions. This finding concurrently indicates that LUCC has a greater effect on the hydrological processes in this region under dry and normal conditions.

Figure 10 shows the spatial heterogeneity and spatial-time variation of the hydrological parameters in the study area from 1990 to 2009 during dry climate conditions. Surface runoff, in sub-basins No. 6, 7, 9, 13, 16, 18, 28, 35, 41, and 78, shows a clear decline for the period from 1990 to 2009. In these sub-basins, the areas of bare land decreased, which is possibly indicative of the main reason that causes a decrease in the surface runoff in the entire watershed. Groundwater increased in sub-basins 13, 29, 35, 41, 52, and 62. In these sub-basins, we observed an increase in farmland area, while in sub-basins 6, 7, 9, 28, 37, 50, 57, and 78, groundwater decreased for the period from 1990 to 2009. In these sub-basins, we observed a decrease in bare land area. Lateral flow increased in sub-basins 13, 41, 53, and 62. In these sub-basins, we observed an increase in farmland area. This may indicate that the main cause of increased lateral flow in the entire watershed is the increase in farmland. We observed increased ET in sub-basins 13, 18, 29, 53, and 62, as well as other sub-basins with increased areas of farmland. Areas of bare land decreased in sub-basins 6, 7, 9, and 28, as well as other sub-basins, which are all characterized by decreasing ET. The main land use change that impacts hydrological processes in the study area is the increasing amount of farmland and the decreasing amount of bare land.

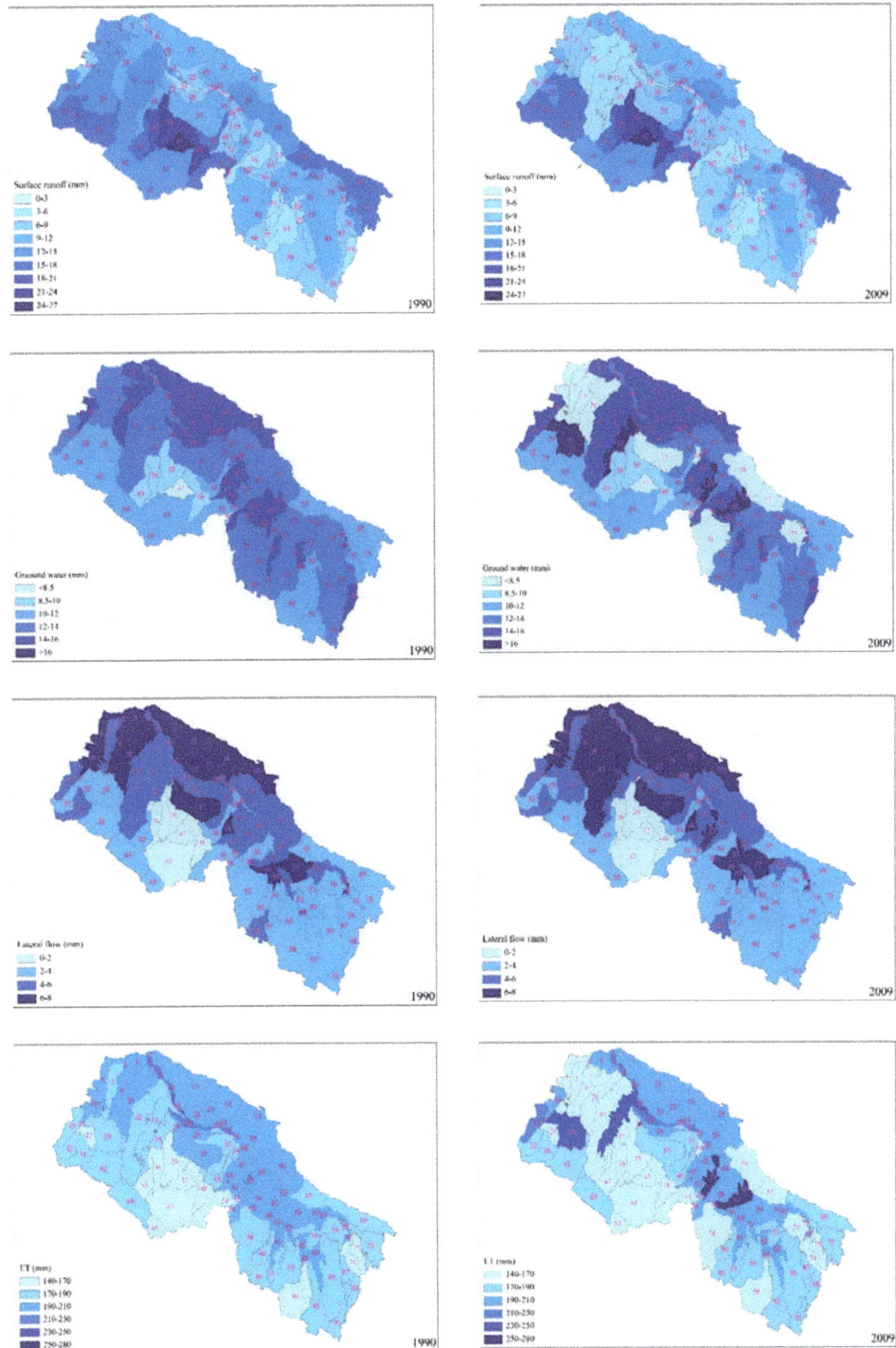

Figure 10. Hydrological parameter spatial variation in the middle reaches of the Heihe River.

Similar studies using the SWAT model reveal the impacts that LUCC has on hydrological processes. However, these studies may fail to reflect year-to-year changes in land use/cover. Some studies [23,33] divided entire simulation periods into uniform time intervals (e.g., 5-yr intervals) and performed interval simulations using land use/cover data from a single year within each interval. An existing study [35] found that, in the SWAT model, HRUs are lumped together and there is no interaction among HRUs in one sub-basin. Therefore, they improved the SWAT model to allow the distribution of HRUs. In this study, HRUs were also lumped, but we have divided large numbers of HRUs based on patches in the overlain multiple year land use maps, which may reduce the defects of lumped HRUs. However, future research requires the true HRU position to avoid defects from the lumped HRUs. Moreover, the number of HRUs in our study is high compared with similar studies. This may add more complexity to hydrological modeling and require more effort to calibrate the model. Setting a proper HRU threshold is a necessary step to solve these problems.

4. Conclusions

In conventional distributed or semi-distributed hydrological models, such as the SWAT, land use/cover type is assumed to remain constant throughout the simulation period, which limits the ability to interpret and predict the effects of LUCC on hydrological processes in a river basin. To overcome this limitation, a modified SWAT (LU-SWAT) was developed that incorporates annual land use/cover data to simulate LUCC effects on hydrological processes under different climatic conditions. To validate this approach, this modified model was applied to the middle reaches of the Heihe River in northwest China. Key findings from this efforts are as follows.

- Implementation of the HWDP and farmland reforestation and regrassing measures in 2000 directly affected land use and cover in this region. From 1990 to 2000, farmland areas increased by 10.65% while grassland areas decreased by 9.13%. From 2000 to 2009, farmland areas continued to increase and grassland areas continued to decrease, but at much slower rates of 3.45% and 0.90%, respectively. Primary land use changes in the study area were from grassland to farmland and from bare land to forest.
- From 1990 to 2009, surface runoff, groundwater runoff, and total water yield showed decreasing trends, while lateral flow and ET volume showed increasing trends under dry, wet, and normal conditions. Under dry, wet, and normal conditions, surface runoff decreased by 75.93%, 23.61%, and 40.44%; groundwater runoff decreased by 45.73%, 11.42%, and 49.55%; and total water yield decreased by 33.74%, 4.23%, and 18.47%; respectively. Lateral flow increased by 99.93%, 30.22%, and 120.21%; and ET volume increased by 35.11%, 13.10%, and 23.60%; respectively. Changes in the various hydrological parameters were most evident under dry and normal climatic conditions.
- The increase in farmland area is the main cause of the increase in lateral flows and evaporation in the middle reaches of Heihe River. The continuous decrease in barren land and grassland areas are the main reasons for the decrease in surface runoff. The increase in farmland area and the corresponding increase in irrigation water are the main causes of the decrease in underground runoff.
- Based on the existing research of the middle reaches of the Heihe River and the performance of SWAT 1 and SWAT 2, the modified LU-SWAT developed in this study outperformed the conventional SWAT when predicting the effects of LUCC on the hydrological processes of river basins. Relative to the conventional SWATs, the proposed LU-SWAT achieved NSE values of 0.75 and 0.82, PBIAS values of 4.43% and 4.43%, and RSR values of 0.50 and 0.42 in the calibration period and NSE values of 0.72 and 0.80, PBIAS values of 7.97% and 7.97%, and RSR values of 0.53 and 0.45 in the validation period when simulating monthly and annual runoff in the middle reaches of the Heihe River, respectively.

The results of this study substantially contribute to the state of knowledge regarding LUCC effects on river basin hydrologic processes. The results of this study also advance the state of practice for hydrologic assessments through the development and validation of a modified SWAT (LU-SWAT) that incorporates annual land use/cover data to simulate LUCC effects on hydrological processes under different climatic conditions. Future research will consider opportunities to enhance or more widely apply the LU-SWAT.

Author Contributions: Model development, X.J.; Y.J.; Formal analysis, X.J.; writing—original draft preparation, X.J.; writing—review and editing, Y.J.; X.M.; supervision, X.M.

Funding: This research was funded by grants from the National Natural Science Foundation of China [No. 41,801,094 & 51669028] and Natural Science Foundation of Qinghai Province [No. 2017-ZJ-961Q & 2019-ZJ-939Q].

Conflicts of Interest: The authors declare no conflict of interest.

References

1. Bonan, G.B. Forests and climate change: Forcings, feedbacks, and the climate benefits of forests. *Science* **2008**, *320*, 1444–1449. [CrossRef] [PubMed]

2. Liu, J.; Dietz, T.; Carpenter, S.R.; Alberti, M.; Folke, C.; Moran, E.; Pell, A.; Deadman, P.; Kratz, T.; Lubchenco, J.; et al. Complexity of coupled human and natural systems. *Science* **2007**, *317*, 1513–1516. [CrossRef]

3. National Research Council. *Advancing Land Change Modeling: Opportunities and Research Requirements*; National Academies Press: Washington, DC, USA, 2014.

4. Hansen, J.; Nazarenko, L.; Ruedy, R.; Sato, M.; Willis, J.; Genio, A.; Koch, D.; Lacis, A.; Lo, K.; Menonn, S.; et al. Earth's Energy Imbalance: Confirmation and Implications. *Science* **2005**, *308*, 1431–1434. [CrossRef] [PubMed]

5. IPCC: Climate Change 2013: The physical science basis. In *Contribution of Working Group I to the Fifth Assessment Report of the Intergovernmental Panel on Climate Change*; Cambridge University Press: New York, NY, USA, 2013.

6. Paeth, H.; Born, K.; Girmes, R.; Podzun, R.; Jacob, D. Regional climate change in tropical and Northern Africa due to greenhouse forcing and land use changes. *J. Clim.* **2009**, *22*, 114–132. [CrossRef]

7. Tilman, D.; Fargione, J.; Wolff, B.; D'Antonio, C.; Dobson, A.; Howwarth, R.; Schindler, D.; Schlesinger, W.; Simberloff, D.; Swackhamer, D. Forecasting Agriculturally Driven Global Environmental Change. *Science* **2001**, *292*, 281–284. [CrossRef] [PubMed]

8. Harr, R.; Fredriksen, R.; Rothacher, J. *Changes in Streamflow Following Timber Harvest in Southwestern Oregon*; USDA Forest Service Research Paper PNW (USA); USDA: Washington, DC, USA, 1979; p. 249.

9. Liu, Z.; Chen, R.; Song, Y.; Han, C.; Yang, Y. Estimation of aboveground biomass for alpine shrubs in the upper reaches of the Heihe River Basin, Northwestern China. *Environ. Earth. Sci.* **2015**, *73*, 5513–5521. [CrossRef]

10. Ngah, M.; Reid, I. The Impact of Land Use Change on Water Yield: The Case Study of Three Selected Urbanised and Newly Urbanised Catchments in Peninsular Malaysia. In *Land Degradation and Desertification: Assessment, Mitigation and Remediation*; Springer: Dordrecht, The Netherlands, 2010; pp. 347–354.

11. Pearce, A.; Rowe, L.; O'Loughlin, C. Water yield changes after harvesting of mixed evergreen forest, Big Bush State Forest, New Zealand. Presented at Soil and Plant Water Symposium, Bulls, New Zealand, 20–27 September 1982.

12. Mao, D.; Cherkauer, K. Impacts of land-use change on hydrologic responses in the Great Lakes region. *J. Hydrol.* **2009**, *374*, 71–82. [CrossRef]

13. Schilling, K.; Jha, M.; Zhang, Y.; Gassman, P.; Wolter, C. Impact of land use and land cover change on the water balance of a large agricultural watershed: Historical effects and future directions. *Water. Resour. Res.* **2008**, *44*, 636–639. [CrossRef]

14. Woodward, C.; Shulmeister, J.; Larsen, J.; Jacobsen, G.; Zawadzki, A. The hydrological legacy of deforestation on global wetlands. *Science* **2014**, *346*, 844–847. [CrossRef]

15. Brown, A.; Zhang, L.; McMahon, T.; Western, A.; Vertessy, R. A review of paired catchment studies for determining changes in water yield resulting from alterations in vegetation. *J. Hydrol.* **2005**, *310*, 28–61. [CrossRef]

16. Choi, W.; Deal, B. Assessing hydrological impact of potential land use change through hydrological and land use change modeling for the Kishwaukee River basin (USA). *J. Environ. Manag.* **2008**, *88*, 1119–1130. [CrossRef] [PubMed]

17. Hundecha, Y.; Bárdossy, A. Modeling of the effect of land use changes on the runoff generation of a river basin through parameter regionalization of a watershed model. *J. Hydrol.* **2004**, *292*, 281–295. [CrossRef]
18. Motevalli, S.; Mahdi Hosseinzadeh, M.; Esmaili, R. Assessing the Effects of Land use Change on Hydrologic Balance of Kan Watershed using SCS and HEC-HMS Hydrological Models—Tehran, IRAN. *Aust. J. Basic Appl. Sci.* **2012**, *6*, 510–519.
19. Bormann, H.; Elfert, S. Application of WaSiM-ETH model to Northern German lowland catchments: Model performance in relation to catchment characteristics and sensitivity to land use change. *Adv. Geosci.* **2010**, *27*, 1–10. [CrossRef]
20. Li, Z.; Liu, W.; Zhang, X.; Zheng, F. Impacts of land use change and climate variability on hydrology in an agricultural catchment on the Loess Plateau of China. *J. Hydrol.* **2009**, *377*, 35–42. [CrossRef]
21. Li, J.; Zhou, Z. Coupled analysis on landscape pattern and hydrological processes in Yanhe watershed of China. *Sci. Total Environ.* **2015**, *505*, 927–938. [CrossRef]
22. Liu, M.; Li, C.; Hu, Y.; Sun, F.; Xu, Y.; Chen, T. Combining CLUE-S and SWAT models to forecast land use change and non-point source pollution impact at a watershed scale in Liaoning Province, China. *Chin. Geogr. Sci.* **2014**, *24*, 540–550. [CrossRef]
23. Zhou, F.; Xu, Y.; Chen, Y.; Xu, C.; Gao, Y.; Du, J. Hydrological response to urbanization at different spatio-temporal scales simulated by coupling of CLUE-S and the SWAT model in the Yangtze River Delta region. *J. Hydrol.* **2013**, *485*, 113–125. [CrossRef]
24. Jin, X.; Zhang, L.; Gu, J.; Zhao, C.; Tian, J.; He, C. Modeling the impacts of spatial heterogeneity in soil hydraulic properties on hydrological process in the upper reach of the Heihe River in the Qilian Mountains, Northwest China. *Hydrol. Process.* **2015**, *29*, 3318–3327. [CrossRef]
25. Luo, Y.; He, C.; Sophocleous, M.; Yin, Z.; Ren, H.; Zhu, O. Assessment of crop growth and soil water modules in SWAT2000 using extensive field experiment data in an irrigation district of the Yellow River Basin. *J. Hydrol.* **2008**, *352*, 139–156. [CrossRef]
26. Vigiak, O.; Malagó, A.; Bouraoui, F.; Vanmaercke, M.; Obreja, F.; Poesen, J.; Habersack, H.; Feher, J.; Groselj, S. Modelling sediment fluxes in the Danube River Basin with SWAT. *Sci. Total Environ.* **2017**, *599–600*, 992–1012. [CrossRef] [PubMed]
27. Arnold, J.; Allen, P.; Volk, M.; Williams, J.; Bosch, D. Assessment of different representations of spatial variability on SWAT model performance. *Trans. ASABE* **2010**, *53*, 1433–1443. [CrossRef]
28. Wang, G.; Liu, J.; Kubota, J.; Chen, L. Effects of land-use changes on hydrological processes in the middle basin of the Heihe River, northwest China. *Hydrol. Process.* **2007**, *21*, 1370–1382. [CrossRef]
29. Wang, C.; Wang, X.; Liu, D.; Wu, H.; Lv, X.; Fang, Y.; Cheng, W.; Luo, W.; Jiang, P.; Shi, J.; et al. Aridity threshold in controlling ecosystem nitrogen cycling in arid and semi-arid grasslands. *Nat. Commun.* **2014**, *5*, 1–8. [CrossRef] [PubMed]
30. Moriasi, D.; Arnold, J.; VanLiew, M.; Binger, R.; Harmel, R.; Veith. Model evaluation guidelines for systematic quantification of accuracy in watershed simulations. *Trans. ASABE* **2007**, *50*, 885–900. [CrossRef]
31. White, K.; Chaubey, I. Sensitivity analysis, calibration, and validations for a multisite and multivariable SWAT model 1. *J. AWRA* **2010**, *41*, 1077–1089. [CrossRef]
32. Lai, Z.; Li, S.; Li, C.; Nan, Z.; Yu, W. Improvement and Applications of SWAT Model in the Upper-middle Heihe River Basin. *J. Nat. Resour.* **2013**, *28*, 1404–1413. (In Chinses)
33. Zhang, L.; Nan, Z.; Wu, Y.; Ge, Y. Modeling Land-Use and Land-Cover Change and Hydrological Responses under Consistent Climate Change Scenarios in the Heihe River Basin, China. *Water. Resour. Manag.* **2015**, *29*, 4701–4717. [CrossRef]
34. Xu, C.; Chen, Y.; Chen, Y.; Zhao, R.; Ding, H. Responses of surface runoff to climate change and human activities in the arid region of Central Asia: A case study in the Tarim River Basin, China. *Environ. Manag.* **2013**, *51*, 926–938. [CrossRef]
35. Meng, F.; Liu, T.; Wang, H.; Luo, M.; Duan, Y.; Bao, A. An Alternative Approach to Overcome the Limitation of HRUs in Analyzing Hydrological Processes Based on Land Use/Cover Change. *Water* **2018**, *10*, 434. [CrossRef]

 sustainability

Article

Temporal-Spatial Variations and Influencing Factor of Land Use Change in Xinjiang, Central Asia, from 1995 to 2015

Qun Liu [1,2], Zhaoping Yang [1], Cuirong Wang [1,*] and Fang Han [1]

1   Xinjiang Institute of Ecology and Geography, Chinese Academy of Sciences, Urumqi 830011, China; liuqun012@163.com (Q.L.); yangzp0124@163.com (Z.Y.); hanfang@ms.xjb.ac.cn (F.H.)
2   University of Chinese Academy of Sciences, Beijing 100049, China
*   Correspondence: wangcr@ms.xjb.ac.cn; Tel: +86-991-788-5354

Received: 11 December 2018; Accepted: 26 January 2019; Published: 29 January 2019

Abstract: In this study, we analyzed the temporal-spatial variations of the characteristics of land use change in central Asia over the past two decades. This was conducted using four indicators (change rate, equilibrium extent, dynamic index, and transfer direction) and a multi-scale correlation analysis method, which explained the impact of recent environmental transformations on land use changes. The results indicated that the integrated dynamic degree of land use increased by 2.2% from 1995 to 2015. The areas of cropland, water bodies, and artificial land increased, with rates of 1047 km^2/a, 39 km^2/a, and 129 km^2/a, respectively. On the other hand, the areas of forest, grassland, and unused land decreased, with rates of 54 km^2/a, 803 km^2/a, and 359 km^2/a, respectively. There were significant increases in cropland and water bodies from 1995 to 2005, while the amount of artificial land significantly increased from 2005 to 2015. The increased areas of cropland in Xinjiang were mainly converted from grassland and unused land from 1995 to 2015, while the artificial land increase was mainly a result of the conversion from cropland, grassland, and unused land. The area of cropland rapidly expanded in south Xinjiang, which has led to centroid position to move cropland in Xinjiang in a southwest direction. Economic development and the rapid growth of population size are the main factors responsible for the cropland increases in Xinjiang. Runoff variations have a key impact on cropland changes at the river basin scale, as seen in three typical river basins.

Keywords: Land use change; temporal-spatial variations; environmental and economic changes; arid region; central Asia

---

1. Introduction

Land cover and land use change is a result of the combined effects of climate change and human activities [1]. Types of land use are different in different climatic conditions. For example, wet areas are dominated by forest [2], arid and semi-arid areas are dominated by grassland [3], while arid areas are dominated by desert [4]. Climate warming and wetting changes will correspondingly result in further land use changes [5].

Global environmental changes are gradually causing changes in land cover and land use [6]. Climate change drives the grasslands vegetation communities to become shrub-encroached grasslands in arid and semi-arid regions, which has attracted the attention of researchers [7–9]. Changes in dry and wet conditions can also lead to changes in land use types [10,11], especially in arid regions, where water resources from river runoff mainly come from glaciers, snow melt, and precipitation in mountainous regions, which are significantly affected by climate change [12]. Global warming has been observed to accelerate in glaciers and snow melt [13], thus runoff will increase in a short period [14], but may not continue. This feature will impact the changes in oasis land use types in arid regions.

In arid areas, the main land use type is unused land, whereas cropland is distributed in oasis regions, while forest land and grassland are mainly located in mountainous areas [15]. In arid and semi-arid regions, there is a consistent performance showed that when cropland increased, the forest and grassland decreased. This occurred in South Asian [16], Central Asian [17], and the Mongolian plateau [18]. The stability of land use structure is relatively fragile and water resources are the basis and key factor for the changes of land use structure in arid regions. When glaciers and snow melt water runoff increase [17], cultivated land will also increase [19]. In the middle and later periods of glacier melt, the river outflow will be reduced and when the amounts of available water cannot support the levels of water use required by croplands, this can be followed by the desertification of cropland, such as in the Shiyang River Basin [20]. At the same time, human activities have also driven land use changes [21,22] and this anthropogenic role will increase with economic development or technology. The impact of human activities on land use types is multifaceted.

Therefore, it is important to study the impacts of climate change and human activities on land use changes in arid regions. In this study, we focused on temporal-spatial variations of land use under climate change, in addition to human activities in arid regions from 1995 to 2015, based on land use data, runoff, climate data, and socioeconomic data.

2. Materials and Methods

2.1. Study Area

Xinjiang region is located in Northwest China, far away from the sea, belonging to the innermost part of Asia. It is largely confined within 34°54′4–9°19′N and 73°44′–96°22′E (Figure 1). The Xinjiang is split by the Tien Shan Mountain, which divides it into two large basins: the Junggar Basin in the north and the Tarim Basin in the south. The Xinjiang has a typical continental subtropical climate, with less precipitation and a large daily temperature range. Therefore, this area is mostly covered with desert and dry grasslands. The dotted oasis is located at the foot of the Tienshan Mountains, the Kunlun Mountains, and the Altai Mountains, because glaciers and snow melt from mountainous areas are the main water sources for agricultural production and social and economic activities in Xinjiang. The Kai-Kong River Basin, the Aksu River Basin, and the Yarkand River Basin are located in Southern Xinjiang (Figure 1) and glaciers and snow melt are main parts of the river outflow [23]. Therefore, the three river basins are typical watersheds in the study area. The downstream areas of the three basins are oasis areas.

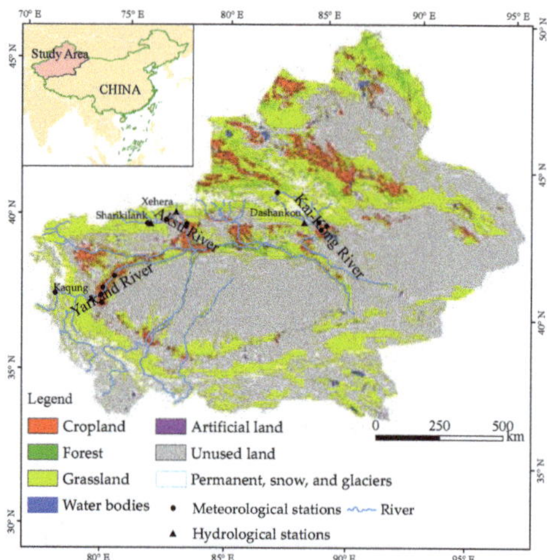

Figure 1. Map of land use in study area. The land use dataset from 2010 was supported by the Data Center for Resources and Environmental Sciences, Chinese Academy of Sciences (RESDC).

2.2. Data

The land use data are provided by the Data Center for Resources and Environmental Sciences, Chinese Academy of Sciences (RESDC), which includes five years: 1995, 2000, 2005, 2010, and 2015, respectively. The years 1995, 2000, 2005, and 2010 are used Landsat ET/ETM images; 2015 used Landsat 8 images. The datasets are gridded at 1 by 1 km. The land use types mainly contain six categories (i.e., cropland, forest, grassland, water bodies, artificial land, and unused land). This is the basic dataset for research on land use changes of the past half century in China [1,24,25].

The temperature and precipitation datasets are from 13 meteorological stations in three river basins (Figure 1), provided by the China Meteorological Administration (CMA) from 1960 to 2012, which are used to analyze the correlation between temperature/precipitation changes and land use changes. The runoff data in this study are derived from hydrological stations of the three typical river basins between 1960 and 2012, which were provided by hydrological bureau of Xinjiang province. The locations of hydrological stations in this study are shown in the Figure 1.

The social and economic data of Xinjiang between 1995 and 2015 were provided by the National Bureau of Statistics of China. In this study, we used the GDP and population size to analyze the impacts of human activities on land use changes in Xinjiang.

2.3. Methods

The sensitivity of temperature effects on runoff is more significant than precipitation in Xinjiang, because the proportion of glaciers and snow meltwater in total runoff is greater than that of rainfall runoff [23]. The multi-scale correlation analysis is very helpful to explain the responses of runoff changes to climate fluctuation [26,27]. In this study, the multi-scale correlation analysis was used to detect the correlation between runoff and climate factors (i.e., precipitation and temperature). The original runoff/temperature/precipitation series that exist create a scale-mixing problem, while the Ensemble Empirical Mode Decomposition (EEMD) method can overcome it [28]. Therefore, we used the EEMD method to separate inter-annual and inter-decadal runoff/precipitation/temperature variation signals from the original data series. The correlation coefficient was estimated by Pearson's method with a two-tailed test. Meanwhile, the Mann-Kendall (M-K) nonparametric trend test [29,30] was used to detect the trend of temperature, precipitation, and runoff. The slope of the trend is estimated by using Sen's nonparametric trend estimator [31].

The river runoff source mainly comes from mountain precipitation and glacier meltwater in Xinjiang [32]. Regional warming and increased precipitation in mountainous areas are the main reasons for the increase in runoff in Xinjiang during the last half century [33]. Therefore, we selected three typical watersheds (the Kai-Kong River Basin, the Aksu River Basin, and the Yarkand River Basin) (Figure 1) to analyze the impacts of climate on cropland changes based on the relationship between temperature, precipitation, runoff, and cropland area.

In this study, we selected four indicators, including land use change rate [25], equilibrium extent [34], dynamic index [35], and transfer direction [35], to describe the characteristics of land use changes in Xinjiang from 1995 to 2015. According to the five periods of land use data, we calculated the centroid position of cropland in different periods using the ArcGIS spatial analysis tool.

The land use change rate is described by a comprehensive index model of land use, as follows:

$$I = 100 \times \sum_{i=1}^{n} (A_i \times C_i) \tag{1}$$

where I represents the comprehensive index of land use change in this study area, A_i represents the classes index of the i class type of land use (unused land is classes 1, grassland and water bodies are classes 2, forest and cropland are classes 3, and artificial land is classes 4), and C_i represents the ratio between the area of the i class land use and total area of land use in the study area.

Therefore, the land use change model (Equation (1)) can be quantitatively expressed using the comprehensive levels and trends of land use:

$$\Delta I_{b-a} = \left(\sum_{i=1}^{n} A_i \times C_{ib} - \sum_{i=1}^{n} A_i \times C_{ia} \right) \times 100 \tag{2}$$

where a and b represent different periods.

The Entropy and the second law of thermodynamics are important indicators to revel land use processes and structures [36,37] and the information entropy has been widely used as an indicator to evaluate the equilibrium balance of land use. In this study, we used the equilibrium extent of a structure of land use [34] to evaluate the structures of land use in the Xinjiang from 1995 to 2015. The range of E is between 0 and 1. The larger the E value, the stronger is the homogeneity of the system. $E = 0$ and $E = 1$ indicates that the land use structure is unbalanced and in an ideal state, respectively.

$$H = - \sum_{i=1}^{n} P_i \log P_i \tag{3}$$

where H represent information entropy and n and P_i represent the number of land use types and the ratio between the area of the i class land use and total area of land use in the study area, respectively.

$$E = - \sum_{i=1}^{n} (P_i \times \log P_i) / \log n \tag{4}$$

where E represents the equilibrium extent of land use structure.

The regional differences of land use change can be expressed by the integrated dynamics index of land use and the dynamic index of single land use types and it can be calculated by the dynamic degree model of land use [1]. The integrated dynamics index of land use is an indicator that can depict the regional difference in land use changes and also reflects the comprehensive influence of human activities on the change in regional land use.

$$S = \left(\sum_{i=1}^{n} (\Delta S_{i-j} / S_i) \right) \times \frac{1}{t} \times 100\% \tag{5}$$

where S represents the integrated dynamic index of land use at time t in the study area, ΔS_{i-j} represents the area of the i class of land use type converted to other types of land use types from the beginning to the end of the monitoring period, S_i represents the total area of the i class of land use type at the monitoring start time, and t is the time in years.

The dynamic index of single land use types describes the change rates and amplitude of different land use types during a certain time.

$$k_i = \frac{S_{it_2} - S_{it_1}}{S_{it_1}} \times \frac{1}{t_2 - t_1} \times 100\% \tag{6}$$

where k_i represents the dynamics index of the i class of land use type between periods t_1 and t_2, and S_{it1} and S_{it2} represent the area of the i class of land use type during the period of t_1 and t_2, respectively.

The transfer direction of land use can be described by the land use change transfer matrix. The land use state transfer matrix comprehensively and concretely depicts the structural characteristics of regional land use change and reflects the direction of land use change, to better reveal the spatiotemporal evolution process of land use change.

$$S_{ij} = \begin{bmatrix} S_{11} & S_{12} & \cdots & S_{1n} \\ S_{21} & S_{22} & \cdots & S_{2n} \\ \cdots & \cdots & \cdots & \cdots \\ S_{n1} & S_{n2} & \cdots & S_{nn} \end{bmatrix} \tag{7}$$

where S_{ij} is the land use state at the beginning and end of the study period, n is the type of land use, and the vector in the state transfer matrix of the land use is the land use area in this study.

3. Results

3.1. The Temporal-Spatial Patterns of Land Use Change

Between 1995 and 2015, the rate of land use change in Xinjiang increased by 2.2%. Figure 2 shows that the equilibrium extent of land use structure increased with the increase in the land use change rate. It indicated that the structure equilibrium balance of land use decreased over the past two decades.

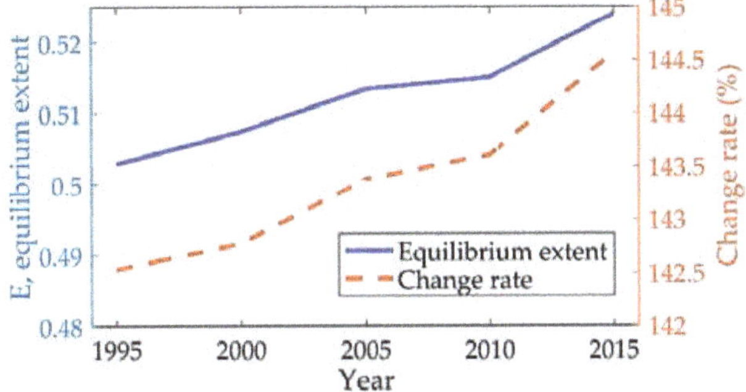

Figure 2. The equilibrium extent of land use structure and land use change rate of land use in Xinjiang from 1995 to 2015.

The area of cropland, water bodies, and artificial land increased in Xinjiang during the period of 1995–2015, while forest, grassland, and unused land decreased (Table 1). Among them, the increase in artificial areas was large, with a rate of 7.92%/a, followed by cropland, with a rate of 3.86%/a, and water bodies, with a rate of 0.81%/a. The forest, grassland, and unused land decreased, with decline rates of −0.26%/a, −0.36%/a, and −0.07%/a, respectively. The characteristics of land use changes during different periods (Table 1) include a large increase in cropland and water bodies from 1995 to 2005 and a large increase in artificial areas during 2005–2015. The decrease in forest mainly occurred during the period of 2005–2015 and the grassland decreased obviously between 1995 and 2005. Therefore, the changes in land use types in different periods require further analysis.

Table 1. Land use change rates in Xinjiang from 1995 to 2015 (units: %/a).

	1995–2005	2005–2015	1995–2015
Cropland	1.93	1.61	3.86
Forest	−0.05	−0.21	−0.26
Grassland	−0.23	−0.13	−0.36
Water bodies	0.76	0.05	0.81
Artificial land	2.71	4.10	7.92
Unused land	−0.01	−0.06	−0.07

The characteristics of the transfer direction of land use changes are also different throughout the years. Table 2 and Figure 3 show how cropland was mainly converted into grassland and artificial areas from 1995–2005, with the conversion areas being 835 km^2 and 481 km^2, respectively. Meanwhile, the amount of forest converted into cropland was 676 km^2, the amount of grassland converted to cropland was 9808 km^2, and the amount of grassland transformed into artificial land was 177 km^2. The amounts of unused land converted into cropland and artificial land were 2147 km^2 and 333 km^2, respectively. Figure 3 shows that the conversion of land use changes mainly occurred in oasis, due to local concentrated human activities.

Table 2. The transfer matrix of land use change from 1995 to 2005 (units: km^2).

	Cropland	Forest	Grassland	Water Bodies	Artificial Land	Unused Land
Cropland	53,770	139	835	86	481	407
Forest	676	37,096	146	49	16	55
Grassland	9808	390	472,849	694	177	2298
Water bodies	87	25	275	11,647	16	318
Artificial land	10	1	3	1	3687	4
Unused land	2147	187	842	837	333	999,879

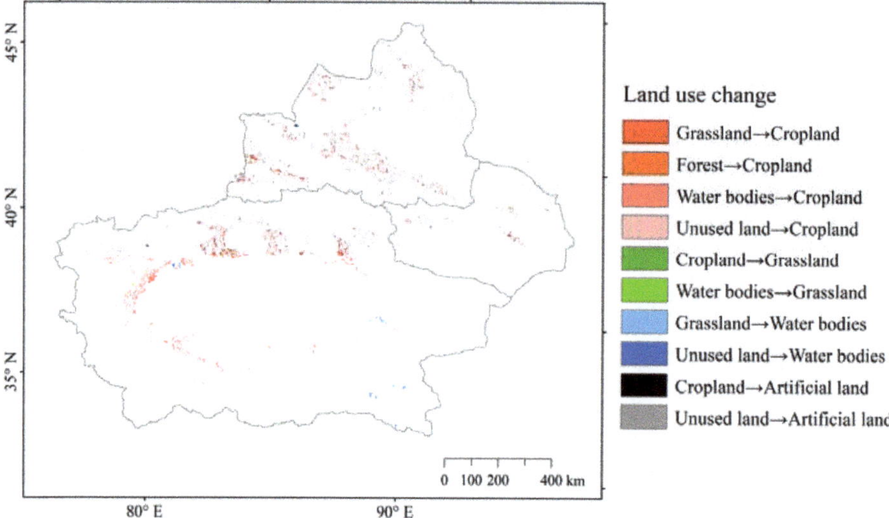

Figure 3. The spatial transfer direction of land use change in Xinjiang from 1995 to 2005.

The results in Table 3 and Figure 4 show that the areas of cropland that were transformed into artificial land and grassland were 566 km^2 and 501 km^2, from 2005 to 2015, respectively. And the amount of forest that was mainly transformed into cropland was around 695 km^2. Meanwhile, grassland were mainly converted into cropland, with an amount of around 6724 km^2, followed by conversion to artificial land, totaling 470 km^2. The unused land was mainly converted into cropland and artificial land, in the amount of 4276 km^2 and 893 km^2, respectively. Therefore, the area of artificial land rapidly increased in Xinjiang during this period, because of the acceleration of urbanization.

Table 3. The transfer matrix of land use change from 2005 to 2015 (units: km^2).

	Cropland	Forest	Grassland	Water Bodies	Artificial Land	Unused Land
Cropland	65,357	19	501	27	566	28
Forest	695	36,884	181	24	38	15
Grassland	6724	56	467,273	249	470	143
Water bodies	126	1	423	12,659	8	97
Artificial land	31	0	4	9	4666	0
Unused land	4276	73	491	330	893	996,859

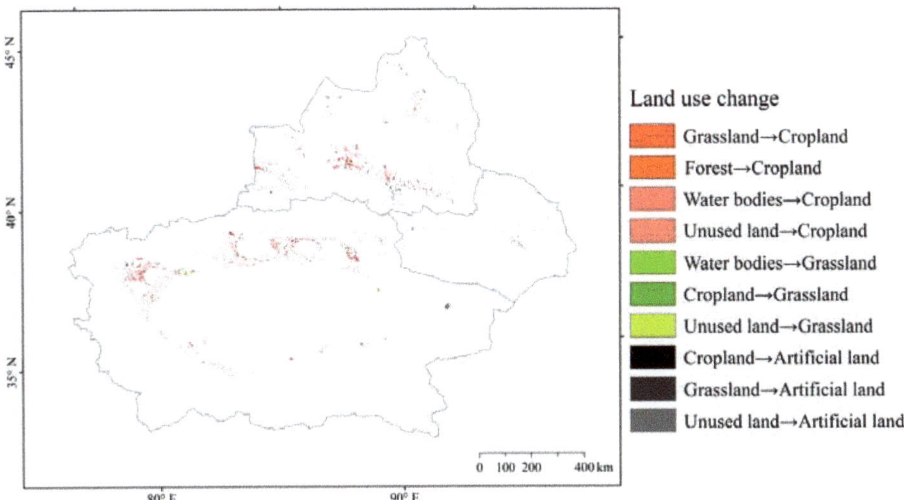

Figure 4. The spatial transfer direction of land use changes in Xinjiang from 2005 to 2015.

From 1995 to 2015, results in Table 4 and Figure 5 showed that the cropland was mainly transformed into artificial land and grassland. Forest was mainly converted into cropland, with a total area of around 1335 km^2, and grassland was converted into cropland and artificial land, with totals of 16,266 km^2 and 752 km^2, respectively. Meanwhile, water bodies converted into grassland and cropland were 475 km^2 and 212 km^2, respectively. Unused land converted into cropland, grassland, water bodies, and artificial land were in the amount of 6294 km^2, 1308 km^2, 1003 km^2, and 1118 km^2, respectively. From 1995–2015, the area of cropland and artificial land increased the most. The area of cropland was mainly transformed from grassland and unused land, while the type of artificial land was mainly transformed from cropland, grassland, and unused land. At the same time, this also shows that Xinjiang's social and economic development directly impacts the expansion of the Oasis scale and the acceleration of urbanization speed.

Table 4. The transfer matrix of land use change from 1995 to 2015 (units: km^2).

	Cropland	Forest	Grassland	Water Bodies	Artificial Land	Unused Land
Cropland	53,070	145	1045	102	1002	354
Forest	1335	36,191	320	64	54	73
Grassland	16,266	422	465,719	822	752	2200
Water bodies	212	20	475	11,305	54	302
Artificial land	32	1	6	2	3661	4
Unused land	6294	254	1308	1003	1118	994,209

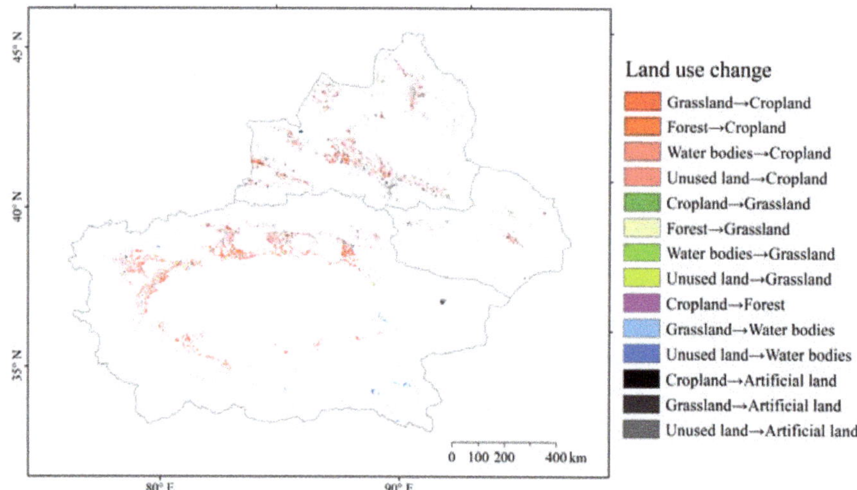

Figure 5. The spatial transfer direction of land use change in Xinjiang from 1995 to 2015.

The centroid of cropland migrated to the southwest of Xinjiang, because of the rapid expansion of cropland in the south of Xinjiang. The results indicated that the migration direction of the centroid position of cropland moves southward and then to the southwest (Figure 6). This also shows that the increase in area of cropland in Southern Xinjiang is faster than the increases in Northern and Eastern Xinjiang. The area of cropland rapidly increasing in South Xinjiang was mainly a result of global warming-accelerated glacier and snow melt, which lead to runoff increases, a driving factor. On the other hand, social and economic development, which has promoted the increased reclamation of other land use types into cropland, is the economic driving factor.

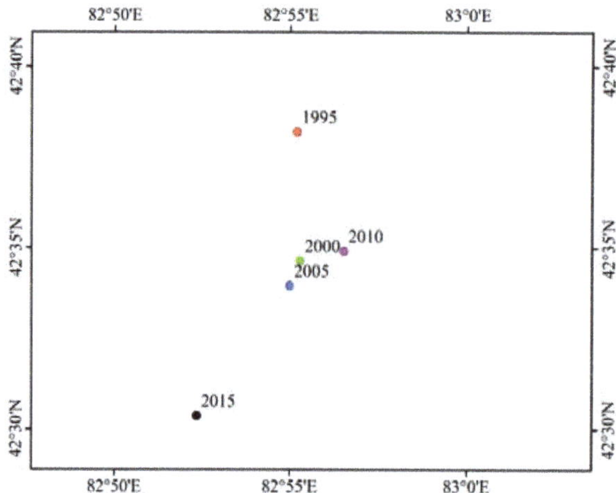

Figure 6. The migration direction of the centroid position of cropland in Xinjiang from 1995 to 2015.

3.2. The Influence of Recent Climate Change on Land Use Change

The results of multi-scale correlation analyses between runoff and precipitation/temperature in the Kai-Kong River Basin shows how runoff has a positive correlation to precipitation and temperature at inter-annual and inter-decadal scales (Table 5). We find that the correlation between runoff and precipitation at inter-annual scale is higher than that at inter-decadal scale, while the correlation between runoff and temperature at inter-annual scale is lower than that at inter-decadal scale. It is very interesting to find that runoff has a negative correlation to precipitation at inter-annual vs. decadal scale. The runoff of the Aksu River Basin has a positive correlation to precipitation at inter-decadal vs. inter-annual, while a negative correlation at inter-annual scale (Table 5). Furthermore, runoff has a positive correlation to temperature at inter-decadal. In the Yarkand River Basin, the runoff has a negative correlation to precipitation, but a positive correlation to temperature at inter-decadal scale (Table 5). The multi-scale correlation between runoff and climate factors are complex, because other than precipitation and temperature, there are many factors that impact runoff changes, such as meteorological station location, glaciers, snow cover, a 0 °C level high (FLH), and temperature lapse rate [38,39].

Table 5. Correlation between runoff and climate factors from 1960 to 2012.

River Basins	Time Scale	Precip. vs. Runoff	Temp. vs. Runoff
	Inter-annual	0.703 ***	0.382 ***
Kai-Kong River Basin	Inter-annual vs. inter-decadal	−0.273 **	−0.076
	Inter-decadal vs. inter-annual	−0.177	−0.111
	Inter-decadal	0.542 ***	0.516 ***
	Inter-annual	−0.297 **	−0.060
Aksu River Basin	Inter-annual vs. inter-decadal	0.178	−0.128
	Inter-decadal vs. inter-annual	0.255 *	−0.119
	Inter-decadal	−0.061	0.494 ***
	Inter-annual	−0.060	0.214
Yarkand River Basin	Inter-annual vs. inter-decadal	0.073	−0.044
	Inter-decadal vs. inter-annual	0.069	0.027
	Inter-decadal	−0.594 ***	0.393 ***

*** correlation is significant at the 0.01 level; ** correlation is significant at the 0.05 level; * correlation is significant at the 0.1 level.

The increase in cropland area in the Kai-Kong River Basin was the fastest, around 29%/decade, followed by the Aksu River Basin and the Yarkand River Basin (Figure 7). Meanwhile, the relationship between climate factors (temperature, precipitation, and runoff) and cropland changes has been analyzed in the three typical watersheds. The results are shown in Figure 7a,b, which indicates how the relationship between temperature, precipitation, and cropland are not very clear due to the influence of temperature and precipitation changes on cropland alterations as a result of the impact on runoff change. Figure 7c shows a large increase rate in cropland responses when there is a large increase in runoff rate. Therefore, the quantity of runoff is closely related to the cropland area in oasis regions in Central Asia.

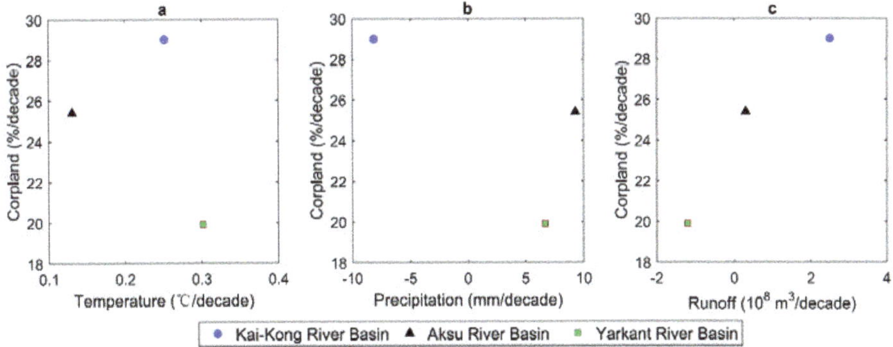

Figure 7. The relationship between climate factors and cropland changes in the three typical river basins from 1995 to 2015 (**a**) is temperature, (**b**) precipitation, and (**c**) runoff. The cropland change rate is calculated by Equation (6).

3.3. The Influence of Social and Economic Factors on Land Use Change

Figure 8a shows that social and economic factors have developed rapidly in Xinjiang during the past 20 years, especially since 2005. The GDP development can be simulated by an exponential model ($y = 648.79 \times e^{0.1316 \times x}$); it is shown that GDP increased with a rate of 649×10^8 RMB/a. Meanwhile, the population size has been rapidly expanding ($y = 34.69 \times x + 1623.59$) and increased by approximately seven million people over the past 20 years (Figure 8a). With rapid growth of the economic development and population size, more cropland areas are needed to support Xinjiang. Moreover, there are obvious changes in land use types in Xinjiang from 1995 to 2015 (Figure 8b), of which the increase rate of cropland was the largest, with a rate of over 1000 km^2/a, followed by the artificial land increase of about 130 km^2/a. However, the forest, grassland, and unused land showed a decreasing trend, in which grassland had the largest reduced rate of about 800 km^2/a. Human activities may have been an important factor in driving Xinjiang cropland increases from 1995 to 2015. Furthermore, there is a need to consider the impact of the migration processes, as rural area depopulation and city growth are related with this an abandonment of cropland. Obviously, it is necessary to deeply analyze impact of economic development on land use change in our next step research agenda.

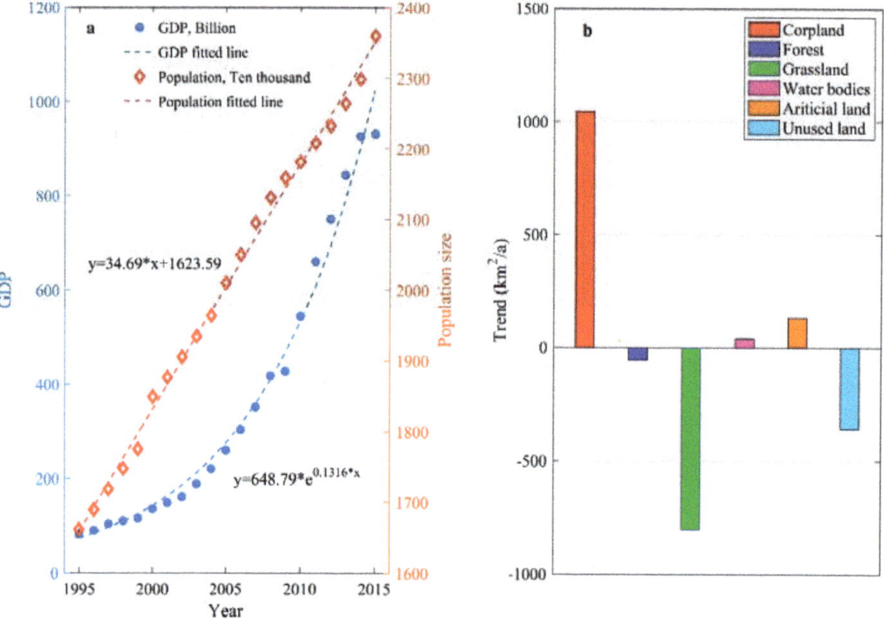

Figure 8. (a) GDP and population size of Xinjiang from 1995 to 2015, with the units of GDP being RMB. (a) Land use changes in Xinjiang from 1995 to 2015.

4. Discussion

Water conditions are the main limiting factor of agricultural in arid areas [40], such as Xinjiang. The water sources in arid areas mainly come from river, lake, and underground water, which are dominated by glaciers and snow meltwater. As a result, croplands are mainly distributed in the middle and lower course of rivers, which are rich in surface and groundwater resources with adequate access to irrigation water. Warmer and wetter climates [41] will contribute to the cropland expanding. During the past half century, runoff significantly increased in Xinjiang [38], which was caused by regional warming that accelerated the glaciers and snow melt. Thus, the cropland rapidly expanded (Figure 8).

However, with glacier storage decreasing, the glaciers and snow meltwater will also decrease, which will then lead to a reduction in runoff [5,12,42]. At this time, the runoff decrease will impact oasis development and perhaps some cropland will gradually experience desertification [17]. Therefore, the determining factor of cropland area in oasis is the glaciers and snow storage, as well as the annual ablation rate in mountainous areas; otherwise, it will lead to oasis development being unsustainable.

Climate change is one of the main factors that drives land-use change [43], especially in arid areas. The change in cropland is significantly affected by climate change. The water use of cropland is mainly provided by river runoff in Xinjiang [44]. The increased rate of runoff is largest in the Kai-Kong River Basin, when compared with other river basins. Meanwhile, the increase in cropland is also largest in the Kai-Kong River Basin (Figure 7). Since 2000, the runoff has been decreasing in the Kai-Kong River Basin [39], but the cropland area is still large, so it will place a large amount of pressure on irrigation water sources. If this situation continues for a long time, it will lead to groundwater being overdrawn and the water level falling, in addition to the lake drying and other environmental problems. The largest annual runoff in the Aksu River Basin is compared with the Kai-Kong River Basin and Yarkand River Basin, although a slight warming rate in the Aksu River Basin has resulted in the rate of runoff increasing at a smaller rate than the Kai-Kong River Basin. However, runoff showed a decreasing trend in the Yarkand River Basin, while temperature increased, perhaps due to the temperature data provided by meteorological stations being in low altitude regions. Then, we analyzed the change of 0 °C level high (FLH), which showed a decrease in the head of the Yarkand River Basin [39], which was limited by the ablation rate of glaciers and snow and then led to a reduction in runoff.

Human activities are important factors that may induced regional land use changes [32], mainly including the economy, while population size plays a dominant role in land use changes in Xinjiang [45–48], especially cropland changes [49]. The social and economic development has driven the unused land transfer into cropland in the middle and lower reaches of the river basin. However, if croplands expand than water usage will rise, and there will be a crowding out of ecological water, leading to the shrinkage of riparian forest in arid regions. Driven by economic interest, it is difficult to find a balance point between irrigation and ecological water use. Therefore, our next focus will be to systematically analyze the impact of human activities on land use change, expound the role of human activities in the relationship between humans and nature in arid areas from an economic view, and put forward a reasonable development strategy to maintain a sustainable development relationship between humans and nature in arid areas.

This paper has taken the equilibrium extent of land use to evaluate land use changes (Figure 2). However, as the land use types were only classified into six categories (i.e., cropland, forest, grassland, water bodies, artificial land, and unused land), the equilibrium extent index can reflect macrostate structural changes of land use, while it remains difficult to reveal the microstate structural changes of land use [50]. Therefore, for our future study, we will consider the use of configurational entropy [51,52], which is calculated using Fragstats software. It captures the microstate changes of the land use in a specific river basin, such as the Kai-Kong River Basin, the Aksu River Basin, or the Yarkand River Basin.

5. Conclusions

Our study has summarized the changes in land use in Xinjiang between 1995 and 2015. These changes were based on datasets and analyzed the influences of climate change on cropland transformation in three typical river basins. The main results are as follows:

Between 1995 and 2015, the largest increase in artificial land was 7.92%/a, followed by an increase of 3.86%/a in cropland and an increase of 0.81%/a in water bodies. The decreases in forestland, grassland, and unused land were −0.26%/a, −0.36%/a, and −0.07%/a, respectively. There were different transfer directions of land use changes during different periods. The forest, grassland, and unused land were transferred into cropland from 1995 to 2005. While, the area of artificial land increased rapidly due to urbanization, which accelerated between 2005 and 2015. Economic development and population expansion are the main factors for the cropland and artificial land increases in Xinjiang. Cropland expanded rapidly in southern Xinjiang between 1995 and 2015, which led to the migration of the centroid of cropland in the southwest direction.

The temperature and precipitation determined the cropland changes by influencing runoff in arid areas. The results of our multi-scale correlation explains the relationship between runoff and precipitation/temperature in the three river basins. The runoff changes were closely related to the cropland changes. The runoff increase was the largest in the Kai-Kong River Basin, corresponding with the increase in cropland being the fastest, reaching 29%/decade. Meanwhile, human activities maybe important factors for cropland increased in Xinjiang.

Author Contributions: Data curation, C.W. and F.H.; Funding acquisition, C.W.; Methodology, Q.L.; Supervision, Z.Y.; Visualization, Q.L.; Writing—original draft, Q.L.; Writing—review and editing, Z.Y. and F.H.

Funding: The research was supported by the Western Young Scholars Project, Chinese Academy of Sciences (2016-QNXZ-B-18).

Acknowledgments: Special thanks are owed to editors and anonymous reviewers for giving valuable suggestions and comments to improve this article. The authors also wish to express gratitude to the Chinese Meteorology Administration (http://data.cma.cn/) for providing air temperature and precipitation data, the Data Center for Resources and Environmental Sciences, Chinese Academy of Sciences (RESDC) (http://www.resdc.cn) for providing the land use data.

Conflicts of Interest: The authors declare no conflict of interest.

References

1. Liu, J.; Liu, M.; Zhuang, D.; Zhang, Z.; Deng, X. Study on spatial pattern of land-use change in China during 1995–2000. *Sci. China Ser. D Earth Sci.* **2003**, *46*, 373–384.

2. Nemani, R.R.; Keeling, C.D.; Hashimoto, H.; Jolly, W.M.; Piper, S.C.; Tucker, C.J.; Myneni, R.B.; Running, S.W. Climate-driven increases in global terrestrial net primary production from 1982 to 1999. *Science* **2003**, *300*, 1560–1563. [CrossRef] [PubMed]

3. Zelikova, T.J.; Williams, D.G.; Hoenigman, R.; Blumenthal, D.M.; Morgan, J.A.; Pendall, E. Seasonality of soil moisture mediates responses of ecosystem phenology to elevated CO_2 and warming in a semi-arid grassland. *J. Ecol.* **2015**, *103*, 1119–1130. [CrossRef]

4. Jeong, S.J.; Ho, C.H.; Brown, M.E.; Kug, J.S.; Piao, S. Browning in desert boundaries in Asia in recent decades. *J. Geophys. Res. Atmos.* **2011**, *116*, D02103. [CrossRef]

5. Aizen, V.B.; Aizen, E.M.; Melack, J.M.; Dozier, J. Climatic and hydrologic changes in the Tien Shan, central Asia. *J. Clim.* **1997**, *10*, 1393–1404. [CrossRef]

6. Ojima, D.S.; Galvin, K.A.; Turner, B.L. The global impact of land-use change. *BioScience* **1994**, *44*, 300–304. [CrossRef]

7. Foley, J.A.; DeFries, R.; Asner, G.P.; Barford, C.; Bonan, G.; Carpenter, S.R.; Chapin, F.S.; Coe, M.T.; Daily, G.C.; Gibbs, H.K.; et al. Global consequences of land use. *Science* **2005**, *309*, 570–574. [CrossRef]

8. Havstad, K.M.; James, D. Prescribed burning to affect a state transition in a shrub-encroached desert grassland. *J. Arid Environ.* **2010**, *74*, 1324–1328. [CrossRef]

9. Li, X.Y.; Zhang, S.Y.; Peng, H.Y.; Hu, X.; Ma, Y.J. Soil water and temperature dynamics in shrub-encroached grasslands and climatic implications: Results from Inner Mongolia steppe ecosystem of north China. *Agric. For. Meteorol.* **2013**, *171*, 20–30. [CrossRef]

10. Chen, L.; Li, H.; Zhang, P.; Zhao, X.; Zhou, L.; Liu, T.; Hu, H.; Bai, Y.; Shen, H.; Fang, J. Climate and native grassland vegetation as drivers of the community structures of shrub-encroached grasslands in Inner Mongolia, China. *Landsc. Ecol.* **2015**, *30*, 1627–1641. [CrossRef]

11. Fu, R.; Li, W. The influence of the land surface on the transition from dry to wet season in Amazonia. *Theor. Appl. Climatol.* **2004**, *78*, 97–110. [CrossRef]

12. Sorg, A.; Bolch, T.; Stoffel, M.; Solomina, O.; Beniston, M. Climate change impacts on glaciers and runoff in Tien Shan (Central Asia). *Nat. Clim. Chang.* **2012**, *2*, 725–731. [CrossRef]

13. Chen, Y.; Li, W.; Deng, H.; Fang, G.; Li, Z. Changes in central Asia's water tower: Past, present and future. *Sci. Rep.* **2016**, *6*, 35458. [CrossRef] [PubMed]

14. Deng, H.; Chen, Y.; Li, Y. Glacier and snow variations and their impacts on regional water resources in mountains. *J. Geogr. Sci.* **2019**, *29*, 84–100. [CrossRef]

15. Farinotti, D.; Longuevergne, L.; Moholdt, G.; Duethmann, D.; Mölg, T.; Bolch, T.; Vorogushyn, S.; Güntner, A. Substantial glacier mass loss in the Tien Shan over the past 50 years. *Nat. Geosci.* **2015**, *8*, 716–722. [CrossRef]

16. Ram, B.; Kolarkar, A. Remote sensing application in monitoring land-use changes in arid Rajasthan. *Int. J. Remote Sens.* **1993**, *14*, 3191–3200. [CrossRef]

17. Li, Y.; Zhao, M.; Motesharrei, S.; Mu, Q.; Kalnay, E.; Li, S. Local cooling and warming effects of forests based on satellite observations. *Nat. Commun.* **2015**, *6*, 6603. [CrossRef]

18. Nendel, C.; Hu, Y.; Lakes, T. Land-use change and land degradation on the Mongolian Plateau from 1975 to 2015—A case study from Xilingol, China. *Land Degrad. Dev.* **2018**, *29*, 1595–1606.

19. Xu, C.; Chen, Y.; Chen, Y.; Zhao, R.; Ding, H. Responses of surface runoff to climate change and human activities in the arid region of Central Asia: A case study in the Tarim River Basin, China. *Environ. Manag.* **2013**, *51*, 926–938. [CrossRef]

20. Ma, Z.; Kang, S.; Zhang, L.; Tong, L.; Su, X. Analysis of impacts of climate variability and human activity on streamflow for a river basin in arid region of northwest China. *J. Hydrol.* **2008**, *352*, 239–249. [CrossRef]

21. Meyer, W.B.; Turner, B.L. Human population growth and global land-use/cover change. *Annu. Rev. Ecol. Syst.* **1992**, *23*, 39–61. [CrossRef]

22. Lambin, E.F.; Meyfroidt, P. Global land use change, economic globalization, and the looming land scarcity. *Proc. Natl. Acad. Sci. USA* **2011**, *108*, 3465–3472. [CrossRef] [PubMed]

23. Fan, Y.; Chen, Y.; Liu, Y.; Li, W. Variation of baseflows in the headstreams of the Tarim River Basin during 1960–2007. *J. Hydrol.* **2013**, *487*, 98–108. [CrossRef]

24. Liu, J.; Liu, M.; Deng, X.; Zhuang, D.; Zhang, Z.; Luo, D. The land use and land cover change database and its relative studies in China. *J. Geogr. Sci.* **2002**, *12*, 275–282.

25. Liu, J.; Liu, M.; Tian, H.; Zhuang, D.; Zhang, Z.; Zhang, W.; Tang, X.; Deng, X. Spatial and temporal patterns of China's cropland during 1990–2000: An analysis based on Landsat TM data. *Remote Sens. Environ.* **2005**, *98*, 442–456. [CrossRef]

26. Bai, L.; Chen, Z.; Xu, J.; Li, W. Multi-scale response of runoff to climate fluctuation in the headwater region of Kaidu River in Xinjiang of China. *Theor. Appl. Climatol.* **2016**, *125*, 703–712. [CrossRef]

27. Chen, Z.; Chen, Y.; Bai, L.; Xu, J. Multiscale evolution of surface air temperature in the arid region of Northwest China and its linkages to ocean oscillations. *Theor. Appl. Climatol.* **2017**, *128*, 945–958. [CrossRef]

28. Wu, Z.; Huang, N. Ensemble empirical mode decomposition: A noise-assisted data analysis method. *Adv. Adapt. Data Anal.* **2009**, *1*, 1–41. [CrossRef]

29. Hirsch, R.; Slack, J. A nonparametric trend test for seasonal data with serial dependence. *Water Resour. Res.* **1984**, *20*, 727–732. [CrossRef]

30. Hamed, K. Trend detection in hydrologic data: The Mann–Kendall trend test under the scaling hypothesis. *J. Hydrol.* **2008**, *349*, 350–363. [CrossRef]

31. Sen, P. Estimates of the Regression Coefficient Based on Kendall's Tau. *J. Am. Stat. Assoc.* K **1968**, *63*, 1379–1389. [CrossRef]

32. Chen, Y.; Xu, C.; Chen, Y.; Li, Z.; Pan, Z. Response of snow cover to climate change in the periphery mountains of Tarim river basin, China, over the past four decades. *Ann. Glaciol.* **2008**, *49*, 166–172.

33. Wang, Y.; Chen, Y.; Ding, J.; Fang, G. Land-use conversion and its attribution in the Kaidu–Kongqi River Basin, China. *Quat. Int.* **2015**, *380*, 216–223. [CrossRef]

34. Chen, Y.; Liu, J. An index of equilibrium of urban land-use structure and information dimension of urban form [in Chinese with English abstract]. *Geogr. Res.* **2001**, *20*, 146–152.

35. Liu, J.; Zhang, Z.; Xu, X.; Kuang, W.; Zhou, W.; Zhang, S.; Li, R.; Yan, C.; Yu, D.; Wu, S.; et al. Spatial patterns and driving forces of land use change in China during the early 21st century. *J. Geogr. Sci.* **2010**, *20*, 483–494. [CrossRef]

36. Wu, J.; Marceau, D. Modeling complex ecological systems: An introduction. *Ecol. Model.* **2002**, *153*, 1–6. [CrossRef]

37. Vranken, I.; Baudry, J.; Aubinet, M.; Visser, M.; Bogaert, J. A review on the use of entropy in landscape ecology: Heterogeneity, unpredictability, scale dependence and their links with thermodynamics. *Landsc. Ecol.* **2015**, *30*, 51–65. [CrossRef]

38. Deng, H.; Chen, Y.; Wang, H.; Zhang, S. Climate change with elevation and its potential impact on water resources in the Tianshan Mountains, Central Asia. *Glob. Planet. Chang.* **2015**, *135*, 28–37. [CrossRef]

39. Chen, Z.; Chen, Y.; Li, W. Response of runoff to change of atmospheric 0 C level height in summer in arid region of Northwest China. *Sci. China Earth Sci.* **2012**, *55*, 1533–1544. [CrossRef]

40. Deng, X.P.; Shan, L.; Zhang, H.; Turner, N.C. Improving agricultural water use efficiency in arid and semiarid areas of China. *Agric. Water Manag.* **2006**, *80*, 23–40. [CrossRef]

41. Shi, Y.F.; Shen, Y.P.; Hu, R.J. Preliminary study on signal, impact and foreground of climate shift from warm–dry to warm–humid in northwest China [in Chinese with English abstract]. *J. Glaciol. Geocryol.* **2002**, *3*, 219–226.

42. Li, Z.; Wang, W.; Zhang, M.; Wang, F.; Li, H. Observed changes in streamflow at the headwaters of the Urumqi River, eastern Tianshan, central Asia. *Hydrol. Processes* **2010**, *24*, 217–224. [CrossRef]

43. Liu, Q.; Yang, Z.; Han, F.; Wang, Z.; Wang, C. NDVI-based vegetation dynamics and their response to recent climate change: A case study in the Tianshan Mountains, China. *Environ. Earth Sci.* **2016**, *75*, 1189. [CrossRef]

44. Yang, X.; Chen, C.; Luo, Q.; Li, L.; Yu, Q. Climate change effects on wheat yield and water use in oasis cropland. *Int. J. Plant Prod.* **2011**, *5*, 83–94.

45. Chen, Z.; Chen, Y.; Li, W. Land use/cover change and their driving forces in Hotian river basin of Xinjiang [in Chinese with English abstract]. *J. Desert Res.* **2010**, *30*, 326–333.

46. Zhou, L.; Tian, Y.; Myneni, R.B.; Ciais, P.; Saatchi, S.; Liu, Y.Y.; Piao, S.; Chen, H.; Vermote, E.F.; Song, C.; et al. Widespread decline of Congo rainforest greenness in the past decade. *Nature* **2014**, *509*, 86–90. [CrossRef] [PubMed]

47. Feng, X.; Fu, B.; Piao, S.; Wang, S.; Ciais, P.; Zeng, Z.; Lü, Y.; Zeng, Y.; Li, Y.; Jiang, X.; et al. Revegetation in China's Loess Plateau is approaching sustainable water resource limits. *Nat. Clim. Chang.* **2016**, *6*, 1019–1022. [CrossRef]

48. Kang, C.; Zhang, Y.; Wang, Z.; Liu, L.; Zhang, H.; Jo, Y. The Driving Force Analysis of NDVI Dynamics in the Trans-Boundary Tumen River Basin between 2000 and 2015. *Sustainability* **2017**, *9*, 2350. [CrossRef]

49. Liu, J.; Hertel, T.W.; Lammers, R.B.; Prusevich, A.; Baldos, U.L.C.; Grogan, D.S.; Frolking, S. Achieving sustainable irrigation water withdrawals: Global impacts on food security and land use. *Environ. Res. Lett.* **2017**, *12*, 104009. [CrossRef]

50. Cushman, S.A. Calculation of configurational entropy in complex landscapes. *Entropy* **2018**, *20*, 298. [CrossRef]

51. Gao, P.; Zhang, H.; Li, Z. A hierarchy-based solution to calculate the configurational entropy of landscape gradients. *Landsc. Ecol.* **2017**, *32*, 1133–1146. [CrossRef]

52. Cushman, S.A. Calculating the configurational entropy of a landscape mosaic. *Landsc. Ecol.* **2016**, *31*, 481–489. [CrossRef]

 *sustainability* 

Article

Modelling Development, Territorial and Legislative Factors Impacting the Changes in Use of Agricultural Land in Slovakia

Lucia Palšová [1,*], Katarína Melichová [2] and Ina Melišková [1]

[1]   Department of Law, Faculty of European Studies and Regional Development, Slovak University of Agriculture in Nitra, Tr. A. Hlinku 2, 949 76 Nitra, Slovakia

[2]   Department of Public Administration, Faculty of European Studies and Regional Development, Slovak University of Agriculture in Nitra, Tr. A. Hlinku 2, 949 76 Nitra, Slovakia

*   Correspondence: lucia.palsova@uniag.sk; Tel.: +421-37-641-5079

Received: 31 May 2019; Accepted: 15 July 2019; Published: 17 July 2019

Abstract: The conflict of interests in agricultural land use based on the diversity of needs of private and public interest is the main problem of the current protection of agricultural land in Slovakia. Therefore, the aim of the paper is to identify factors affecting the withdrawal of agricultural land, i.e., conversion of the agricultural land to non-agricultural purposes, and to initiate a professional discussion on the concept of protection and use of the agricultural land in Slovakia. Through panel regression models, the developmental, territorial, and legislative factors affecting land withdrawal for the purpose of housing, industry, transport, mining, and other purposes were analyzed. Research has shown that developmental factors, compared to legislative ones, affect the total volume of agricultural land withdrawn in bigger scope. From the perspective of the conflict of interests between the individuals and state regarding land protection, the private interest prevails over the public one. As a consequence, agricultural land is withdrawn in suburbanized and attractive areas, where the land of the highest quality is mostly located. In accordance with the precautionary principle, the state should adopt a long-term conceptual document defining the areas of agricultural land use taking into account the impact of the developmental factors on the land protection.

Keywords: withdrawal of agricultural land; contributions; developmental factors; territorial factors; legislative factors

---

1. Introduction

The competition for land resources creates serious risks of geopolitical imbalances in the European Union (hereinafter as "EU") and worldwide [1]. For the following reasons, it is currently emphasized to ensure effective tools of land policy. Its complexity lies in the fact that the management of land nearly always faces a trade-off between various social, economic, and environmental needs [2]. Land use for non-agricultural purposes has resulted in impermeable coverage of the land as an irreversible process leading to the loss of soil resources [3–8].

Research on land use is now being redefined. Until now, the emphasis had been on the impermeable coverage of land in general [9–11], currently more important is the examination of the factors causing the land withdrawal. Indeed, the withdrawal of agricultural land for non-agricultural purposes will often irreversibly depreciate the most productive soil in the area, and therefore there is an emphasis put on the support for the rational decision-making process of the administrative authorities [12–14].

The process of agricultural land withdrawal is predominant in suburban areas [7,8,15]. In some countries, due to the rapid urbanization process and the increasing value of land, developers started to focus on the construction of housing units on low quality land [16].

The issue of urbanization and the withdrawal of agricultural land is a worldwide phenomenon, and while it draws attention to the economic growth, the majority of the agricultural land is withdrawn for non-agricultural purposes (e.g., up to 70% in Pakistan), resulting in a lack of food and nutrition for the growing population and an increase of socio-economic and infrastructure issues [17]. China is another example, where they are experiencing severe depletion of agricultural land in several urban agglomerations. This situation underlines the urgent need for the government to develop an effective policy for the protection of agricultural land [18–21]. An elaboration of studies on the use and withdrawal of agricultural land can be seen as an important tool for policy makers and future generations and for maintaining ecological well-being of inhabitants [22]. A study conducted in selected EU cities has demonstrated that an integrated approach, along with other sustainability objectives instead of isolated programs, reduces the withdrawal of land and subsequent impermeable coverage [5].

The specificities of the land use conflicts in the EU are also linked to the implementation of the Common Agricultural Policy (hereinafter as "CAP"), which affects the regional land market by disrupting agricultural land prices. Research in Poland has shown that in the case of the most urbanized regions, the contribution of the CAP to the agricultural land prices was relatively low, leading to a strong incentive for farmers to sell their land for non-agricultural purposes and thus to land use conflicts [23]. Contrary, in Italy, the risks associated with changes in the use of agricultural land are the most significant. Its consumption is mainly related to the urbanization pressure, calling for the highly efficient decision-making tools, such as municipal town planning schemes [24,25]. All EU countries are currently addressing the withdrawal of agricultural land related to the urban expansion, with the objective to help authorities and policy makers understand the effects of urban expansion and related land use change on agriculture [26,27]. Research in this area has shown that the CAP subsidies have an impact on reducing urbanization and reducing agricultural land use [28].

Despite the importance of joint action to protect the area and quality of the agricultural land in the EU, responsibility for land policy and legislation and its implementation remain a national competence of the EU Member States [29].

Slovakia is the EU country with the smallest total area of agricultural land (48.58% of agricultural land, of which 50% is arable land [30]). The majority of the agricultural land belongs to the quality categories 5–7 (out of the total 9 quality categories ordered from the most to the least productive [31]).

The conflict between satisfying the private and public needs is addressed by the state through the state legislative instruments. The protection of agricultural land is defined as the basic constitutional obligation of the state [32], which was subsequently amended in the sectoral Act No 220/2004 Coll. on the protection and use of agricultural land and amending Act No 245/2003 Coll. on integrated pollution prevention and control and in amendments to certain acts (lex generalis).

From a qualitative point of view, the Act transfers the state's responsibility for protecting the land to the owner, user or land manager. Although the law expressis verbis determines the obligations of the parties concerned, enforcement for the infringement is weak [33]. The reason is the legal and factual problem related to the complexity of the whole issue (causality and its proving, reversibility of damages, forms of remedy) on the one hand and the small societal and political interest in increasing the enforceability of legal obligations on the other hand [34].

From a quantitative point of view, the Act regulates the withdrawal of land for non-agricultural purposes according to the framework areas of the occupation, but does not reflect the purpose of the land. The Act is limited to economic instruments, such as contributions for the withdrawal of agricultural land for non-agricultural purposes and fines for breaching of land area protection obligations. The contributions are based on the principle that the person in whose interest the land is withdrawn is obliged to pay a specified amount [5]. Therefore, they have a motivation function, as it should lead the applicant to withdraw the land to limit the required occupation, to choose the withdrawal of land of less quality, or choose an alternative solution [34,35].

Also for this reason, contributions have become an effective tool of the state to pursue state land policy objectives. The development of legislative changes implies that the state has abolished, respectively modified, the contributions according to the expected investment development. With the accession of Slovakia to the EU in 2004, the contributions were abolished in order to open up the internal market [36]. As the legislator's expectation of a positive impact on landowners or users has not been met, in 2009 contributions were reintroduced for the agricultural land of the highest quality [37]. In 2013, contributions were extended to withdrawal of agricultural land of the highest quality in each cadastral area [38]. The introduced change has brought a positive effect on the protection of the land area and balanced development within the regions [39]. Constitutional change in 2017, brought a special protection of agricultural land [32].

The application of the Act is left to the decision-making competence of the employees in state administration on the district level, who lack methodological tools and clear guidelines on how to solve conflict in land use [40]. Izakovičová et al. [2] provide such a methodical approach to the integrative assessment of the land use conflicts specifically under conditions currently present in Slovakia.

In Slovakia, the territorial concept of the use of agricultural land and the planning of strategic activities has been absent for a long time. This assumes an analysis of factors affecting the withdrawal of agricultural land and the subsequent establishment of long-term priorities, which takes into account indicators of soil quality, climate change and development [41,42].

As the impacts of the state land policy on the sustainable use of agricultural land have never been comprehensively examined in Slovakia, the aim of the article is to identify development, territorial, and legislative factors affecting the loss of volume of agricultural land in Slovakia due to its withdrawal.

2. Materials and Methods

To achieve the aim, first, we conducted an analysis of the spatial distribution of the withdrawal of agricultural land in Slovak districts for the period of 2007–2016, due to data availability. To better comprehend trends in the agricultural land withdrawal, a Delphi method was employed. The panel of experts consisted of 28 respondents from the relevant scientific fields as well as from practice (Ministry of Agriculture and Rural Development of the Slovak Republic, Soil Science and Conservation Research Institute, Research Institute of Agriculture and Food Economics, Central Control and Testing Institute in Agriculture, Slovak University of Agriculture in Nitra, chairman of an agricultural cooperative and representatives of the Land and Forestry Departments of district offices located in the seats of 8 NUTS III regions of the Slovak Republic).

Conducting the interviews with listed experts allowed us to confront the empirical findings with the information provided by the experts. Consequently, this confrontation, together with the theoretical assumptions discussed in detail in the first part of the paper, helped us to narrow down the factors affecting the volume of agricultural land withdrawal, as well as to formulate the final statistical models used to quantify their effects statistically. For this purpose, in the final part of the paper, a panel data set containing the information on the amount of land withdrawn from the agricultural land fund (for housing development purpose, industrial development purpose, mining, transportation and other purposes) for the period of 2007–2016 and spanning 41 districts of Slovakia was used, meaning 410 observations in total. The withdrawal of agricultural land is an administrative process that results in the conversion of agricultural land into one of the alternative uses listed above.

The decision to extend the analysis to panel data (as opposed to a simple time series) was reached mainly due to the findings generated by the interviews with experts, especially the need to take into account regional and local specificities. Indeed, interviews with representatives of the district authorities revealed that the factors and circumstances that caused the intensity of the withdrawal of agricultural land in individual districts varied spatially. However, this decision eliminated a group of potentially important factors, for which it is not possible to obtain relevant data on the district level. In addition, some important factors are invariable or do not vary sufficiently, resulting in an arbitrary decision to use a panel regression model with random errors. The dependent variable of the individual models is the volume of agricultural land withdrawn, calculated as a proportion to the population of the respective district in a given year. We examine the effect of relevant factors not only on the total volume withdrawn but also on the volume withdrawn for individual purposes, since it is justified to assume that the circumstances affecting the intensity of land withdrawal will be different. Data on the withdrawal of agricultural land were obtained from the Electronic Land Service Yearbook and from the Ministry of Agriculture and Rural Development of the Slovak Republic.

Individual potentially relevant developmental, territorial, and legislative factors enter the models as explanatory variables and have been chosen based on the study of relevant literature in the first part of the article and based on the results of interviews carried out with the panel of experts.

All input data were standardized by calculation of z-scores using the formula:

$$I^s_{ij} = \frac{I_{ij} - \overline{I_j}}{\sigma}$$

where I^s_{ij} is the new normalized value of indicator I in district i in year j, $\overline{I_j}$ is the average value of indicator I in the year j and σ is the standard deviation.

In terms of the developmental factors, we examine the impact of foreign direct investment flows—FDI_POP (measured as foreign direct investments in euros per capita; data were obtained from the National Bank of Slovakia), and domestic business growth—DBUS_POP (measured as the number of newly established enterprises per capita; data were obtained from the Register of Organizations). Additionally, income level of the population—AMNW (measured as the average monthly nominal wage), the number of immigrants—IMMIG (measured as the absolute number of immigrants), the net migration—NETMIG and NETMIG_POP (measured as a migration balance in absolute terms and relative to the population in the base year) were included into the models and obtained from the Statistical Office of the Slovak Republic. In the case of migration, all three indicators need to be taken into account, as each reflects a different aspect of the impact of population movements.

In the case of territorial factors, the most important characteristic is usually the settlement structure, not only because it affects the demand for the use of space, but also because it is closely related to the developmental factors. In our analyses, we examine two: The average size of settlements in the district—AVER_SET (measured as the arithmetic average of the population of all municipalities and cities in a given district) and the size of the central city—CENT_CITY (measured as the population of the largest city in the district) for which the data were obtained from the Statistical Office of the Slovak Republic. In addition, we also consider the territorial characteristics of the land fund itself, namely the fragmentation of land. We also analyze two factors: Fragmentation of arable land—FRAG_ARABLE (measured as the average size of land blocks identified as arable land) and fragmentation of agricultural land—FRAG_AGRI (measured as the average size of land blocks identified as agricultural land). A shapefile of Land Parcel Identification System (LPIS) blocks was used to generate the variables, and it was obtained from the Ministry of Agriculture and Rural Development of the Slovak Republic. It should be noted that the land fragmentation data is only available for the present, so these variables have only a spatial dimension.

In the case of legal factors, while constructing the model, we rely mainly on the information provided by the panel of experts. Dummy and categorical variables quantifying potentially relevant legislative milestones were used. Dummy variable—EXEMP_VINEYARD was designed to introduce contributions on the withdrawal of vineyards as well as the introduction of exemptions for family homes, where 0 means the absence of the regulations in question and 1 means that the regulations are introduced/in force. These two legislative changes were introduced in the same year and were in effect until the end of the analyzed timeframe, so we cannot separate their effects. A categorical variable CON was used to measure the impact of contributions imposed for the withdrawal of agricultural land because in addition to being introduced in the given period, the contributions were also significantly altered (by adding land quality categories 5 to 9 later on). The reference value of the variable acquires a value of 0 if contributions were not levied for the withdrawal of land, 1 for the years when they were paid only for quality categories 1 through 4 and 2, for the years when contributions were paid for withdrawal of all quality categories. In this case, it should be noted that the variables reflecting the legislative milestones have only a temporal dimension as they apply equally to all spatial units. For every withdrawal purpose we also examine all the factors together, with the resulting panel regression model having the following form:

$$\begin{aligned} zWTD_{it}^{p} = \ & \alpha + \beta_1 zFDI\_POP_{it} + \beta_2 zDBUS\_POP_{it} + \beta_3 AMNW_{it} + \beta_4 zIMMIG_{it} \\ & + \beta_5 zNETMIG_{it} + \beta_6 zNETMIG\_POP_{it} + \beta_7 zFRAG\_ARABLE_i \\ & + \beta_8 zFRAG\_AGRI_i + \beta_9 zAVER\_SET_{it} + \beta_{10} zCENT\_CITY_{it} + \beta_{11} CON(1\_4)_t \\ & + \beta_{12} CON(1\_9)_t + \beta_{13} EXEMP\_VINEYARD_t + u_t + \varepsilon_{it} \end{aligned}$$

where $zWTD_{it}^{p}$ is the value of the dependent variable in district i in year j (normalized value of volume of withdrawn agricultural land for purpose p), αi denotes the value of intercept, μ_i introduces a time-invariant district-specific unobserved component, while ε_{it} is idiosyncratic error term.

The proposed approach is original mainly in its scope, both geographic and in terms of comprehensiveness of the impacts tested. In Slovakia specifically, analyzing the land use conflicts and subsequent changes relies on conducting case studies [2] or on using a specific set of indicators, like legal aspects [39]. We believe that analyzing both development and legislative factors, while taking territorial conditions into account, will help to determine how the individual entities and public policies interact in land use conflicts, thus providing insights into what needs to be done to increase the effectiveness of the legal framework aimed at achieving sustainable use of land resources.

3. Results and Discussion

In Slovakia, due to historical and territorial circumstances (the emergence of of separate state, socialism), emphasis was placed on the farming on agricultural land to the detriment of ownership relations with the aim of increasing production. As a result, despite the social and political changes in 1989, more than 90% of the agricultural land is currently leased, causing the loss of owners' relationship to the land and loss of their motivation to implement sustainable agriculture practices [41]. Therefore, when the demand for investment activities and the convergence to the EU-15 standard of living occurred, after the accession of Slovakia to the EU in 2004, landowners were not interested in maintaining the land in the agricultural land fund. In addition, through legislation, the state had created a leeway for land use for non-agricultural purposes for both private and public needs [43].

The trend of withdrawing of the agricultural land for non-agricultural purpose was fluctuating in the given period (Figure 1). The most pronounced conflict in alternative use of the agricultural land has been proven between both agriculture and residential use and agriculture and industrial development. The most notable extreme in amount of withdrawn agricultural land happened in 2008, which is the last year when the contributions for withdrawal were not levied on those that withdrew agricultural land of top four quality categories. An increase was recorded in withdrawal for all listed purposes, the most significant, however, occurred due to the conversion to residential and industrial use, as well as the accompanying infrastructure development.

Almost 5 thousand hectares of agricultural land were withdrawn in one year alone, clearly indicating that both private and public entities reacted to the upcoming legislative amendment introducing the contributions in the following year. In the following years, the trend of land withdrawal steadily declined, mostly not exceeding the level of one thousand hectares per year. Over the course of analyzed time period, two more rises occurred. The first one was in 2012, in the year that also coincided with relatively major legislative change, namely the extension of the contribution obligation for withdrawal of agricultural land for all quality categories that came into force the following year. This, however, seemed to affect mostly the withdrawal for transport infrastructure development. This is usually planned years ahead, so the causative relationship between the mentioned contribution extension and a prior increase in land withdrawal might not be straightforward. The second increase occurred in 2016, which was caused by a marked rise in the conversion of agricultural land to industrial sites and it coincided with the latest major foreign investment in Slovakia – Jaguar Land Rover located in the Nitra district.

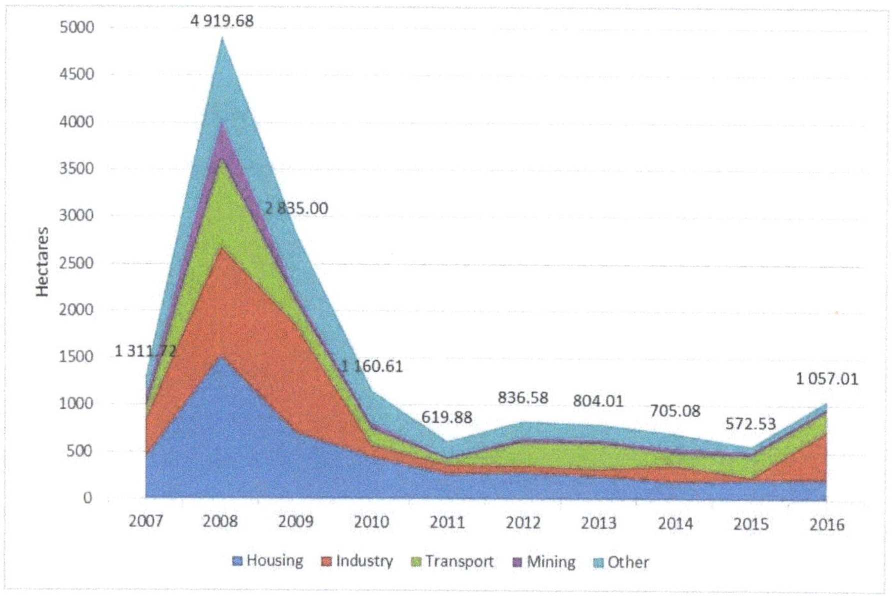

Figure 1. Volume of withdrawn agricultural land according to the purpose of withdrawal in the Slovak Republic in 2007–2016 (in ha). Source: Own processing based on the Electronic Land Service Yearbook and data from the Ministry of Agriculture and Rural Development of the Slovak Republic.

Overall, the lowest amount of the agricultural land was lost for mining purposes, which seems to be phasing out almost completely in recent years.

Based on the total amount of the withdrawn agricultural land for the analyzed ten-year period (Figure 2) we can conclude that the most intensive conflict in land use arises in the western and north-western part of the country, supporting the assumptions presented in Reference [42]. The greatest losses of agricultural land occurred in the districts with a heavy presence of foreign investments (Trnava, Nitra, and Galanta) located close to the capital city Bratislava. Over the ten-year period, the withdrawal in these districts was caused mainly by the expansion of industrial areas, most notably the automobile sector and expansion of the related economic activities, which was confirmed by the experts. Although the amount of the withdrawn agricultural land in the Bratislava district is relatively low, all neighboring districts experienced severe pressure in alternative use of agricultural land, with a most pronounced increase in demand for the residential space. Although the capital city of Bratislava, and the most developed region in Slovakia, exerts positive economic spatial spillovers on the surrounding regions, it also creates intense conflicts in the use of land, resulting in permanent losses of the agricultural land of the highest quality. We can observe a similar influence of Košice—the second largest city in Slovakia, located in the eastern part of the country, although the land takes intensity in its neighboring districts is of a lesser magnitude. Overall, in the eastern districts, aside from the development of industrial and commercial sites, residential use seems to be the biggest rival to the agricultural sector in use of land.

In the tables below (Tables 1–6), for each withdrawal purpose under analysis, we introduce several models representing different combinations of factors defined in the Materials and Methods section of the article. Specifically, Model I represents the estimates of the impacts of developmental factors on the withdrawal of agricultural land, model II represents the estimates of the impacts of territorial characteristics, model III the impact of significant legislative changes, and model IV the combination of all the mentioned factors. Statistically significant regression coefficients are indicated by asterisks in the table, while standard errors are given in brackets below the regression coefficients. Due to the standardization of input data, the coefficients are dimensionless and can be compared.

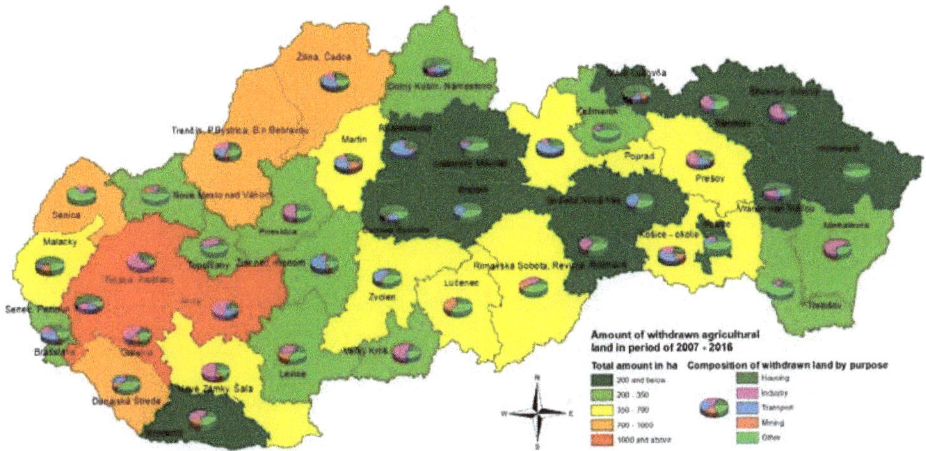

Figure 2. Spatial aspects of the agricultural land withdrawal in Slovakia in 2007–2016. Source: Own processing based on the Electronic Land Service Yearbook and data from the Ministry of Agriculture and Rural Development of the Slovak Republic.

From Table 1 it is clear that all factor groups have a statistically significant effect on the total volume of the agricultural land withdrawal (all models are statistically significant, as evidenced by Wald's χ^2). By comparing the R^2 coefficients of determination, it can be concluded that the developmental factors have the greatest impact on the total volume of the agricultural land withdrawn. Among these, foreign direct investment, income level, as well as mechanical population movement (i.e., migration) are statistically significant. This confirms not only the theoretical assumptions but also the statements of several interviewed experts and previously conducted studies [44–46].

In the case of absolute value of the migration balance, we have identified contradictory effects. The increase of the absolute value of the migration balance leads to the reduction in the volume of the land withdrawal. However, when we recalculate the migration balance as a share of the total population in the base year, we find the effect has opposite direction. With the increase in the net migration per capita, the volume withdrawn from the agricultural land fund increases significantly.

Although suburbanization and de-urbanization processes are already taking place in Slovakia, the urbanized regions tend to have the highest migration balance. In many cases, these are the regions where the starting point of the agricultural land area has already been very low. It can, therefore, be assumed that the increasing demands for the alternative use of space caused by immigration are being addressed to a greater extent by adapting areas, other than agricultural land, to new functions for which demand has increased (e.g., former industrial districts, increasing the number of floors in residential areas, etc.).

However, when assessing the net migration balance relative to the baseline population number, the highest values of this indicator are recorded by the regions that previously had a lower population density but became more attractive to migrants at that time. Often these are the regions that are the target of suburbanization processes. Thus, we have also statistically confirmed that it is precisely in these regions that the most significant changes in the use and functions of the land are taking place (as existing capacities are not sufficient given the growing demand). Charles and Guna [47] affirm that migration has a negative impact on the city's limited land and it is necessary to propose solutions for its sustainable use strategy.

Table 1. Results of panel regression models where the dependent variable is the total volume of agricultural land withdrawn.

Factor:	Model I	Model II	Model III	Model IV
zFDI_POP	0.2757 *			0.2059
	(0.1267)			(0.1431)
zAMNW	−0.1983 **			−0.0096
	(0.0616)			(0.0938)
zIMMIG	−0.1836			−0.3306
	(0.1248)			(0.3405)
zNETMIG	−0.3769 *			−0.1941
	(0.1641)			(0.2320)
zNETMIG_POP	0.8424 ***			0.6486 ***
	(0.1548)			(0.175)
zDBUS_POP	−0.039			−0.0868
	(0.0455)			(0.057)
zFRAG_ARABLE		0.2172 **		0.0713
		(0.0665)		(0.0522)
zFRAG_AGRI		0.0706		−0.0123
		(0.0724)		(0.0535)
zAVER_SET		0.0419		0.1377
		(0.2355)		(0.2167)
zCENT_CITY		−0.185		−0.0846
		(0.229)		(0.2051)
CON (1–4)			−0.4024 **	−0.3149 **
			(0.1224)	(0.1323)
CON (1–9)			−0.5938 **	−0.5659 **
			(0.1730)	(0.1984)
EXEMP_VINEYARD			−0.0049	0.0025
			(0.1631)	(0.1719)
Intercept	-6.72×10^{-17}	-1.90×10^{-16}	0.3999 **	0.3515 **
	(0.0439)	(0.0647)	(0.1161)	(0.1236)
N° of observations	410	410	410	410
R^2				
within	0.0830	0.0107	0.0000	0.0971
between	0.7397	0.3248	0.0000	0.7713
overall	0.2228	0.0709	0.0477	0.2445
sigma_u	0.0000	0.2895	0.3783	0.0000
sigma_e	0.8928	0.9208	0.9046	0.8841
rho	0.0000	0.09	0.1488	0.0000
Wald χ^2	115.52 ***	16.74 **	23.85 ***	128.17 ***

Note: standard errors are given in brackets, * indicates significance level $\alpha < 0.05$, ** $\alpha < 0.02$, and *** $\alpha < 0.001$. Source: own calculation.

At first sight, a statistically significant inverse relationship between the average monthly nominal wage and the volume of withdrawn agricultural land is unclear. It can be seen as a reflection of the fact that income levels in Slovakia are higher in the urbanized regions, where the baseline area of the agricultural land per capita is lower. However, we cannot exclude the possibility that this relationship is more direct. The income level often reflects the overall level of financial capital in a given territory. In this context, this may mean that entities in such regions have the opportunity to use alternative resources for areas with new features; because they have the resources that they can use e.g., for brownfield revitalization. This is considerably more expensive than building on the green field, but it reduces the pressure on the agricultural land [48].

Within the territorial characteristics, only the rate of arable land fragmentation was statistically significant, and with increasing parcel size, the withdrawal rate is increasing. We also explain this relationship through the financial costs, specifically transaction costs. In the case of any construction on the agricultural land, it is much cheaper for an investor, an owner or other relevant entity to obtain a unified land plot than to negotiate with several original owners. The Ministry of Agriculture and Rural Development of the Slovak Republic in 2019 identified an average of 12 co-owners per 1 land plot, while one owner owns 23 parcels [49].

In terms of legislative changes, the introduction of contributions on vineyards, or the introduction of exemptions for family houses does not statistically significantly affect the withdrawal rate. On the other hand, we also showed a positive impact of introducing the contributions on reducing the volume of the withdrawn agricultural land. For comparison, when withdrawal contributions were not collected, extending their collection to all quality groups (i.e., from 1 to 9) had a noticeably greater effect on reducing the volume of the agricultural land withdrawn than when they were collected only for the withdrawal of quality groups 1 to 4. As confirmed by Sobocká [41], the legislative change has increased the protection of the agricultural land from 21 to 37% for the highest quality agricultural land in each cadastral area.

When taking into account all factors (Model IV), only the positive impact of the legislative changes remains significant together with the migration balance per capita, which has a negative effect on the total volume of the land withdrawn.

In order to comprehensively understand the cause-and-effect relationships between the analyzed factors and the intensity of the withdrawal of agricultural land, we need to carry out more detailed analyses. Therefore, in the following sections, we estimate the relationship between all the factors analyzed and the volume of the land withdrawn for individual purposes.

3.1. The Volume of Agricultural Land Withdrawn for the Housing Purposes

As shown by the models' results (Table 2), only developmental and legislative factors are significant in the case of land withdrawal.

In terms of developmental factors, the highest withdrawal intensity has been demonstrated in the suburbanized areas and in the areas with the inflow of foreign direct investment. Contrarily, in the urbanized areas, there is lower land availability resulting in the lower occupation of the agricultural land. Urbanized areas offer alternative and more expensive housing options, which are, however, allowed by the higher income level of the population. Conversely, the lower-income population is to a greater extent migrating to urbanized areas and looking for housing opportunities in suburbanized areas, where the pressure on the land withdrawal is consequently higher [50].

This is also evidenced by a statistically significant positive coefficient of the share of migration balance per capita. Even in the case of land withdrawal for housing purposes, there has been a decreasing trend after the introduction of contributions, with a more substantial effect on this reduction being the extension of the obligation to pay contributions and for the withdrawal of agricultural land from lower quality groups, which was predicted by Sobocká [41].

Legislative factors have a significant impact on the withdrawal of land, but according to the opinion of experts, it is not a decisive factor in the choice of housing.

Table 2. Results of panel regression models where the dependent variable is the volume of agricultural land withdrawn for housing purposes.

Factor:	Model I	Model II	Model III	Model IV
zFDI_POP	−0.0843 (0.1264)			−0.2356 (0.1429)
zAMNW	−0.1171 (0.0615)			0.0516 (0.0936)
zIMMIG	0.1061 (0.1246)			−0.0992 (0.3399)
zNETMIG	−0.1046 (0.1638)			0.1288 (0.2316)
zNETMIG_POP	0.5529 *** (0.1545)			0.3932 * (0.1747)
zDBUS_POP	0.0222 (0.0454)			−0.0054 (0.0569)
zFRAG_ARABLE		0.1519 (0.0794)		−0.0503 (0.0522)
zFRAG_AGRI		0.1019 (0.0864)		0.0534 (0.0534)
zAVER_SET		−0.0574 (0.2811)		0.1257 (0.2164)
zCENT_CITY		−0.0446 (0.2733)		0.0529 (0.2048)
CON (1–4)			−0.4054 ** (0.1199)	−0.3500 ** (0.1320)
CON (1–9)			−0.5181 ** (0.1696)	−0.4418 * (0.1980)
EXEMP_VINEYARD			−0.0371 (0.1599)	−0.0671 (0.1716)
Intercept	-5.25×10^{-18} (0.0438)	-7.42×10^{-17} (0.0773)	0.3805 ** (0.1185)	0.3369 ** (0.1233)
N° of observations	410	410	410	410
R^2				
within	0.0227	0.0003	0.0000	0.0426
between	0.8193	0.1599	0.0000	0.8532
overall	0.2258	0.0408	0.0404	0.2471
sigma_u	0.0000	0.4024	0.4277	0.0000
sigma_e	0.8860	0.9064	0.8869	0.8723
rho	0.0000	0.1646	0.1887	0.0000
Wald χ^2	117.51 ***	6.80	20.99 ***	129.98 ***

Note: standard errors are given in brackets, * indicates significance level $\alpha < 0.05$, ** $\alpha < 0.02$, and *** $\alpha < 0.001$. Source: own calculation.

3.2. The Volume of Agricultural Land Withdrawn for Industry Development Purposes

Land withdrawal for industrial purposes is significantly affected by all the examined factors. Regarding legislative changes, there is a comparable effect of introducing contributions on the withdrawal of agricultural land to the volume of the land withdrawn for industrial development, as was the case with the already analyzed models.

In terms of territorial factors, arable land fragmentation significantly affects the volume of the withdrawn land. As the average land size increases, the volume of the withdrawn agricultural land also increases. This further confirms the assumption that investors, who prefer to occupy larger parcels of land, are searching for sites where they can acquire land with low transaction costs, whose amount is significantly affected by fragmentation and complex ownership relationships. At the same time, the experts underlined that higher land fragmentation, as a negative phenomenon, often hindering the usability of the land for large scale farming as well.

Table 3. Results of panel regression models where the dependent variable is the volume of agricultural land withdrawn for industry development purposes.

Factor:	Model I	Model II	Model III	Model IV
zFDI_POP	0.2893 (0.1631)			0.3798 ** (0.1641)
zAMNW	−0.2045 ** (0.0706)			−0.1446 (0.1071)
zIMMIG	−0.1498 (0.1661)			−0.1663 (0.391)
zNETMIG	−0.2039 (0.2148)			−0.1844 (0.2659)
zNETMIG_POP	0.4519 * (0.2023)			0.3019 (0.2012)
zDBUS_POP	−0.0044 (0.0484)			−0.0046 (0.0606)
zFRAG_ARABLE		0.2785 *** (0.0515)		0.2552 *** (0.0624)
zFRAG_AGRI		0.0368 (0.0561)		−0.0212 (0.064)
zAVER_SET		−0.3499 (0.1825)		−0.4297 (0.2555)
zCENT_CITY		0.2669 (0.1775)		0.2957 (0.2421)
CON (1–4)			−0.2998 ** (0.1275)	−0.2232 (0.1419)
CON (1–9)			−0.5258 ** (0.1804)	−0.3993 (0.2145)
EXEMP_VINEYARD			0.0834 (0.1700)	0.1831 (0.1840)
Intercept	-3.91×10^{-17} (0.0629)	-6.43×10^{-17} (0.0501)	0.3052 ** (0.1142)	0.1941 (0.1371)
N° of observations	410	410	410	410
R^2 within	0.0452	0.0093	0.0000	0.0549
between	0.1677	0.4844	0.0000	0.4996
overall	0.0636	0.0854	0.0292	0.1318
sigma_u	0.2715	0.1078	0.3003	0.1573
sigma_e	0.9383	0.9537	0.9429	0.9366
rho	0.0773	0.0126	0.0921	0.0274
Wald χ^2	23.6 ***	33.93 ***	13.44 **	52.55 ***

Note: standard errors are given in brackets, * indicates significance level $\alpha < 0.05$, ** $\alpha < 0.02$, and *** $\alpha < 0.001$. Source: own calculation.

In the case of the developmental factors, we observe a similar impact of immigration on the requirements for the use of space and on the withdrawal of agricultural land. However, we must acknowledge that the relationship between these two processes is likely to work in the opposite direction (the inflow of investors will cause the withdrawal of agricultural land and at the same time it attracts labor force to the region). We also identify a statistically significant effect of higher income levels on the reduction of the volume of agricultural land withdrawn for the industrial development. We explain this by the fact that in the most developed regions of Western Slovakia, the local economy has already been "saturated" in the context of the development of industrial sectors. Those are mainly services that do not have the same space requirements, which are being developed. Additionally, industries with lower added value, typical for their location in less developed regions, are more demanding regarding the occupation of space compared to the light industry sectors, which are more concentrated in the more developed regions [51].

Foreign direct investments are a statistically significant factor after taking into account all analyzed factor groups (Model IV). In the same analysis, no legislative changes have been shown to be significant, indicating the failure of legislative measures in protecting the agricultural land when conflicts between the use of space for the agricultural production and the interest of investors arise.

When assessing the impact of the legislative factors, land withdrawal is differentiated by the industry size. While the contribution obligation is negatively affecting smaller investors, large strategic investments are exempted from paying the contributions. As several experts have stated, this approach of the state is not justifiable from an environmental point of view. The negative side is that the large investors do not have any additional compensatory obligation in relation to the withdrawn land.

3.3. The Volume of Agricultural Land Withdrawn for Transport Development Purposes

So far, the weakest interactions of the analyzed factors are recorded in case of the land withdrawal for the purpose of the transport development (Table 4).

The effects of neither territorial nor developmental factors on the volume of the withdrawn land for the transport development purposes are significant as a whole.

The only factor that has a statistically significant impact on the intensity of the land withdrawal for the purpose of transport is the introduction of contributions for its withdrawal, while even the extending of categories of land, for which these contributions are compulsory, does not affect this intensity as in previous cases. The explanation for these empirical findings is most likely that the development and construction of the transport network (especially the most important transport routes) is governed by completely different mechanisms than the rest of the withdrawal purposes. Planning of the development of the transport network and major routes is given by the Concept of Territorial Development of the Slovak Republic [52], which is designed and managed centrally. For these reasons, this process is very little influenced by the examined factors. Experts add that, in order to reduce the land use conflicts, cooperation between the agricultural land protection authorities, architects and authorizing authorities is essential.

Table 4. Results of panel regression models where the dependent variable is the volume of agricultural land withdrawn for transport development purposes.

Factor:	Model I	Model II	Model III	Model IV
zFDI_POP	−0.1497 (0.1404)			−0.2387 (0.1635)
zAMNW	−0.0028 (0.0683)			0.1639 (0.1069)
zIMMIG	0.0684 (0.1384)			−0.0029 (0.3893)
zNETMIG	0.2081 (0.1819)			0.3338 (0.2650)

Table 4. *Cont.*

Factor:	Model I	Model II	Model III	Model IV
zNETMIG_POP	0.0012			−0.0907
	(0.1716)			(0.2001)
zDBUS_POP	0.0225			0.0035
	(0.0505)			(0.0628)
zFRAG_ARABLE		0.0370		−0.0618
		(0.0610)		(0.0608)
zFRAG_AGRI		0.0053		−0.0155
		(0.0664)		(0.0623)
zAVER_SET		−0.1995		−0.0705
		(0.2163)		(0.2507)
zCENT_CITY		0.1649		0.1132
		(0.2103)		(0.2374)
CON (1–4)			−0.3841 **	−0.4374 **
			(0.1319)	(0.1463)
CON (1–9)			−0.3246	−0.4157
			(0.1865)	(0.2203)
EXEMP_VINEYARD			−0.0639	−0.1669
			(0.1758)	(0.1900)
Intercept	2.74×10^{-17}	2.66×10^{-17}	0.3026 **	0.3913 **
	(0.0486)	(0.0594)	(0.1114)	(0.1388)
N° of observations	410	410	410	410
R^2				
within	0.0120	0.0456	0.0000	0.0324
between	0.2886	0.0285	0.0000	0.3847
overall	0.0449	0.0043	0.0233	0.0768
sigma_u	0.0000	0.2155	0.1842	0.1068
sigma_e	0.9821	0.9626	0.9749	0.9620
rho	0.0000	0.0477	0.0344	0.0122
Wald χ^2	18.96 **	1.31	10.02 **	30.96 **

Note: standard errors are given in brackets, * indicates significance level $\alpha < 0.05$, ** $\alpha < 0.02$, and *** $\alpha < 0.001$. Source: own calculation.

3.4. The Volume of Agricultural Land Withdrawn for Mining Purposes

There are very similar interactions between the analyzed factors and the rate of withdrawal of agricultural land for the mining purposes as in the case of the total volume of the withdrawn land. This effect is even more visible in the case of the population movements measured by the migration balance. At the same time, we assume that the similar causal consequences that we considered at the beginning of this chapter also apply to the withdrawal for the mining purposes. Regarding the statistically significant inverse relation of the average monthly nominal wage and the volume of agricultural land for mining purposes, analogous to the previous interpretations, we assume that it is due to the nature of the localization behavior of the mining industries. These, of course, depend primarily on the spatial distribution of mineral resources, but they also tend to concentrate in less developed regions. These activities are space-demanding in the area they occupy. This was also confirmed by the finding that the larger average land size (namely arable land) leads to an increased land withdrawal frequency for mining purposes. In case of the withdrawal of land for the mining purposes, the relatively consistent impact can clearly be attributed to the foreign direct investments. What is positive, however, is that the introduction of contributions, as well as their extension to other land quality categories, has led to a reduction in the volume of the land withdrawal for the mining purposes. The experts confirmed that the increase in the extraction of gravel and other minerals had a negative impact on the extent of the agricultural land withdrawn. Therefore, it is necessary that the mining industry, as well as the state, take responsibility for land use conflicts in the field of mining.

Table 5. Results of panel regression models where the dependent variable is the volume of agricultural land withdrawn for mining purposes.

Factor:	Model I	Model II	Model III	Model IV
zFDI_POP	0.4329 **			0.4730 **
	(0.1561)			(0.1711)
zAMNW	−0.1942 **			−0.0558
	(0.068)			(0.1110)
zIMMIG	−0.230			−0.2785
	(0.1587)			(0.4080)
zNETMIG	−0.7248 ***			−0.6681 **
	(0.2053)			(0.2767)
zNETMIG_POP	1.032 ***			0.8992 ***
	(0.1934)			(0.2106)
zDBUS_POP	−0.0288			0.0034
	(0.0468)			(0.0584)
zFRAG_ARABLE		0.1825 **		0.1069
		(0.0653)		(0.0688)
zFRAG_AGRI		0.0497		−0.0409
		(0.0711)		(0.0706)
zAVER_SET		0.1355		0.0895
		(0.2313)		(0.2766)
zCENT_CITY		−0.2386		−0.1782
		(0.225)		(0.2626)
CON (1–4)			−0.5727 ***	−0.5349 ***
			(0.1237)	(0.1378)
CON (1–9)			−0.6313 ***	−0.5460 **
			(0.1749)	(0.2103)
EXEMP_VINEYARD			0.0352	0.1068
			(0.1649)	(0.1782)
Intercept	6.64×10^{-17}	-1.91×10^{-16}	0.3423 **	0.4004 **
	(0.0598)	(0.0636)	(0.1128)	(0.1388)
N° of observations	410	410	410	410
R^2				
within	0.0858	0.0087	0.0000	0.1120
between	0.3722	0.259	0.0000	0.4683
overall	0.1331	0.0501	0.0405	0.1786
sigma_u	0.2453	0.2758	0.2740	0.2347
sigma_e	0.8794	0.9413	0.9448	0.8729
rho	0.0722	0.079	0.0776	0.0674
Wald χ^2	50.95 ***	12.26 **	18.58 ***	71.77 ***

Note: standard errors are given in brackets, * indicates significance level $\alpha < 0.05$, ** $\alpha < 0.02$, and *** $\alpha < 0.001$. Source: own calculation.

3.5. The Volume of Agricultural Land Withdrawn for Other Purposes

Non-agricultural purposes, for which agricultural land may be withdrawn from the agricultural land fund on request and which are not included in the categories of withdrawal we have already analyzed, are included in the group of other purposes. In order to comprehensively analyze the factors that affect the volume of withdrawn agricultural land, but also to ensure that all factors are correctly identified, we also analyze this withdrawal category (Table 6). All models are significant as a whole, and a set of factors that have been identified as statistically significant is almost identical to a set of factors that determine the total volume of the withdrawn agricultural land. It should be noted, however, that in the case of land withdrawal for the non-categorized purposes, the positive effects of legislative changes (namely the introduction of contributions) on the intensity of withdrawal of agricultural land have only been manifested after their extension to the all land quality groups.

Table 6. Results of panel regression models where the dependent variable is the volume of agricultural land withdrawn for other purposes.

Factor:	Model I	Model II	Model III	Model IV
zFDI_POP	0.3990 **			0.3348 *
	(0.1489)			(0.1649)
zAMNW	−0.2353 **			−0.0549
	(0.0684)			(0.1076)
zIMMIG	−0.2361			−0.7014
	(0.1492)			(0.3929)
zNETMIG	−0.4617 **			−0.1370
	(0.1944)			(0.2672)
zNETMIG_POP	0.7046 ***			0.4604 *
	(0.1832)			(0.2022)
zDBUS_POP	−0.003			−0.0361
	(0.0486)			(0.0610)
zFRAG_ARABLE		0.1549 **		0.1348 *
		(0.0584)		(0.0627)
zFRAG_AGRI		0.0306		−0.0103
		(0.0636)		(0.0642)
zAVER_SET		0.2619		0.4441
		(0.2072)		(0.2565)
zCENT_CITY		−0.3574		−0.1257
		(0.2015)		(0.2431)
CON (1–4)			−0.3138 **	−0.2780
			(0.1278)	(0.1427)
CON (1–9)			−0.4661 **	−0.4365 *
			(0.1807)	(0.2158)
EXEMP_VINEYARD			−0.101	−0.0182
			(0.1704)	(0.1852)
Intercept	6.04×10^{-17}	2.00×10^{-17}	0.3423 **	0.2913 *
	(0.0542)	(0.0569)	(0.1128)	(0.1378)
N° of observations	410	410	410	410
R^2				
within	0.0541	0.0117	0.0000	0.0631
between	0.2933	0.2796	0.0000	0.4167
overall	0.0917	0.0441	0.0405	0.1198
sigma_u	0.1765	0.1984	0.2740	0.1570
sigma_e	0.9456	0.9599	0.9448	0.9463
rho	0.0337	0.041	0.0776	0.0268
Wald χ^2	36.5 ***	13.45 **	18.58 ***	48.53 ***

Note: standard errors are given in brackets, * indicates significance level $\alpha < 0.05$, ** $\alpha < 0.02$, and *** $\alpha < 0.001$. Source: own calculation.

Within V4 countries, similar comprehensive analysis was conducted in Poland by Sroka et al. [53] with somewhat comparable results. Although lacking the legislative aspects, with application of regression trees, they confirmed the high importance of population density and migration on differences in share of agricultural land in the overall surface of the municipal area, while the most important factor was the intensity of entrepreneurial activity, which we did not confirm. It should be noted, however, that they did not differentiate between foreign investments and local business activities.

4. Conclusions

The main problem of the current protection of the agricultural land in Slovakia is the conflict of interests in agricultural land use based on the diversity of needs of individuals on the one hand and the interest of the state and society on the other hand. The policy-making of the state is based on the assumption that the land protection can be ensured in particular by maintaining its area.

By examining the legislative development of land protection, the state mainly uses economic instruments to introduce and modify the withdrawal contribution obligation and regulates exemptions from it based on the state's current interests.

The research has shown that developmental factors for investment and development activities directly affect the total volume of the withdrawn agricultural land. From the aspect of the conflict of interests between individual and state regarding land protection, the private interest prevails over the public one. As a consequence, agricultural land is withdrawn in suburbanized and attractive areas where the land of the highest quality category is mostly located.

Aside from the industrial development, great conflicts arise between agricultural and residential use of the land. In many municipalities in Slovakia, there is still housing stock available, but nevertheless, most new residential areas are built on the agricultural land that has been withdrawn.

In accordance with the precautionary principle, the state should adopt a long-term conceptual document defining areas of agricultural land use taking into account the impact of developmental factors on the land protection. Similarly, the state should introduce new tools for land protection intended for entities not currently influenced by the introduced legislative instruments.

Author Contributions: Conceptualization, I.M., L.P., K.M.; methodology, K.M.; validation, I.M., L.P., K.M.; formal analysis, I.M.; investigation, I.M., L.P., K.M.; resources, I.M.; data curation, I.M, K.M.; writing—original draft preparation, I.M., L.P., K.M.; writing—review and editing, I.M., L.P., K.M.; visualization, I.M.; supervision, I.M., L.P., K.M.; project administration, L.P.; funding acquisition, L.P.

Funding: This research was funded by the Scientific Grant Agency project "Protection of conservation of agricultural land acreage in Slovakia", No. V-18-015-00, with the support of the Ministry of Education, Science, Research and Sport of the Slovak Republic and the Slovak Academy of Sciences.

Conflicts of Interest: The authors declare no conflict of interest. The sponsors had no role in the design, execution, interpretation, or writing of the study.

References

1. European Commission. The Implementation of the Soil Thematic Strategy and Ongoing Activities COM/2012/046 Final. Report from the Commission to the European Parliament, the Council, the European Economic and Social Committee and the Committee of the Regions. Available online: https://eur-lex.europa.eu/LexUriServ/LexUriServ.do?uri=COM:2012:0046:FIN:EN:HTML (accessed on 15 April 2019).

2. Izakovičová, Z.; Miklós, L.; Miklósová, V. Integrative Assessment of Land Use Conflicts. *Sustainability* **2018**, *10*, 3270. Available online: https://www.mdpi.com/2071-1050/10/9/3270/pdf (accessed on 15 April 2019).

3. Haase, D. The Rural-to-Urban Gradient and Ecosystem Services. In *Atlas of Ecosystem Services: Drivers, Risks, and Societal Responses*; Schröter, M., Bonn, A., Eds.; Springer: Basel, Switzerland, 2019; pp. 141–146.

4. Ronchi, S. *Ecosystem Services for Spatial Planning: Innovative Approaches and Challenges for Practical Applications*; Springer: Basel, Switzerland, 2018.

5. Naumann, S.; Frelih-Larsen, A.; Prokop, G.; Ittner, S.; Reed, M.; Mills, J.; Morari, F.; Verzandvoort, S.; Albrecht, S.; Bjuréus, A.; et al. Land Take and Soil Sealing—Drivers, Trends and Policy (Legal) Instruments: Insights from European Cities. In *International Yearbook of Soil Law and Policy 2018*; Ginzky, H., Dooley, E., Eds.; Springer: Basel, Switzerland, 2018; pp. 83–112.

6. Yi, Y.; Zhao, Y.; Ding, G.; Gao, G.; Shi, M.; Cao, Y. Effects of Urbanization on Landscape Patterns in a Mountainous Area: A Case Study in the Mentougou District, Beijing, China. *Sustainability* **2016**, *8*, 1190. Available online: https://www.mdpi.com/2071-1050/8/11/1190 (accessed on 21 June 2019).

7. Rimal, B.; Zhang, L.; Stork, N.; Sloan, S.; Rijal, S. Urban Expansion Occurred at the Expense of Agricultural Lands in the Tarai Region of Nepal from 1989 to 2016. *Sustainability* **2018**, *10*, 1341. Available online: https://www.mdpi.com/2071-1050/10/5/1341 (accessed on 21 June 2019).

8. Radwan, T.M.; Blackburn, G.A.; Whyatt, J.D.; Atkinson, P.M. Dramatic Loss of Agricultural Land Due to Urban Expansion Threatens Food Security in the Nile Delta, Egypt. *Remote Sens.* **2019**, *11*, 332. Available online: https://www.mdpi.com/2072-4292/11/3/332 (accessed on 21 June 2019).

9. Report on Best Practices for Limiting Soil Sealing and Mitigating Its Effects. Technical Report-2011-50. Environment Agency Austria—Umweltbundesamt. Available online: http://ec.europa.eu/environment/archives/soil/pdf/sealing/Soil%20sealing%20-%20Final%20Report.pdf (accessed on 21 June 2019).

10. Janků, J.; Sekáč, P.; Baráková, J.; Kozak, J. Land Use Analysis in Terms of Farmland Protection in the Czech Republic. *Soil Water Res.* **2016**, *11*, 20–28. Available online: https://www.agriculturejournals.cz/publicFiles/173200.pdf (accessed on 21 June 2019).

11. Urban Land Take. Environmental Indicator Report 2018. European Environment Agency. Available online: https://www.eea.europa.eu/airs/2018/natural-capital/urban-land-expansion#tab-based-on-data (accessed on 21 June 2019).

12. Bandlerová, A.; Palšová, L.; Schwarcz, P.; Roháčiková, O.; Viti, D.; Celi, G.; Sadowski, A.; Przygodzka, R.; Bueno-Armijo, A.; Mansberger, R.; et al. *The Land Management Manual of the EU*; Slovak University of Agriculture in Nitra: Nitra, Slovak Republic, 2017.

13. Busko, M.; Szafranska, B. Analysis of Changes in Land Use Patterns Pursuant to the Conversion of Agricultural Land to Non-Agricultural Use in the Context of the Sustainable Development of the Malopolska Region. *Sustainability* **2018**, *10*, 136. Available online: https://www.mdpi.com/2071-1050/10/1/136/htm (accessed on 21 June 2019).

14. Cosentino, C.; Amato, F.; Murgante, B. Population-Based Simulation of Urban Growth: The Italian Case Study. *Sustainability* **2018**, *10*, 4838. Available online: https://www.mdpi.com/2071-1050/10/12/4838 (accessed on 21 June 2019).

15. Rondhi, M.; Pratiwi, P.; Handini, V.; Sunartomo, A.; Budiman, S. Agricultural Land Conversion, Land Economic Value, and Sustainable Agriculture: A Case Study in East Java, Indonesia. *Land* **2018**, *7*, 148. Available online: https://www.mdpi.com/2073-445X/7/4/148/htm (accessed on 21 June 2019).

16. Alam, J. Rapid urbanization and changing land values in mega cities: Implications for housing development projects in Dhaka, Bangladesh. *Bdg. J. Glob. South* **2018**, *5*, 1–19. Available online: https://link.springer.com/article/10.1186/s40728-018-0046-0 (accessed on 21 June 2019).

17. Peerzado, M.B.; Magsi, H.; Sheikh, M.J. Land use conflicts and urban sprawl: Conversion of agriculture lands into urbanization in Hyderabad, Pakistan. *J. Saudi Soc. Agric. Sci.* **2018**, *17*. Available online. https://www.sciencedirect.com/science/article/pii/S1658077X17303818 (accessed on 21 June 2019). [CrossRef]

18. Xiao, R.; Liu, Y.; Huang, X.; Shi, R.; Yu, W.; Zhang, T. Exploring the driving forces of farmland loss under rapidurbanization using binary logistic regression and spatial regression: A case study of Shanghai and Hangzhou Bay. *Ecol. Indic.* **2018**, *95*, 455–467. Available online: https://www.sciencedirect.com/science/article/pii/S1470160X18305909 (accessed on 21 June 2019).

19. Cui, Y.; Liu, J.; Xu, X.; Dong, J.; Li, N.; Fu, Y.; Lu, S.; Xia, H.; Si, B.; Xiao, X. Accelerating Cities in an Unsustainable Landscape: Urban Expansion and Cropland Occupation in China, 1990–2030. *Sustainability* **2019**, *11*, 2283. Available online: https://www.mdpi.com/2071-1050/11/8/2283 (accessed on 21 June 2019).

20. Zhang, C.; Zhong, S.; Wang, X.; Shen, L.; Liu, L.; Liu, Y. Land Use Change in Coastal Cities during the Rapid Urbanization Period from 1990 to 2016: A Case Study in Ningbo City, China. *Sustainability* **2019**, *11*, 2122. Available online: https://www.mdpi.com/2071-1050/11/7/2122 (accessed on 21 June 2019).

21. Zhang, C.; Zhong, S.; Wang, X.; Shen, L.; Liu, L.; Liu, Y. Urban Land Revenue and Sustainable Urbanization in China: Issues and Challenges. *Sustainability* **2018**, *10*, 2111. Available online: https://www.mdpi.com/2071-1050/10/7/2111 (accessed on 21 June 2019).
22. Song, M.; Huntsinger, L.; Han, M. How Does the Ecological Well-Being of Urban and Rural Residents Change with Rural-Urban Land Conversion? The Case of Hubei, China. *Sustainability* **2018**, *10*, 527. Available online: https://www.mdpi.com/2071-1050/10/2/527 (accessed on 21 June 2019).
23. Milczarek-Andrzejewska, D.; Zawalińska, K.; Czarnecki, A. Land-use conflicts and the Common Agricultural Policy: Evidence from Poland. *Land Use Policy* **2018**, *73*, 423–433. Available online: https://www.sciencedirect.com/science/article/pii/S0264837717313935 (accessed on 21 June 2019).
24. Mazzocchi, C.; Corsi, S.; Sali, G. Agricultural Land Consumption in Periurban Areas: A Methodological Approach for Risk Assessment Using Artificial Neural Networks and Spatial Correlation in Northern Italy. *Appl. Spat. Anal. Policy* **2017**, *10*, 3–20. Available online: https://link.springer.com/article/10.1007%2Fs12061-015-9168-9 (accessed on 21 June 2019).
25. Romano, B.; Zullo, F.; Marucci, A.; Fiorini, L. Vintage Urban Planning in Italy: Land Management with the Tools of the Mid-Twentieth Century. *Sustainability* **2018**, *10*, 4125. Available online: https://www.mdpi.com/2071-1050/10/11/4125 (accessed on 21 June 2019).
26. Policy Brief: Soil Sealing and Land Take. The RECARE Project. Available online: https://www.ecologic.eu/sites/files/publication/2018/2730_recare_soil-sealing_web.pdf (accessed on 21 June 2019).
27. Opportunities for Soil Sustainability in Europe. EASAC Policy Report 36. European Academies' Science Advisory Council. Available online: https://easac.eu/fileadmin/PDF_s/reports_statements/EASAC_Soils_complete_Web-ready_210918.pdf (accessed on 21 June 2019).
28. Ustaoglu, E.; Williams, B. Determinants of Urban Expansion and Agricultural Land Conversion in 25 EU Countries. *Environ. Manag.* **2017**, *60*, 717–746. Available online: https://link.springer.com/article/10.1007/s00267-017-0908-2 (accessed on 21 June 2019).
29. Stankovics, P.; Tóth, G.; Tóth, Z. Identifying Gaps between the Legislative Tools of Soil Protection in the EU Member States for a Common European Soil Protection Legislation. *Sustainability* **2018**, *10*, 2886. Available online: https://www.mdpi.com/2071-1050/10/8/2886/pdf (accessed on 21 June 2019).
30. The Geodesy, Cartography and Cadastre Authority of the Slovak Republic. Available online: http://www.skgeodesy.sk/files/slovensky/ugkk/kataster-nehnutelnosti/sumarne-udaje-katastra-podnom-fonde/statisticka-rocenka-2018.pdf (accessed on 15 April 2019).
31. Representation of Agricultural Land Quality Groups. Soil Service Portal. Available online: www.podnemapy.sk/portal/reg_pod_infoservis/kvalita/kvalita.aspx (accessed on 21 June 2019).
32. Explanatory Memorandum to the Amendment to the Constitution of the Slovak Republic. Available online: https://www.nrsr.sk/web/Dynamic/Download.aspx?DocID=437404 (accessed on 15 April 2019).
33. Act No. 220/2004 Coll. on the Protection and Use of Agricultural Land and Amending Act No. 245/2003 Coll. On Integrated Pollution Prevention and Control and in Amendments to Certain Acts. Available online: https://www.zakonypreludi.sk/zz/2004-220 (accessed on 15 April 2019).
34. Damohorský, M.; Drobník, J.; Smolek, M. *Právo Životního Prostředí*, 3rd ed.; C.H. Beck: Praha, Czech Republic, 2010.
35. Pérez, A.P.; Sánchez, S.P. *European Achievements in Soil Remediation and Brownfield Redevelopment*; Publications Office of the European Union: Brussel, Belgium, 2017.
36. Explanatory Memorandum to the Act No. 220/2004 Coll. On the Protection and Use of Agricultural Land and Amending Act No. 245/2003 Coll. On Integrated Pollution Prevention and Control and in Amendments to Certain Acts. Available online: www.mpsr.sk/download.php?pkfID=2024 (accessed on 15 April 2019).
37. Explanatory Memorandum to the Act No. 219/2008 Coll. Amending Act No. 220/2004 Coll. on the Protection and Use of Agricultural Land and Amending Act No. 245/2003 Coll. On Integrated Pollution Prevention and Control and in Amendments to Certain Acts. Available online: http://www.epi.sk/dovodova-sprava/Dovodova-sprava-k-zakonu-c-219-2008-Z-z.htm (accessed on 15 April 2019).
38. Explanatory Memorandum to the Act No. 57/2013 Coll. Amending Act No. 220/2004 Coll. On the Protection and Use of Agricultural Land and Amending Act No. 245/2003 Coll. On Integrated Pollution Prevention and Control and in Amendments to Certain Acts. Available online: https://www.nrsr.sk/web/Dynamic/Download.aspx?DocID=373171 (accessed on 15 April 2019).

39. Sobocká, J.; Bezák, P.; Skalský, R. Ochrana pôdy na Slovensku a návrhy na novelizáciu súčasnej legislatívy. In *Pestovateľské Technológie a ich Význam pre Prax*; Malovcová, Ľ., Babulicová, M., Sekerková, M., Eds.; Centrum Výskumu Rastlinnej Výroby Piešťany: Piešťany, Slovak Republic, 2012; pp. 8–17.

40. Bielek, P. *Pôdoznalectvo pre Enviromanažérov*; Slovak University of Agriculture in Nitra: Nitra, Slovak Republic, 2017.

41. Sobocká, J. *Dlhodobé Výzvy Klimatických Zmien a Význam Pôdno-Klimatických Podmienok*; Manažment Rizík v Agrosektore: Bratislava, Slovak Republic, 2018.

42. Schwarcz, P.; Bandlerová, A.; Dimitrova, V.; Floriš, N.; Grigorova, Z.; Hristov, L.; Iliev, T.; Ivanova, M.; Kováčik, M.; Levkov, K.; et al. *European Agricultural and Environmental Policy*; Slovak University of Agriculture in Nitra: Nitra, Slovak Republic, 2016.

43. Bandlerová, A.; Bielek, P.; Schwarcz, P.; Palšová, L. *EU Land Policy, The Pathway Towards Sustainable Europe*; Slovak University of Agriculture in Nitra: Nitra, Slovak Republic, 2016.

44. Wyrwa, J. Foreign Direct Investments and Poland's Economic Development—Current Situation and Development Prospects. *Acta Oeconomica Univ. Selye* **2018**, *7*, 188–200. Available online: http://acta.ujs.sk/docs/Acta%207_2%20print.pdf (accessed on 27 April 2019).

45. Food and Agriculture Organization of the United Nations. *Foreign Direct Investment—Win-Win or Land Grab*; World Summit on Food Security: Rome, Italy, 2009.

46. Häberli, C. Foreign Direct Investment in Agriculture: Land Grab or Food Security Improvement? In *Economic Analysis of International Law*; 2014; pp. 283–303. Available online: https://ssrn.com/abstract=2556244 (accessed on 20 April 2019).

47. World Economic Forum. Migration and Its Impact on Cities—An Insight Report. Available online: http://www3.weforum.org/docs/WEF_Migration_Report_Embargov.pdf (accessed on 25 April 2019).

48. Ignjatić, J.; Nikolić, B.; Rikalović, A. Trends and Challenges in Brownfield Revitalization: A Gis Based Approach. In Proceedings of the XVII International Scientific Conference on Industrial Systems (IS'17), Novi Sad, Serbia, 4–6 October 2017.

49. Ministry of Agriculture and Rural Development of the Slovak Republic. Available online: http://www.mpsr.sk/?navID=1&id=14097 (accessed on 24 April 2019).

50. The Ministry of Labour, Social Affairs and Family of the Slovak Republic. Available online: https://www.employment.gov.sk/files/slovensky/ministerstvo/analyticke-centrum/sprava-socialnej-situacii-obyvatelstva-slovenskej-republiky-za-rok-2017.pdf (accessed on 24 April 2019).

51. Slovak Business Agency. Available online: http://www.sbagency.sk/sites/default/files/analyza_podnikatelskeho_prostredia_v_regionoch_sr_final1.pdf (accessed on 18 April 2019).

52. The Concept of Territorial Development of the Slovak Republic. Available online: https://www.mindop.sk/ministerstvo-1/vystavba-5/uzemne-planovanie/dokumenty/koncepcia-uzemneho-rozvoja-slovenska-kurs2001/textova-cast (accessed on 28 April 2019).

53. Sroka, W.; Mikolajczyk, J.; Wojewodzic, T.; Kwoczynska, B. Agricultural Land vs. Urbanisation in Chosen Polish Metropolitan Areas: A Spatial Analysis Based on Regression Trees. *Sustainability* **2018**, *10*, 837. [CrossRef]

Article

Assessment of Soil Suitability for Improvement of Soil Factors and Agricultural Management

Sameh Kotb Abd-Elmabod [1,2,3,4,*], Noura Bakr [3], Miriam Muñoz-Rojas [5,6], Paulo Pereira [7], Zhenhua Zhang [1,*], Artemi Cerdà [8], Antonio Jordán [4], Hani Mansour [9], Diego De la Rosa [10] and Laurence Jones [2]

1   Institute of Agricultural Resources and Environment, Jiangsu Academy of Agricultural Sciences, Nanjing 210014, China
2   Centre for Ecology & Hydrology (CEH-Bangor), Environment Centre Wales, Deiniol Road, Bangor LL57 2UW, UK; lj@ceh.ac.uk
3   Soil and Water Use Department, Agricultural and Biological Research Division, National Research Centre, Cairo 12622, Egypt; nourabakr@yahoo.com
4   MED_Soil Research Group, Department of Crystallography, Mineralogy and Agricultural Chemistry, Seville University, 41012 Seville, Spain; ajordan@us.es
5   Centre for Ecosystem Science, School of Biological, Earth and Environmental Sciences, University of New South Wales, Sydney, NSW 2052, Australia; m.munoz-rojas@unsw.edu.au
6   School of Biological Sciences, The University of Western Australia, Perth, WA 6009, Australia
7   Environmental Management Centre, Mykolas Romeris University, Ateities, Vilnius LT-08303, Lithuania; paulo@mruni.eu
8   Soil Erosion and Degradation Research Group, Department of Geography, Valencia University, 46010 Valencia, Spain; artemio.cerda@uv.es
9   Water Relations and Field Irrigation Department, Agricultural and Biological Research Division, National Research Centre, Cairo 12622, Egypt; mansourhani2011@gmail.com
10   Earth Sciences Section, Royal Academy of Sciences, 41012 Seville, Spain; diego.delarosa@rasc.es
*   Correspondence: sk.abd-elmabod@nrc.sci.eg (S.K.A.-E.); zhenhuaz70@hotmail.com (Z.Z.)

Received: 12 February 2019; Accepted: 12 March 2019; Published: 15 March 2019

Abstract: The dramatic growth of the world's population is increasing the pressure on natural resources, particularly on soil systems. At the same time, inappropriate agricultural practices are causing widespread soil degradation. Improved management of soil resources and identification of the potential agricultural capability of soils is therefore needed to prevent further land degradation, particularly in dryland areas such as Egypt. Here, we present a case study in the El-Fayoum depression (Northern Egypt) to model and map soil suitability for 12 typical Mediterranean crops. Two management scenarios were analyzed: the current situation (CS) and an optimal scenario (OS) of soil variables. The Almagra model was applied to estimate soil suitability under CS and OS. Management options based on the CS assessment were proposed to reduce some limiting factors: a fixed value of 2 dSm^{-1} for soil salinity and 5% for sodium saturation; these defined the OS. Under optimal management, the OS scenario showed potential, where a notable increase of the area covered by a high suitability class (around 80%) for annual and semi-annual crops was observed. There was also a marked increase (about 70% for CS and 50% for OS) for perennial crops shifting from the marginal to moderate soil suitability class. The results reveal the importance of proper management to massively alter soil suitability into better states in order to achieve sustainable land use in this fertile agro-ecosystem.

Keywords: sustainable agriculture; MicroLEIS DSS; land-use planning; soil reclamation

---

1. Introduction

The dramatic increase of the world's population is inducing enormous pressure on natural resources [1,2]. This pressure causes multiple environmental problems for land and water systems [3], and therefore, appropriate land-use and management strategies are needed to reduce the magnitude of these human impacts [4,5]. Agricultural activities have a direct effect on soils' physical, chemical, and biological properties [6], resulting in environmental problems such as soil degradation [7], waterlogging [8], salinization/alkalization [9], and contamination [10,11]. These environmental problems affect soil quality and crop productivity, reducing food production capacity and food security.

Water scarcity is also a serious constraint for agriculture in drylands, where land is highly vulnerable to land degradation due to aridity [9]. At the global scale, Africa is highly affected by desertification, with over 45% of the land exposed to this process [12]. This issue calls for the urgent improvement of agricultural practices and water use efficiency [13] in order to reduce environmental problems and prevent further degradation [14]. Irrigation is often used to alleviate water scarcity, and often inappropriate irrigation practices induce soil salinization and soil sodicity, which are major causes of land degradation [15]. The excess of salts may cause clay dispersion and create soil crusts, decreasing permeability and soil productivity, but suitable agricultural land management may also improve conditions [16]. Although drainage problems occur naturally, agricultural land management can exacerbate it [16]. Improving soils' physical properties facilitates sodium leaching, and decreases salinity and sodicity effects in soils [17].

Agriculture is the main economic activity in Egypt and supports the livelihoods of approximately 55% of the population, contributing to around 20% of foreign exchange earnings, and approximately 30% of Egypt's commodity exports [18]. Agricultural land in Egypt represents approximately 3.8 Mha and the main crops cultivated are wheat, cotton, maize, sunflower, clovers, tomatoes, aromatic and medicinal plants, mangoes, olives, and citrus [18]. In the largely fertile Nile valley, soil productivity is restricted by salinity as result of irrigation and by urban sprawl over productive soils. Such unsustainable management results in land degradation, with implications for soil productivity, food production, and food security [19–21].

Land evaluation involves determination of the land potential for agricultural purposes [22,23] and its main objective is to manage and improve land in a sustainable way to increase its potential for human uses [22,24]. Land suitability status is based on intrinsic properties of soils (e.g., parent materials, soil texture and depth) and characteristics that can be altered by human management (e.g., drainage, salinity, nutrient concentration and vegetation cover) [25,26]. Several land evaluation guidelines have been created in the last decades (e.g., [22,25–31]).

The evolution of technology and models capable of analyzing a large number of variables has increased the sophistication of land suitability analyses [32,33]. Furthermore, the development of geographical information systems (GIS) and geostatistical techniques has allowed improved spatial processing of information on the variables that affect land degradation and land suitability, which are important for sustainable use of agricultural areas [6]. Soil and land evaluation models, hypothetical scenarios of agriculture management, and spatial analyses are valuable tools for land managers and decision makers to achieve sustainable land-use planning and management for targeted areas [34–36]. The Micro Land Evaluation Information System has been extensively used worldwide for land suitability assessment (MicroLEIS, [29,30,37]). The MicroLEIS focuses on an integrated system of soil, climate, and agriculture management databases for land evaluation and contains two sets of models related to land suitability and land vulnerability. These models have been widely used for: (i) land degradation evaluation and prediction of optimal land-use and management practices, and (ii) assisting decision makers in solving agro-ecological problems (e.g., [36,38–44]). The Almagra model is one of the main components of the MicroLEIS DSS that was designed for land suitability evaluation [45].

Given the importance of the agricultural sector in Egypt, there is a strong need for assessments of the agricultural potential of the existing soils. This is especially important in an area extremely vulnerable to land degradation and where soil productivity is important for food security. In this study, the Almagra model was used to evaluate soil suitability in the El-Fayoum depression (Northern Egypt). The main objectives were to: (i) assess soil suitability in relation to cultivation of 12 Mediterranean crops (annual, semi-annual, and perennial) under current scenario (CS) and optimal scenarios (OS) of soil management, and (ii) to obtain the spatial distribution of soil suitability under the proposed scenarios based on physiographic units.

2. Material and Methods

2.1. Study Area

The El-Fayoum depression is located within the El-Fayoum region (western desert of Egypt) at coordinates 29°02′–29°35′ N and 30°23′–31°05′ E (Figure 1). Monthly mean temperature and precipitation inside and outside of the depression are shown in Figure S1.

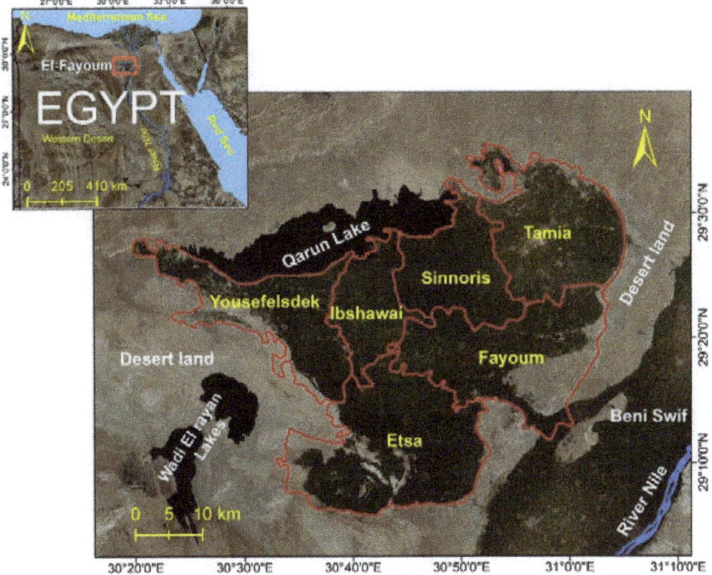

Figure 1. Location of the El-Fayoum depression within Egypt.

The El-Fayoum altitude ranges between −52 to 141 m in relation to mean sea level and has a flat topography (Figure 2A,B). According to Reference [46], the soils of the El-Fayoum depression belong to the Aridisol and Entisol soil groups. The dominant soil subgroup is the Vertic Torrifluvent that covers an area of 43% of the study area; the other subgroups are Typic Haplocalcids, Typic Torrifluvents, Typic Haplogypsids, Typic Haplosalids, and Typic Torripsamments [47]. The depression is linked to the Nile River by the canal of Hawara that transports the water to the depression. The study area involves six districts (Tamia, Sinnoris, Ibshawai, Fayoum, Yousef El Sadik, and Itsa). Agriculture is the main activity in the study area and the total cultivated land is 177,802 ha [18].

2.2. Soil and Climate Databases

Twenty-nine soil profiles were selected from the studies of [48,49] to analyze the soil characteristics in the current situation (CS). Each of these profiles was the average of three individual points (Figure 2C). For this study, we selected the most important soil variables (model input) to evaluate soil quality, e.g., electrical conductivity (EC), exchangeable sodium percentage (ESP%) pH, carbonate content (CaCO$_3$), texture, organic matter (OM), and cation exchange capacity (CEC). Agroclimatic parameters such as evapotranspiration and aridity index (ARi, numbers of arid months in which the actual precipitation was lower than the evapotranspiration) were calculated from El-Fayoum weather station data from the period 1962–2006 [50]. Water supply from the Nile for irrigation (2.64×10^9 m^3/year) was also considered in the models. The monthly distribution of these parameters is shown in Figure S1.

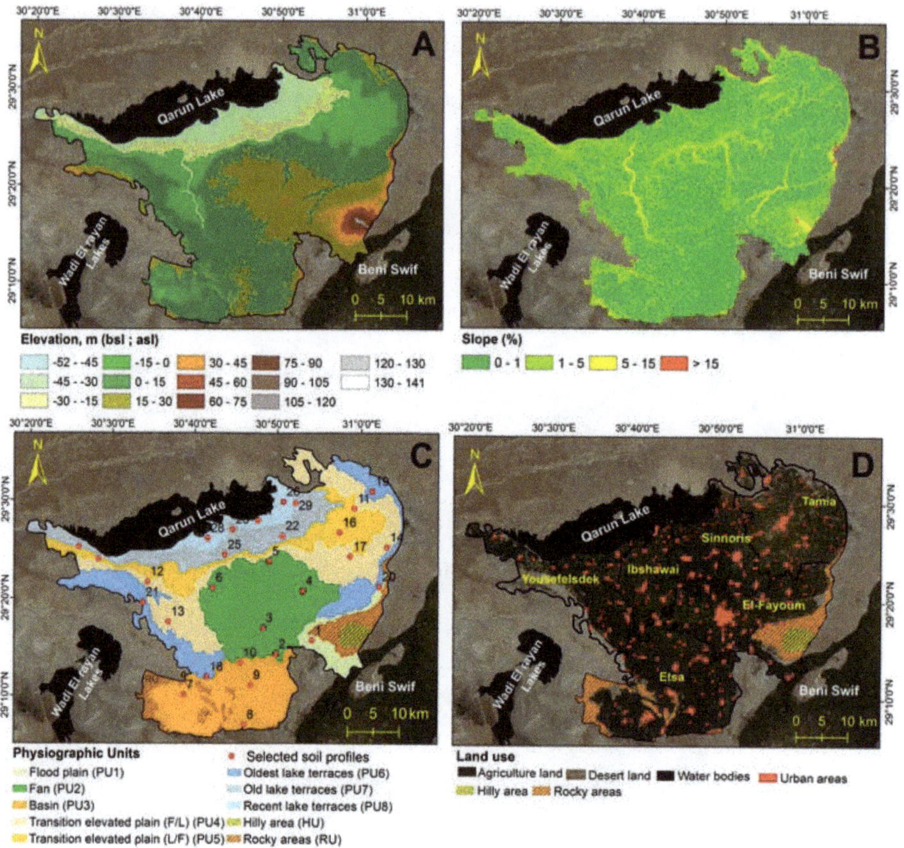

Figure 2. Study area: (**A**) elevation, (**B**) slope, (**C**) sampled sites and physiographic units, and (**D**) land use.

2.3. Soil Suitability Evaluation

Almagra Model

This model is considered as a qualitative biophysical evaluation model that uses the diagnostic criteria of soil variables and the favorable conditions for crop growth within MicroLEIS ([30,45,51], Figure S2). This model is more practical to use than other land suitability models [52]. The Almagra model has been used and calibrated in previous studies in Mediterranean regions [37,41,53–55]. In this study, the Almagra model defined soil suitability within five different classes (Figure S3): optimum (S1), high (S2), moderate (S3), marginal (S4), and not suitable (S5) for twelve traditional crops that included annuals, e.g., wheat, sunflower, sugar beet, melon, potato, soybean, cotton, and corn; semi-annuals, e.g., alfalfa; and perennials, e.g., olive, citrus, and peach. The level of generalization of the variables was dependent on the crop requirement for each soil variable using the "most limiting factor" approach to identify the soil suitability classes. Here, the Almagra model was applied to evaluate the CS of soil suitability for the twelve Mediterranean crops considering all the soil limiting factors in the El-Fayoum depression. The optimal scenario was based on the soil variables that could be managed, such as EC, ESP%, and drainage, without taking into account the interaction between them. Soil depth and texture were not considered as these variables are not easily modified. The soil suitability classes were further divided into 18 subclasses depending on the number of limiting factors in each soil suitability class (Figure S3).

The suggested optimal scenario (OS) was determined according to the following equation:

$$OS = CS - URs$$

where; OS: optimal scenario; CS: current situation; URs: units of reduction.

The units of reduction were defined based on the CS assessment to meet the proposed fixed value of the OS in order to increase final soil suitability to S2 (high soil suitability). The proposed fixed values of each manageable soil factor are highlighted in green boxes in Figure S2. Accordingly, when soil suitability under CS is S5 or S4 (marginal and not suitable, respectively), these will require higher URs compared to S3 (moderate), which need lower reduction units to meet the fixed value of OS for each soil variable. For soil salinity, the OS was proposed to decrease ES classes (ranged from slightly saline to high saline) to a fixed value of 2 dSm^{-1} (non-saline class). In terms of ESP%, the intended value of the OS was 5%. Finally, as the drainage factor is a qualitative parameter in the Almagra model, the OS was aimed to enhance the drainage from "very poor", "poor", and "moderate", to "good" drainage status.

2.4. Spatial Analysis

A Landsat 8 satellite image (path/raw: 177/40 acquired in July 2016) was used as a base map for the spatial analysis of the studied variables. The image was masked to the study area based on the administrative boundaries of the El-Fayoum depression using ArcGIS 10.4 [56].

The Shuttle Radar Topographic Mission (SRTM) digital elevation model (DEM) produced by NASA (version 3, 30 m resolution, updated on 6 August 2015) was downloaded from the US Geological Survey website using EarthExplorer. The DEM of the El-Fayoum depression was subset to the study area to accurately allocate the physiographic units [49,57]. The slope percentage was calculated based on the DEM using the spatial analysis tools in ArcMap and the slope classes were generated based on the Soil Survey Manual [58,59].

In order to obtain more reliable results, the largest extended urban areas (>10 hectare) were excluded from the total area coverage. A supervised classification procedure by maximum likelihood classifier was used to separate the urban areas from other land cover classes. All area coverages (km^2 and %) for the physiographic units and urban areas were calculated in the attribute table using ArcMap.

To display the spatial distribution of soil salinity and sodium saturation percentage of the soil profiles under CS and URs, these variables were interpolated using ArcMap. This interpolation considered the values of the soil input factor as well as the DEM using Co-Kriging. Additionally, inverse distance weighting (IDW) was executed to spatially represent the distribution of soil suitability for the twelve studied crops under CS and OS, as well as the total average of soil suitability.

2.5. Statistical Analysis for Soil Factors

All soil variables analyzed were tested for normality using the Shapiro–Wilk normality test. Data were log transformed as necessary for analysis, but all presented data were non-transformed for ease of interpretation. Differences among physiographical units for each soil property were tested using a one-way ANOVA test. The SPSS 23 software [60] was used at a 95% confidence interval (significance level of $\alpha = 0.05$). The main descriptive statistical parameters (maximum, minimum, and mean) were also generated.

3. Results and Discussion

3.1. Spatial Analysis and Soil Physiochemical Properties

3.1.1. Physiographic Units

Eight physiographic units were distinguished in the area besides the hilly and rocky areas (Figure 2C, Table 1). When considering the entire study area (with the urban area excluded), the fan physiographic unit represented one fifth of the study area (20%). The transition elevated plains (fluvial/lacustrine) covered an area of 30% and around the same area percentage (30%) was allocated to lake terraces (oldest, old, and recent). The uncultivated areas: rocky area (RU), and hilly area (HU) covering 105 and 22 km^2 respectively, were excluded from the geometric area calculation for the evaluation process. The total urban area obtained was 115 km^2 (Figure 2D) which represents around 6% of the area coverage.

Table 1. The main physiographic units (PU), soil taxonomy according to the United States Department of Agriculture (USDA, 2014) and area coverage by km^2 and % for the studied area. VTFE, Vertic Torrifluvents; TTFE, Typic Torrifluvents; TTPE, Typic Torripsamments; THCI, Typic Haplocalcids; THGI, Typic Haplogypsids; THIS, Typic Haplosalids.

Unit	Taxonomy (USDA, 2014)	Area			
		Urban Included		Urban Excluded	
		km^2	%	km^2	%
PU1-Flood plain	VTFE	59.47	3.33	55.5	3.32
PU2-Fan	TTFE and VTFE	374.51	20.98	336.04	20.13
PU3-Basin	THCI and VTFE	295.16	16.54	283.43	16.98
PU4-Transition A	THCI, THGI, and VTFE	347.5	19.47	325.87	19.52
PU5-Transition B	VTFE and THCI	182.81	10.24	166.81	9.99
PU6-Oldest lake terraces	TTPE and THCI	234.32	13.13	226.47	13.57
PU7-Old lake terraces	TTFE and THCI	203.03	11.38	192.85	11.55
PU8-Recent lake terraces	THIS	87.87	4.92	82.58	4.95
Total		**1784.67**	100	**1669.54**	100

3.1.2. Soil Properties

Almost all physiographic units were represented by four profiles except the flood plain unit (PU1), which was represented by only one soil profile, and the fan unit (PU2), which was represented by five. Table 2 illustrates the mean values of the soil factors in each physiographic unit. The elevated mean value of the EC was assigned to the recent lake terraces unit (PU8), adjacent to Qarun Lake, with a value of 31.9 dSm^{-1}. The elevated EC in the soil close to Qarun lake could be caused by the percolation of the saline lake water, which can eventually lead to a rising water table within the soil profile depth [14]. Generally, the soils' pH ranged from 7.54 to 8.87 (from slightly to strongly alkaline). The highest mean value of CaCO$_3$ content in the El-Fayoum soils was 19.71%, which was observed in the elevated plain (Lacustrine/Fluvial) unit (PU5), while the lowest mean value (3.96%) was observed in the PU1-Flood plain. Several soil texture classes were found in the study area, from sand (>90% sand) to clay (around 65% clay). The heavy texture (clay content of 45%–60%) was primarily found in the PU2-fan physiographic unit. The mean values of exchangeable sodium percentage ranged from 8.72 to 19.83% based on the type of the physiographic unit, while the highest CEC value occurred within the fan physiographic unit with highest amount of clay. Generally, the El-Fayoum depression soils were poor in organic matter, with around 80% of the studied area having less than 1.5%. Deeper profiles (120–150 cm) were observed in the flood plain, fan, and basin physiographic units. Sand, silt, and clay contents differed significantly ($p < 0.05$) among physiographic units, but the rest of the variables (e.g., EC, pH, CaCO$_3$, OM, and ESP) were not significantly different.

Table 2. The mean values of the soil factors that were used in the suitability model for each physiographic unit. PU1, Flood plain; PU2, Fan; PU3, Basin; PU4, Transition elevated plain A; PU5, Transition elevated plain B; PU6, Oldest lake terraces; PU7, Old lake terraces; PU8, Recent lake terraces; EC, electric conductivity; OM, organic matter; ESP, exchangeable sodium percentage; CEC, cation exchange capacity.

Physiographic Unit	Particle Size Distribution, %			OM, %	EC, dS/m	pH	$CaCO_3$, %	ESP, %	CEC, meq/100 g
	Sand	Silt	Clay						
PU1	43.97	24.17	31.86	2.05	2.56	7.55	3.69	8.72	24.96
PU2	19.13	28.74	52.13	1.59	5.79	8.21	4.9	14.55	41.16
PU3	57.54	19.85	22.61	1.54	5.92	7.94	13.45	12.03	19.8
PU4	32.25	34.92	32.83	1.28	2.86	8.07	14.81	17	26.92
PU5	43.74	21.8	34.45	1.21	10.86	8.11	19.71	10.1	26.02
PU6	77.66	8.73	13.61	0.72	2.43	7.72	7.18	8.87	13.34
PU7	35.95	26.33	37.72	1.17	8.98	8.08	18.33	14.1	35.75
PU8	42.6	16.68	40.73	2.2	31.89	8.41	11.38	19.83	30

3.2. Soil Factors under Current and Optimal Scenarios

3.2.1. Soil Salinity

Around 70% of the study area had a slight-to-moderate soil salinity with EC ranging between 2 and 6 dSm^{-1} under CS (Table 3). Approximately 5% of the study area showed extreme salinity (>16 dSm^{-1}) (Figure 3a).

Figure 3. Highly saline soil (**a**), sodic soil (**b**), and very poorly drained soil (**c**) compared with the highly suitable soil, planted with sugar beet and corn (**d**) in El-Fayoum.

Table 3. Area (km^2 and %) of each soil salinity class and sodium saturation in the study area.

Soil Salinity (dSm^{-1}) [a]	Area		Sodium Saturation (%)	Area	
	km^2	%		km^2	%
1.65–2	0.79	0.05	5.36–7	1.27	0.08
2–4	586.37	35.12	7–10	282.72	16.93
4–6	592.08	35.46	10–12	326.89	19.58
6–8	213.74	12.80	12–15	658.23	39.43
8–10	84.68	5.07	15–17	215.60	12.91
10–16	106.46	6.38	17–20	113.84	6.82
>16	85.42	5.12	20–22	41.88	2.51
Total	**1669.54**	**100.00**	22–25	29.09	1.74
			>25	0.01	0.00
			Total	**1669.54**	**100.00**

[a] 0–2: non-saline; 2–4: slightly saline; 4–8: moderately saline; 8–16: highly saline; >16: very highly saline.

Under the optimal scenario, all salinity classes were decreased to 2 dSm^{-1} (non-saline soil) with more than two-thirds of the study area assigned as slight-to-moderately saline soil (Table 3). The maximum soil salinity values (>16 dSm^{-1}) were observed in PU8 under CS and OS (Figure 4A,B). On the other hand, the non-saline soil (<2 dSm^{-1}) resulted in no reduction and the highest salinity classes (10–16 dSm^{-1}) in 8 to 14 units of reduction (Figure 4A,B).

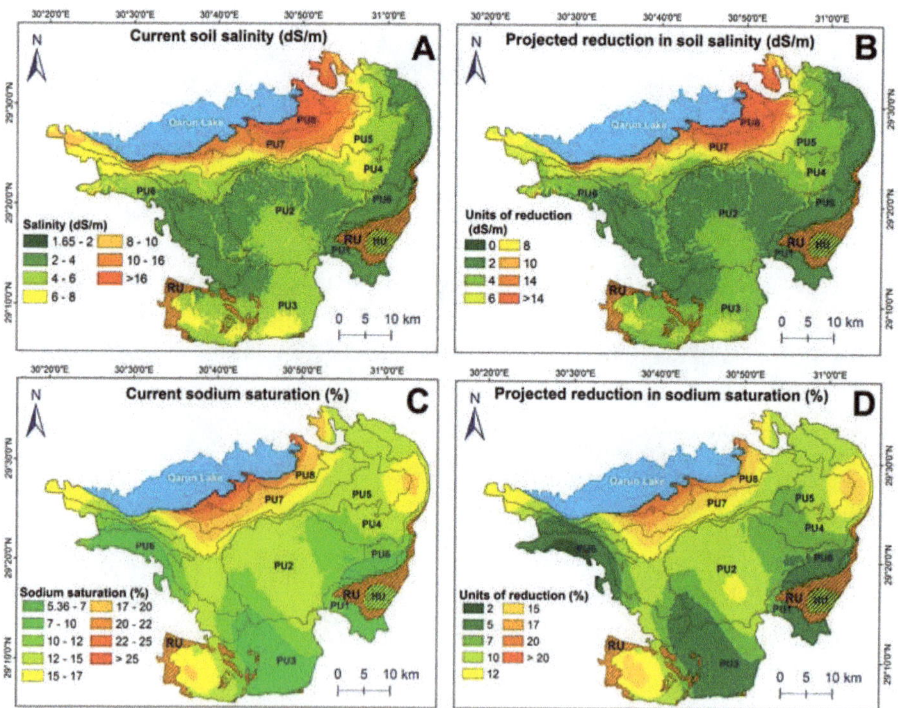

Figure 4. Spatial distribution of soil salinity under the current situation (CS) and the projected reduction units in the optimal scenario (OS) scenario (**A,B**); and the spatial distribution of sodium saturation under the current situation and the projected reduction in the OS scenario units (**C,D**).

Different soil management options have been suggested to reduce soil salinity, such as removal of salts from the soil root zone through a leaching process via low salinity water [61]. However, the growth rates of plants under salt stress often vary between cultivars and strongly differ between plant species [62,63].

3.2.2. Sodium Saturation

In around 39% of the study area, sodium saturation ranged from 12 to 15% under CS (Table 3). Low-sodium saturation (5–12%) represented around 37% of the study area, and high-sodium saturation (20–25%) covered <5% of the area (Figure 3b). Therefore, 37% of the study area can be improved to the intended value (5%) with 5–7 reduction units, and 7–10 reduction units are needed to enhance 39% of the study area. The spatial distribution of sodium saturation under CS and the projected reduction units of sodium saturation for each class are shown in (Figure 4C,D).

The increased sodium saturation and salinization in arid and semiarid regions negatively affect soils' physical properties such as soil structure and hydraulic conductivity, and consequently adversely impact crop yield, and if the exchangeable sodium percentage passes 15, these soils are classified as sodic soils [64,65]. On the other hand, high-sodium saturation could be improved by the addition of gypsum [16,17,66]. The addition of gypsum leads to the replacement of sodium with calcium on soil particles, which directly affects soil aggregation and reduces pH [17,67]. Thus, the incorporation of calcium using gypsum has proved to enhance soil particle aggregation, thus creating an adequate soil physical condition for nutrient uptake in sodic soils [68].

3.2.3. Drainage

Agriculture drainage in the studied area ranged from excessive (PU6) to very poor (PU8) (Figure 3c), and the dominant drainage status was the poor class under CS. Adequate drainage is essential to discharge leached salts from locations that have suffered from salinity problems [69,70]. In El-Fayoum, the main source of irrigation is Bahr Youssif (an old branch of the River Nile) with a water salinity value < 0.4 dSm^{-1}. An optimum drainage status may be achieved by the addition of OM and gypsum, that would directly enhance soils' physiochemical properties and consequently improves soil suitability [71,72]. Additionally, reducing tillage in sandy soils [73] and adding sand in the cases of clayey soil with very poor drainage [74,75], could lead to increased agriculture suitability under OS. An improved irrigation and drainage system could additionally help to recover and enhance soil properties, although over-irrigation could lead to decreasing soil suitability [76].

3.3. Soil Suitability

3.3.1. General Evaluation of Current Situation (CS) and Optimal Scenario (OS)

Table 4 shows how the Almagra model classified the three soil profile examples, and also illustrates the improvement in suitability classifications under the OS compared with the CS. Generally, the non-suitable class (S5) occurred mainly in the physiographic unit PU8 (Figure 5). For the perennial crops (citrus, peach, and olive), the main area assigned was the marginal suitability class (S4), and for the rest of the studied crops, the dominant classes were moderate (S3) and marginal (S4). High salinity, elevated sodium saturation, poor drainage, and heavy texture were observed as limiting factors in the physiographic units PU4 and PU5.

According to the spatial distribution (Figure 5) under CS, the average of soil suitability for each crop was classified as: cotton > sugar beet > wheat, melon, potato, soybean sunflower, and alfalfa > corn > olive > citrus, and peach. Soil suitability assessments for annual and semi-annual crops under CS were mostly classified as moderate in the adjacent south-western areas of Qarun Lake to not suitable for almost all crops in the south-eastern soils (Figure 5).

Table 4. Soil suitability classifications according to the Almagra model. Each soil profile was classified on a scale from 1 (best) to 5 (worst) for each subclass, based on the specific requirements for each crop. The final classification was determined by the worst subclasses (in red), which was indicated by their letter. The crops were: I, wheat; II, corn; III, melon; IV, potato; V, soybean; VI, cotton; VII, sunflower; VIII, sugar beet; IX, alfalfa; X, peach; XI, citrus fruits; XII, olive. CS, Current Situation; OS, Optimal Scenario.

Profile Code	Soil Factors/Classification	Crops											
		I	II	III	IV	V	VI	VII	VIII	IX	X	XI	XII
F10	Useful depth (p)	1	1	1	1	1	1	1	1	1	1	1	1
	Texture (t)	1	1	2	2	1	2	1	1	1	2	2	3
	Drainage (d)	3	2	2	2	3	2	2	3	3	4	4	4
	Carbonate (c)	2	1	1	1	2	1	2	2	2	1	1	2
	Salinity (s)	1	1	2	2	2	1	2	1	2	2	2	2
	Sodium sat (a)	2	2	2	2	2	1	2	1	2	2	2	2
	Profile dev (g)	1	1	1	1	1	1	1	1	1	2	2	1
	CS classification	S3d	S2da	S2tdsa	S2tdsa	S3d	S2td	S2dcsa	S3d	S3d	S4d	S4d	S4d
	OS classification	S2c	S1	S2t	S2t	S2c	S1	S2c	S2c	S2c	S2t	S2t	S2t
F22	Useful depth (p)	1	1	1	1	1	1	1	1	1	1	1	1
	Texture (t)	1	1	2	2	1	2	1	1	1	2	2	3
	Drainage (d)	1	1	1	1	1	1	1	1	1	2	2	2
	Carbonate (c)	2	1	1	1	2	1	2	2	2	1	1	2
	Salinity (s)	4	4	4	4	4	3	4	3	3	5	5	3
	Sodium sat (a)	2	3	2	2	2	1	2	1	2	2	2	2
	Profile dev (g)	1	1	1	1	1	1	1	1	1	2	2	1
	CS classification	S4s	S4s	S4s	S4s	S4s	S3s	S4s	S3s	S3s	S5s	S5s	S3ts
	OS classification	S2c	S1	S2t	S2t	S2c	S2c	S2c	S2c	S2c	S2tcg	S2tdg	S3t
F26	Useful depth (p)	1	1	2	1	2	2	2	2	2	3	3	3
	Texture (t)	2	2	2	2	2	2	2	2	2	4	4	4
	Drainage (d)	4	3	3	3	4	3	3	4	4	5	5	5
	Carbonate (c)	1	2	2	2	1	2	1	1	1	2	2	1
	Salinity (s)	5	5	5	5	5	5	5	5	5	5	5	5
	Sodium sat (a)	3	4	3	3	3	3	3	3	3	3	3	3
	Profile dev (g)	2	2	2	2	2	2	2	2	2	2	2	2
	CS classification	S5s	S5s	S5s	S5s	S5s	S5s	S5s	S5s	S5s	S5s	S5s	S5s
	OS classification	S5s	S5s	S5s	S5s	S5s	S5s	S5s	S5s	S5s	S5s	S5s	S5s

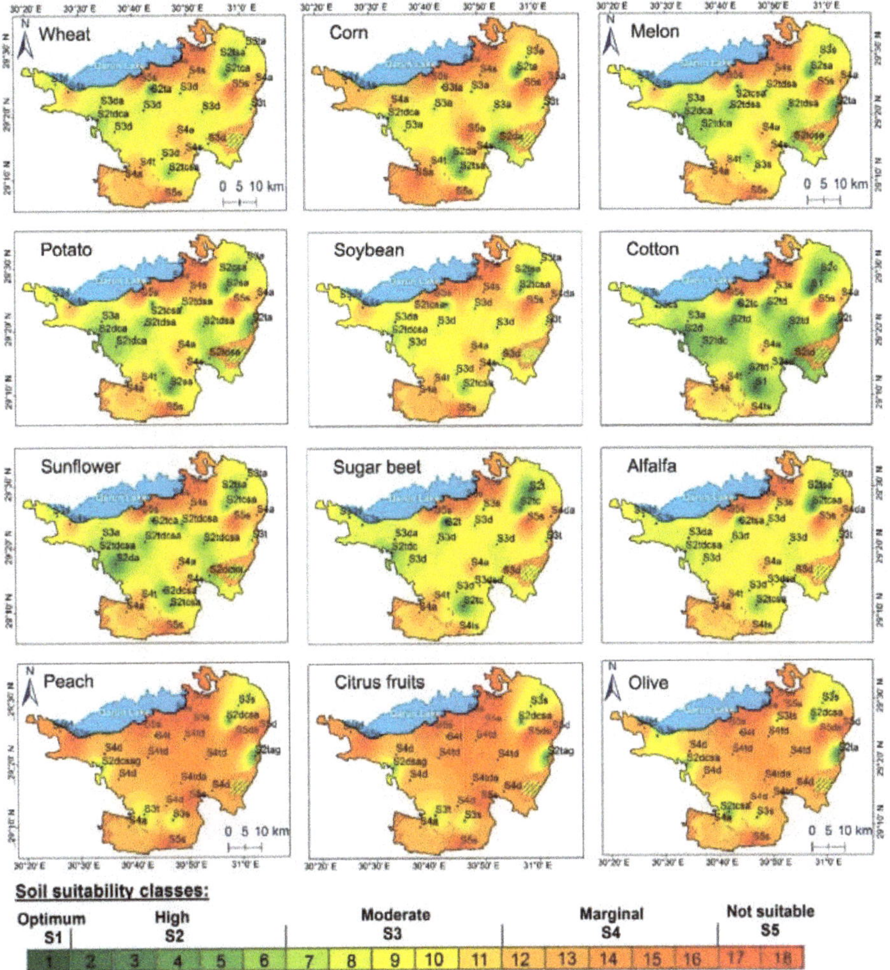

Figure 5. Spatial distribution of the soil suitability classes under the current situation of soil factors, according to the application of the Almagra model in the El-Fayoum depression. The Main limiting factors; s, salinity; d, drainage; a, sodium saturation; t, texture; c, carbonate content; g, profile development.

While under the OS, the marginal and moderated classes changed to high suitability (Figures 5 and 6). Importantly, there was at least one factor limiting soil suitability and preventing classification as optimum (Table 4). Although for all annual and semi-annual crops, the dominating soil suitability class was S2 (high), where the moderate and marginal suitability classes were predominant under perennials crops due to the difficulty of managing or improving soil factors, such as soil texture, high carbonate content, and soil depth (Figure 6).

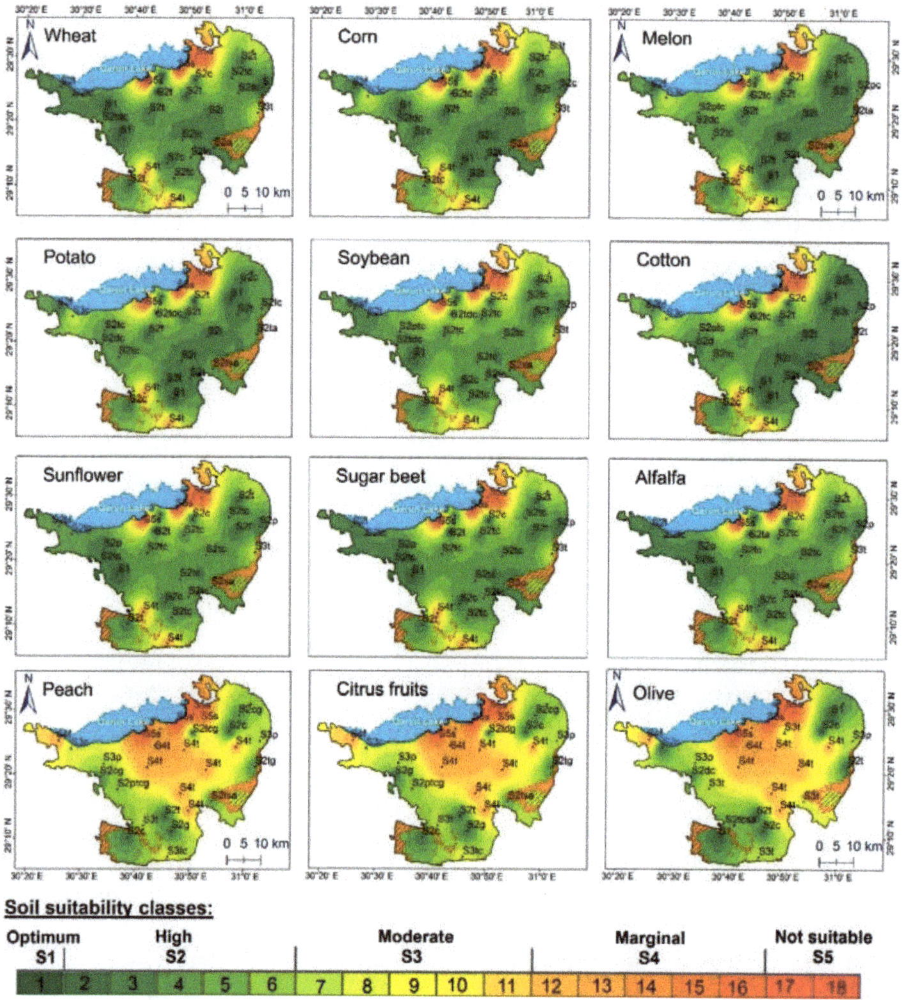

Figure 6. Spatial analysis of soil suitability classes under the projected improvement of soil factors, according to the application of the Almagra model in the El-Fayoum depression. The main limiting factors: t, texture; c, carbonate content; p, soil depth; g, profile development s, salinity; d, drainage; a, sodium saturation.

3.3.2. Soil Suitability for Studied Crops

El-Fayoum soils are inherently very fertile, but with poor management and environmental conditions, the salinity levels and the ESP have increased in the area, which significantly affects the suitability of the studied crops [77,78]. Reducing the severity of the manageable soil limiting factors (EC, ESP, and drainage), where possible, resulted in an increase of soil suitability for all the studied crops under OS. Only areas with extreme salinity and shallow soil profile depths were excluded from the OS, as the modification of these factors would be unfeasible (Figure 7). Under CS, the main soil suitability subclasses were represented by subclasses 7 to 17, which covers the suitability classes of S3 and S4 for the most evaluated crops (Figure 4). With the application of the OS, the moderate and marginal suitability classes (S3 and S4, respectively) shifted to the high suitability class (S2) (Table 5).

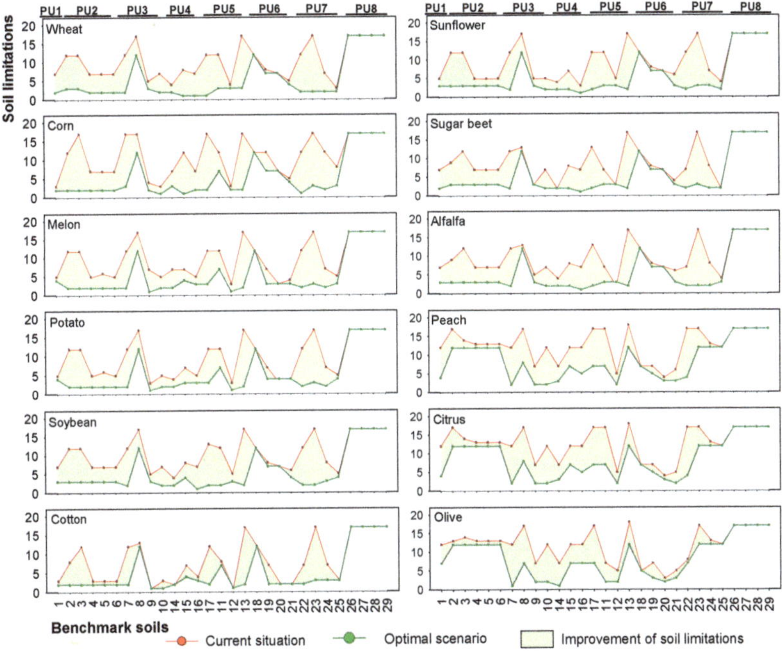

Figure 7. Improvement degree of soil suitability evaluation under the OS compared with CS for each benchmark soil for twelve Mediterranean crops. Lower values on the *y*-axis represent low soil limitations and better suitability. PU1, Flood plain; PU2, Fan; PU3, Basin; PU4, Transition elevated plain A; PU5, Transition elevated plain B; PU6, Oldest lake terraces; PU7, Old lake terraces; PU8, Recent lake terraces.

The studied annual and semi-annual crops were wheat, corn, melon, potato, soybean, cotton, sunflower, sugar beet, and alfalfa. Under CS, the S3 and S4 soil suitability classes for wheat covered 95% of the study area (1187 km^2 and 403 km^2, respectively), whereas under OS, 79% (1314 km^2) of the area improved to S2. For the corn crops, around 93% of the study area was S3 and S4 classes (881 km^2 and 664 km^2, respectively) under CS. In OS, the S2 class occupied 78%, 1298 km^2 (Figure 8). Regarding melon crop, the S3 and S4 classes covered around 91% of the study area (1150 km^2 and 362 km^2, respectively) under CS. However, with the application of OS, the area of the S2 class increased substantially from 119 km^2 to 1350 km^2, which represented 81% of the study area. Under the CS, 88% of the study area was classified as S3 and S4 for potato crop (1150 km^2 and 317 km^2, respectively), which was enhanced under the OS to the S2 class covering an area of 1342 km^2. The S3 and S4 classes covered 90% of the study area (1175 km^2 and 431 km^2, respectively) for soybean under CS. However, with the implementation of the OS, the S2 class improved to cover an area of 1287 km^2 (77%) (Table 5, Figures 5, 6 and 8). Cotton is one of the strategic economic crops in Egypt [79]. In the El-Fayoum depression, the soil suitability for cotton under CS ranged from S2, S3, and to S4, with a coverage of 29%, 58%, and 12%, respectively. Applying the OS enhanced the area coverage for S2 to represent 81% (1357 km^2). The S3 and S4 classes for sunflower under CS represented 90% of the study area (70% and 21%, respectively), whereas with OS, the S2 class represented 1299 km^2 (78%) of the study area. For the sugar beet crop, 93% of the study area was allocated for the S3 and S4 classes (77% and 16%, respectively), while under the OS, the area of S2 increased to 1299 km^2 (78%). Finally, the S3 and S4 classes for alfalfa under CS covered 96% of the study area (79% and 17%, respectively); however, with OS, the area coverage for S2 was improved to cover 1299 km^2 (Table 5, Figures 4, 5 and 7). Regarding subclasses, the subclass 4 represented the large area for almost all annual and semi-annual crops studied, except for cotton and sunflower the subclass 3 represented the maximum extension.

Table 5. Area (%) of soil suitability classes for the 12 crops under the CS (A) and OS (B).

Class/Scenarios	Crops																							
	Wheat		Corn		Melon		Potato		Soybean		Cotton		Sunflower		Sugar Beet		Alfalfa		Peach		Citrus Fruits		Olive	
	A	B	A	B	A	B	A	B	A	B	A	B	A	B	A	B	A	B	A	B	A	B	A	B
S2-high	2.48	78.7	3.77	77.8	7.14	80.8	9.92	80.4	1.43	77.1	29.1	81.3	7.38	77.8	5.17	77.8	2.58	77.8	0.68	27.3	0.84	27.9	1.39	31.9
S3-moderate	71.1	15.8	52.8	16.8	68.9	13.8	68.9	14.1	70.4	17.2	57.6	13.4	69.7	16.6	77.3	16.5	78.8	16.6	20.7	50.4	21.1	49.9	31.9	44.3
S4-marginal	24.1	4.8	39.8	4.81	21.7	4.72	19	4.78	25.8	5.01	11.9	4.66	20.7	4.93	15.9	4.94	17	4.92	70.7	21.1	70.1	21	62.8	22.5
S5-not suitable	2.28	0.68	3.69	0.67	2.3	0.67	2.21	0.68	2.38	0.71	1.38	0.66	2.22	0.69	1.55	0.69	1.61	0.69	7.95	1.22	7.93	1.22	3.84	1.32
Total	100	100	100	100	100	100	100	100	100	100	100	100	100	100	100	100	100	100	100	100	100	100	100	100

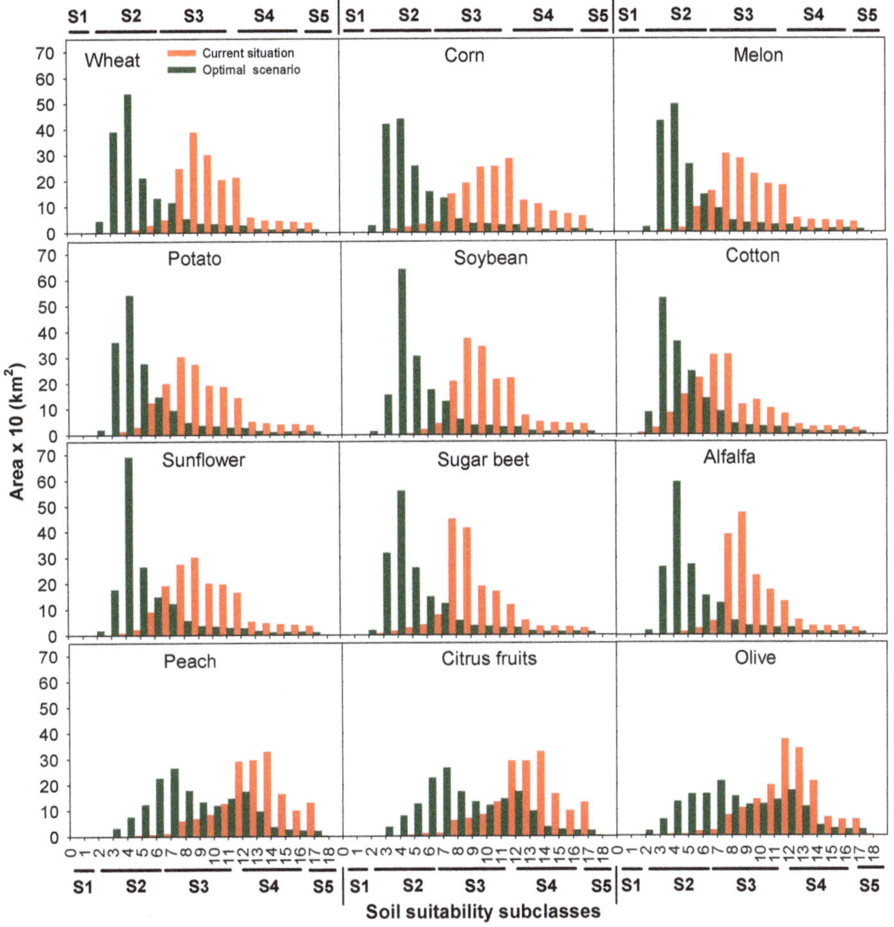

Figure 8. Soil suitability subclasses area (km^2) under CS and OS.

Three perennials crops were studied; peach, citrus, and olive. Under CS, around 91% of the study area was covered with S3 and S4 classes (21% and 71%, respectively) for peach and citrus. After implementing the OS, the area of S2 and S3 increased to around 28% and 50%, respectively. The S3 and S4 classes for olive under CS covered 95% of the study area (32% and 63%, respectively), which enhanced to S2 and S3 classes and covered areas of 533 km^2 (32%) and 44% (739 km^2), respectively (Table 5, Figures 4, 5 and 7).

Under CS, the average of soil suitability for the 12 studied crops was assigned to moderate and marginal suitability classes representing 71% and 25% of the study area, respectively (Figure 9). At the subclass level, the maximum area coverage of 378 km^2 (23%) was occupied by subclass 10 (S3 class). Remarkably, with the application of the OS, a notable improvement was observed, as approximately 75% (1248 km^2) of the study area was highly suitable for all evaluated crops. Conversely, the marginal suitability class was assigned to <5% of the area. The subclass 5 (S2 class) had a maximum area of 622 km^2 (37%) compared to the rest of soil suitability subclasses.

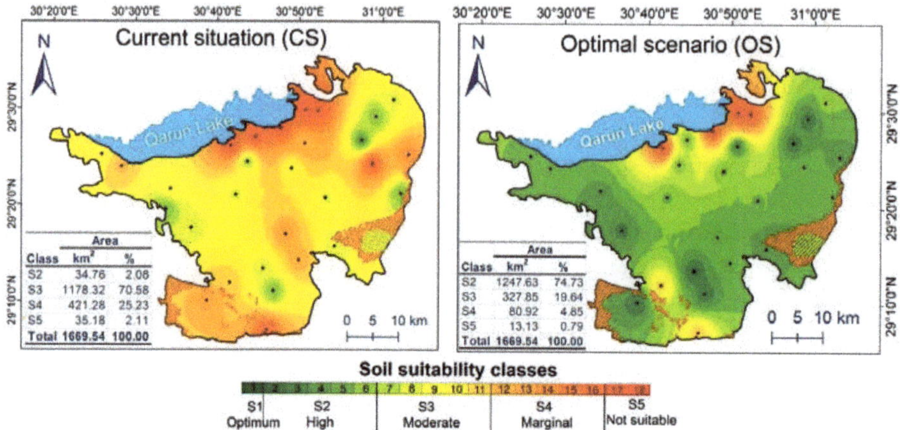

Figure 9. The total average of soil suitability for the 12 studied crops under current and optimal scenarios.

Successful sustainable agricultural management programs rely on the best choice of suitable crops for specific land in specific environmental conditions. There is robust interaction between soil characteristics and crop requirements. Thus, when the soil qualities fail to meet the requirements of the crop (soil suitability assigned as marginal or not suitable), this can be amended in some cases by improving soil management [80]. Enhancing agricultural management practices via improving soil physiochemical characteristics helps to raise soil suitability for optimal production of maize, melon, and olive [81–83]. Improving soil suitability for potato, maize, and alfalfa needs substantial efforts of soil management to increase crop production [84]. Reference [85] found that the most limiting factors of soil suitability for wheat production were elevated soil salinity and alkalinity. Additionally, they applied a qualitative model for evaluating soil suitability to assist decision makers for sustainable agriculture planning and economic productivity of soil resources in Iran. While, in the north-western region of Libya, Reference [86] reported that the main limiting factors of soil suitability for wheat, maize, alfalfa, sunflower, soybean, potato, citrus, and olives crops were soil salinity, soil texture, alkaline pH, calcium carbonate, and soil depth. Their findings are consistent with this research's observations.

Several scientists have utilized a computer/GIS-based model for mapping soil suitability [45,85,87–89]. The spatial analysis of soil suitability has a vital role in carrying out agriculture management processes [85,90]. Reference [91] used a soil evaluation model to estimate the land suitability of wheat, sugar beet, potato, and alfalfa in Iran. On the other hand, Reference [92] introduce a Bayesian network model, ALECA (Agroecological Land Evaluation for *Coffea arabica* L.), to evaluate land suitability for coffee production. Other scientists have utilized multi-criteria decision-making processes to provide accurate estimations of land suitability for different crops such as citrus [93], tobacco [94], and tea [95].

4. Conclusions

In this study, the application of the Almagra soil suitability model, a component of the MicroLEIS, allowed for predictions of agriculture soil suitability for twelve Mediterranean crops. The model identified the most suitable crops and the optimal spatial distribution under current and optimal management scenarios. The assessment of soil suitability under the current situation (CS) of soil factors helps to distinguish the most limiting soil factors. Likewise, applying the soil suitability model under the improvement of manageable limiting factors such as soil salinity, sodium saturation, and drainage assists in predicting the degree of improvement in soil suitability under the proposed optimal scenario (OS).

Consequently, the Almagra model assisted in identifying which of the studied crops were suitable to grow and where under the CS and the OS. The soil suitability classes varied spatially from optimum to non-suitable classes depending on the crop type and the possibility of soil factor improvements in case of the OS. Generally, under the CS of soil factors, the dominant soil suitability classes in the El-Fayoum depression were the moderate and marginal classes. On the other hand, the highly suitable class was dominant under the projected improvement of the manageable soil factors. Cotton was the most suitable current crop, while under optimal management, the suitability for all crops improved except for the perennial crops (i.e., peach, citrus fruits, and olive trees), where the most limiting factors for these crops are soil texture, depth, soil profile development, and carbonate content which are inflexible to modification. The total average of soil suitability for the twelve Mediterranean crops under CS were cotton > sugar beet > wheat, melon, potato, soybean sunflower, and alfalfa > corn > olive > citrus, and peach.

The assessment of soil suitability can help decision makers recognize the most limiting soil factors. Assessing the potential for improvement of manageable limiting factors such as soil salinity, sodium saturation, and drainage may assist in predicting the degree of improvement in soil suitability under the proposed OS. Mapping current soil suitability in this study and its improvement potential therefore provides valuable information to decision makers for appropriate land-use planning and sustainable development in the El-Fayoum depression.

Supplementary Materials: The following are available online at http://www.mdpi.com/2071-1050/11/6/1588/s1.

Author Contributions: Conceptualization, S.K.A.-E., N.B. and M.M.; Introduction, A.C., H.M., Z.Z.; Methodology, D.D. and A.J.; Results and discussions, S.K.A.-E., N.B., P.P. and M.M.; Reviewing and conclusions, L.J. and P.P. Funding Acquisition, Z.Z.

Funding: This research was funded by Talented Young Scientist Program (TYSP), China Science and Technology Exchange Center), grant number P150I3315, and the Independent Innovation Fund of Agricultural Science and Technology, Jiangsu Province, grant number CX (17)1001, and The APC was funded by Independent Innovation Fund of Agricultural Science and Technology, Jiangsu Province.

Conflicts of Interest: The authors declare no conflict of interest.

References

1. Santana-Cordero, A.M.; Ariza, E.; Romagosa, F. Studying the historical evolution of ecosystem services to inform management policies for developed shorelines. *Environ. Sci. Policy* **2016**, *64*, 18–29. [CrossRef]

2. Hanh, H.Q.; Azadi, H.; Dogot, T.; Ton, V.D.; Lebailly, P. Dynamics of Agrarian Systems and Land Use Change in North Vietnam. *Land Degrad. Dev.* **2017**, *28*, 799–810. [CrossRef]

3. Tengberg, A.; Radstake, F.; Zhang, K.; Dunn, B. Scaling up of Sustainable Land Management in the Western People's Republic of China: Evaluation of a 10-Year Partnership. *Land Degrad. Dev.* **2016**, *27*, 134–144. [CrossRef]

4. Lal, R. Soils and sustainable agriculture: A review. *Agron. Sustain. Dev.* **2009**, *28*, 57–64. [CrossRef]

5. Brevik, E.C. The potential impact of climate change on soil properties and processes and corresponding influence on food security. *Agriculture* **2013**, *3*, 398–417. [CrossRef]

6. Brevik, E.; Calzolari, C.; Miller, B.; Pereira, P.; Kabala, C.; Baumgarten, A.; Jordán, A. Historical perspectives and future needs in soil mapping, classification and pedological modelling. *Geoderma* **2016**, *264*, 256–274. [CrossRef]

7. Panagos, P.; Borrelli, P.; Poesen, J.; Ballabio, C.; Lugato, E.; Meusburger, K.; Montanarella, L.; Alewell, C. The new assessment of soil loss by water erosion in Europe. *Environ. Sci. Policy* **2015**, *54*, 438–447. [CrossRef]

8. Singh, A. Soil salinization and waterlogging: A threat to environment and agricultural sustainability. *Ecol. Indic.* **2016**, *57*, 128–130. [CrossRef]

9. Velmurugan, A.; Swarnam, T.P.; Ambast, S.K.; Kumar, N. Managing waterlogging and soil salinity with a permanent raised bed and furrow system in coastal lowlands of humid tropics. *Agric. Water Manag.* **2016**, *168*, 56–67. [CrossRef]

10. Sam, K.; Coulon, F.; Prpich, G. Working towards an integrated land contamination management framework for Nigeria. *Sci. Total Environ.* **2016**, *571*, 916–925. [CrossRef] [PubMed]

11. Chartzoulakis, K.; Bertaki, M. Sustainable Water Management in Agriculture under Climate Change. *Agric. Agric. Sci. Procedia* **2015**, *4*, 88–98. [CrossRef]

12. Longjun, C. UN Convention to Combat Desertification. In *Encyclopedia of Environmental Health*; Nriagu, J.O., Ed.; Elsevier: Burlington, NJ, USA, 2011; pp. 504–517. ISBN 9780444522726.

13. Zhou, D.; Wang, X.; Shi, M. Human driving forces of oasis expansion in Northwestern china during the last decade—A case of study of the Heihe river basin. *Land Degrad. Dev.* **2017**, *28*, 412–420. [CrossRef]

14. Ali, M. Management of salt-affected soils. In *Practices of Irrigation & On-Farm Water Management*; Springer: New York, NY, USA, 2011; Volume 2, pp. 271–325.

15. Daliakopoulos, I.N.; Tsanis, I.K.; Koutroulis, A.; Kourgialas, N.N.; Varouchakis, A.E.; Karatzas, G.P.; Ritsema, C.J. The threat of soil salinity: A European scale review. *Sci. Total Environ.* **2016**, *573*, 727–739. [CrossRef] [PubMed]

16. FAO (Food and Agriculture Organization). *Salt-Affected Soils and Their Management*; Soils Bulletin 39; FAO: Rome, Italy, 1988.

17. Chi, C.M.; Zhao, C.W.; Sun, X.J.; Wang, Z.C. Reclamation of saline-sodic soil properties and improvement of rice (*Oriza sativa* L.) growth and yield using desulfurized gypsum in the west of Songnen Plain, northeast China. *Geoderma* **2012**, *187*, 24–30. [CrossRef]

18. CAPMAS. Egypt Statistical Yearbook Population. 2015. Available online: http://www.capmas.gov.eg/Pages/Publications.aspx?page_id=5104&YearID=23011 (accessed on 12 July 2017).

19. Gouda, A.A.; Hosseini, M.; Masoumi, H.E. The Status of Urban and Suburban Sprawl in Egypt and Iran. *GeoScape* **2016**, *10*, 1–15. [CrossRef]

20. Negm, A.M.; Saavedra, O.; El-Adawy, A. Nile Delta Biography: Challenges and Opportunities. In *The Nile Delta, The Handbook of Environmental Chemistry*; Negm, A.M., Ed.; Springer: Cham, Switzerland, 2016; pp. 1–16.

21. Mahmoud, H.; Divigalpitiya, P. Modeling Future Land Use and Land-Cover Change in the Asyut Region Using Markov Chains and Cellular Automata. In *Smart and Sustainable Planning for Cities and Region*; Springer: Cham, Switzerland, 2017; pp. 99–112.

22. FAO (Food and Agriculture Organization). *Land Evaluation: Towards a Revised Framework*; Land and Water Discussion Paper 6; FAO: Rome, Italy, 2007; p. 107.

23. Anaya-Romero, M.; Abd-Elmabod, S.K.; Muñoz-Rojas, M.; Castellano, G.; Ceacero, C.J.; Alvarez, S.; Méndez, M.; De la Rosa, D. Evaluating soil threats under climate change scenarios in the Andalusia region. Southern Spain. *Land Degrad. Dev.* **2015**, *26*, 441–449. [CrossRef]

24. Rossiter, D.G. A Theoretical Framework for Land Evaluation. *Geoderma* **1996**, *72*, 165–202. [CrossRef]

25. FAO (Food and Agriculture Organization). *Guidelines: Land Evaluation for Irrigated Agriculture*; Soils Bulletin 55; FAO: Rome, Italy, 1985.

26. FAO (Food and Agriculture Organization). *Guidelines for Land-Use Planning*; FAO Development Series 1; FAO: Rome, Italy, 1993; p. 96.

27. Sys, C.; Van Ranst, E.; Debaveye, J. *Land Evaluation. Part 3: Crop Requirements*; Agricultural Publications, General Administration of Development Cooperation: Belgium, Brussels, 1993; p. 199.

28. Rossiter, D.G. ALES (Automated Land Evaluation System): A framework for land evaluation using a microcomputer. *Soil Use Manag.* **1990**, *6*, 7–20. [CrossRef]

29. De la Rosa, D.; Moreno, J.A.; Garcia, L.V.; Almorza, J. MicroLEIS: A microcomputer-based Mediterranean land evaluation information system. *Soil Use Manag.* **1992**, *8*, 89–96. [CrossRef]

30. De la Rosa, D.; Mayol, F.; Diaz-Pereira, E.; Fernandez, M. Aland evaluation decision support system (MicroLEIS DSS) for agricultural soil protection. *Environ. Model. Softw.* **2004**, *19*, 929–942. [CrossRef]

31. Wenkel, K.O.; Berg, M.; Mirschel, W.; Wieland, R.; Nendel, C.; Köstner, B. LandCaRe DSS—An interactive decision support system for climate change impact assessment and the analysis of potential agricultural land use adaptation strategies. *J. Environ. Manag.* **2013**, *127*, S168–S183. [CrossRef] [PubMed]

32. Pereira, P.; Brevik, E.; Munoz-Rojas, M.; Miller, B.; Smetanova, A.; Depellegrin, D.; Misiune, I.; Novara, A.; Cerda, A. Soil mapping and process modelling for sustainable land management. In *Soil Mapping and Process Modelling for Sustainable Land Use Management*; Pereira, P., Brevik, E., Munoz-Rojas, M., Miller, B., Eds.; Elsevier: Amsterdam, The Netherland, 2017; pp. 29–60. ISBN 9780128052006.

33. Pereira, P.; Brevik, E.; Trevisani, S. Mapping the Environment. *Sci. Total Environ.* **2018**, *610*, 17–23. [CrossRef] [PubMed]

34. Muñoz-Rojas, M.; Doro, L.; Ledda, L.; Francaviglia, R. Application of CarboSOIL model to predict the effects of climate change on soil organic carbon stocks in agro-silvo-pastoral Mediterranean management. *Agric. Ecosyst. Environ.* **2015**, *202*, 8–16. [CrossRef]

35. Muñoz-Rojas, M.; Jordán, A.; Zavala, L.M.; De la Rosa, D.; Abd-Elmabod, S.K.; Anaya-Romero, M. Impact of land use and land cover changes on organic carbon stocks in Mediterranean soils (1956–2007). *Land Degrad. Dev.* **2015**, *26*, 168–179.

36. Muñoz-Rojas, M.; Abd-Elmabod, S.K.; Zavala, L.M.; De la Rosa, D.; Jordán, A. Climate change impacts on soil organic carbon stocks of Mediterranean agricultural areas: A case study in Northern Egypt. *Agric. Ecosyst. Environ.* **2017**, *238*, 142–152. [CrossRef]

37. De la Rosa, D. *Soil Survey and Evaluation of Guadalquivir River Terraces, in Sevilla Province*; Cent. Edaf. Cuarto Pub.: Seville, Spain, 1974.

38. Farroni, A.; Magaldi, D.; Tallini, M. Total sediment transport by the rivers of Abruzzi (Central Italy): Prediction with the RAIZAL model. *Bull. Eng. Geol. Environ.* **2002**, *61*, 121–127.

39. Erdogan, H.E.; Yüksel, M.; De La Rosa, D. Evaluation of Sustainable Land Management Using Agro-Ecological Evaluation Approach in Ceylanpınar State Farm (Turkey). *Turk. J. Agric. For.* **2003**, *27*, 15–22.

40. López García, J.; Acosta, R.; Bojórquez Serrano, J.I. Aptitud relativa agrícola del municipio de Tuxpan, Nayarit, utilizando el modelo Almagra del Sistema MicroLEIS. *Investig. Geogr.* **2006**, *59*, 59–73. [CrossRef]

41. Darwish, K.M.; Wahba, M.M.; Awad, F. Agricultural Soil Suitability of Haplo-Soils for Some Crops in Newly Reclaimed Areas of Egypt. *Appl. Sci. Res.* **2006**, *12*, 1235–1243.

42. Bakr, N.; Bahnassy, M.H.; El-Badawi, M.M.; Ageeb, G.W.; Weindorf, D.C. Land capability evaluation in newly reclaimed areas: A case study in Bustan 3 area, Egypt. *Soil Sur. Horiz.* **2009**, *51*, 90–95. [CrossRef]

43. Shahbazi, F.; De la Rosa, D.; Anaya-Romero, M.; Jafarzade, A.; Sarmadian, F.; Neyshaboury, M.; Oustam, S. Land use planning in Ahar area (Iran) using MicroLEIS DSS. *Int. Agrophys.* **2010**, *22*, 277–286.

44. Muñoz-Rojas, M.; Jordán, A.; Zavala, L.M.; González-Peñaloza, F.A.; De la Rosa, D.; Pino-Mejias, R.; Anaya-Romero, M. Modelling soil organic C stocks in climate change scenarios: A CarboSOIL model application. *Biogeosciences* **2013**, *10*, 8253–8268.

45. Abd-Elmabod, S.K.; Jordán, A.; Fleskens, L.; Phillips, J.D.; Muñoz-Rojas, M.; Van der Ploeg, M.; Anaya-Romero, M.; De la Rosa, D. Modelling agricultural suitability along soil transects under current conditions and improved scenario of soil factors. In *Soil Mapping and Process Modeling for Sustainable Land Use Management*; Elsevier: Amsterdam, The Netherland, 2017; pp. 193–219. ISBN 9780128052006.

46. ASRT (Academy of Scientific Research and Technology). *Preparation of Land Data Base for Agriculture Use" Fifth Report*; ASRT: Cairo, Egypt, 2009.

47. Soil Survey Staff. *Keys to Soil Taxonomy*, 12th ed.; United States Department of Agriculture, Natural Resources Conservation Service: Washington, DC, USA, 2014.

48. Haroun, O.R. Soil Evaluation Systems as a Guide to Identify an Economical Feasibility Study for Agricultural Purposes in El-Fayoum Province, Egypt. Ph.D. Thesis, Faculty of Agriculture, El-Fayoum Cairo University, El-Fayoum, Egypt, 2004.

49. Ali, R.R. Geomatics Based Soil Mapping and degradation risk Assessment of the Cultivated Land in El-Fayoum Depression Egypt. *J. Soil Sci.* **2005**, *45*, 349–360.

50. CLAC (Central Laboratory for Agricultural Climate). *The Climatic Normals-El Fayoum Station*; Annual Report; Central Laboratory for Agricultural Climate CLAC: Cairo, Egypt, 2010.

51. De la Rosa, D.; Cardona, F.; Paneque, G. Evaluación de suelos para diferentes usos agrícolas. Un sistema desarrollado para regiones mediterráneas. *Anales de Edafología y Agrobiología* **1977**, *36*, 1100–1112.

52. Manna, P.; Basile, A.; Bonfante, A.; De Mascellis, R.; Terribile, F. Comparative Land Evaluation approaches: An itinerary from FAO framework to simulation modelling. *Geoderma* **2009**, *150*, 367–378. [CrossRef]

53. De la Rosa, D.; Anaya-Romero, M.; Diaz-Pereira, E.; Heredia, N.; Shahbazi, F. Soil-Specific Agro-Ecological Strategies for Sustainable Land Use-A Case Study by Using MicrolEIS DSS in Sevilla Province (Spain). *Land Use Policy* **2009**, *26*, 1055–1065. [CrossRef]

54. Aldabaa, A.A.; HaiLin, Z.; Shata, A.; El-Sawey, S.; Abdel-Hameed, A.; Schroder, J.L. Land suitability classification of a desert area in Egypt for some crops using MicrolEIS program. *Am.-Eurasian J. Agric. Environ. Sci.* **2010**, *8*, 80–94.

55. Darwish, K.M.; Kawy, W.A. Land suitability decision support for assessing land use changes in areas west of Nile Delta, Egypt. *Arab. J. Geosci.* **2014**, *7*, 865–875. [CrossRef]

56. ESRI (Environmental Systems Research Institute). *ArcGIS Desktop: Release 10.4*; Environmental Systems Research Institute: Redlands, CA, USA, 2016.

57. Shendi, M.M. Pedological Studies on Soils Adjacent to Qarun Lake, Fayoum Governorate, Egypt. Master's Thesis, Faculty of Agriculture, Cairo University, Giza, Egypt, 1984.

58. Soil Survey Division Staff. Soil survey manual. In *U.S. Department of Agriculture Handbook 18*; Soil Conservation Service: Washington, DC, USA, 1993.

59. IUSS (International Union of Soil Sciences); ISRIC (ISRIC World Soil Information); FAO (Food and Agriculture Organization). *World Reference Base for Soil Resources and Communication*; World Soil Resources Report 103; International Union of Soil Sciences, ISRIC World Soil Information, FAO: Rome, Italy, 2006.

60. IBM SPSS. *IBM SPSS Statistics for Windows, Version.23.0*; IBM Corp.: Armonk, NY, USA, 2015.

61. Zalacáin, D.; Martínez-Pérez, S.; Bienes, R.; García-Díaz, A.; Sastre-Merlín, A. Salt accumulation in soils and plants under reclaimed water irrigation in urban parks of Madrid (Spain). *Agric. Water Manag.* **2019**, *213*, 468–476. [CrossRef]

62. Qadir, M.; Schubert, S. Degradation processes and nutrient constraints in sodic soils. *Land Degrad. Dev.* **2002**, *13*, 275–294. [CrossRef]

63. Jacobsen, S.E.; Jensen, C.R.; Liu, F. Improving crop production in the arid Mediterranean climate. *Field Crops Res.* **2012**, *128*, 34–47. [CrossRef]

64. Balks, M.R.; Bond, W.J.; Smith, C.J. Effects of sodium accumulation on soil physical properties under an effluent-irrigated plantation. *Soil Res.* **1998**, *36*, 821–830. [CrossRef]

65. Paes, J.L.; Ruiz, H.A.; Fernandes, R.B.; Freire, M.B.; Barros, M.D.; Rocha, G.C. Hydraulic conductivity in response to exchangeable sodium percentage and solution salt concentration. *Revista Ceres* **2014**, *61*, 715–722. [CrossRef]

66. Rasouli, F.; Pouya, A.K.; Karimian, N. Wheat yield and physico-chemical properties of a sodic soil from semi-arid area of Iran as affected by applied gypsum. *Geoderma* **2013**, *193*, 246–255. [CrossRef]

67. Temiz, C.; Cayci, G. The effects of gypsum and mulch applications on reclamation parameters and physical properties of an alkali soil. *Environ. Monit. Assess.* **2018**, *190*, 347. [CrossRef] [PubMed]

68. Anikwe, M.A.N.; Ibudialo, E.A.N. Influence of lime and gypsum application on soil properties and yield of cassava (*Manihot esculenta* Crantz.) in a degraded Ultisol in Agbani, Enugu Southeastern Nigeria. *Soil Till. Res.* **2016**, *158*, 32–38. [CrossRef]

69. Abrol, I.P.; Jai Singh, P.Y.; Massoud, F.I. *Salt-Affected Soils and Their Management*; No. 39; Food & Agriculture Org.: Rome, Italy, 1988.

70. Chang, X.; Gao, Z.; Wang, S.; Chen, H. Modelling long-term soil salinity dynamics using SaltMod in Hetao Irrigation District, China. *Comput. Electron. Agric.* **2019**, *156*, 447–458. [CrossRef]

71. Hu, Z.Y.; Zaho, F.J.; McGrath, S.P. Sulphur fractionation in calcareous soils and bioavailability to plants. *Plant Soil* **2005**, *268*, 103–109. [CrossRef]

72. Yazdanpanah, N.; Mahmoodabadi, M.; Cerdà, A. The impact of organic amendments on soil hydrology, structure and microbial respiration in semiarid lands. *Geoderma* **2016**, *266*, 58–65. [CrossRef]

73. Assefa, T.; Jha, M.; Reyes, M.; Worqlul, A.W. Modeling the Impacts of Conservation Agriculture with a Drip Irrigation System on the Hydrology and Water Management in Sub-Saharan Africa. *Sustainability* **2018**, *10*, 19. [CrossRef]

74. Park, T.W.; Kim, H.J.; Tanvir, M.T.; Lee, J.B.; Moon, S.G. Influence of coarse particles on the physical properties and quick undrained shear strength of fine-grained soils. *Geomech. Eng.* **2018**, *14*, 99–105.

75. Feng, W.Q.; Li, C.; Yin, J.H.; Chen, J.; Liu, K. Physical model study on the clay–sand interface without and with geotextile separator. *Acta Geotech.* **2019**, 1–17. [CrossRef]

76. Valipour, M. Drainage, waterlogging, and salinity. *Arch. Agron. Soil Sci.* **2014**, *60*, 1625–1640. [CrossRef]

77. Ali, R.R.; Abdel Kawy, W.A. Land degradation risk assessment of El Fayoum depression, Egypt. *Arab J. Geosci.* **2013**, *6*, 2767–2776. [CrossRef]

78. Abd-Elmabod, S.K. Evaluation of Soil Degradation and Land Capability in Mediterranean Areas under Climate and Management Change Scenarios: (Andalusia Region, Spain and El-Fayoum Province, Egypt). Ph.D. Thesis, Universidad de Sevilla, Sevilla, Spain, 2014.

79. Amin, A.A.; Gergis, M.F. Integrated management strategies for control of cotton key pests in middle Egypt. *Agron. Res.* **2006**, *4*, 121–128.

80. Manrique, L.A.; Uehara, G. A Proposed Land Suitability Classification for Potato: I. Methodology 1. *Soil Sci. Soc. Am. J.* **1984**, *48*, 843–847. [CrossRef]

81. Abagyeh, S.O.; Idoga, S.; Agber, P.I. Land suitability evaluation for maize (*Zea mays*) production in selected sites of the Mid-Benue valley, Nigeria. *IJAPR* **2016**, *4*, 46–51.

82. Akinrinde, E.A.; Bello, O.S.; Ayegboyin, K.O.; Iroh, L. Added benefits of combined organic and mineral phosphate fertilizers applied to maize and melon. *J. Food Agric. Environ.* **2005**, *3*, 75.

83. Kavvadias, V.; Papadopoulou, M.; Vavoulidou, E.; Theocharopoulos, S.; Repas, S.; Koubouris, G.; Psarras, G.; Kokkinos, G. Effect of addition of organic materials and irrigation practices on soil quality in olive groves. *J. Water Clim. Chang.* **2018**, *9*, 775–785. [CrossRef]

84. Jafarzadeh, A.A.; Abbasi, G. Qualitative land suitability evaluation for the growth of onion, potato, maize, and alfalfa on soils of the Khalat pushan research station. *Biologia* **2006**, *61*, S349–S352. [CrossRef]

85. Ashraf, S.; Afshari, H.; Munokyan, R.; Ebadi, A.G. Multicriteria land suitability evaluation for barley by using GIS in Damghan plain (Northeast of Iran). *J. Food Agric. Environ.* **2010**, *8*, 626–628.

86. El-Aziz, S.H. Evaluation of land suitability for main irrigated crops in the North-Western Region of Libya. *Eurasian J. Soil Sci.* **2018**, *7*, 73–86. [CrossRef]

87. Doolittle, J.; Dobos, R.; Peaslee, S.; Waltman, S.; Benham, E.; Tuttle, W. Revised Ground-Penetrating Radar Soil Suitability Maps. *J. Environ. Eng. Geophys.* **2010**, *15*, 111–118. [CrossRef]

88. Ennaji, W.; Barakat, A.; El Baghdadi, M.; Oumenskou, H.; Aadraoui, M.; Karroum, L.A.; Hilali, A. GIS-based multi-criteria land suitability analysis for sustainable agriculture in the northeast area of Tadla plain (Morocco). *J. Earth Syst. Sci.* **2018**, *127*, 14. [CrossRef]

89. Salkovic, E.; Djurovic, I.; Knezevic, M.; Popovic-Bugarin, V.; Topalovic, A. Digitization and Mapping of National Legacy Soil Data of Montenegro. *Soil Water Res.* **2018**, *13*, 83–89. [CrossRef]

90. Liu, Y.-S.; Wang, J.-Y.; Guo, L.-Y. GIS-based assessment of land suitability for optimal allocation in the Qinling Mountains, China. *Pedosphere* **2006**, *16*, 579–586. [CrossRef]

91. Safari, Y.; Esfandiarpour-Boroujeni, I.; Kamali, A.; Salehi, M.H.; Bagheri-Bodaghabadi, M. Qualitative land suitability evaluation for main irrigated crops in the shahrekord plain, Iran: A geostatistical approach compared with conventional method. *Pedosphere* **2013**, *23*, 767–778. [CrossRef]

92. Lara-Estrada, L.D.; Rasche, L.; Schneider, U. Modeling land suitability for *Coffea arabica* L. in Central America. *Environ. Model. Softw.* **2017**, *95*, 96–209. [CrossRef]

93. Zabihi, H.; Ahmad, A.; Vogeler, I.; Said, M.N.; Golmohammadi, M.; Golein, B.; Nilashi, M. Land suitability procedure for sustainable citrus planning using the application of the analytical network process approach and GIS. *Comput. Electron. Agric.* **2015**, *117*, 114–126. [CrossRef]

94. Zhang, J.; Su, Y.; Wu, J.; Liang, H. GIS based land suitability assessment for tobacco production using AHP and fuzzy set in Shandong province of China. *Comput. Electron. Agric.* **2015**, *114*, 202–211. [CrossRef]

95. Bo, L.; Zhang, F.; Zhang, L.W.; Huang, J.F.; Zhi-Feng, J.I.; Gupta, D.K. Comprehensive suitability evaluation of tea crops using GIS and a modified land ecological suitability evaluation model. *Pedosphere* **2012**, *22*, 122–130.

Article

Land Use Changes and Their Driving Forces in a Debris Flow Active Area of Gansu Province, China

Songtang He [1,2,3] , **Daojie Wang** [1,2,*], **Yong Li** [1,2] **and Peng Zhao** [1,2,3]

1   Key Laboratory of Mountain Hazards and Earth Surface Processes, Chinese Academy of Sciences, Chengdu 610041, China; hst1529568372@126.com (S.H.); ylie@imde.ac.cn (Y.L.); dilikexuezp@126.com (P.Z.)
2   Institute of Mountain Hazards and Environment, Chinese Academy of Sciences, Chengdu 610041, China
3   University of Chinese Academy of Sciences, Beijing 100049, China
*   Correspondence: wangdj@imde.ac.cn

Received: 27 June 2018; Accepted: 2 August 2018; Published: 4 August 2018

Abstract: Land use change is extremely sensitive to natural factors and human influence in active debris flow. It is therefore necessary to determine the factors that influence land use change. This paper took Wudu District, Gansu Province, China as a study area, and a systemic analysis of the transformational extent and rate of debris flow waste-shoal land (DFWSL) was carried out from 2005 to 2015. The results show that from 2005 to 2015, cultivated land resources transformed to other types of land; cultivated lands mainly transformed to grassland from 2005 to 2010 and construction land from 2010 to 2015. Moreover, the growth rate of construction land from 2005 to 2010 was only 0.11%, but increased to 6.87% between 2010 and 2015. The latter is more than 60 times the former. This increase was brought about by natural disasters (debris flow, earthquakes, and landslides) and anthropogenic factors (national policies or strategies), which acted as driving forces in debris flow area. The former determines the initial use type of the DFWSL while the latter only affects the direction of land use and transformation.

Keywords: debris flow waste-shoal land; land use and transformation; driving forces analysis; territorial development; marginal land resources

---

1. Introduction

Land use is an anthropogenic activity in which the natural characteristics of the land are identified, and bio-technological means are adopted to periodically harness and manage the land for socioeconomic benefits [1]. Land cover refers to the natural and biophysical properties of the Earth's surface [2]. Land use and land cover changes result from interactions among socioeconomic, institutional, and environmental factors [3]. In 1995, the International Geosphere-Biosphere Program (IGBP) and the Human and Environmental Plan (HDP) jointly launched the land use and land cover change (LUCC) research program to further research on land use and transformation, especially with regard to global changes. Studies on LUCC over the past decades have focused on the analysis of factors driving land use, dynamic processes of land use, and the interrelation between land use and land cover changes [4–6]. Understanding the dynamic system of driving forces is key to revealing the sources of land use change. In the past, research has been conducted to understand the following: (1) Change in land cover due to human activities over the past 300 years; (2) anthropogenic factors influencing land use and transformation in different historical stages and geographic units [7–9]; (3) impact of land use and transformation on land cover over the next 50–100 years [10,11]; (4) impacts of anthropogenic and biophysical factors on the carrying capacities of different land types [12]; and (5) impacts of climate and global biogeochemistry on land use and land cover [13,14]. As urbanization prevails and the global population surges, the food supply is predicted to increase by 60% by 2050 [15], implying that the area of agricultural land (arable area) may require an increase of 10–26% [16]. In extreme cases, farmland will need to increase by 42% and pastures by 15% to meet the growing demands [17]. This raises questions about the extent to which technology can help improve the capacity of land, how land can be utilized effectively, and whether current land resources will meet future demands [18].

Considering the above-mentioned scenarios, as well as intense human activities and severe natural disasters, the capacity of the available land resources (including cultivated land, forests, industrial and mining land, etc.) in ecologically fragile areas is likely to decrease. Moreover, the possibility of an increase in quality or quantity is limited. Additionally, ecological security serves as the core of land resource sustainability. Reshaping the relationship between land use and the ecological environment, and maintaining the renewability of land resources, are important for the restoration of damaged land ecosystems [19]. Consequently, in order to increase the available area, and to provide an opportunity to achieve a multi-win of soil and water conservation, environmental governance and economic development, research perspectives should shift toward exploring land resources that were previously deemed unavailable. Moreover, past studies have paid special attention to areas with vulnerable ecosystems [20,21] and especially those characterized by sensitivity and instability [22]. For instance, the successful management and use of desert lands in northwestern China [23], reuse of abandoned mining land [24], land development and use in karst areas [25], and the proper use of torrential floods in arid and semi-arid areas [26]. These lands are important reserves of resources and not only provide sufficient space for industrialization and urbanization but also work as effective buffers for ecological construction. Therefore, it is essential to address the dilemma of protecting cultivated land while pursuing development in mountainous regions [27]. To explore the possibilities of developing unused land in mountainous areas, this study focused on the use of debris flow waste-shoal land (DFWSL), cover changes in mountainous areas, and the driving forces behind these changes.

DFWSL is the products of debris flows, which frequently occur and generally develop in the lower reaches of valleys [28–32]. Some scholars refer to them as 'bajada' [33], 'floodplain', or 'alluvial fans' [34–36] because of their fan-shaped structures when formed [30,37,38]. Wang Daojie [39] pointed out that the raw debris flow fan is alkaline with unreasonable soil hierarchy (no clear plough layer, plow pan and subsoil) and poor physical structure (compact structure). The different textures can be further categorized as boulder clay, silty soil, and gravel. In addition, its organic content is under 1%, and the contents of available nitrogen, phosphorus, as well as potassium, are presented in moderate levels or lower. Through soil improvement, the soil layer of the debris flow fan can reach 35 cm, with a gravel content of only 1%. The soil texture can also be significantly improved from silty soil to silty loam soil. Other scholars deemed that there were distinctive characteristics between the debris flow fan and other piedmonts, such as surface features, and the stratum. The sub-characteristics, such as the characters of grooving, dikes, ditches, and the shape or tip of the debris flow fan were also used to identify the debris flow fan (Figure 1a,b) [30,37,38,40]. Valley-type debris flow fans have features such as a gentle topography (about 5°), abundant water sources, convenient transportation, and easy access to irrigation [41]. Therefore, agricultural activities are carried out on some DFWSL, while some towns and villages are also built on the fans [31,42]. Additionally, some settlements also improve access to the fans to enable conversion to commercial or recreational sites with infrastructure like golf courses [43]. However, unlike conventional land resources, the use of DFWSL is subject to greater spatial-temporal variability [44–46]. Several studies have looked at dynamic land use and its driving forces; however, analyses of land use and transformation, as well the driving forces in mudslide-prone areas are limited [47]. Additionally, land use is subject to limitations of the natural environment and intervention of human activities. Focusing on the driving forces behind use and transformation patterns is a major approach to elucidate the general relationship between the socioeconomic and natural environments [48,49]. Therefore, it is necessary to analyze the transformation of DFWSL as well as the driving forces behind the transformation to lay the foundation for balanced, efficient, and safe development and use of DFWSL.

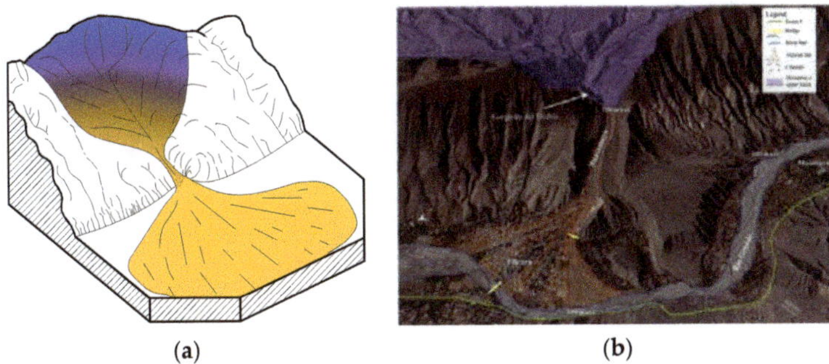

Figure 1. Various debris flow fans. (**a**) The original debris flow fan; (**b**) Tilcara alluvial fan in Argentina [31].

Land use in developed regions of western China has received significant attention [50–54]; however, the study of land use and driving forces in underdeveloped areas of western China, especially in ecologically fragile areas, requires further attention [20]. Debris flow active areas are typical ecologically fragile areas. These areas are prone to natural factors (e.g., debris flow and landslide). Moreover, the ecological equilibrium of these areas is vulnerable to social, economic, and technological development activities concerning land cover changes [22]. Longnan City in Gansu Province, China, has been affected by debris flows for a long time; however, it is also a model with mature DFWSL use experience. Agriculture and townships in Longnan City are developed along the Bailongjiang River. The debris flow area in the middle of Bailongjiang River accounts for 52.07% of the flat area in the valley [55]. The development and use of the debris flow area has initiated positive outcomes with regard to serious water and soil losses and shortage of land resources.

This study analyzed the evolution of DFWSL use and cover changes in Longnan City from 2005 to 2015 and investigated the influence of natural factors (natural disasters) and socioeconomic factors (policies) on the transformation of land use patterns. The objectives were to explore new development paths for the development and use of land resources in mountainous regions and to find new ways to accelerate the transition to sustainable development [56], especially in ecologically fragile areas where soil erosion has taken place. Besides simultaneously addressing poverty alleviation and soil erosion prevention, this study also aimed to solve the problem of dislocation between ecologically sustainable construction and farmers' income through comprehensive management of small basins. This is likely to promote the evolution of a harmonious development zone and a land ecological experimental zone [57]. On one hand, these measures will help control the degree of mountain exploitation; on the other hand, they will improve the efficiency of land use in ecologically fragile areas, optimize the development and management of wasteland, ensure the minimum area of arable land, and finally, achieve sustainable use of land resources in mountainous areas. All these measures can contribute to rural population retention and rural prosperity.

2. Materials and Methods

2.1. Study Area

Wudu District is located in southeastern Gansu Province, China and is situated along the middle reaches of the Bailongjiang River, a tributary of the Jialing River in the Yangtze River Basin. Its geographical coordinates are 32°47′~33°42′ N and 104°34′~105°38′ E (Figure 2). Wudu District is a debris flow-stricken area. According to incomplete statistics, the district has been affected by debris flow nearly 240 times [58], causing significant casualties and property losses. Field investigations [56] have indicated 172 debris flow gullies in the Zhouqu and Wudu districts along the middle reaches of the Bailongjiang River to form DFWSL areas of various sizes. Eighty-six of these are scattered from the

Shawan section to the Gushuizi section, with a total acreage of 2550.91 hm². The number of DFWSL areas on the left bank is 52, with an area of 1619 hm²; the other 34 have formed on the right bank, covering a total area of 931.91 hm². The largest debris flow accumulation fan in this section is the Gouba fan, with a maximum width of 2.3 km and a maximum length of 2.4 km, to cover a total area of 3.01 km². Fifty-nine DFWSLs exist from the Gushuizi section to the Bikou section and cover a total area of 155.99 hm². Of these, 33 DFWSL areas are found on the left bank and 26 on the right back, covering 75.64 hm² and 8.35 hm², respectively. The single largest DFWSL area in this section, the Jugan DFWSL, is about 410 m wide and 650 m long, with an area of about 15.89 hm². The smallest, the Yangsiba debris flow land, is 100 m long, 80 m wide, and 0.32 hm² in area. According to preliminary data, exploited areas of the Bailongjiang River basin total 2808.2 hm², accounting for 93% of the flat area.

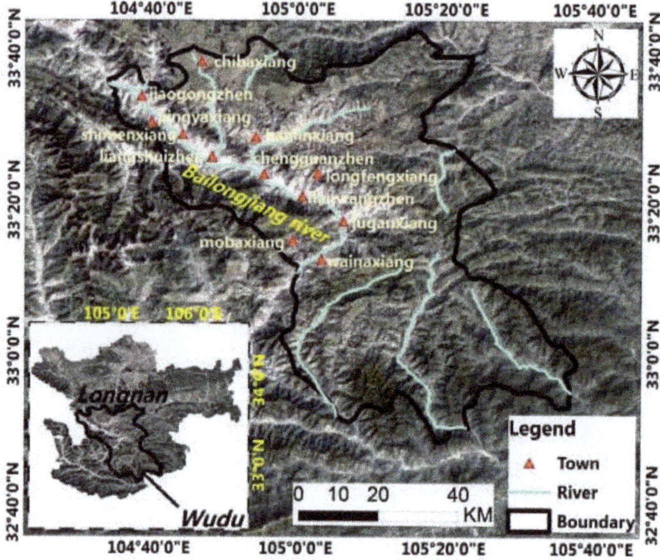

Figure 2. Location of Wudu District.

Cinnamon soil, entisol, loessial soil, and rocky soil (soil classification is based on the World Reference Base) are the four main soil types in Wudu District. Cinnamon soil and entisol are mainly distributed in valley terraces with low altitudes. The two soils are high yield soil types because of moderate texture and high potential fertility. Grain crops include corn, wheat, potato and rapeseed, while economic crops include vegetables, watermelons, olives, Chinese prickly ash, and grapes. The local economy of Wudu District mainly thrives on the sale of crops such as olives, peppers, grapes, potatoes, and canola. According to a year-long field investigation, the olive yield is about 2250 kg per ha and the purchase price is about 10 CNY/kg. Thus, income is about CNY 22,500/ha. The Chinese prickly ash yield is about 750 kg per ha with a purchase price of about 100 yuan/kg, providing an income about CNY 75,000 per ha. The rapeseed yield is about 2250~3000 kg per ha and its purchase price is about CNY 4 /kg, yielding an income per ha of about CNY 9000–12,000. The potato yield is about 600 kg per ha, its purchase price is about CNY 2/kg, and the income per ha is about CNY 1200.

At present, apart from farmland and forests, the DFWSL areas in Wudu District have also been transformed into construction land and township development sites where site conditions are appropriate (Figure 3).

Figure 3. Different types of debris flow waste-shoal lands use in Wudu District: (**a**) Primitive debris flow waste-shoal land; (**b**) overgrown with weeds; (**c,d**) developed into cultivated land; (**e,f**) villages or towns; and, (**g,h**) traffic network construction.

2.2. Data Collection and Methodology

Satellite remote sensing has been proven to be a suitable approach for detecting and monitoring land use transformations [59]. This study used land use raster data with a spatial resolution of 1 km to study land use changes in the Wudu District in 2005, 2010, and 2015 (Tables 1 and 2, Figure 3). A 2.15 m (no deviation) resolution image of Chenjiaba village in 2014, and again 2017, was obtained from Google Maps. The images were digitized using ArcGIS 9.3. The features in the area were divided into four categories by visual interpretation: Cultivated land, woodland, water, and construction land (Figure 4). Furthermore, the land area was constructed as a land use transfer matrix (Tables 3–5) using the method of Braimoh [60]. Additionally, the current situation and changes of various land use types were quantitatively analyzed. To quantitatively characterize the rate of land use change in the study area within a given time interval, the dynamic degree of land use was used [21]. It is described as follows:

$$K = \frac{U_2 - U_1}{U_1} \times \frac{1}{T} \times 100\%$$

(1)

where K is the dynamic degree of land use; U_1 and U_2 are the area of a land type at the beginning and the end of a period, respectively; and T is time interval (years). This equation can be used to analyze and compare the rates of change among different land use types in the study area.

3. Results

3.1. Temporal and Spatial Patterns of Land Use Transformation in DFWSL

3.1.1. Overall Conditions of Land Use Transformation in Wudu District

The distribution of land use types in Wudu District, according to the classified remote sensing images is shown in Figure 4; the ratios are shown in Table 1. Wudu District has developed along the Bailongjiang River and belongs to the mountain-valley land type. The land use pattern in mountainous areas is limited by site conditions and the climate. The middle and high mountain areas are dominated by grasslands and woodlands, which accounted for about 52% and 22% of the total area, respectively. The lowland flat areas, mainly constituted cultivated land and accounted for about 24% of the total area (Table 1). The transformation in land use types from 2005 to 2015 (Table 2) suggests that the area of cultivated land has decreased (a decrease of 36.5395 km^2 from 2005 to 2015) and transformation to other land types has largely shown an increasing trend.

Table 1. Land use in Wudu District in 2005, 2010, and 2015.

Type	2005		2010		2015	
	Acreage (km^2)	Ratio (%)	Acreage (km^2)	Ratio (%)	Acreage (km^2)	Ratio (%)
Grassland	2388.8834	0.5218	2395.0128	0.5232	2406.3738	0.5257
Cultivated land	1106.5583	0.2417	1098.9520	0.2400	1070.0188	0.2337
Woodland	1018.3564	0.2223	1019.4410	0.2225	1020.5569	0.2228
Water	15.4206	0.0034	15.7698	0.0034	16.4690	0.0036
Construction land	40.1770	0.0088	40.2202	0.0088	54.0303	0.0118
Unused land	8.4034	0.0018	8.4034	0.0018	10.3504	0.0023

Table 2. Area statistics on land use changes in Wudu District (km^2).

Type	2005	2010	2015	2005–2010	2010–2015	2005–2015
Grassland	2388.8834	2395.0128	2406.3738	6.1294	11.361	17.4904
Cultivated land	1106.5583	1098.952	1070.0188	−7.6063	−28.9332	−36.5395
Woodland	1018.3564	1019.441	1020.5569	1.0846	1.1159	2.2005
Water	15.4206	15.7698	16.469	0.3492	0.6992	1.0484
Construction land	40.177	40.2202	54.0303	0.0432	13.8101	13.8533
Unused land	8.4034	8.4034	10.3504	0	1.947	1.947

In particular, after 2010, the growth rate of construction land was at its highest, showing an increase of 13.8533 km^2 from 2005 to 2015 to account for 25.64% of the total construction land. The growth rate of construction land from 2010 to 2015 was 6.87%, while the growth rate from 2005 to 2010 was only 0.11%. This rate brought about an increase in the area of construction land, which can also be verified from the number of patches indicated by red circles in Figure 4.

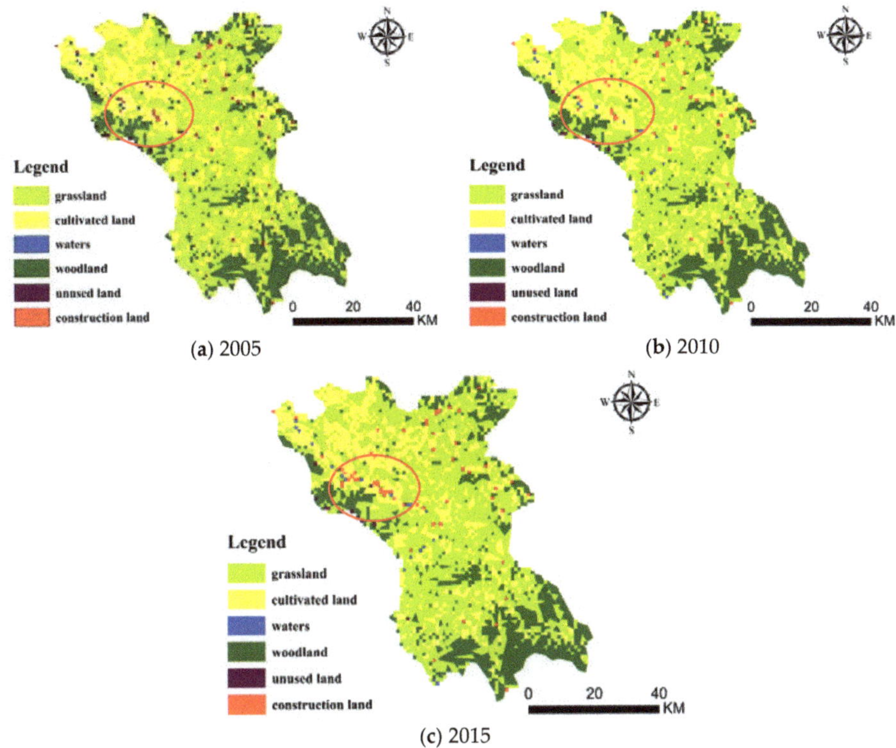

Figure 4. Distribution of land use types in Wudu District in (a) 2005, (b) 2010, and (c) 2015.

To illustrate the mutual transformation of land types on DFWSL and clear land transformation trends, and analyze the spatial-temporal change patterns of DFWSL [61], ArcGIS was used to calculate matrices of land use transformation from 2005 to 2010 and from 2010 to 2015 (Tables 3 and 4).

As indicated in Table 3, the overall change from 2005 to 2010 was slight and represented by transfer-ins and transfer-outs of cultivated land, with transfer-outs outnumbering transfer-ins. Nearly 10.4233 km^2 of the cultivated land was converted into grassland, 1.2984 km^2 into woodland, and a very small part (0.3791 km^2) was converted into water area, totaling 12.1001 km^2. The transfer-ins were mainly from grasslands, which accounted for 4.2508 km^2. The remaining areas were converted from woodland and water, which accounted for 0.2137 km^2 and 0.0299 km^2, respectively. As a result of the transformation progress, the cultivated land outflowed to other land types with a net outflow of 7.6063 km^2.

Table 3. Land uses transformation matrix for Wudu District from 2005 to 2010 (km^2).

Type	Grassland	Cultivated Land	Woodland	Water	Construction Land	Unused Land	2005
Grassland	2384.4447	4.2508	0	0	0.188	0	2388.8834
Cultivated land	10.4233	1094.4575	1.2984	0.3791	0	0	1106.5583
Woodland	0	0.2137	1018.1427	0	0	0	1018.3564
Water	0	0.0299	0	15.3907	0	0	15.4206
Construction land	0.1449	0	0	0	40.0322	0	40.177
Unused land	0	0	0	0	0	8.4034	8.4034
2010	2395.0128	1098.9520	1019.4410	15.7698	40.2202	8.4034	4577.7992

A comparison of results from 2010 to 2015 (Table 4) with those from 2005 to 2010 indicates that although the decrease of arable land was still the prevailing trend, an increase in either the total quantity or transformation rate was more prominent. Although the outflow of arable land was still dominated by grassland (15.4292 km^2), the outflow to woodland, water, and unused land increased exponentially from 2005 to 2010. Additionally, the area of arable land transforming to construction land significantly increased. It surged from near zero in 2010 to 13.998 km^2 in 2015 and the increased acreage accounted for 34.80% of the acreage of construction land in 2010.

Table 4. Land use transformation matrix for Wudu District from 2010 to 2015 (km^2).

Type	Grassland	Cultivated Land	Woodland	Water	Construction Land	Unused Land	2010
Grassland	2389.1242	3.8906	0.8162	0	0.2071	0.9747	2395.0128
Cultivated land	15.4292	1065.2159	2.3967	0.9400	13.9980	0.9722	1098.9520
Woodland	1.8204	0.2467	1017.3440	0	0.0299	0	1019.4410
Water	0	0.1809	0	15.5290	0.0598	0	15.7698
Construction land	0	0.4847	0	0	39.7355	0	40.2202
Unused land	0	0	0	0	0	8.4034	8.4034
2015	2406.3738	1070.0188	1020.5569	16.4690	54.0303	10.3503	4577.7992

3.1.2. Land Use and Transformation in a Typical DFWSL

To illustrate the land use transformation of DFWSL, this study narrowed its study scope by selecting Chenjiaba village as a research site. Chenjiaba village is completely constructed on converted DFWSL (Figure 5). Land use maps for 2014 and 2017 were selected, together with Google images (Figure 6); an analysis was conducted in terms of land use and transformation. As shown in Figure 5, in 2014 the DFWSL in the village was mainly developed into cultivated land and a small area of construction land. However, in 2017, most of the cultivated land had outflowed and transformed into construction land, which significantly increased the acreage of construction land. Figure 5 shows detailed land use and transformation trends in 2014 (a) and 2017 (b).

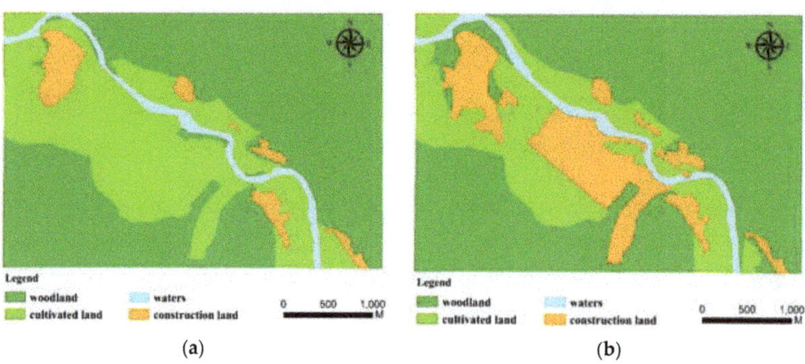

Figure 5. Land use and transformation maps of Chenjiaba village: (**a**) Status of land use type in 2014; (**b**) status of land use type in 2017.

Figure 6. Google images of land use transformation of DFWSL in Chenjiaba village: (**a**) Status of land use in 2014; (**b**) status of land use in 2017. Red circles represent arable land in 2014, most of which became construction land (residential area) in 2017; yellow circles show woodland (olive-planting sites) in 2014, which was converted into construction land in 2017.

The DFWSL developed along the riverside showed features of flat terrain, abundant water and heat, and easy access to traffic. Therefore, the use of this land is goal-oriented and well targeted. Woodlands were transformed into artificial forests for growing economic crops, such as olives; and some were also converted into arable land. Before 2014, woodland and arable land accounted for 92.06% of the land in the village. In 2017, the woodland remained largely unchanged. However, almost a third of the arable land resources outflowed and transformed into construction land, which tripled the percentage of construction land from 4.6% to 14.59% (Table 5). As a representative of the DFWSL use and transformation model of the Bailongjiang River Basin, the pattern of land transformation in Chenjiaba village reflects the trend of conversion of land use type. Most of the arable land in Wudu District was transformed from DFWSL along the river. Therefore, the increased construction land was transformed from DFWSL in the low-relief terrain.

Table 5. Land use and transformation matrix of DFWSL in Chenjiaba village, Wudu District from 2014 to 2017 (km^2).

Type	Woodland	Cultivated Land	Waters	Construction Land	2014
Woodland	6.8687	0.04947	0.03942	0.06428	7.0219
Cultivated land	0.2928	2.3796	0	1.0981	3.7704
Waters	0	0.0066	0.3776	0	0.3842
Construction land	0	0	0	0.5440	0.5440
2017	7.1615	2.4356	0.4170	1.7063	11.7205

3.2. Driving Forces of Land Use and Land Changes of DFWSL

Traditional geography focuses on morphology and patterns and pays little attention to underlying processes, especially the influences of patterns on the process. Therefore, this study revealed the underlying factors driving DFWSL use and transformation through change patterns.

3.2.1. Natural Driving Forces

DFWSL areas are located in the active debris flow area. They are extremely sensitive to vulnerable habitats and easily prone to natural disasters. Therefore, land use and changes of DFWSL in Wudu District resulted from debris flows and earthquakes. The annual precipitation in Wudu District is relatively large and the temporal and spatial distributions are uneven. Precipitation is largely in the form of heavy and torrential rains that often occur from May to September. The rainfall during this period accounts for 79.2% of the whole year. Since 1951, heavy rains with daily precipitation of over 25 mm has occurred on average 2.7 times per year; daily precipitation of over 50 mm has occurred once per year. In the past 20 years, the number of heavy and torrential rain events has increased significantly; for example, extremely heavy rainstorm events occur on average 4.3 times a year. Each event exceeds the criticality for debris flows. According to data from the meteorological department, in the event of torrential rains, intensity exceeds 8 mm within 10 min.

According to historical records, in Houba village, Wudu District, large-scale debris flows have occurred five times, in 1956, 1967, 1984, 1998 [62], and 2005. Moreover, on 21 June 2006, the village was struck by a 5.0 magnitude earthquake with an epicenter at 105° E, and 33.1° N. This earthquake adversely impacted the Niwangou gully and banks. The village was also affected by the Wenchuan earthquake which occurred in Sichuan on 12 May, 2008. Parts of Niwangou gully collapsed and mudslides blocked the gully while cracks were widened. The most serious blockages happened in the middle and lower reaches of the Niwangou gully, where slumps from both sides of the gully bank silted up the trench. The volume of debris flow from solid loose material was about 20×104 m^3.

Houba village is built on the DFWSL and contains 520 households with a total of 2600 residents. Five large-scale mudslides have occurred (i.e., those in 1956, 1967, 1984, 1998, and 2005); these destroyed more than 50 private residences and over 150 acres of farmland, burying 15 livestock, damaging a 200 m streth of long state highway, and resulting in direct economic losses of more than CNY 4.20 million. The damaged houses could not be reconstructed on the original sites and the buried farmlands were unsuitable for cultivation over a short period of time. For these reasons, these lands were developed into woodland, which increased water and soil conservation. In summary, frequent natural disasters have caused uncertainties in the use of DFWSL.

The frequent occurrence of debris flow hazards in the Wudu District is closely related to the condition of soil and vegetation in the area. The middle-high mountains mainly consist of loessial and rocky soils. They belong to entisol, with a large quantity of gravel, and their structure is loose. There is almost no bedrock distribution, suggesting that the soil is unstable and highly erodible. In addition, the vegetation in the area is unevenly distributed, with forests as the main type. The vegetation coverage at an altitude of 1800 m is less than 50%. The vegetation at 1400 m altitude comprises various grasses with a coverage rate of less than 20%. The vegetation coverage in the valleys on both sides of the mainstream of the Bailongjiang River (debris flow beach concentrated area) is low (NDVI < 0.3). Within this range, the capacity of the vegetation to retain water and soil is poor; therefore, it is prone to mudslides. About 65.53% of the total area is a high-risk mudslide-intensive area [63]. In summary, the harsh environment of the underlying surface cannot withstand natural disasters; this results in frequent changes in land use in the debris flow area.

Land use in the debris flow area, especially the development and use of DFWSL, is initially cultivated land, woodlands, or construction land and is largely dependent on the presence of frequent and catastrophic mudslides. If land is relatively stable with good site conditions, it will be developed into construction or arable land; otherwise, it will be converted into woodland in order to preserve water and soil. This is very different from the choice of land use types in ecologically non-vulnerable areas. In general, the development of ecologically non-vulnerable areas assumes that the impacts of natural disasters are non-existent or negligible because the priority is to balance economic benefits and ecological conservation. Therefore, in the debris flow active area, land use and transformation are strongly affected by natural factors, which significantly impact land planning, resource management, and risk control.

3.2.2. Characteristics of Land Transformation and Demographic Migration

According to the Regional Statistical Yearbook (Wudu district), the population was 530,000 in 2005, 544,480 in 2010, and 592,000 in 2015 indicating a growing trend, especially from 2010 to 2015. Along with population growth, land use and transformation also changed across the same period [64]. From 2010 to 2015, woodland and construction land noticeably increased while cultivated land decreased. This change was brought about by transformation of land use of the DFWSL along the river. Only a small region in the middle-high mountainous area was affected by mudslides; because the underlying surface in this area is seldom affected, there remain largely unchanged land types. Moreover, owing to different topographies, construction land and cultivated land underwent few changes.

To understand the correlation between population growth and the use and transformation of DFWSL, the population of Chenjiaba village (a village completely built on DFWSL in Wudu District) from 2005 to 2015 was assessed. The population was close to 300 in 2005, and had increased to 560 in 2010, and 1156 in 2015. Land use and transformation in Chenjiaba village were characterized by the outflow of cultivated land to construction land (Figures 5 and 6). The relationship between population and land use and transformation is as follows: A negative correlation between population growth and the reduction of arable land was observed while the correlation between population growth and an increase in construction land was positive. The growth of the population in Chenjiaba village was due to inward migration of people from Pingya village, which is located in the middle-high mountains. Therefore, the land originally used for residential areas was converted to forests, grasslands or cultivated land in Pingya village. Consequently, land transformation conditions in Pingya village were the opposite to those in Chenjiaba village.

Therefore, the relationship between land transformation and population change results from the impact of population distribution on land use. Population distribution and land use type changes are also affected by scale and range. At large scales, they may not be related while at small and medium scales, local population changes will lead to distinctive changes in land use.

3.2.3. Changes in Land Use Policies and Strategies

Anthropogenic activity and regional development profoundly influence ecosystems and land use patterns [65]. According to the Regional Land Resources Bulletin (Wudu district), since the implementation of the policy of returning farmland to forests in 1999, China has attached great importance to the recovery of forests and, in particular, the recovery of ecologically fragile areas. The forest areas in Wudu District increased from 2005 to 2015. Additionally, China implemented the policy of "direct subsidies to grain producers" in 2004, canceled agricultural taxes in 2005, issued farmland protection funds in 2009, and has gradually provided skills training to rural laborers. The implementation of these policies has promoted the development and use of DFWSL to turn unavailable land into useful land resources. At present, in response to the initiative of rural revitalization [66] and the country's poverty alleviation policies, the development and use of DFWSL is approaching the goal of building a nation that boasts "productive and compact producing spaces, comfortable and cozy living spaces, and picturesque ecological scenes" [67].

The period from 2016 to 2020 is a decisive stage for China with regard to building a moderately prosperous society. Poverty alleviation has been the focus of rural construction work in recent years. Although the amount of traditional arable land continues to decrease, recent years have witnessed the effective adjustment of agricultural structure and the industrial layout structure has begun to take shape. In particular, the adoption of intensive and efficient land use patterns for the cultivation of pepper, walnuts, olives, vegetables, and Chinese herbal medicines have been explored and modified through greenhouse agriculture and multi-dimensional eco-agriculture to significantly increase the income of farmers. In 2010, the household operating income of farmers in Wudu District reached CNY 1250 and reflected an increase of CNY 502 over that in 2005 (CNY 748). This is an average annual increase of 8.03% and accounts for 56.43% of total net income. Moreover, urbanization also provides a broader perspective for people to develop secondary and tertiary industries within their households, especially in the catering, transportation, traffic, and construction industries, all of which have not only provided employment for farmers but also increased farmer incomes. For example, the Lanyu railway and the Lanhai expressway have facilitated traffic in Wudu District. Moreover, the Hanwang 330 KV transformer substation has been completed and put into production, stimulating investment in exponential growth and subsequent industry development. At the same time, this substation also created employment and provided opportunities for increasing farmer incomes. The development of catering and real estate industries and increased traffic require more land which is often categorized as land for construction. Therefore, the acreage of construction land increased dramatically after 2010, and arable land decreased rapidly, especially in development-oriented areas (for example, the land near Chenjiaba, as shown in Figures 5 and 6). The increased land for construction also includes space for various types of infrastructure such as drainage systems and blockage measures.

In summary, in the area where mudslides occurred, the country's policies or strategies in different periods influenced the diversity of certain types of land use; however, the complete transformation of one land type to another could not be determined. For any country or region, land use should comply with the overall standards of being "safe, harmonious, open, coordinated, competitive, sustainable, and beautiful". This highlights the difference between the influence of policy or strategy and natural factors on land use and transformation. In general, natural driving forces can promote complete transformation of land use types, while policy or strategy does not play a decisive role in the transformation.

4. Discussion

4.1. A Comparison between Our Study and Other Similar Area

The Pokhara valley is an intermountain fluvial basin occupying the midsection of the Seti River in the Lesser Himalayas of Nepal. It holds a large volume of layered clastic deposits of gravel, silt and clay of Quaternary age [68,69]. This area, like Wudu, is located in an ecologically fragile area and is subject to frequent landslides and debris flow. Both these areas are sensitive to natural factors and human influence. The land cover changes are characterized by increses in construction land. In addition, land use structure is becoming increasingly diverse, expanding to include not only construction and forest land, but also water areas and grassland. At present, much of the city's rapid growth in population has been accommodated in temporary settlements that are at high risk of environmental impairments and hazards in Pokhara [70]. However, Wudu district has adopted a more rational settlement for population growth for example, moving people away from danger zones to a safe area (Pingya village to Chenjiaba village, as shown in our study). Furthermore, in Wudu district, in addition to land use and land use changes, emphasis is also placed on the measures to conserve soil and water and ecological protection (Figure 7), such as cultivating vegetation for slope protection (Figure 7a,b), and geotechnical engineering (Figure 7c,d).

In recent years, marginal lands have attracted widespread concern due to their potential for exploitation [71,72]. Caserta [73] highlighted that this is likely because many abandoned regions have latent resources of water, energy, and vegetation (e.g., commercial agriculture) which, if properly exploited, could lead to a flourishing modern rural system. It is worth mentioning that the Wudu district pays more attention to maximizing the efficiency of land resources use. Debris flow waste-shoal lands are developed into available land without disrupting the natural environment. While the objectives and ultimate aim of reclamation may differ in the face of individual needs and natural conditions, the essential ideas remain the same: Take ecological restoration and construction as the main lines, pursue the maximization of economic and social benefits from land and achieve sustainable development. Hence, although Wudu is in the debris flow area and experiences fast land use transformation, the risks from natural disasters are also accounted for in the form of corresponding prevention and control measures.

Figure 7. Various measures for ecological preservation: (**a,b**) Slope ecological protection measures; (**c,d**) drainage groove with check dam; (**e,f**) ecological engineering protection in gully.

4.2. Development Mode Based on Local Conditions

Moving out of environmentally fragile areas where the environment is harsh and disasters occur frequently is a common sentiment. China is a mountainous country with a large population. The mountainous area accounts for two-thirds of the country's total land and population in these areas exceeds 1.3 billion. Since abandoning these places is unlikely, it is beneficial to take advantage of unique mountainous resources. Although debris flows cause significant damage and losses to the natural environment and humans alike, fans shaped by debris flows are advantageous for human development. At present, there are two types of development models worth recommending: Greenhouse agriculture and multi-dimensional eco-agriculture.

Considering the climate of Wudu District and its site conditions, efficient multi-dimensional eco-agriculture has resulted in intercropping of olives, grapes, walnuts, and pasture to utilize different levels of photothermal resources. It has also equipped the region with supporting facilities such as reservoirs, irrigation systems, pig houses, chicken houses, and biogas digesters. Forage is used as feed for chickens, pigs, and fish; the manure of livestock, branches and leaves of walnuts and grapes, and branches of crops can be added to biogas digesters for fermentation; the biogas produced can be used as energy and the remaining waste can be used as fertilizer; the deposits of fish ponds can be used as fertilizer for plant and crops; and the flowers of walnut and grape crops can be used for honey-making during the flowering period (Figure 8). This model, through the recycling of materials and energy flows, forms a number of intertwined industrial chains and organically links the biological environment within the multi-dimensional agricultural system. It is also likely to promote agricultural tourism such as flower viewing in spring and fruit gathering in autumn. Certain locations may also have the potential to develop rural tourism and operate farm stays. Booming rural tourism can increase consumption of green foods and encourage more farmers to embrace three-dimensional agriculture, and finally, build a desirable development situation. Primary industry boosts tertiary industry, which in return can promote the development of agriculture.

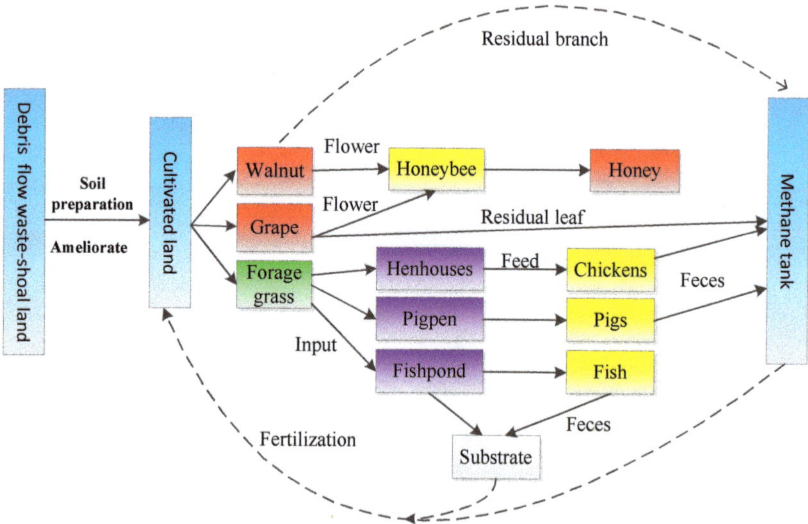

Figure 8. Operational route of multi-dimensional agricultural mode of DFWSL: Recycling of materials and energy flows.

Greenhouse agriculture adopts precise operating and large-scale planting methods. It helps utilize the flat terrain and sufficient water sources to plant non-seasonal fruits and vegetables (Figure 9). Wudu's greenhouse vegetables adjust to market demand and farmers mainly plant white leaf mustard, lettuce, and strawberries. A 2 hm^2 area of land in the Liangshui town of Wudu District has an annual production value of CNY 450,000 per ha. It not only provides employment for local residents, but also increases their income by CNY 20,000–30,000 per year. At the same time, greenhouse agriculture is pollution-free (as it replaces chemical fertilizers with soybeans, which are highly nutritious), and high-tech; for example, using plant sterilization technology such as the garlic bactericidal method and the use of biological insecticides (introducing *Neoseiulus californicus* (Acari: Phytoseiidae) to get rid of red spiders in fruit sheds). Such measures have greatly reduced environmental pollution and improved the taste and safety of vegetables and fruits. Therefore, the two types of development models are worth promoting and developing.

Figure 9. Greenhouse agriculture on DFWSL: (**a,b**) exteriors of greenhouse on DFWSL; (**c,d**) land consolidation and fertilization; (**e,f**) vegetable (lettuce) and fruit (strawberry).

4.3. Relationship between Population Change and Land Use and Transformation

Different schools of thought exist regarding the relationship between population growth and land use type transformation. One opinion is that population increase is directly related to the change in the quantity of woodland and cultivated land. For example, population growth leads to a reduction in tropical forests [74,75] and an increase in the demand for arable land, thus causing excessive reclamation of agricultural land, serious land degradation, and related environmental problems [76]. As urbanization proceeds, population growth can cause a sharp decline in the quantity of arable land and an increase in the quantity of construction land [77]. Another opinion is that the increase in population is not related to land use or transformation; in particular, there is no obvious correlation between population growth and the increase or decrease in woodland and cultivated land [78]. Zhu [79] studied land use changes in areas surrounding the Bohai region of China and found that population growth did not lead to any increase in cultivated land from 1985 to 1995.

The development and use of DFWSL, especially for newly formed lands, are subject to fluctuations due to their instability and vulnerability. Land use type changes are greatly affected by natural disasters. For instance, a flat DFWSL might be used as arable land or construction land; however, owing to certain risks, it must instead be used as water and soil conservation forest. Areas seldom affected by natural conditions are often subject to greater volatility due to strategic policies [70,80]. Under these conditions, woodland is transformed into cultivated land, which is in turn transformed into construction land and may eventually become woodland or cultivated land. In mountainous areas, the growth and migration of populations are oriented to suitable and safe environments. Therefore, there is no direct and obvious correlation between the change of land-use type and population change in DFWSL. The complex relationship among them is co-determined by land use scales, policy shifts, and human decision-making. Based on our findings from the field investigation in the Wudu District, population growth leads to different population distributions that affect further development of land as cultivated land, woodland or construction land. For example, if the population in an area is concentrated and the land is resilient, the area may develop into residential areas and towns, while ecologically sensitive areas would be transformed into cultivated land or woodland. Therefore, changes in land use, especially changes in the quantity of arable land, may result in population increase, population decrease, or unchanged population, indicating no direct or universal correlation.

4.4. Limitations and Recommendations

Although the development and use of DFWSL areas have brought substantial ecological, economic, and social benefits to local farmers, certain limitations still exist. The location of the Wudu District in the gully region of the Loess Plateau, with the presence of mountains, rivers and ravines, frequent natural disasters, unfavorable temporal and spatial distributions of rainfall, and poor irrigation, have limited development and sustainability of agriculture, resulting in high cost and low income from agricultural production. Additionally, financial assistance is insufficient in Wudu District. Lack of investment and agricultural infrastructure, coupled with lack of knowledge and awareness of market competition among farmers, and the lack of new technology and systematic development, makes it difficult to popularize new crops. Although efficient multi-dimensional eco-agriculture and greenhouse technology have been promoted in some regions, current agricultural industrialization continues to lag. Channels for farmers to increase their incomes are not diversified; agricultural products remain at the roughly processed stage without packaging, thus making it challenging for farmers to increase their incomes.

To cope with these limitations, it is recommended to strengthen the construction of industrial bases, expand the scale of production, develop trade and export-oriented agriculture, and upgrade industrial production. It is also suggested that local governments continue the construction of leading industrial bases for fruits, vegetables and livestock, formulate construction standards for industrial bases, and focus on concentrated development. Furthermore, the government should provide support for agricultural cooperatives in terms of technical help, policy, and financial aid. For example, the government should help train various types of agricultural brokers who can give training and guidance to farmers and expand the scope of agriculture development. Additionally, the administration should actively explore new approaches of land circulation in accordance with the principle of "lawful, voluntary and paid" and modify policies to upgrade land circulation.

Local governments should also pursue brand strategies and promote unique local agriculture. The specialties of Wudu District, such as peppers and olives, need resource integration to dominate the market. Efforts should be made to accelerate the establishment and improvement of a standardization system for pollution-free agricultural products and organic foods. Additionally, the formulation and implementation of standards should be strengthened, while continuing to introduce new crop varieties and technologies, and to update knowledge. To help leading enterprises building brand awareness, an increase in investments and active participation in various agricultural product competitions must be encouraged. Winning awards and appraisal can improve the popularity of products. The government should also organize and coordinate the development of local agriculture. For example, the government should classify local specialties such as peppercorns, walnuts, and olives, integrate brands, increase production processing, adopt more technologies, and strengthen resource management.

The exploitation and protection of DFWSL are contradictory. In debris flow active areas, it is imperative to increase awareness and strengthen the ability for disaster prevention and relief. At the same time, it is significant to enhance cultural literacy, strictly observe conservation, and ensure rapid and sustainable socioeconomic development. With these two measures, an overall improvement in the economy, life quality, and environment can be achieved, and an ecological-economic comprehensive development zone can be built.

5. Conclusions

The development and use of DFWSL areas provide potential for relieving conflicts between humans and land in environmentally fragile mountainous areas and address the dilemma of land protection and sustainable development. This study was conducted in Wudu District, which is a typical model for the use of DFWSL. The trend of land use and transformation from 2005 to 2015 was analyzed and the results indicated that the land use type changes from 2005 to 2010 and from 2010 to 2015 was from cultivated land to grassland and construction land. However, the rate and scale of change between 2005 and 2010 were far lower than those during 2010–2015. Additionally, the conversion rate of cultivated land to construction land during 2010–2015 (6.87%) was more than 60 times that of 2005–2010 (0.11%). Finally, the area of construction land increased by 13.38533 km^2 from 2005 to 2015, accounting for 25.64% of the entire construction land area.

Natural driving forces behind changes in land use type are natural disasters and include mudslides, landslides, and earthquakes, while anthropogenic factors are mainly the transformation of national policies and strategies. Moreover, in terms of land-use change in the debris flow area, the natural driving force has a decisive effect on the type of the initial development of DFWSL areas, while national policies or strategies in different periods only affect the direction of transformation in that specific period.

Author Contributions: S.H. and D.W. conceived and designed the framework of this article; S.H. and P.Z. collected and analyzed the data; All the authors contributed to the writing of this paper.

Funding: This work was supported by the National Natural Science Foundation of China (Grant No. 41790434) and the National Sci-tech Support Plan of China (2014BAL05B01).

Acknowledgments: We are grateful for the comments and criticisms of an early version of this manuscript by our colleagues and the journal's reviewers.

Conflicts of Interest: The authors declare no conflict of interest.

References

1. Fu, B. *Ecosystem Services and Ecological Security*; Higher Education Press: Beijing, China, 2013.
2. Wang, X.; Bao, Y. Study on the methods of dynamic change of Land use. *Adv. Geogr.* **1999**, *181*, 81–87.
3. Lesschen, J.P.; Verburg, P.H.; Staal, S.J. *Statistical Methods for Analysing the Spatial Dimension of Changes in Land Use and Farming Systems*; Wageningen University: Wageningen, The Netherlands, 2005.
4. Chen, Y.; Yang, P. New progress in the study of land use/land cover change in the world. *Econ. Geogr.* **2001**, *21*, 95–100.
5. Stern, P.C.; Young, O.R.; Druckman, D. *Global Environmental Change: Understanding the Human Dimension*; National Research Council: Washington, DC, USA, 1992.
6. Tumer, B.L.; Meyer, W.B.; Skole, D.L. Global land use/land cover change: Toward an integrated program of study. *Ambio* **1994**, *23*, 91–95.
7. Lawler, J.J.; Lewis, D.J.; Nelson, E.; Andrew, J.P.; Stephen, P.; John, C.; Withey, D.P.; Helmers, S.M.; Derric, P.; Volker, C.R. Projected land-use change impacts on ecosystem services in the United States. *Proc. Natl. Acad. Sci. USA* **2014**, *111*, 7492–7497. [CrossRef] [PubMed]
8. Curebal, I.; Efe, R.; Soykan, A.; Sonmez, S. Impacts of anthropogenic factors on land degradation during the anthropocene in Turkey. *J. Environ. Biol.* **2015**, *36*, 51–58. [PubMed]
9. Pan, H.; Deal, B.; Chen, Y.; Geoffrey, H. A Reassessment of urban structure and land-use patterns: Distance to CBD or network-based? Evidence from Chicago. *Reg. Sci. Urban Econ.* **2018**, *70*, 215–228. [CrossRef]
10. Brovkin, V.; Boysen, L.; Arora, V.K.; Boisier, J.P.; Cadule, P.; Chini, L.; Claussen, M.; Friedlingstein, P.; Gayler, V.; van den Hurk, B.J.J.M.; et al. Effect of anthropogenic land-use and land-cover changes on climate and land carbon storage in CMIP5 projections for the twenty-first century. *J. Clim.* **2013**, *26*, 6859–6881. [CrossRef]
11. Wu, K.Y.; Ye, X.Y.; Qi, Z.F.; Zhang, H. Impacts of land use/land cover change and socioeconomic development on regional ecosystem services: The case of fast-growing Hangzhou metropolitan area, China. *Cities* **2013**, *31*, 276–284. [CrossRef]

12. Vanloocke, A. The Impact of Land-Use And Global Change on Water-Related Agro-Ecosystem Services in the Midwest US. Ph.D. Thesis, University of Illinois at Urbana–Champaign, Champaign, IL, USA, 2012.

13. Luyssaert, S.; Jammet, M.; Stoy, P.C.; Stephan, E.; Julia, P.; Eric, C.; Galina, C.; Axel, D.; KarlHeinz, E.; Morgan, F.; et al. Land management and land-cover change have impacts of similar magnitude on surface temperature. *Nat. Clim. Chang.* **2014**, *4*, 389–393. [CrossRef]

14. Turner, B.L.; Skole, D.; Sanderson, S.; Fischer, G.; Fresco, L.; Leemans, R. *Land-Use and Land-Cover Change: Science/Research Plan*; International Geosphere-Biosphere Programme: Stockholm, Sweden, 1995.

15. Alexandratos, N.; Bruinsma, J. *World Agriculture towards 2030/2050: The 2012 Revision*; ESA Working paper Rome; FAO: Rome, Italy, 2012.

16. Schmitz, C.; van Meijl, H.; Kyle, P.; Nelson, G.C.; Fujimori, S.; Gurgel, A.; Havlik, P.; Heyhoe, E.; d'Croz, D.M.; Popp, A.; et al. Land-use change trajectories up to 2050: Insights from a global agro-economic model comparison. *Agric. Econ.* **2014**, *45*, 69–84. [CrossRef]

17. Bajzelj, B.; Richards, K.S.; Allwood, J.M.; Smith, P.; Dennis, J.S.; Curmi, E.; Gilligan, C.A. Importance of food-demand management for climate mitigation. *Nat. Clim. Chang.* **2014**, *4*, 924–929. [CrossRef]

18. Benton, T.G.; Bailey, R.; Froggatt, A.; King, R.; Lee, B.; Wellesley, L. Designing sustainable landuse in a 1.5° world: The complexities of projecting multiple ecosystem services from land. *Curr. Opin. Environ. Sustain.* **2018**, *31*, 88–95. [CrossRef]

19. Jiang, Z. Strategic framework for sustainable use of land resources in China. *Rural Econ. Technol.* **2003**, 13–16, (In Chinese with English Abstract).

20. Wang, X.H.; Zheng, D.; Shen, Y.C. Land use change and its driving forces on the Tibetan Plateau during 1990–2000. *Catena* **2008**, *72*, 56–66. [CrossRef]

21. Li, X.Y.; Ma, Y.J.; Xu, H.Y.; Wang, J.H.; Zhang, D.S. Impact of landuse and land cover change on environmental degradation in Lake Qinghai watershed, northeast Qinghai-Tibet Plateau. *Land Degrad. Dev.* **2010**, *20*, 69–83. [CrossRef]

22. Glasby, G.P. Sustainable development: The need for a new paradigm. *Environ. Dev. Sustain.* **2002**, *4*, 333–345. [CrossRef]

23. Cao, S.; Liu, Y.; Yu, Z. China's successes at combating desertification provide roadmap for other nations. *Environ. Sci. Policy Sustain. Dev.* **2018**, *60*, 16–24. [CrossRef]

24. Liu, H.; Liu, Y.; Bi, R.; Xu, Y.; Wang, S. Reuse type judgment of mining wasteland based on land use competitiveness. *Trans. Chin. Soc. Agric. Eng.* **2016**, *32*, 258–266.

25. Chen, T.T.; Peng, L.; Liu, S.Q.; Wang, Q. Land cover change in different altitudes of Guizhou-Guangxi karst mountain area, China: Patterns and drivers. *J. Mt. Sci.* **2017**, *14*. [CrossRef]

26. Zhang, J.; He, C.; Chen, L.; Cao, X. Improving food security in China by taking advantage of marginal and degraded lands. *J. Clean Prod.* **2018**, *171*, 1020–1030. [CrossRef]

27. Cai, Y. It is possible to turn unused land into construction land. *Chin. Land* **2010**, *8*, 27–31.

28. Dorn, R.I. Desert rock coatings. In *Geomorphology of Desert Environments*; Springer: Dordrecht, The Netherlands, 2009; pp. 153–186.

29. Frankel, K.; Dolan, J. Characterizing arid region alluvial fan roughness with airborne laser swath mapping digital topographic data. *J. Geophys. Res.* **2007**, *112*. [CrossRef]

30. Blair, T.; McPherson, J. Processes and forms of alluvial fans. In *Geomorphology of Desert Environments*; Parsons, A., Abrahams, A., Eds.; Springer: New York, NY, USA, 2009; pp. 413–467.

31. Marcato, G.; Bossi, G.; Rivelli, F.; Borgatti, L. Debris flood hazard documentation and mitigation on the Tilcara alluvial fan (Quebrada de Humahuaca, Jujuy province, North-West Argentina). *Nat. Hazard Earth Syst. Sci.* **2012**, *12*, 1873–1882. [CrossRef]

32. Scheinert, C.; Wasklewicz, T.; Staley, D. Alluvial fan dynamics—Revisiting the field. *Geogr. Compass* **2012**, *6*, 752–775. [CrossRef]

33. Blackwelder, E. Desert plains. *J. Geol.* **1931**, *39*, 133–140. [CrossRef]

34. Bull, W.B. The alluvial fan environment. *Prog. Phys. Geogr.* **1977**, *1*, 222–270. [CrossRef]

35. Tsujimoto, Y.; Inusah, B.; Katsura, K.; Fuseini, A.; Dogbe, W.; Zakaria, A.I.; Yoichi, F.; Masato, O.; Jun-Ichi, S. The effect of sulfur fertilization on rice yields and nitrogen use efficiency in a floodplain ecosystem of northern ghana. *Field Crop Res.* **2017**, *211*, 155–164. [CrossRef]

36. Haas, T.D.; Densmore, A.L.; Stoffel, M.; Suwa, H.; Imaizumi, F.; Ballesteros-Cánovas, J.A.; Wasklewiczf, T. Avulsions and the spatio-temporal evolution of debris-flow fans. *Earth Sci. Rev.* **2018**, *177*, 53–57. [CrossRef]

37. Staley, D.M.; Wasklewicz, T.A.; Blaszczynsky, J.S. Surficial patterns of debris-flow deposition on alluvial fans in Death Valley, CA using airborne laser swath mapping data. *Geomorphology* **2006**, *74*, 152–163. [CrossRef]

38. Cavalli, M.; Marchi, L. Characterization of the surface morphology of an alpine alluvial fan using airborne LiDAR. *Nat. Hazard Earth Syst. Sci.* **2008**, *8*, 323–333. [CrossRef]

39. Wang, D.; Cui, P.; Zhu, B.; Wei, F. Characteristics of High Sand flow and improvement of debris flow waste-shoal land in Jiangjia gully, Yunnan, China. *Mt. Res.* **2003**, *6*, 745–751.

40. Bull, W.B. *Geomorphology of Segmented Alluvial Fans in Western Fresno County, California*; US Government Printing Office: Washington, DC, USA, 1964.

41. Cui, P.; Ge, Y.; Zhuang, J.; Wang, D. Soil evolution features of debris flow waste-shoal land. *J. Mt. Sci.* **2009**, *6*, 181–188. [CrossRef]

42. Sancho, C.; Pena, J.L.; Rivelli, F.; Rhodes, E.; Munoz, A. Geomorphological evolution of the Tilcara alluvial fan (Jujuy Province, NW Argentina): Tectonic implications and palaeoenvironmental considerations. *J. South Am. Earth Sci.* **2008**, *26*, 68–77. [CrossRef]

43. Okunishi, K.; Suwa, H. Assessment of debris-flow hazards of alluvial fans. *Nat. Hazards* **2001**, *23*, 259–269. [CrossRef]

44. Curry, G.N.; Koczberski, G.; Selwood, J. Cashing out, cashing in: Rural change on the south coast of Western Australia. *Aust. Geogr.* **2001**, *32*, 109–124. [CrossRef]

45. Petit, S. The dimensions of land use change in rural landscapes: Lessons learnt from the GB Countryside Surveys. *J. Environ. Manag.* **2009**, *90*, 2851–2856. [CrossRef] [PubMed]

46. Rudel, T.K. Tree farms: Driving forces and regional patterns in the global expansion of forest plantations. *Land Use Policy* **2009**, *26*, 545–550. [CrossRef]

47. Poyatos, R.; Latron, J.; Llorens, P. Land use and land cover change after agricultural abandonment: The case of a Mediterranean mountain area (Catalan Pre-Pyrenees). *Mt. Res. Dev.* **2003**, *23*, 362–368. [CrossRef]

48. Zhou, Z. Landscape changes in a rural area in China. *Landsc. Urban Plan.* **2000**, *47*, 33–38.

49. Burgi, M.; Russell, E.W.B. Integrative methods to study landscape changes. *Land Use Policy* **2001**, *18*, 9–16. [CrossRef]

50. Weng, Q.H. Land use change analysis in the Zhujiang Delta of China using satellite remote sensing, GIS and stochastic modeling. *J. Environ. Manag.* **2002**, *64*, 273–284. [CrossRef]

51. Chen, L.D.; Messing, I.; Zhang, S.R.; Fu, B.J.; Ledin, S. Land use evaluation and scenario analysis towards sustainable planning on the Loess Plateau in China-case study in a small catchment. *Catena* **2003**, *54*, 303–316. [CrossRef]

52. Lu, L.; Li, X.; Cheng, G.D. Landscape evolution in the middle Heihe River Basin of North-West China during the last decade. *J. Arid Environ.* **2003**, *53*, 395–408. [CrossRef]

53. Jia, B.Q.; Zhang, Z.Q.; Ci, L.J.; Ren, Y.P.; Pan, B.R.; Zhang, Z. Oasis land-use dynamics and its influence on the oasis environment in Xinjiang, China. *J. Arid Environ.* **2004**, *56*, 11–26. [CrossRef]

54. Zhao, H.L.; Zhao, X.Y.; Zhou, R.L.; Zhang, T.H.; Drake, S. Desertification processes due to heavy grazing in sandy rangeland, Inner Mongolia. *J. Arid Environ.* **2005**, *62*, 309–319. [CrossRef]

55. Liu, J.; Huang, J.; Ou, G.; Fan, J. The developing potentiality of slope farmland on the alluvial fans of debris flow in the midstream of Bailongjiang River. *Res. Soil Water Conserv.* **2011**, *18*, 92–96.

56. Future Earth. *Future Earth 2025 Vision*; Future Earth Interim Secretariat: Paris, France, 2014.

57. Liu, Y. Strategies to Guarantee Land Resources Safety in China. *Proc. Chin. Acad. Sci.* **2006**, *21*, 379–384.

58. Records of Wudu County, Gansu Province, China. In *Compilation Committee of Local Records of Wudu County*; Sanlian Bookstore: Beijing, China, 1998.

59. Longley, P.A. Geography: Will development in urban remote sensing and GIS lead to better urban geography? *Prog. Hum. Geogr.* **2002**, *26*, 231–239. [CrossRef]

60. Braimoh, A.K. Random and systematic land-cover transitions in northern Ghana. *Agric. Ecosyst. Environ.* **2006**, *113*, 254–263. [CrossRef]

61. Wang, J.; Fu, B. The impact of land use on spatial and temporal distribution of soil moisture on the Loess Plateau. *Acta Geogr. Sin.* **2000**, *55*, 84–91.

62. Huang, Z.; Yu, Y. Distribution and Developmental characteristics of debris flow in Longnan area of the Upper reaches in Yangtze River. *People Yangtze River* **1998**, *29*, 42–44.

63. Ning, N.; Tian, L.; Zhang, P.; Qi, S.; Ma, J. Risk Assessment of debris flow in Wudu area, South Gansu Province, China. *Mt. Res.* **2013**, *31*, 601–609.

64. Wang, D.; Meng, X.; Guo, P.; Guo, J.; Tan, L. Dynamic changes of land use before and after returning farmland to forest in Wudu, Gansu Province, China. *Guizhou Agric. Sci.* **2012**, *40*, 227–229.

65. Nagashima, K.; Sands, R.; Whyte, A.G.D.; Bilek, E.M.; Nobukazu, N. Regional landscape change as a consequence of plantation forestry expansion: An example in the Nelson rejoin, New Zealand. *For. Ecol. Manag.* **2002**, *163*, 245–261. [CrossRef]

66. Liu, Y.; Li, Y. Revitalize the world's countryside. *Nature* **2017**, *548*, 275–277. [CrossRef] [PubMed]

67. Shi, Z.; Deng, W.; Zhang, S. Spatio-temporal pattern changes of land space in hengduan mountains during 1990–2015. *J. Geogr. Sci.* **2018**, *28*, 529–542. [CrossRef]

68. Yamanaka, H.; Yoshida, M.; Arita, K. Terrace landform and Quaternary deposit around Pokhara Valley, central Nepal. *J. Nepal Geol. Soc.* **1982**, *2*, 95–112.

69. Gautam, P.; Panta, S.R.; Ando, H. Mapping of subsurface karst structure with gamma ray and electrical resistivity profiles: A case study from Pokhara valley, central Nepal. *J. Appl. Geophys.* **2000**, *45*, 97–110. [CrossRef]

70. Rimal, B.; Baral, H.; Stork, N.; Kiran, P.; Sushila, R. Growing City and Rapid Land Use Transition: Assessing Multiple Hazards and Risks in the Pokhara Valley, Nepal. *Land* **2015**, *4*, 957–978. [CrossRef]

71. Tilman, D.; Hill, J.; Lehman, C. Carbon-negative biofuels from low-input high-diversity grassland biomass. *Science* **2006**, *314*, 1598–1600. [CrossRef] [PubMed]

72. Robertson, G.P.; Dale, V.H.; Doering, O.C.; Hamburg, S.P.; Melillo, J.M.; Wander, M.M.; Parton, W.J.; Adler, P.A.; Barney, J.N.; Cruse, R.M.; et al. Sustainable biofuels redux. *Science* **2008**, *322*, 49–50. [CrossRef] [PubMed]

73. Caserta, G. Reclamation of abandoned lands in Mediterranean countries through the contribution of biomass and other renewable energy sources. A metodological proposal. In Proceedings of the Conference on Responsible Coastal Zone Management; 2000; pp. 275–280. Available online: https://www.researchgate.net/publication/295998952_Reclamation_of_abandoned_lands_in_Mediterranean_countries_through_the_contribution_of_biomass_and_other_renewable_energy_sources_A_metodological_proposal (accessed on 3 August 2018).

74. Kammerbauer, J.; Ardon, C. Land use dynamics and landscape change pattern in a typical watershed in the hillside region of central Honduras. *Agric. Ecosyst. Environ.* **1999**, *75*, 93–100. [CrossRef]

75. Verburg, P.H.; Veldkamp, T.A.; Bouma, J. Land use change under conditions of high population pressure: The case of Java. *Glob. Environ. Chang.* **1999**, *9*, 303–312. [CrossRef]

76. Nemes, I. Land reclamation works applied in the amelioration perimeter in territorial agrarian fond otelec-uivar, Timiş County, Romania. *Res. J. Agric. Sci.* **2009**, *41*, 472–476.

77. Xu, Y.; Tang, Q.; Fan, J.; Bennett, S.J.; Li, Y. Assessing construction land potential and its spatial pattern in china. *Landsc. Urban. Plan.* **2011**, *103*, 207–216. [CrossRef]

78. Lambin, E.F.; Turner, B.L.; Geist, H.J.; Samuel, B.A.; Arild, A.; John, W.B.; Oliver, T.C.; Rodolfo, D.; Günther, F.; Carl, F.; et al. The causes of land-use and land-cover change: Moving beyond the myths. *Glob. Environ. Chang.* **2001**, *11*, 261–269. [CrossRef]

79. Zhu, H.; He, S.; Zhang, M. Driving forces analysis of land use change in Bohai Rim. *Geogr. Res.* **2001**, *20*, 669–678.

80. Deal, B.; Pan, H. Discerning and Addressing Environmental Failures in Policy Scenarios Using Planning Support System (PSS) Technologies. *Sustainability* **2016**, *9*, 13. [CrossRef]

 *sustainability*

Article

Comprehensive Land Carrying Capacities of the Cities in the Shandong Peninsula Blue Economic Zone and their Spatio-Temporal Variations

Guangming Cui [1], Xuliang Zhang [2,*], Zhaohui Zhang [3], Yinghui Cao [2] and Xiujun Liu [2]

[1] School of Environmental Science and Engineering, Qingdao University, Qingdao 266071, Shandong, China; qducgm@163.com
[2] School of Tourism and Geography Science, Qingdao University, Qingdao 266071, Shandong, China; yinghui.cathy@gmail.com (Y.C.); kevinvernal@126.com (X.L.)
[3] The First Institute of Oceanography, Ministry of Natural Resources, Qingdao 266071, Shandong, China; zhang@fio.org.cn
* Correspondence: Geo_zhang@163.com

Received: 8 December 2018; Accepted: 9 January 2019; Published: 16 January 2019

Abstract: The comprehensive land carrying capacities of seven cities in the Shandong Peninsula Blue Economic Zone between 2007–2014 were assessed using a multi-criterion comprehensive evaluation approach and an index of 27 indicators, and cluster analysis was conducted to identify the spatial-temporal variations of the cities' comprehensive land carrying capacities. The results showed that the carrying capacity of the water and soil resources of the cities had declined except Dongying City; in contrast, the carrying capacities of the eco-environment, the social resources and the economy and technology of the seven cities had all arisen. The carrying capacities of social resources of Dongying and Weihai were markedly higher than the other five cities. The carrying capacities of economy and technology of Qingdao and Dongying were high, the capacities of Weihai and Yantai were moderate, and the capacities of Weifang, Rizhao, and Binzhou were low. In general, the comprehensive land carrying capacities of the eastern cities were higher than those of the western cities, which was similar to the spatial pattern of the economy development of those cities. In addition, positive correlations were identified between the comprehensive land carrying capacity and the per capita land for construction, areal proportion of wetland to total land, percentage of green space to build up area, per capita public green space, comprehensive utilization rate of industrial solid waste residues, urbanization rate, area of per capita urban road, per capita GDP, economy density, fixed-assets investment per area, etc. However negative correlations were discovered between the comprehensive land carrying capacity and the discharge of industrial waste water per 10,000 Yuan RMB GDP, as well as the proportion of added value of the primary industry to total GDP. Finally, we discussed measures to improve the comprehensive land carrying capacities of the cities, such as elevating the intensive land utilization and economic development, decreasing the proportion of added value of the primary industry to total GDP, promoting energy saving and emission reduction, etc.

Keywords: comprehensive land carrying capacity; multi-criterion comprehensive evaluation; analytic hierarchy process; standard deviation; weight; spatial variation

---

1. Introduction

Population explosion and land resource shortage are two important constraints on current economic development of the world [1]. By the early 19th century, scientists started to realize that the relationship of social economic activities with the land resources should be coordinated, and that is when the line of research on land carrying capacity began. During the period of 1970s to 1980s, research on land carrying capacity was mainly based on the ecological limiting factors, potential land capacity for natural production, grain yield and relationship of man and grain to confirm the land's population carry capacity. For example, the research of land carrying capacity for Australia on the perspective of the restriction imposed on population by all types of resources [2], the determination of land population carrying capacity by using Agricultural Ecology Zone method proposed by FAO in 1997 [3,4]. The enhancement of carrying capacity options (ECCO) put forward by Slessor in the early 1980s took thorough consideration of the relation between population change and land carrying capacity [5]. Furthermore, research of land resource production capacity, especially grain output and land's population carrying capacity of China were also carried out by the Commission for Integrated Survey of Natural Resources, China [6].

Since the 1990s, with the progress of economic globalization, the land carrying capacity research on the relationship between human and grain showed its limitation in effectively explaining and guiding human activities. Research on land carrying capacity evaluation gradually developed from the application of single grain index to comprehensive index, from the evaluation of land's population carrying capacity to the evaluation of comprehensive land carrying capacity by considering all kinds of human activities rather than merely focusing on population. The index composed by cost and benefit indicators, which belong to four subsystems of water and soil resources, eco-environment, society, economy, and technology, is put forward for evaluating comprehensive land carrying capacity of eastern China [1,7–12]. The comprehensive land carrying capacity of major grain-producing areas, Heilongjiang Province in northeastern China and its risk factors were also researched [13]. The weights of the indicators were mainly constructed using only objective weighting methods, and short of the evaluator's professional experience and subjective judgment, the methods of system dynamic, ecological footprint, projection pursuit, pair analysis, variables fuzzy assessment method began been used in the research [1,7–15]. In this period, the research on land's population or livestock carrying capacity, water resources carrying capacity, and ecological carrying capacity were also carried out [14,16–19], however the aims of and index used by these studies were different from the evaluation of comprehensive land carrying capacity. The comprehensive land carrying capacity is the threshold of human activities that the ecological-economic-social system formed mainly by land can bear at a particular state, or limiting factors of land reached their maximum value or minimum value. It is composed of the carrying capacity of water and soil resources, the carrying capacity of eco-environment, the carrying capacity of social resources and the carrying capacity of economy and technology.

"The Plan to Develop the Shandong Peninsula Blue Economic Zone" approved by the state council of China in 2011 indicated that Shandong Peninsula Blue Economic Zone takes the accumulation area of modern marine industry with high international competitiveness, the world's leading core area of marine science and technology and marine education, the national marine economic reform and open zone and the marine ecological civilization demonstration area as its construction goals [20]. The total population of the area had increased to 3691.4×10^4 by 2014. Such large population and shortage of land resource per capita gave high pressure on the land, which has become the main obstacles for the region. This research aims to provide theoretical basis for the development of marine industry, the coordinated development of inland and coastal area, and the adjustment and optimization of land utilization, as well as improve the quality of regional eco-environment by studying the spatio-temporal variations and the influences of the comprehensive land carrying capacity in the Shandong Peninsula Blue Economic Zone. In regard to research methods, determining the weight of indicators is the key technology of evaluating the comprehensive land carrying capacity. Previous studies showed defects on subjective or objective empowerment of evaluating indicators, therefore this paper tries to determine the weight of indicators for evaluation by using the multi-criterion comprehensive evaluation model to compensate the shortage of subjective evaluation or objective evaluation, and in order to get more scientific and accurate results.

2. Research methods

2.1. Research Area

Shandong Peninsula Blue Economic Zone includes the 6 cities of Qingdao City, Dongying City, Yantai City, Weifang City, Weihai City, Rizhao City, and Wudi County and Zhanhua County in Binzhou City of Shandong Province as well as the regional offshore waters, it has land area of 6.4×10^4 km^2, sea area of 15.95×10^4 km^2 and coastline of 3345 km in length. The warm temperate continental monsoon climate of the research area is characterized by precipitation concentrated in summer and autumn, the temperature of spring lower than that of autumn. The terrain is characterized by higher altitudes in the southern and northern regions than that in the central region, and the regional average altitude is less than 300 m. The land area comprises mainly mountains and hills of granite, as well as alluvial plains and marine-deposition plains partially. The soils are classified as brown soil, cinnamon soil, fluvo-aquic soil, cultivated soil, alluvial-salty soil, saline soil, etc. The zonal vegetation of the region is temperate deciduous broad-leaved forest. With excellent location, rich marine resources and favorable eco-environment, Shandong Peninsula Blue Economic Zone has seen rapid regional social and economic development. By 2014, the population of the cities of Qingdao, Dongying, Yantai, Weifang, Weihai, Rizhao and Binzhou were 904.62×10^4, 209.91×10^4, 700.23×10^4, 924.72×10^4, 280.92×10^4, 287.05×10^4, and 383.96×10^4, and each had increased 7.86, 5.43, 0.11, 4.65, 0.34, 5.42, and 4.59% respectively. The total population had increased 4.27% during the period of 2007–2014. By 2014, the regional gross domestic product (GDP) reached 2.77×10^{12} Yuan RMB and gross ocean product (GOP) reached 1.04×10^{12} Yuan RMB, accounting for 46.64% and 17.50% of the GDP and GOP of Shandong Province respectively. The per capita GDP of the research area was 81,656.38 Yuan RMB, which was 34.51% higher than that of Shandong Province.

2.2. Establish Evaluation Index System

With the principles of comprehensiveness, hierarchy and regional particularity, the evaluation index system of comprehensive land carrying capacity of Shandong Peninsula Blue Economic Zone based on multi-objective comprehensive evaluation model is composed of three layers of target, criterion and index [10,18,21,22]. The target layer is the comprehensive land carrying capacity of Shandong Peninsula Blue Economic Zone, the criterion layer is composed by four supporting land carrying capacity subsystems, namely the carrying capacity of the water and soil resources, the carrying capacity of eco-environment, the carrying capacity of social resources and the carrying capacity of economy and technology, the index layer includes 22 benefit indictors and 5 cost indictors [23], the benefit indictors include cultivated land area per capita, effective irrigation rate of cultivated land, and water resources per capita, etc., which are positively correlated with the comprehensive land carrying capacity, and the cost indictors negatively related to the comprehensive land carrying capacity, which include the proportion of value added of primary industry to total GDP, discharge of industrial waste water per 10 thousand Yuan RMB GDP, density of population, registered urban unemployment rate, and natural population growth rate.

2.3. Data Source and Processing Method

In order to make the indictors used in the research comparable, the 7 cities of Rizhao, Qingdao, Weihai, Yantai, Weifang, Dongying, and Binzhou were selected as evaluation units; the original data of the 27 indictors of the cities in 2007–2014 for evaluation were collected from China's City Statistical Almanacs, Shandong Statistical Almanacs, and the Statistical Almanacs of the seven cities during the period of 2008–2015.

Due to data of the indicators of the seven cities is not comparable for the difference of dimensions and units, the original data should be normalized, the method of normalizing original data is to bring the original data of benefit indictors and cost indictors in the index layer of the seven cities in the period of 2007–2014 into formula (1) or (2), then the normalized value of the indicators is obtained.

In the data series of a given benefit indictor in 2007–2014, the largest normalized value of the seven cities is 1, the smallest normalized value is 0, and the cost indicators are reversed.

The benefit indictor is

$$Z_{ij} = \frac{x_{ij} - x_{j\min}}{x_{j\max} - x_{j\min}},\tag{1}$$

The cost indictor is

$$Z_{ij} = \frac{x_{j\max} - x_{ij}}{x_{j\max} - x_{j\min}},\tag{2}$$

where x_{ij} is the original data of the indicators, i is the sample number of the indicators, j is the number of the indicators, Z_{ij} is the normalized value of the jth indicator, $0 \leq Z_{ij} \leq 1$; $X_{j\min}$ and $X_{j\max}$ are the minimum values and maximum values respectively of the original data of the jth indicator.

2.4. Determining the Weight of the Indicators

There are three methods of determining the weights of the indicators for carrying capacity evaluation, which include subjective weighting method, objective weighting method, and multi-criterion comprehensive weighting method. The subjective weighting method determines the weights of the indicators according to the evaluator's professional experience and the relative importance of the indicators, and the weights have some subjective meaning; the objective weighting method determines the weights of the indicators according the amount of information contained in the indicators by using mathematical model, but the evaluator's subjective understanding and professional experience were ignored [24–26]. The multi-criterion comprehensive weighting method possesses the advantages of both subjective weighting method and objective weighting method, and avoids their disadvantages. The weights of indicators reflect both the evaluator's subjective understanding and the amount of information content of the indicators [27]. The process of calculating the indicators' weights using the multi-criterion comprehensive weighting method is as follows:

The weight of each indicator was determined by the analytic hierarchy process and the standard deviation method. Analytic hierarchy process is a combination of subjective method and objective method for determining weights of indicators, to determine the weights of indicators by using analytic hierarchy process, we firstly establish hierarchical structure model of index, and a judgment matrix is constructed to calculate the weights of indicators of all levels, coupled comparing and ranking all indicators in each level respectively, and then the consistency test of the indicators' weights is performed, the indicators in each level are sorted declining according their weights, and ranking the total indicators on their weights [28]. However, the standard deviation method only uses objective mathematical model to determine the weights of indicators, the average value and standard deviation of the indicators are calculated by using formula (3) or (4) respectively, and then normalize the standard deviations of each indicators to obtain the weights of the indicators by using formula (5) [29].

The average value of the indicators is calculated by

$$\overline{Z_j} = \frac{1}{n}\sum_{i=1}^{n} Z_{ij},\tag{3}$$

The standard deviation of the indicators is

$$\sigma_j = \sqrt{\sum_{i=1}^{n} (Z_{ij} - \overline{Z_j})},\tag{4}$$

The weight of the indicators is

$$q_j = \frac{\sigma_j}{\sum\limits_{j=1}^{m} \sigma_j},\tag{5}$$

After calculating the weights of the indicators by using the analytic hierarchy process and standard method separately, the paper combines the two results of calculation, the weight of every indicator is further determined by using multi-criterion comprehensive method. Assume the comprehensive weight of the *j*th indicator is w_j, the value of w_j can be calculated using formula (6).

The comprehensive weight of the *j*th indicator is

$$w_j = k_1 p_j + k_2 q_j,\tag{6}$$

where p_j and q_j are the weight of the *j*th indicator calculated by using analytic hierarchy process and standard deviation method respectively, k_1 and k_2 are the undetermined constant.

Establishing objective function in formula (7), when the value of the objective function is maximum and the formula (7) meet the demands of $k_1 > 0$, $k_2 > 0$ and $k_1^2 + k_2^2 = 1$, we obtain k_1 and k_2 by using formulae (8) and (9) according to the Lagrange extreme value principle

$$\sum_{i=1}^{n} y_i = \sum_{i=1}^{n} \sum_{j=1}^{m} (k_1 p_j + k_2 q_j) Z_{ij},\tag{7}$$

$$k_1 = \frac{\sum_{i=1}^{n} \sum_{j=1}^{m} p_j Z_{ij}}{\sqrt{(\sum_{i=1}^{n} \sum_{j=1}^{m} p_j Z_{ij})^2 + (\sum_{i=1}^{n} \sum_{j=1}^{m} q_j Z_{ij})^2}},\tag{8}$$

$$k_2 = \frac{\sum_{i=1}^{n} \sum_{j=1}^{m} q_j Z_{ij}}{\sqrt{(\sum_{i=1}^{n} \sum_{j=1}^{m} p_j Z_{ij})^2 + (\sum_{i=1}^{n} \sum_{j=1}^{m} q_j Z_{ij})^2}},\tag{9}$$

The results of the calculation are $k_1 = 0.6956$ and $k_2 = 0.7184$, substituting the value of k_1, k_2, and the weights of the indicators calculated by using analytic hierarchy process and the standard method respectively into formula (6), calculate their normalization to get the weight of the indicators by using the multi-criterion comprehensive method (Table 1).

Table 1. Weight of the indicators for assessing the comprehensive land carrying capacities of Shandong Peninsula Blue Economic Zone.

Target Layer	Criterion Layer	Index Layer	Subjective Weight Calculated by Analytic Hierarchy Process	Objective Weight Calculated by Standard Deviation Method	The Weight Calculated by Multi-Criterion Comprehensive Method
Comprehensive land carrying capacities of cities in Shandong Peninsula Blue Economic Zone	Carrying capacity of water and soil resources/C_1	Cultivated land per capita C_{11}/(hm^2/capita)	0.0770	0.0376	0.0570
		Effective irrigation rate of cultivated land C_{12}/(%)	0.0376	0.0354	0.0365
		Water resource per capita C_{13}/(m^3/capita)	0.0199	0.0367	0.0284
		Intensity of land exploitation C_{14}/(%)	0.0247	0.0357	0.0303
		Land for construction per capita C_{15}/(m^2/capita)	0.0550	0.0367	0.0457
		Proportion of wetlands to total land area C_{16}/(%)	0.0177	0.0348	0.0264
		Coordination of water and land C_{17}/(%)	0.0422	0.0375	0.0398

Table 1. *Cont.*

Target Layer	Criterion Layer	Index Layer	Subjective Weight Calculated by Analytic Hierarchy Process	Objective Weight Calculated by Standard Deviation Method	The Weight Calculated by Multi-Criterion Comprehensive Method
	Carrying capacity of eco-environment/C_2	Percentage of green space to built up area C_{21}/(%)	0.0365	0.0339	0.0352
		Public green space per capita C_{22}/(m^2/capita)	0.0252	0.0357	0.0305
		Forest coverage rate C_{23}/(%)	0.0712	0.0357	0.0532
		Proportion of environment protection investment to GDP C_{24}/(%)	0.0116	0.0406	0.0263
		Comprehensive utilization rate of industrial solid waste residues C_{25}/(%)	0.0134	0.0411	0.0275
		Urban sewage treatment rate C_{26}/(%)	0.0180	0.0353	0.0268
		Discharge of industrial wastewater per 10,000 Yuan RMB GDP C_{27}/(t)	0.0180	0.0414	0.0299
	Carrying capacity of social resources/C_3	Population density C_{31}/(capita/m^2)	0.0537	0.0312	0.0423
		Natural population growth rate C_{32}/(%)	0.0105	0.0394	0.0252
		Registered urban unemployment rate C_{33}/(%)	0.0195	0.0395	0.0297
		Urbanization rate C_{34}/(%)	0.0314	0.0402	0.0359
		Proportion of technology and education investment to total GDP C_{35}/(%)	0.0105	0.0315	0.0212
		Urban road area per capita C_{36}/(m^2)	0.0074	0.0390	0.0235
	Carrying capacity of economy and technology/C_4	GDP per capita C_{41}/($\times 10^4$ Yuan RMB)	0.1142	0.0359	0.0744
		Economy density C_{42}/($\times 10^4$ Yuan RMB/m^2)	0.0792	0.0348	0.0566
		Growth rate of GDP C_{43}/(%)	0.0615	0.0415	0.0513
		Fixed assets investment per area C_{44}/($\times 10^4$ Yuan RMB/hm^2)	0.0549	0.0397	0.0472
		Yield of grain per area C_{45}/(kg/hm^2)	0.0398	0.0374	0.0386
		Proportion of value added of the primary industry to total GDP C_{46}/(%)	0.0211	0.0414	0.0314
		Proportion of value added of the secondary industry to total GDP C_{47}/(%)	0.0284	0.0305	0.0295

Note: The indicators of C_{11}, C_{13}, C_{15}, C_{31}, and C_{41} are calculated according to the permanent residents, C_{14}, C_{21}, C_{22}, and C_{36} are the statistical data of urban areas, among them, C_{14} is the proportion of land for construction to total urban land, and C_{24} is the proportion of the budget of environment protection investment to GDP, C_{35} is the proportion of technology and education investment to GDP in the fiscal budgets of the cities, and C_{44} is the ratio of total fixed assets investment to acreage of agricultural and construction land.

2.5. Calculation of the Indicators of Comprehensive Land Carrying Capacity

Putting the normalized value of the indicators and the weights of the indicators calculated by using multi-criterion comprehensive method into formula (10), we get the indicators of comprehensive land carrying capacity of each evaluating unit y_i ($0 \leq y_i \leq 1$) (Table 2). The value of the indicator is bigger, the comprehensive land carrying capacity is higher.

$$y_i = \sum_{j=1}^{m} w_j Z_{ij} = \sum_{j=1}^{m} (k_1 p_j + k_2 q_j) Z_{ij}, \tag{10}$$

where y_i is the indicator of the comprehensive land carrying capacity, Z_{ij} is the normalized value of the indicators, w_j is the weight of the indicators.

Then analyze the overall spatial difference of the comprehensive land carrying capacities of the seven cities in Shandong Peninsula Blue Economic Zone in 2007–2014 by using cluster analysis method.

3. Results of Evaluation and Analysis

3.1. Annual Changes and Spatial Differences of the Supporting Land Carrying Capacity Subsystems

We obtain the indicators' value of comprehensive land carrying capacity of the seven cities in Shandong Peninsula Blue Economic Zone in the period of 2007–2014 (Table 2) and analyze the annual changes and spatial differences of the supporting land carrying capacity subsystems.

Table 2. Comprehensive land carrying capacities of the seven cities in Shandong Peninsula Blue Economic Zone in the period of 2007–2014

City	2007	2008	2009	2010	2011	2012	2013	2014
Qingdao	0.424	0.435	0.456	0.482	0.508	0.506	0.513	0.521
Dongying	0.500	0.506	0.514	0.555	0.590	0.612	0.626	0.652
Yantai	0.417	0.405	0.423	0.468	0.475	0.477	0.474	0.451
Weifang	0.311	0.331	0.369	0.384	0.392	0.418	0.433	0.428
Weihai	0.555	0.515	0.526	0.591	0.584	0.608	0.577	0.590
Rizhao	0.345	0.370	0.362	0.386	0.414	0.429	0.422	0.422
Binzhou	0.351	0.371	0.404	0.428	0.460	0.471	0.476	0.471

With the economy development and the population growth in the period of 2007–2014, the land carrying capacity of soil and water resources of six cities except for Dongying showed a slow decrease due to the decrease of land and water resource per capita, the land carrying capacity of soil and water resources of Dongying, Weifang, and Weihai were higher than those of the other four cities (Figure 1a).

The land carrying capacity of eco-environment of the seven cities rose significantly because the increase of percentage of green space to urban built up area, the public green space per capita, the forest coverage rate, the proportion of environmental protection investment to GDP, the comprehensive utilization rate of industrial solid waste residues and the urban sewage treatment rate, as well as the decrease of discharge of industrial wastewater per 10,000 Yuan RMB GDP. The carrying capacity of eco-environment of Weihai was in high level, those of Qingdao Rizhao and Yantai were in middle level, and those of Weifang, Dongying, and Binzhou were low (Figure 1b).

The land carrying capacity of social resource of the seven cities had increased to a small extent, and the increase of Dongying and Weihai was obvious higher than the others five cities, due to the cities showed increase of the population density, natural population growth rate, registered urban unemployment rate, urbanization rate, proportion of technology and education investment of total GDP, the urban road area per capita, etc., however, the growth rate of the three benefit indicators, such as the urbanization rate were higher than that of the three cost indicators which include the population density (Figure 1c).

The cultivated land per capita of the cities had decreased, however with the improvement of agricultural technology, the yield of grain per area had increased significantly, which partly makes up for the decrease of cultivated land per capita and improves the population carrying capacity of land resource; the GDP per capita, economy density and fixed assets investment per area of the seven cities had been constantly increasing, the land use intensification has been continuously improved, the carrying capacity of economy and technology of the seven cities had increased. The carrying capacity of economy and technology of Qingdao and Dongying were in high level, those of Weihai and Yantai were in middle level, and those of Rizhao, Weifang, and Binzhou were low (Figure 1d).

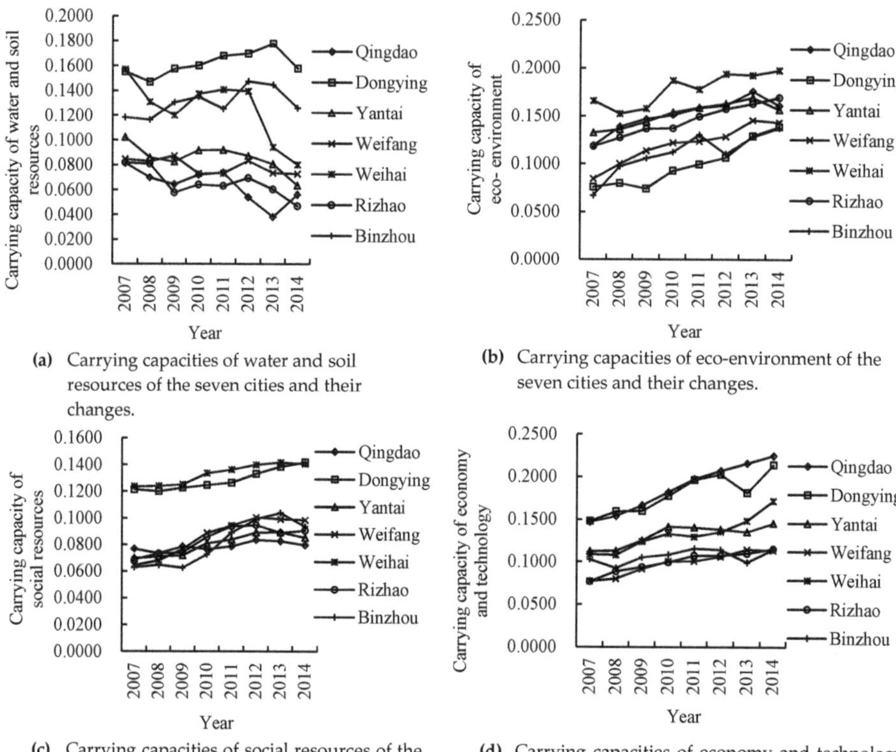

(a) Carrying capacities of water and soil resources of the seven cities and their changes.

(b) Carrying capacities of eco-environment of the seven cities and their changes.

(c) Carrying capacities of social resources of the seven cities and their changes.

(d) Carrying capacities of economy and technology of the seven cities and their changes.

Figure 1. The sub-systems' indexes of land carrying capacities of the seven cities in Shandong Peninsula Blue Economic Zone and their spatiotemporal variations in the period of 2007–2014.

3.2. Annual Changes of the Comprehensive Land Carrying Capacity

The comprehensive land carrying capacity of the seven cities had kept been continual or fluctuant rising in the period of 2007–2014, which reveal the seven cities attach great importance to the management of land use, and constantly improve the type structure and spatial distribution of land use, overall planning of land utilization, as well as improve the intensity of land exploitation, strengthen prime cropland construction, protect cultivated land, and coordinate the relationship between land use and protection of eco-environment. Since 2011, the growth rate of comprehensive land carrying capacities of the 6 cities except for Dongying had decreased because the growth of economy slowed down and the carrying capacities of water and soil resources decreased (Figure 2).

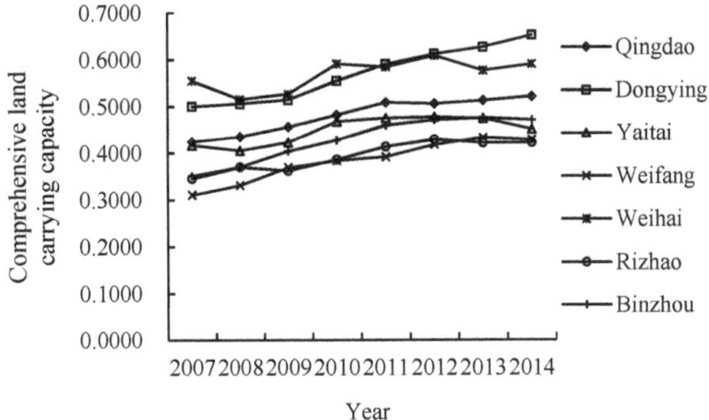

Figure 2. Comprehensive land carrying capacities of the seven cities in Shandong Peninsula Blue Economic Zone and their changes in the period of 2007 and 2014.

3.3. Spatial Differences of the Comprehensive Land Carrying Capacity of the Seven Cities

In order to understand the spatial differences of comprehensive land carrying capacity of the seven cities in Shandong Peninsula Blue Economic Zone in the period of 2007–2014, the software of SPSS 19.0 and clustering methodology were used for calculating the comprehensive land carrying capacities of the seven cities in the period of 2007–2014, and the comprehensive land carrying capacities of the cities were divided into three grades of I, II, and III from high to low. The average situation of the comprehensive land carrying capacity of the seven cities in the period of 2007–2014 shows roughly the spatial difference characteristics of east-high and west-low which similar to the regional differences of economy growth: the east Weihai and central Dongying belonged to the region of grade I, their comprehensive land carrying capacities were high; Qingdao and Yantai belonged to the region of grade II, their comprehensive land carrying capacities were intermediate, the west cities of Weifang, Rizhao, and Binzhou belonged to the region of grade III, and their comprehensive land carrying capacity level were low (Figure 3).

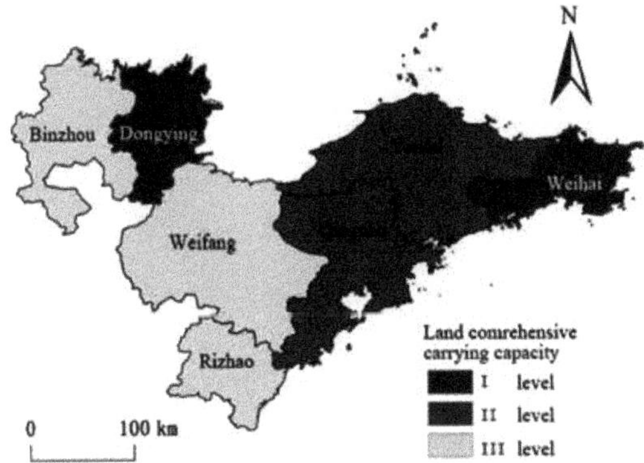

Figure 3. Grades of comprehensive land carrying capacities and their spatial difference of the seven cities in Shandong Peninsula Blue Economic Zone in the period of 2007–2014.

The region of grade I includes Dongying and Weihai. The comprehensive land carrying capacities of the two cities were higher than those of the other five cities, the comprehensive land carrying capacity of Dongying was the highest in the seven cities. The indexes of all supporting land carrying capacity subsystems were relatively high, and the indicators were well coordinated, or some indicators have significant advantages to compensate for the disadvantages of other indicators. For example, the proportion of environment protection investment to GDP and forest coverage rate of Dongying were relatively poor, which caused a low regional carrying capacity of eco-environment, however the carrying capacity of water and soil resources, the carrying capacity of economy and technology of the city were relatively high.

The region of grade II includes Qingdao and Yantai. The comprehensive land carrying capacities of the two cities were slightly lower than those of the two cities of grade I. The indexes of all supporting land carrying capacity subsystems were intermediate, and the supporting subsystems were well coordinated, for example, Qingdao had developed economy, effective eco-environment governance, high efficiency of input and output in the process of land resource exploitation, and relatively high indicators of carrying capacity of eco-environment, carrying capacity of social resources as well as carrying capacity of economy and technology. However, the indicators of carrying capacity of soil and water resources were relatively low limited by the large population, shortage of cultivated land, and water resource per capita.

The region of grade III includes Rizhao, Weifang, and Binzhou. The comprehensive land carrying capacities were low, the supporting subsystems lacked prominent advantages and were poorly coordinated with each other. For example, the carrying capacity of soil and water resources and carrying capacity of social resources of Binzhou were at a medium level, and more seriously, its carrying capacity of eco-environment and carrying capacity of economy and technology were low.

4. Conclusions and Discussion

4.1. Conclusions

The comprehensive land carrying capacity of the seven cities in Shandong Peninsula Blue Economic Zone from 2007 to 2014 was evaluated by using multi-criterion comprehensive evaluation method, and the spatio-temporal variations and the factors which influenced the evaluating result were analyzed.

The comprehensive land carrying capacities of the seven cities had been rising either continuously or with a fluctuation. In addition, it and shows a great regional difference that cities in the east demonstrated higher comprehensive land carrying capacity than the cities in the west during 2007–2014. The seven cities were consequently divided into three grades, among which Dongying and Weihai belonged to the region of grade I and their comprehensive land carrying capacities were high; Qingdao and Yantai belonged to the region of grade II, the comprehensive land carrying capacities of the two cities were slightly lower than those of Dongying and Weihai; Rizhao, Binzhou, and Weifang belonged to the region of grade III, their comprehensive land carrying capacities were lower than those of Qingdao and Yantai.

4.2. Discussions

The index used in this paper was adapted from references [7–10], however it is optimized by adding new effective indicators, such as intensity of land exploitation, proportion of wetlands to total land area, urbanization rate, etc. Additionally, several ineffective indicators in the previous index were removed. The indicators' weights were determined by using multi-criterion comprehensive weighting method, and the evaluator's subjective thoughts and the indicators intrinsic information were significantly reflected, avoiding the disadvantages of both subjective and objective weighting method. There is still a shortage of missing the data in the latest two years; however, after a preliminary analysis of the new data of 2015 and 2016, we adding data of the two years would not overt change of the research conclusion.

It has to be noted that even clean service economics, such as tourism, would have negative impacts on the land system [30]. It seemed that one of the research results, the comprehensive land carrying capacities of the seven cities should had been decreasing for many economic indicators had been increasing. However, the truth is that the capacities had kept been increasing. The main reason is that we regarded most economic indicators as benefit indicators which have positive impacts on comprehensive land carrying capacity; however, we also accounted for their negative impacts in the evaluation index by using the five cost indicators and a few benefit indicators, maybe we underestimate negative impacts of economic indicators on land.

Due to the marine industry and marine eco-environment play an important role on the regional economic development, therefore the indicators for evaluation which reflect the marine economy development, the status of marine resources exploitation and the characteristics of marine eco-environment can be added to further optimize the index system of evaluation, in order that the results of evaluation can reflect the impact of marine industry development and marine eco-environment on the comprehensive land carrying capacity.

We should take countermeasures such as improving intensive land use, strengthening regional eco-environment governance, enhancing urban economy growth rate, improving the energy use efficiency, reducing the proportion of added value of the primary industry to total GDP, as well as reducing greenhouse gas emissions to improve the comprehensive land carrying capacities of the cities in research area.

Author Contributions: X.Z. and Z.Z. conceived and designed the assessing mythology and indicators; G.C. performed the data collection and model simulation; Y.C. and X.L. contributed the analysis tools and English spell check; X.Z. wrote the paper.

Funding: The National Key Research and Development Plan of China (2016YFC0503503) and the Project of Natural Science Foundation of Shandong Province, China (ZR2016DM11).

Acknowledgments: The author is grateful to the members of Advanced Institute of Culture & Tourism, Qingdao University for their valuable advice.

Conflicts of Interest: The authors declare no conflict of interest.

References

1. Guo, S.; Li, C.; Liu, S.; Zhou, K. Land carrying capacity in rural settlements of three gorges reservoir based on the system dynamic model. *Nat. Resour. Model.* **2018**, *31*, e12152. [CrossRef]

2. Millington, R.; Gifford, R. *Energy and How We Live*; Australian UNESCO Seminar, Committee for Man and Biospher: Adelaide, Australia, 1973.

3. Jin, X.M.; Chen, L. Paradigm shift in the study of land carrying capacity: An overview. *J. Nat. Resour.* **2018**, *33*, 526–540.

4. Higgins, G.M.; Kassan, A.H.; Naiken, L.; Shah, M.M. *Potential Population Supporting Capacities of Lands in the Developing World*; Food & Agriculture Organization: Washington, DC, USA, 1982.

5. Slessor, M. *Enhancement of Carrying Capacity Options ECCO*; The Resource Use Institute: London, UK, 1990.

6. Chen, B.M. An outline of the research method of the project "the productivity and population carrying capacity of the land resource in China". *J. Nat. Res.* **1991**, *6*, 197–205.

7. Wang, S.H.; Mao, H.Y. Design and evaluation on the indicator system of land comprehensive carrying capacity. *J. Nat. Resour.* **2001**, *16*, 248–254.

8. Wang, S.H.; Mao, H.Y.; Zhao, M.H. Thinking on the index system design to the land comprehensive carrying capacity—A case study: Coastal region of China. *Hum. Geogr.* **2001**, *16*, 57–61.

9. Zhang, B.; Xue, J.B.; Fan, Q.; Jia, K.J. An evaluation of regional land comprehensive carrying capacity and spatial disparity analysis—A case of Jiaxing City. In Proceedings of the Zhejiang Province. International Symposium on Humanistic Management and Development of New Cities and Towns, Hangzhou, China, 31 October–2 November 2014; pp. 174–180.

10. Yu, G.H.; Sun, C.Z. Land carrying capacity spatiotemporal differentiation in the Bohai Sea coastal areas. *Acta Ecol. Sin.* **2015**, *35*, 4860–4870.

11. Jiang, Q.X.; Fu, Q.; Meng, J.; Wang, Z.L.; Zhao, K. Comprehensive evaluation of land resources carrying capacity under different scales based on RAGA-PPC. In *Computer and Computing Technologies in Agriculture VIII (CCTA 2014), Beijing, China, 16–19 September 2014*; Li, D., Chen, Y., Eds.; IFIP Advances in Information and Communication Technology; Springer International Publishing: Cham, Switzerland, 2015; Volume 452, pp. 200–209.

12. Guo, H.H.; Li, B.; Hou, Y. Research on the capacity of land resource based on land function: Haidian as an example. *J. Beijing Normal Univ. (Nat. Sci.)* **2011**, *47*, 424–427.

13. Cheng, K.; Fu, Q.; Cui, S.; Li, T.X.; Pei, W.; Liu, D.; Meng, J. Evaluation of the land carrying capacity of major grain-producing areas and the identification of risk factors. *Nat. Hazards* **2017**, *86*, 263–280. [CrossRef]

14. Shi, Y.S.; Wang, H.F.; Yin, C.Y. Evaluation method of urban land population carrying capacity based on GIS: A case of Shanghai, China. *Comput. Environ. Urban Syst.* **2013**, *39*, 27–38. [CrossRef]

15. He, R.W.; Liu, S.Q.; Liu, Y.W. Application of SD model in analyzing the cultivated land carrying capacity: A case study in Bijie Prefecture, Guizhou Province, China. *Procedia Environ. Sci.* **2011**, 1985–1991. [CrossRef]

16. Zhang, Y.Z.; Chang, L.P.; Zhang, B.; Zhang, S.W.; Huang, T.Q.; Liu, Y.Q. Land resources survey by remote sensing and analysis of land carrying capacity for population in Tumen river region. *Chin. Geogr. Sci.* **1996**, *6*, 342–350. [CrossRef]

17. Thapa, G.B.; Paudel, G.S. Evaluation of the livestock carrying capacity of land resources in the Hills of Nepal based on total digestive nutrient analysis. *Agric. Ecosyst. Environ.* **2000**, *78*, 223–235. [CrossRef]

18. Yang, Z.Y.; Song, J.X.; Cheng, D.D.; Xia, J.; Li, Q.; Ahamad, M.I. Comprehensive evaluation and scenario simulation for the water resources carrying capacity in Xi'an city, China. *J. Environ. Manag.* **2019**, *230*, 221–233. [CrossRef] [PubMed]

19. Peng, B.H.; Wang, Y.Y.; Elahi, E.; Wei, G. Evaluation and prediction of the ecological footprint and ecological carrying capacity for Yangtze River urban agglomeration based on the Grey Model. *Int. J. Environ. Res. Public Health* **2018**, *15*, 2543. [CrossRef] [PubMed]

20. National Development and Reform Commission. *Shandong Peninsula Blue Economic Zone Development Plan*; National Development and Reform Commission: Beijing, China, 2012.

21. Yu, S.; Wang, M.Y. Comprehensive evaluation of scenario schemes for multi-objective decision-making in river ecological restoration by artificially recharging river. *Water Resour. Manag.* **2014**, *28*, 5555–5571. [CrossRef]

22. Sun, Y.; Li, X.G. The research on the coordinated development degree of urban land comprehensive carrying capacity system in Shandong province. *China Popul. Resour. Environ.* **2013**, *23*, 123–129.

23. Zhang, X.L.; Zhang, Z.H.; Su, W.X. Comprehensive assessment of the ecological carrying capacity of the Yellow River delta. *J. Saf. Environ.* **2015**, *15*, 364–369. [CrossRef]

24. Liu, Y.; Zeng, C.; Cui, H.; Song, Y. Sustainable land urbanization and ecological carrying capacity: A spatially explicit perspective. *Sustainability* **2018**, *10*, 3070. [CrossRef]

25. Zhou, S.H.; Chen, G.Q.; Fang, L.G.; Nie, Y.W. GIS-based integration of subjective and objective weighting methods for regional landslides susceptibility mapping. *Sustainability* **2016**, *8*, 334. [CrossRef]

26. Herva, M.; Roca, E. Review of combined approaches and multi-criteria analysis for corporate environmental evaluation. *J. Clean. Prod.* **2013**, *39*, 355–371. [CrossRef]

27. Lu, B.H. *Study on Comprehensive Carrying Capacity of Land Resources in Linan Based on Multi-Index System*; Zhejiang University: Hangzhou, China, 2014.

28. Xue, R.; Wang, C.; Liu, M.L.; Zhang, D.; Li, K.L.; Li, N. A new method for soil health assessment based on analytic hierarchy process and meta-analysis. *Sci. Total Environ.* **2019**, *650*, 2771–2777. [CrossRef] [PubMed]

29. Zhang, Y.R.; Ma, J.Z.; Qi, Z. Human activities, climate change and water resources in Shiyang Basin. *Resour. Sci.* **2012**, *34*, 1922–1928.

30. Koens, K.; Postma, A.; Papp, B. Is overtourism overused? Understanding the impact of tourism in a city context. *Sustainability* **2018**, *10*, 4384. [CrossRef]

Article

National Green GDP Assessment and Prediction for China Based on a CA-Markov Land Use Simulation Model

Yuhan Yu [1], Mengmeng Yu [1], Lu Lin [2], Jiaxin Chen [1], Dongjie Li [3], Wenting Zhang [1,4,*] and Kai Cao [5,*]

1   College of Resources and Environment, Huazhong Agricultural University, Wuhan 430070, China; hzyh199418@163.com (Y.Y.); yumm.gis@gmail.com (M.Y.); jiaxinc1998@gmail.com (J.C.)
2   School of Economics and Management, Academy of Chinese Energy Strategy, China University of Petroleum, Beijing 102249, China; linlu26@hotmail.com
3   Tourism College, Hunan Normal University, Changsha 410006, China; zhouwulian0408@163.com
4   Laboratory of urban Land Resources Monitoring and Simulation, Ministry of Land and Resources, Shenzhen 518000, China
5   Department of Geography, National University of Singapore, Singapore 117570, Singapore
*   Correspondence: wentingzhang@mail.hzau.edu.cn (W.Z.); geock@nus.edu.sg (K.C.)

Received: 13 December 2018; Accepted: 18 January 2019; Published: 22 January 2019

Abstract: Green Gross Domestic Product (GDP) is an important indicator to reflect the trade-off between the ecosystem and economic system. Substantial research has mapped historical green GDP spatially. But few studies have concerned future variations of green GDP. In this study, we have calculated and mapped the spatial distribution of the green GDP by summing the ecosystem service value (ESV) and GDP for China from 1990 to 2015. The pattern of land use change simulated by a CA-Markov model was used in the process of ESV prediction (with an average accuracy of 86%). On the other hand, based on the increasing trend of GDP during the period of 1990 to 2015, a regression model was built up to present time-series increases in GDP at prefecture-level cities, having an average value of R square (R^2) of approximately 0.85 and significance level less than 0.05. The results indicated that (1) from 1990 to 2015, green GDP was increased, with a huge growth rate of 78%. Specifically, the ESV value was decreased slightly, while the GDP value was increased substantially. (2) Forecasted green GDP would increase by 194,978.29 billion yuan in 2050. Specifically, the future ESV will decline, while the rapidly increased GDP leads to the final increase in future green GDP. (3) According to our results, the spatial differences in green GDP for regions became more significant from 1990 to 2050.

Keywords: Green GDP; Ecosystem service value; Gross Domestic Product; Land Use; CA-Markov

1. Introduction

China's economic development has improved people's living standards. The GDP in China increased from 367.87 billion yuan to 74 trillion yuan during the period of 1978 to 2016 [1]. With the rapid economic growth, ecosystem degradation has been aggravated and has severely affected the ecosystem function. The contradiction between economic growth and ecosystem carrying capacity has become an important limiting factor for sustainable development [2]. Therefore, a set of evaluation indicators was proposed and supposed to properly reveal the relationship between socio-economic development and ecosystem services. Those indicators provided a basis for the formulation of human social and economic sustainable development decisions. Among all of those indicators, green GDP is the most popular.

The United Nations first proposed the concept of green GDP in 1993 and released the system of ecosystem and economic accounting to reflect the relationship between economic development and preservation of natural resources [3,4]. Based on the concept proposed by the United Nations, numerous academic studies have been carried out to measure, evaluate and display green GDP for different regions at varying scales. For example, Sutton and Costanza [5] first attempted to map the ESV including green GDP accounting and estimated global market and non-market economic value and presented an integrated indicator called the subtotal ecological-economic product (SEP) that includes ecosystem services products (ESPs) and GDP. Kultida et al. [6] calculated the green GDP for Thailand from 2015 to 2020 by considering three types of environmental costs. Lin et al. [7] calculated the green GDP for China from 2006 to 2010 by considering energy intensity and pollution intensity. Talberth and Bohara [8] compared the gaps between green GDP increase and GDP increase. Li and Lang [9] implemented green GDP accounting for China with the purpose of illustrating the ongoing conflict between environmental preservation goals and economic growth goals. Li and Fang [10] mapped and estimated national green GDP globally by considering the economic value of ecosystem services and GDP. Yang and Poon [11] calculated the green GDP for China in 2007 and found that once considering green GDP, China's geography of income inequality is much less obvious.

It is obvious that green GDP accounting has become increasingly popular. Green GDP can be calculated by various methods concerning energy intensity, pollution intensity [7], environmental costs [6] and ecosystem service value [9,12,13]. Until now, using different methods to measure green GDP, comparing differences between green GDP and GDP and mapping the spatial distribution of green GDP have been widely studied concerning the historical value of green GDP. However, even if substantial studies have focused on the temporal trends of GDP [14], due to the difficulties in predicting spatial variation of environmental cost or value by statistical models, few studies have predicted the future changes in green GDP, especially the spatial change. Being aware of the future green GDP in space is significant for government to control local economic development and ecological preservation.

On the other hand, the method used to predict future ecosystem service value in response to land use change has been well developed and fully implemented. For example, Arowolo et al. [15] evaluated changes in ESV in response to land-use/land-cover dynamics in Nigeria. Xu and Ding [16] focused on the desertification in North China and its impact on the regional ESV from 1981 to 2010. Anderson et al. [17] calculated the ESV for South Africa according to land use datasets. Numerous studies have been carried out to evaluate ESV changes by considering land use changes. Obviously, the response of ESV to land use change has been proved and applied in calculating ESV. In addition, among all these studies, the cellular automata (CA) approach integrated with the Markov model (CA-Markov) is the most popular for simulating land use [18,19]. To be specific, CA is a spatiotemporal dynamic simulation model with discretization in time, space and state [20,21]. It focuses mainly on the local interactions of cells with distinct temporal and spatial coupling features. It has been widely used in simulating pedestrian dynamics [22,23], traffic flow [24] and fire spreading [25]. CA was first proposed in geographical modeling by Tobler [26]. Now, CA has become a popular method to simulate spatial distribution of urban sprawl for urban planning [27,28]. On the other hand, the Markov method is commonly used in predicting geographical characteristics lacking after-effect events. The integration of CA and Markov can provide a reasonable approach for future land use pattern prediction. For example, Etemadi et al. [29] used CA-Markov algorithms to monitor and predict future land use changes in coastal mangrove forests in Iran; Kamusoko [21] and Moghadam and Helbich [30] simulated the spatiotemporal urbanization process of one megacity in India using CA-Markov.

In response to the gaps in green GDP prediction and significance of being aware of future green GDP and given the advantages of CA-Markov in land use change simulation, this study attempted to account for the green GDP for the entirety of China from 1990 to 2015 at the prefectural level and combined the land use simulating approach, CA-Markov, to the green GDP prediction model to predict the future variations of ESV in China. In the second part, our study area and data sources are introduced and the third part carefully presents the methods for green GDP accounting and prediction.

The fourth part presents the spatial results of green GDP from 1990 to 2050. The last part is the conclusions and discussion section.

2. Study Area and Data Sources

2.1. Study Area

With an approximately 9.6 million square kilometer territory, China accommodated a population of 1374.62 million in 2015, with an annual population growth rate of 0.9676% from 1978 to 2015 [1]. Due to the reform and opening up policy, China has experienced drastic economic growth, with the GDP value increasing from 367.87 billion yuan to 74,114 billion yuan from 1978 to 2016 [1]. However, along with the rapid development of the economy and population, the ecological environment has been depressed [31]. Therefore, it is essential to evaluate the economic growth along with the environmental degradation.

2.2. Data Sources

The land use dataset with a spatial resolution of 1 km for the mainland China for the period of 1990 to 2015 was downloaded from the Data Center of the Institute of Geographic Sciences and Natural Resources Research, Chinese Academy of Sciences (http://www.resdc.cn/). Six land use types are included: cultivated land, forest land, grassland, waterbody, construction land and unused land.

The national GDP values at the prefectural level from 1990 to 2015 were obtained from provincial and prefectural statistical yearbooks. Some of the regional data were missing due to the change in administrative boundaries at the prefectural level and the missing data were set to Nodata. Meanwhile, Taiwan, Hong Kong and Macao were not considered in this study.

The outputs of major crops used as coefficients in ESV evaluation were obtained from the China statistical yearbook of 2016 [1]. The major crops in China include rice, wheat, corn and soybeans. The average purchase prices of major crops were provided by the China agricultural product price survey yearbook of 2016 [32]. A summary of all the datasets used in this study is presented in Table 1.

Table 1. Description of the datasets used in this study.

Data	Time	Scale/Form/Resolution	Source
Land use	From 1990 to 2015	National, 1 km resolution	Institute of Geographic Sciences and Natural Resources Research Chinese Academy of Sciences (http://www.resdc.cn/)
GDP	From 1990 to 2015	National, spreadsheet	National, provincial and prefectural Statistical Yearbook
Outputs of major crops	2015	National, spreadsheet	National Statistical Yearbook
Crops purchase price	2015	National, spreadsheet	National Statistical Yearbook
Administrative Boundaries	2007	National, vector shapefile	Institute of Geographic Sciences and Natural Resources Research Chinese Academy of Sciences (http://www.resdc.cn/)

3. Methods

We first used the coefficient method proposed by Costanza et al. [33] to evaluate the ESV in response to land use changes for China from 1990 to 2015 and meanwhile, the GDP at the prefectural level was obtained from provincial and prefectural statistic yearbooks. After accounting for the green GDP for China, we used the CA-Markov model to simulate the land use changes from 2015 to 2050 and the increasing trend of GDP was built by linear regression. The framework of this study is presented in Figure 1.

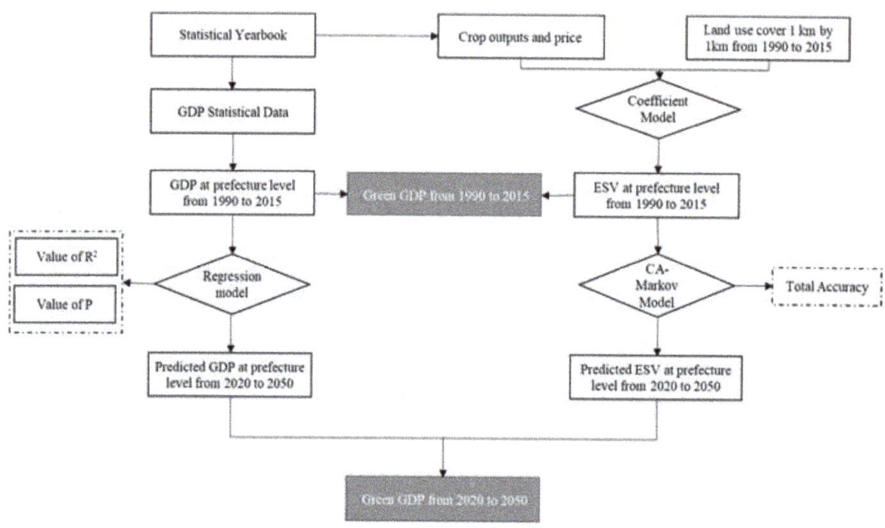

Figure 1. The framework.

3.1. Estimation of the Ecosystem Service Value

This study referred to the coefficient method proposed by Costanza et al. [33] to calculate the ESV at the prefectural level for the entirety of China (see Equation (1)):

$$ESV = \sum_{i=1}^{n} A_i \times E_i, \tag{1}$$

where ESV is the total value of the regional ecosystem service value; A_i is the area of the i-th land-use; E_i is the ESV coefficient for the i-th land use, which indicate the ESV per unit area of the i-th land-use type (yuan/hm^2); and n is the number of land use types. E_i is vital in ESV calculations and can be determined by Equation (2).

$$E_i = e_i \times E_a \tag{2}$$

e_i is the equivalent factor of the unit price of the ESV of the i-th land-use type (see Table 2), which can be determined according to the estimation method of the ecosystem service function value coefficient proposed by Constanza et al. [33] and Xie et al. [34]. E_a represents the economic value of the food production function for per unit area of farmland and can be determined using Equation (3).

$$E_a = \frac{1}{7} \sum_{j=1}^{k} \frac{m_j p_j q_j}{M}, \quad (i = 1, \cdots, k) \tag{3}$$

Table 2. Equivalent factor of the unit price of the ecosystem service value (ESV) for land use types in 2015 (Units: yuan/hm^2).

Land Use Types	e_i
Cultivated land	$16{,}538.88 \times 10^8$
Forest land	$108{,}095.26 \times 10^8$
Grass land	$86{,}873.69 \times 10^8$
Waterbody	$54{,}264.14 \times 10^8$
Unused land	7357.97×10^8

j is the j-th type of crop and k is the number of crop types. P_j is the national average price of crops in the j-th crop (yuan/ton) and q_j is the yield of the j-th crop (ton/hm^2). m_j is the planting area of the j-th crop (hm^2). M is the total crop area of all crops (hm^2) and 1/7 refers to one-seventh of the economic value of the function of farmland providing food production unit area. According to the status of the main crops in China, wheat, corn, rice and cotton were selected as the major crops. In addition, the P_j can be achieved from a China Yearbook of Agricultural Price Survey [31] for each major crop, and, on the other hand, q_j and m_j can be achieved from the provincial statistical yearbook of 2016. The ESVs from 1990 to 2015 were calculated based on the E_a evaluated by Equation (3) and were expressed in national average prices of crops for 2015 and can reflect the variation in ecosystem service caused by land use change directly.

Since the ESV is a total index for the whole city, we defined the UESV as the ESV in a unit area for a specific city (see Equation (4)).

$$\text{UESV}_m = \frac{ESV_m}{Area_m} \tag{4}$$

UESV$_m$ is the value of the ecological services of the m-th city on its per unit area (yuan/hm^2) and $Area_m$ is the area of the m-th city (hm^2).

3.2. Estimation of Green GDP

Green GDP was proposed to reflect the development of economy and ecological preservation simultaneously and it can be measured as Equation (5) shows.

$$G_m = TESV_m + GDP_m \tag{5}$$

G_m refers to the value of green GDP for the m-th city. $TESV_m$ is the ecological service value for the m-th city. GDP_m is the gross domestic product for the m-th city.

3.3. ESV Forecasting Based on the CA-Markov Model

Because ESV is measured according to land use pattern, we simulated the future land use pattern by the CA-Markov model first and then forecasted the ESV in the future. CA can be expressed by Equation (6).

$$\delta_i + 1 = f(St, N) \tag{6}$$

δ is a finite and discrete set of states of the cell. N is the neighborhood of the cell, t and $t+1$ refer to a different time of iteration. f is the transformation rule of the state of the local space cell [35].

The Markov method uses Markov chain theory and probability theory to analyze the law of random event changes. This method was used to predict a long-term prediction method in the future state of change in each period [36].

$$S_{(t+1)} = P_{ij} \times S_{(t)} \tag{7}$$

$S(t+1)$ is the state of the land use system at time $t+1$. P_{ij} is the state of the land use system at time t. $S(t)$ is the transfer matrix.

Both the CA model and the Markov model are dynamic models with discrete states. The CA-Markov model combines the advantages of the CA model to simulate spatial variations of complex systems and the advantages of the Markov model for quantitative prediction [27,37]. With the combined advantages of CA and Markov, the CA-Markov model has been widely used in land use space simulation applications [38].

To reflect the regional characteristics and regional differences in land use changes, according to the existing research [39–41], the entire country was divided into four regions—eastern, central, western and northeast—in the process of future land use simulation based on the 16th National Congress of the Communist Party of China. After the land use simulation for each region separately, we used the merging tool in ArcGIS to combine the results so as to obtain the final land use pattern for the entirety of China. Detailed descriptions of the four regions are listed in Table 3. To carry out the simulation, the CA-Markov module in IDRISI software was used.

Table 3. The descriptions of the four sections in the CA-Markov land use simulation.

Region Index	Name	Zone	Abbreviation
Z1	Eastern China	Beijing, Tianjin, Hebei, Shandong, Jiangsu, Shanghai, Zhejiang, Fujian, Guangdong, Hainan, Taiwan, Macao and Hong Kong	EC
Z2	Northeast China	Heilongjiang, Jilin and Liaoning	NEC
Z3	Western China	Inner Mongolia, Ningxia, Shaanxi, Chongqing, Guizhou, Guangxi, Yunnan, Sichuan, Gansu, Qinghai, Xinjiang and Tibet	WC
Z4	Central China	Shanxi, Henan, Anhui, Hubei, Hunan and Jiangxi	CC

3.4. Gross Domestic Product Forecasting Based on the Regress Model

According to the statistical yearbook from 1990 to 2016, the GDP at the prefectural level was obtained. A linear regression model was built up to predict the GDP for 2020-2050. Specifically, years were considered as the independent variable and GDP at the prefectural level was set as the dependent variable in the regression, as Equation (8) shows:

$$Y = \alpha x + \beta, \tag{8}$$

where Y is the dependent variable of GDP; α and β are the regression coefficients; and x is the independent variable of years.

We also defined the UGDP to indicate the GDP value on per unit area in a specific city (see Equation (9)):

$$UGDP_m = \frac{GDP_m}{Area_m}, \tag{9}$$

where $UGDP_m$ is the GDP value for per unit area of the m-th city (thousand yuan/hm^2) and $Area_m$ is the area of the m-th land use (hm^2).

4. Results

4.1. Temporal and Spatial Variations of ESV from 1990 to 2015

According to the national 11th Five-Year Plan, the whole nation can be classified into four major sections: the eastern region, central region, western region and northeastern region. To display and analyze the differences in ESV, GDP and green GDP in the regions in detail, we divided the four major regions into eight major economic zones by referring to existing studies [42,43]. Descriptions of the eight major economic zones are listed in Table 4. The differences, variations and trends for ESV, GDP and green GDP were analyzed at the economic zone level.

As the results showed, UESV gradually increased from northwest to southeast (see Figure 2). This pattern was consistent with the spatial distribution of major forests in China. The better the hydrothermal conditions, the higher the value of ecosystem services would be. The UESV has been analyzed for 2015 (see Table 5). The economic zones with high values of UESV mainly included the SCEZ, with a UESV value of 420.35 yuan/hm^2 and the ECEZ, with a UESV value of 354.41 yuan/hm^2. The high values of UESV in the SCEZ and ECEZ are probably due to better hydrothermal conditions and higher vegetation coverage, such as for Foshan city in Guangdong province, with a UESV of 831.49 yuan/hm^2 and Suzhou city in Jiangsu province, with a UESV value of 854.65 yuan/hm^2, whereas the lowest values of UESV, 246.22 yuan/hm^2 and 137.95 yuan/hm^2, were for Xiamen city and Yancheng city, respectively. The NEEZ, MYTREZ and SWEZ experienced the second highest UESV. To be specific, the UESV values in the NEEZ, MYTREZ and SWEZ were 325.14 yuan/hm^2, 323.10 yuan/hm^2 and 320.29 yuan/hm^2, respectively. In particular, in the SWEZ, the high value of

UESV, 506.71 yuan/hm^2, was concentrated in Nujiang city in Yunnan province. Usually, in the SWEZ, the places with rich hydrothermal conditions in Yunnan province tended to maintain high values of UESV. On the other hand, Ziyang city in Sichuan suffered the lowest USEV value of 105.62 yuan/hm^2 in the SWEZ.

Table 4. Descriptions of China's eight economic zones and abbreviations.

Region Index	Name	Zone	Abbreviation
A	Northeast Economic Zone	Liaoning, Jilin, Heilongjiang	NEEZ
B	Northwest Economic Zone	Gansu, Qinghai, Ningxia, Tibet and Xinjiang	NWEZ
C	Middle Reaches of the Yellow River Economic Zone	Shaanxi, Shanxi, Henan and Inner Mongolia	MYREZ
D	Northern Coastal Economic Zone	Beijing, Tianjin, Hebei and Shandong	NCEZ
E	Eastern Coastal Economic Zone	Shanghai, Jiangsu and Zhejiang	ECEZ
F	Middle Yangtze River Economic Zone	Hubei, Hunan, Jiangxi and Anhui	MYTREZ
G	Southwest Economic Zone	Yunnan, Guizhou, Sichuan, Chongqing and Guangxi	SWEZ
H	Southern Coastal Economic Zone	Fujian, Guangdong and Hainan	SCEZ

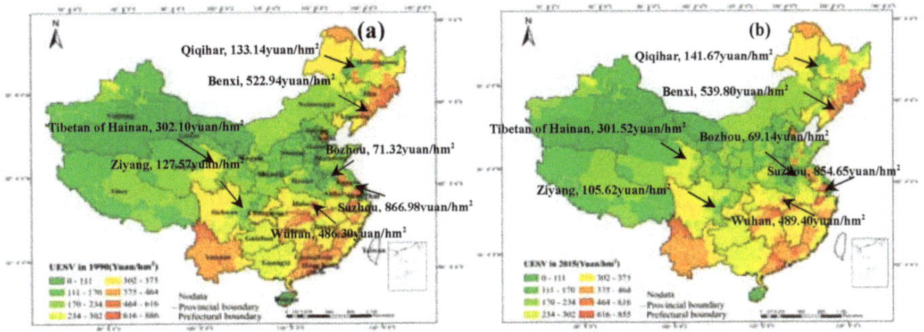

Figure 2. Spatial pattern of UESV: (**a**) UESV in 1990 and (**b**) UESV in 2015.

Table 5. The maximum and minimum values of UESV in 2015 (Units: yuan/hm^2).

	The Maximum Value of UESV	The City with the Maximum Value of UESV	The Minimum Value of UESV	The City with the Minimum Value of UESV	The UESV Value
NEEZ	539.80	Benxi	141.67	Qiqihar	325.14
NWEZ	301.52	Tibetan Autonomous Prefecture of Hainan	44.92	Jiuquan	140.13
MYREZ	347.42	Hulun Buir	51.85	Alxa League	195.25
NCEZ	429.70	Tianjin	65.46	Langfang	179.57
ECEZ	854.65	Suzhou	137.95	Yancheng	354.41
MYTREZ	609.69	Ezhou	69.14	Bozhou	323.10
SWEZ	506.71	Nujiang Lisu Nationality Autonomous	105.62	Ziyang	320.29
SCEZ	831.49	Foshan	246.22	Xiamen	420.35

Typically, in the NEEZ, the high values of UESV are probably due to the fact that there are massive forest reserves in the NEEZ and forests are an important carbon pool in China [44]. Some places in the NEEZ trended to maintain the highest average UESV of 539.80 yuan/hm^2, such as Benxi city in Liaoning province. However, some places in the NEEZ also experienced lower UESV, such as Qiqihar city in Heilongjiang province, with a UESV value of merely 141.67 yuan/hm^2. In the MYTREZ, the high values of UESV were distributed in the highly developed central cities, such as Wuhan city in Hubei province and Changsha city in Hunan province, which have well-developed waterbodies and relatively rich forest lands, with UESVs of up to 489.40 yuan/hm^2 and 360.67 yuan/hm^2, respectively. On the other hand, the lowest UESV was located in Bozhou city in Anhui province for the MYTREZ, which had few forest lands, with a UESV of only 69.14 yuan/hm^2. As for the SWEZ, the places with rich hydrothermal conditions in Yunnan tended to maintain high values of UESV. On the other hand, Ziyang city suffered the lowest USEV value of 105.62 yuan/hm^2 in the SWEZ.

The NWEZ, MYREZ and NCEZ suffered the lowest UESV. The shortage of waterbodies and massive unused land covers resulted in relatively low UESV values in the NWEZ, MYREZ and NCEZ, which were 140.13 yuan/hm^2, 195.25 yuan/hm^2 and 179.57 yuan/hm^2, respectively. The maximum UESV values in the NWEZ, MYREZ and NCEZ were merely 301.52 yuan/hm^2, 347.42 yuan/hm^2 and 429.70 yuan/hm^2 in the Tibetan Autonomous Prefecture of Hainan, Hulun Buir city and Tianjin, respectively.

The total ESV for China experienced an obvious decreasing trend from 1990 to 2015, with the average UESV dropping 1004.64 yuan/hm^2. Most of China has suffered from an approximately 10% decline in ESV (see the green area in Figure 2). Specifically, the UESV in the NWEZ, MYTREZ and SWEZ showed an obvious decreasing trend, with decreases of 1.72 yuan/hm^2, 108.62 yuan/hm^2 and 6.35 yuan/hm^2 from 1990 to 2015, respectively. Among them, the forest land and grassland decreased and the construction land increased over large areas of Hefei city, which suffered the greatest decrease from 427.08 yuan/hm^2 in 1990 to 248.31 yuan/hm^2 in 2015. However, UESV in some areas still showed slight growth. For example, the UESV in the ECEC, MYREZ, NCEZ and SCEZ grew by 22.90 yuan/hm^2, 49.83 yuan/hm^2, 54.72 yuan/hm^2 and 5.23 yuan/hm^2 from 1990 to 2015, respectively. Waterbodies enlarged Jining city in Shandong province and Zhuhai city in Guangdong province, which maintained large increases of 178.35 yuan/hm^2 and 164.31 yuan/hm^2, respectively (see Figure 3).

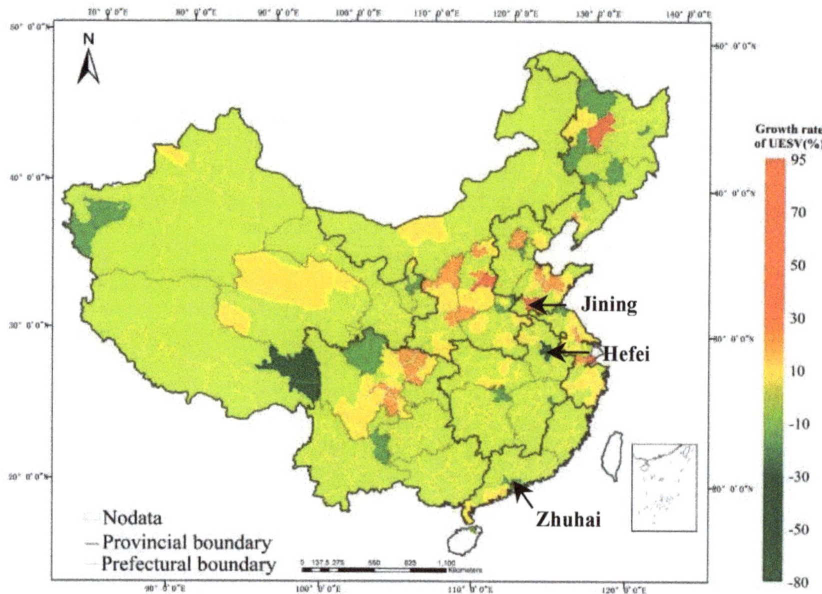

Figure 3. The spatial growth rate of UESV from 1990 to 2015.

4.2. ESV Forecasting Based on Future Land Use Change

4.2.1. Future Land Use Simulation

To reduce the uncertainty in the land use simulation, this study used the CA-Markov model to determine land suitability so as to guide the land use simulation process. The land use characteristics and geographical conditions were different in the four regions. For example, in the northeast, eastern and central regions, cultivated land was the main land use type and the transportation network was dense without steep slopes in regions. However, in the western region, most areas were covered by unused land with a low density of built-up land and transportation networks. Therefore, according to the different natural characteristics of each region, the suitability atlas calculation process differed for different regions, which was set as follows.

(1) Selection of impact factors:

There were many constraints for different land uses. Various impact factors such as the slope, distance to transportation networks and distance to the waterbody can be used to determine the constraints for land uses.

(2) Establishment of suitability maps for land uses

Cultivated land: First, the places covered with cultivated land were set to 1. Otherwise, they would be set to 0. Moreover, the places covered with unused land in Z1, Z2 and Z4 were set to 1, indicating that the unused land in those regions can be transformed into cultivated land. However, the unused land in Z3 was set to 0, since the unused land in Z3 is usually desert, frozen ground or other land uses and cannot be developed. Therefore, the places covered with unused land in Z3 were set to 0. According to the national Regulations on the Work of Water and Soil Conservation [45], cultivated land cannot be located in places with slopes larger than 25 degrees. Therefore, the suitability for cultivated land was set to 0 at places with slopes exceeding 25 degrees.

Forest land: According to the transfer matrix, the grass land can be transferred to forest land. Therefore, the places with grass land were set to 1. Meanwhile, it was then determined that the forest land tended to be located in places close to waterbodies. It is learned that in Z1, 73% of the forest land was located in places within 550 km of a waterbody. Therefore, we selected a threshold of 550 km as a constraint, indicating that a place is suitable for forest land. Otherwise, the places would be set to 0. As for Z2, Z3 and Z4, similarly, the distance thresholds for waterbodies have been set to 450 km, 880 km and 420 km when 71%, 77% and 72% forest land, respectively, is located within those distances from waterbodies. Similar to cultivated land, for forest land, the places with unused land in Z1, Z2 and Z4 were set to 1, indicating that those places are capable of being converted to forest land.

Grass land: According to the land use transfer matrix, forest land can be converted to grass land. Therefore, the places covered by forest land and grass land were set to 1 first. Meanwhile, it is also found out that the grass land tended to be located close to waterbodies. We set the threshold distance such that 70% grassland was located in the distance buffer. Specifically, the threshold distances from a waterbody for Z1, Z2, Z3 and Z4 were 550 km, 420 km, 850 km and 400 km, respectively. Last, the places covered by unused land in Z1, Z2 and Z4 were set to 1.

Waterbody: Transfers and development were not allowed for waterbodies. Therefore, the places with waterbody types were set to 1 and the other places were set to 0.

Construction land: First, the places with construction land were set to 1. It was then determined that the construction land tended to be located in the places close to transportation networks. Therefore, we selected the threshold distances from transportation networks when there was 75% construction land located in those buffer areas. Specifically, the threshold distances were 800 km, 900 km, 1350 km and 1000 km for Z1, Z2, Z3 and Z4, respectively.

The suitability maps for different land uses were obtained at the national scale and are presented in Figure 4.

The land uses in 2015 were simulated based on land use in 2010 and land use changes from 2005 to 2010 were used to verify the model (see Table 6). The Markov model was used to calculate the land use transfer matrix and then quantitatively explained the transformation among land-use types from 2005 to 2010. Then, based on the suitability maps and transfer matrix, the land use spatial pattern was simulated for 2015 (see Figure 5). The simulation results indicated that the simulation accuracies in Z1, Z2, Z3 and Z4 were 84%, 81%, 91% and 88%, respectively. According to existing studies, the accuracies of our model were acceptable [46,47]. Therefore, the model can be used for further forecasting of the spatial distribution.

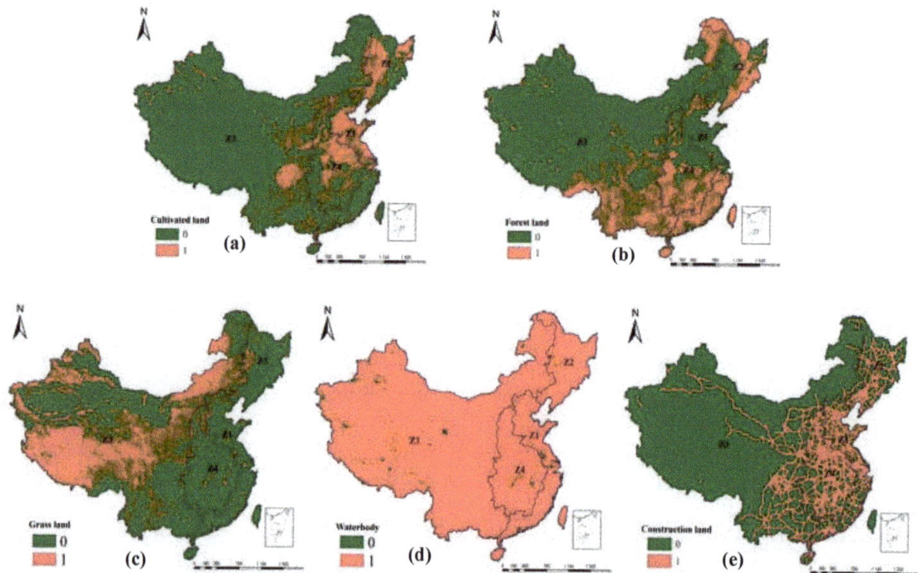

Figure 4. Suitability maps for different land uses based on binary and continuous values in the CA-Markov model (the red raster indicates the places that are suitable for the specific land use, while the green raster indicates the places that are not suitable for the specific land use): (**a**) cultivated land, (**b**) forestland, (**c**) grassland, (**d**) waterbody and (**e**) construction land.

Table 6. Transition probability matrix of land use change from 2005 to 2010.

	Land Use Type	2010					
		Cultivated Land	Forest Land	Grass Land	Waterbody	Construction Land	Unused Land
2005	Cultivated land	0.9939	0.001	0.0007	0.0006	0.004	0.0001
	Forest land	0.0007	0.9983	0.0006	0.0012	0.0005	0.0001
	Grass land	0.0023	0.0007	0.9942	0.0015	0.0002	0.0008
	Waterbody	0.0015	0.0002	0.0014	0.9946	0.0016	0.0017
	Construction land	0.0006	0.0004	0.0034	0.0009	0.9942	0.0003
	Unused land	0.0012	0	0.001	0.0013	0.0002	0.9976

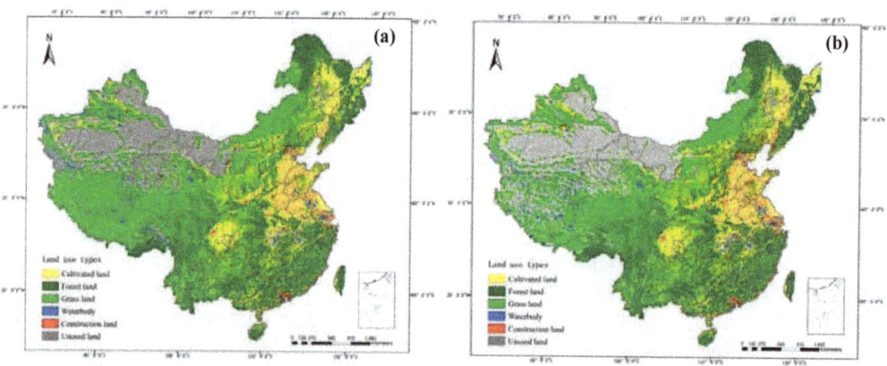

Figure 5. Spatial distribution of land use: (**a**) actual land use pattern and (**b**) prediction of land use change in 2015.

The verified model with land use transfer matrix from 2005 to 2010 was then used to simulate the future land use patterns for Z1, Z2, Z3 and Z4. The results indicated that the areas of cultivated land, grassland and unused land in the next 30 years will dynamically decrease with decline rates of −8.71%, −2.97% and −9.28%, respectively (see Table 7). By contrast, forest land, waterbody and construction land will increase by 11.80%, 3.99% and 96.53%, respectively (see Table 7). Some regions in particular experienced sharp growth in construction land, such as Z1 and Z3, with increases of 6.75×10^6 hm^2 and 15.48×10^6 hm^2, respectively. On the other hand, the areas of cultivated land, grass land and unused land were largely decreased in the western region, those losses being 15.42×10^6 hm^2, 12.50×10^6 hm^2 and 15.24×10^6 hm^2, respectively (see Table 8).

Table 7. Land use change from 1990 to 2050 (Units: 10^6 hm^2).

Year	1990	2015	2020	2025	2030	2035	2040	2045	2050	Growth Rate (%)
					Forecast Data from 2020 to 2050					
Cultivated land	177.18	178.6	181.25	178.43	175.35	172.29	169.4	165.65	161.75	−8.71
Forest land	225.5	224.02	228.99	230.7	232.04	237.24	240.44	244.58	252.1	11.80
Grass land	300.19	295.4	264.93	268.33	267.27	258.39	262.57	276.81	291.28	−2.97
Waterbody	27.05	28.01	26.9	27.41	27.74	27.17	27.08	26.98	28.13	3.99
Construction land	15.86	22.19	25.01	25.46	26.54	27.46	28.63	29.93	31.17	96.53
Unused land	200.89	198.45	219.59	216.34	217.73	224.12	218.55	202.72	182.24	−9.28

Table 8. Land use growth changes in Z1, Z2, Z3 and Z4 from 1990 to 2050 (Units: 10^6 hm^2).

	Cultivated Land	Forest Land	Grass Land	Waterbody	Construction Land	Unused Land
Z1	−6.52	2.81	−3.02	0.01	6.75	−0.03
Z2	0.18	2.36	−2.82	−0.15	3.28	−2.85
Z3	−15.42	26.61	−12.50	1.07	15.48	−15.24
Z4	−11.78	7.80	−3.66	−0.06	9.31	−1.61

4.2.2. Forecasted ESV from 2020 to 2050

Based on the simulated future land use pattern, the ESV from 2020 to 2050 was measured. In terms of spatial variation, it was found that the high values of UESV were mainly distributed in coastal and northeast China, such as the SCEZ and ECEZ (see Figure 6). Specifically, the UESV in the SCEZ would be up to 421.35 yuan/hm^2 in 2050, with the UESV in the ECEZ following at 361.85 yuan/hm^2 in 2050. According to our results, the overall pattern of UESV was ranked as the SCEZ > ECEZ > MYTREZ > NEEZ > SWEZ > MYREZ > NCEZ > NWEZ in 2050.

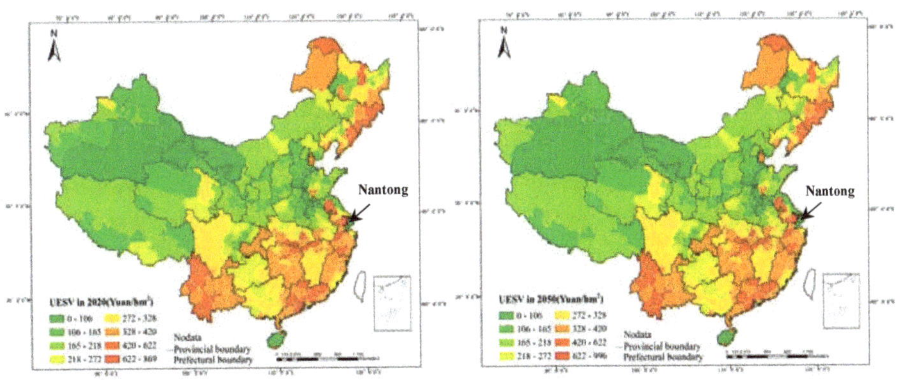

Figure 6. Spatial pattern of UESV: (**a**) forecasting the spatial pattern of UESV in 2020 and (**b**) UESV forecast for 2050.

In terms of temporal variation, the UESV was gradually decreased from 2020 to 2050, with a total reduction of 1108 yuan/hm^2 and rate of growth of -1.21% (Table 9). The ESV in the SWEZ, NCEZ and ECEZ showed a decreasing trend, with reducing rates of -10.53%, -1.968% and -1.512%, respectively. As an example, the maximum decline rate of Nantong city was -43.91% (see Figure 6). The UESV in the NWEZ showed a slight downward trend, with reductions of -0.268%.

Table 9. The variations in UESV in regions from 2020 to 2050 (Units: Yuan/hm^2).

Economic Zone	2020	2025	2030	2035	2040	2045	2050	Growth Rate (%)
NEEZ	319.80	324.64	324.26	320.70	324.79	322.71	320.63	0.259
NWEZ	144.48	144.04	147.06	144.06	144.06	144.59	144.09	−0.270
MYREZ	196.89	197.31	197.32	209.28	197.32	197.37	197.26	0.188
NCEZ	173.76	170.63	171.78	170.42	170.59	170.19	170.34	−1.967
ECEZ	367.40	361.78	361.90	361.97	362.23	362.06	361.85	−1.511
MYTREZ	324.57	325.18	325.19	325.14	317.93	323.50	325.05	0.148
SWEZ	324.31	333.80	324.39	321.37	324.37	324.28	290.16	−10.530
SCEZ	421.33	421.41	421.59	411.44	421.46	421.53	421.35	0.004

A slight upward trend of UESV was detected in the NEEZ, MYREZ, MYTREZ and SCEZ with increases of 0.259%, 0.188%, 0.148% and 0.004%, respectively. For the SCEZ and NEEZ, it is probably because of increasing forest land in these two regions. Specifically, in the SCEZ and NEEZ, the forest land increases by 1.38×10^6 hm^2 and 3.21×10^6 hm^2, respectively. A larger area of forest land (with e_i of $108{,}095.26 \times 10^8$ yuan/hm^2) would lead to a higher value of ecological service. On the other hand, even if the construction land increases by 1.30×10^6 hm^2 and 1.78×10^6 hm^2 in both regions, it trended to cover the other land uses with lower ecological service values, such as unused land (with e_i of 7357.97×10^8 yuan/hm^2) and cultivated land (with e_i of 0 yuan/hm^2). Therefore, an increasing trend has been detected in these regions.

4.3. Temporal and Spatial Variations of GDP from 1990 to 2015

The results indicated that in 2015, the UGDP decreased from the southeast to the northwest and the high values of UGDP were mainly concentrated in the NEEZ, ECEZ and SWEZ (see Figure 7). In addition, it was also found that the UGDP trended to be much higher in one or two cities than the UGDPs of the surrounding cities. For example, the average UGDPs in Shenzhen city (3051.39 thousand yuan/hm^2) and Guangzhou city (877.10 thousand yuan/hm^2) were the highest in Guangdong province, whereas the average UGDP for other cities in Guangdong province was merely 211.18 thousand yuan/hm^2 in the period from 1990 to 2015.

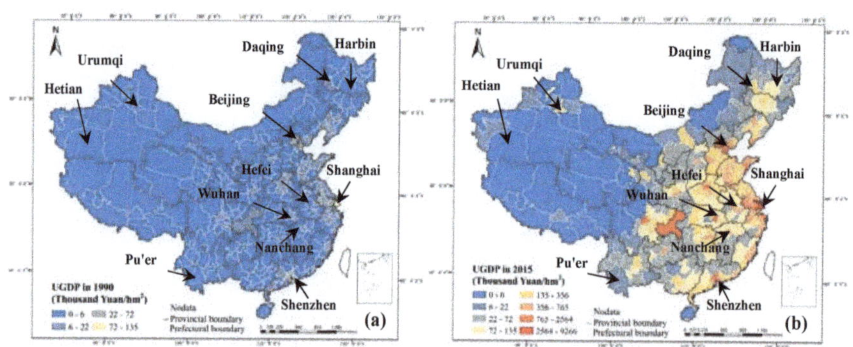

Figure 7. Spatial distribution of UGDP in China: (**a**) UGDP in 1990 and (**b**) UGDP in 2015.

The spatial differences in UGDP in China became more severe from 1990 to 2015. For example, the maximum UGDP of Shenzhen city was up to 72.00 thousand yuan/hm^2 in Guangdong province and the maximum UGDP of Urumqi city in Xinjiang province was merely 4.00 thousand yuan/hm^2 in 1990; that gap was up to 9077.00 thousand yuan/hm^2 in 2015. Different growth rates in different regions led to this spatial difference in UGDP in 2015.

The UGDP showed an obvious upward trend in the ECEZ, SCEZ and NCEZ. For example, Shenzhen city, Shanghai city and Beijing city experienced the largest growth, with increases of 9193.81 thousand yuan/hm^2, 4329.82 thousand yuan/hm^2 and 1369.45 thousand yuan/hm^2 from 1990 to 2015, respectively.

The growth rates of the NEEZ and MYTRZE were moderate. Traditional industrial cities in the NEEZ experienced higher development than other cities in NEEZ. For example, the UGDPs in Harbin city and Daqing city were increased by 106.23 thousand yuan/hm^2 and 133.78 thousand yuan/hm^2, with growth rates of 98.05% and 94.97%, respectively. Capital cities, for example, Wuhan in Hubei province, Hefei in Anhui province and Nanchang city in Jiangxi province in the MYTREZ, experienced greater growth than did other cities in the MYTREZ and accounted for 41.13%, 13.20%, 30.79% and 29.8% of the total UGDP for the whole region of the MYTREZ in 2015.

However, the NWEZ, SWEZ and MYREZ experienced relatively low growth rates. For example, the UGDP of Hetian in Xinjiang province, Pu'er in Yunnan province and Yushu in Qinghai province experienced the lowest growths of 0.94 thousand yuan/hm^2, 0.91 thousand yuan/hm^2 and 0.73 thousand yuan/hm^2 from 1990 to 2015, respectively.

4.4. GDP Forecasting from 2020 to 2050

Based on the temporal changes in GDP, a linear regression was built up for each city to predict the GDP in 2020 to 2050. The values of R square and significant P for the regressions of each city are presented in Figure 8a. It is noted that the values of R square for most of the cities were in the range of 0.72 and 0.99, indicating a high fitting accuracy. Specifically, the R square was larger in the highly developed cities of China, such as Beijing (with R square of 0.95), Shanghai (with R square of 0.98) and Guangzhou (with R square of 0.95). However, R square was lower in those eastern cities suffering depressed development, such as Qitaihe in Heilongjiang (with R square of 0.72), Liupanshui in Guizhou province (with R square of 0.75) and Xiangyang in Hubei province (with R square of 0.76). In statistics, a significant P less than 0.05 and greater than 0.01 indicates statistical significance and P less than 0.01 indicates very significant. However, the significant P exceeded 0.05, indicating that the difference was not significant. The significant P was closer to 0.01 in most of the highly developed cities (see Figure 8b), such as Beijing (with P of 0.007), Shanghai (with P of 0.003) and Guangzhou (with P of 0.007). However, values of P exceeded 0.05 in the economically under-developed areas, for example, Haixi Mongol (with P of 0.266) in Qinghai province, Laibin (with P of 0.154) in the Guangxi Zhuang Autonomous Region and Tongren (with P of 0.157) in Guizhou province. According to the regression model, the GDP value for each city from 2020 to 2050 was predicted.

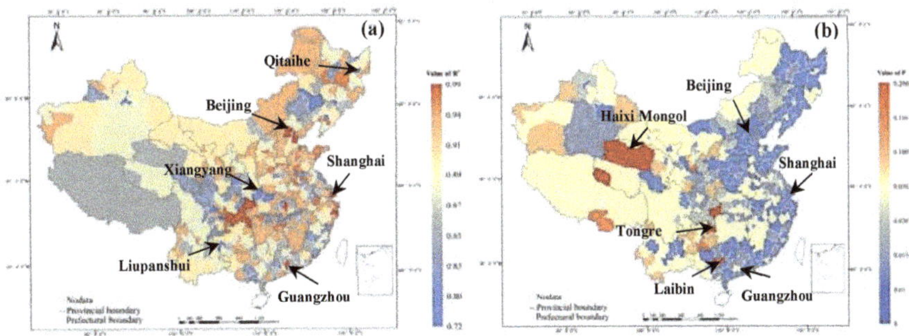

Figure 8. The fitting values of the R square relationship and significant P between the UGDP and the time series: (**a**) R square and (**b**) significant P.

The predicted GDP was presented as the UGDP value for each city in 2020 and 2050 (see Figure 9). According to the results, a huge increase in GDP has been detected—the total national GDP increases from 85,271.13 billion yuan in 2020 to 180,718.28 billion yuan in 2050. The increase is mainly concentrated in the highly developed cities, such as Beijing city, Shanghai city, Guangzhou city and Hangzhou city and the sum of GDP values is as large as 21,847.24 billion yuan, contributing 23.76% of the entire national GDP value in 2050.

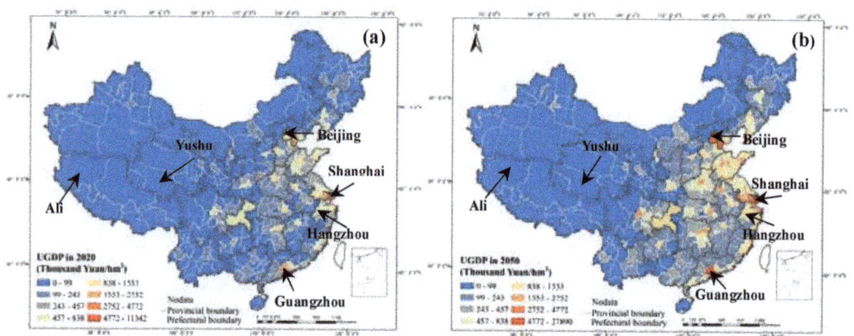

Figure 9. Spatial distributions of UGDP in China: (**a**) UGDP in 2020 and (**b**) UGDP in 2050.

As a result, the spatial differences in UGDP become stronger from 2020 to 2050. The high value regions of UGDP were mainly distributed in the ECEZ and SCEZ, with values of 45,061.62 thousand yuan/hm^2 and 68,205.96 thousand yuan/hm^2, respectively, in 2050. High UGDP values in the ECEZ and SCEZ are primarily found in Shanghai city, Hangzhou city and Guangzhou city; for 2050, their values were 9949.94 thousand yuan/hm^2, 1279.45 thousand yuan/hm^2 and 5405.24 thousand yuan/hm^2, respectively. In contrast, the UGDP values in the NWEZ and SWEZ were only 4363.51 thousand yuan/hm^2 and 16,449.02 thousand yuan/hm^2, respectively, in 2050. In the NWEZ and SWEZ, the lowest UGDPs were located in the blue area in Figure 9, such as Ali in Tibet province and Yushu in Qinghai province, with values of 0.33 thousand yuan/hm^2 and 0.97 thousand yuan/hm^2, respectively, in 2050. In the ECEZ and SCEZ, the difference between the high value of UGDP and the low value of UGDP in the NWEZ and SWEZ was 46,135.58 thousand yuan/hm^2 in 2020 and then the difference was 93,689.98 thousand yuan/hm^2 in 2050; therefore, the spatial difference was greater between 2020 and 2050.

4.5. Temporal and Spatial Variations of Green GDP from 1990 to 2050

The green GDP was measured according to ESV and GDP at the prefectural level and a huge increase has been revealed from 1990 to 2050. It was found that the green GDP increased from 19,623.94 billion yuan in 1990 to 194,978.29 billion yuan in 2050. Even if the ESV was decreasing, due to the rapid economic development growth, the national green GDP was increasing.

In terms of spatial distribution (see Figure 10), high values of green GDP tended to be located in the south China and east coastal regions, such as Beijing city, Shenzhen city and Guangzhou city, having green GDP of 5006.06 billion yuan, 5273.93 billion yuan and 3910.14 billion yuan in 2050. In addition, low values of green GDP tended to be located in west China, such as Jinchang and Jiayuguan in Gansu province, having green GDP of 75.33 billion yuan and 63.50 billion yuan in 2050. To be specific, the economic zones can be ranked as follows according to their green GDP values in 2050: NCEZ > MYTREZ > SWEZ > MYREZ > ECEZ > SCEZ > NEEZ and NWEZ (Table 10). It was obvious that places with higher green GDP values tended to experience stronger spatial differences or spatial inequalities. For example, the Ugreen GDP in Wuhan was up to 3104.02 thousand yuan/hm^2 in 2050 but the Ugreen GDPs for cities surrounding Wuhan city were relatively low, such as for Huanggang city, with a Ugreen GDP of 492.19 thousand yuan/hm^2 in 2050 (see Figure 10).

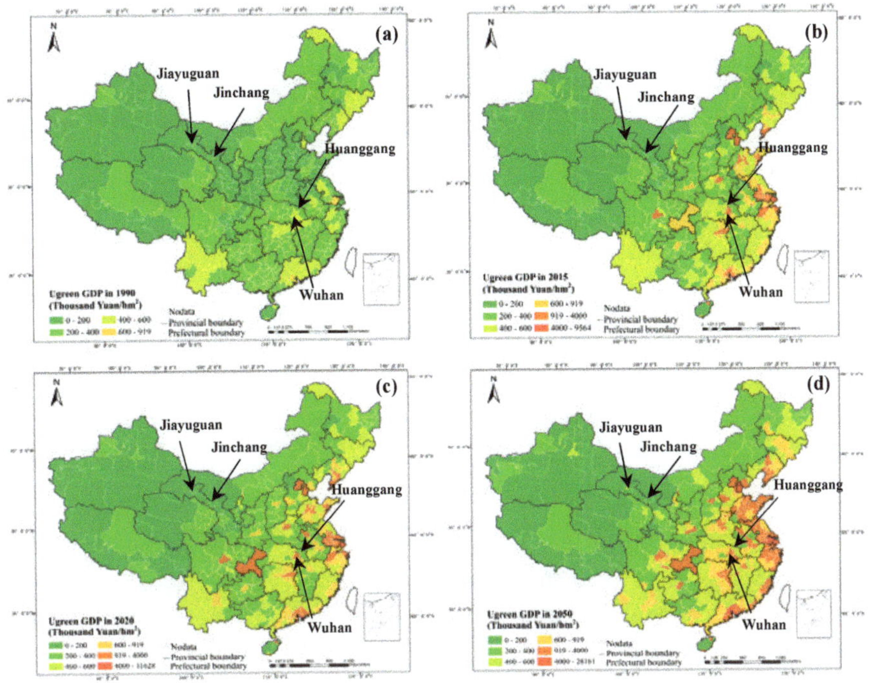

Figure 10. National Ugreen GDP map: (**a**) Ugreen GDP map for 1990, (**b**) Ugreen GDP map for 1995, (**c**) Ugreen GDP map for 2020 and (**d**) Ugreen GDP map for 2050.

Table 10. Statistics of total ESV, GDP and green GDP in eight districts in 1990 and 2050 (Billion Yuan) and the proportions of ESV and GDP in green GDP (%).

	1990					2050				
	ESV	Proportion of ESV in Green GDP	GDP	Proportion of GDP in Green GDP	Value of Green GDP	ESV	Proportion of ESV in Green GDP	GDP	Proportion of GDP in Green GDP	Green GDP
NEEZ	2547.46	93.29	183.10	6.71	2730.56	2489.93	15.66	13,414.89	84.34	15,904.82
NWEZ	3580.29	99.06	33.99	0.94	3614.28	5747.21	47.10	6456.22	52.90	12,203.43
MYREZ	3184.91	97.72	74.33	2.28	3259.24	3268.02	12.66	22,540.06	87.34	25,808.08
NCEZ	599.23	71.15	242.93	28.85	842.16	620.36	1.73	35,244.08	98.27	35,864.44
ECEZ	698.20	71.13	283.33	28.87	981.53	733.46	2.80	25,457.61	97.20	26,191.07
MYTREZ	2264.68	95.78	99.89	4.22	2364.57	2270.84	8.46	24,566.68	91.54	26,837.52
SWEZ	4298.28	97.66	103.05	2.34	4401.33	3920.97	14.71	22,726.93	85.29	26,647.87
SCEZ	1242.60	86.88	187.67	13.12	1430.27	1250.80	4.90	24,270.26	95.10	25,521.06
Sum	18,415.65	–	1208.29	–	19,623.94	20,301.59	–	174,676.73	–	19,4978.29

The huge differences in green GDP values between one city and the surrounding cities may lead to numerous social problems. Being aware of highly spatial differences in green GDP in south China and coastal China and low green GDPs in west China can help the government when making decisions concerning environmental preservation and economic promotion.

The regional differences in the contributions of GDP and ESV to green GDP were analyzed. The proportion of ESV and GDP in green GDP changed significantly in regions (see Table 10 and Figure 11). In 1990, for most cities in China, the ESV accounted for 80% to 95% of green GDP. However, when it comes to 2015, due to the drastic economic growth in China, the contribution of ESV has been hugely diminished (see Figure 11).

In regions with high green GDP, the contribution of GDP to green GDP was increased gradually, such as for the SCEZ, ECEZ, NCEZ and the MYTREZ. For example, in the NCEZ, the proportion of GDP accounting for the green GDP had a large increase from 28.85% in 1990 to 98.27% to 2050.

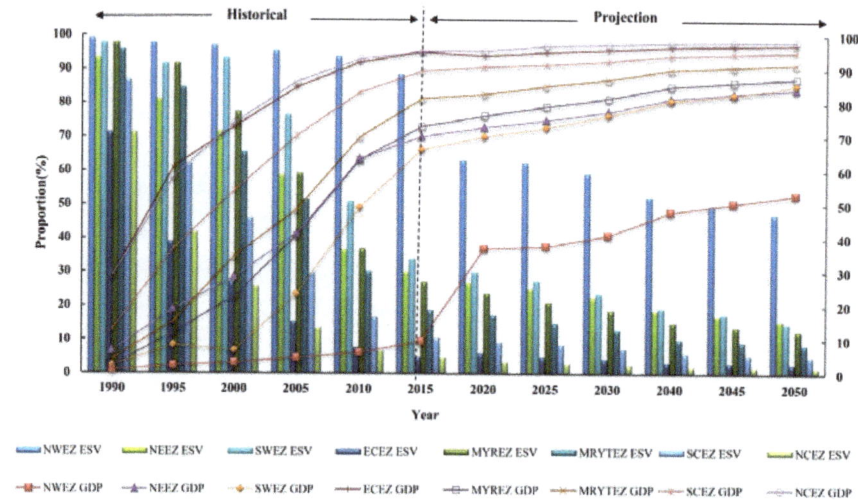

Figure 11. ESV and GDP as a percentage of green GDP from 1990 to 2050.

Cities with reducing ESV values and dramatically increasing GDP experienced a huge growth in green GDP, such as Shenzhen, Chongqing and Beijing (see Figure 12). Taking Shenzhen as an example, it demonstrated that the proportion of GDP in green GDP was 98.30%.

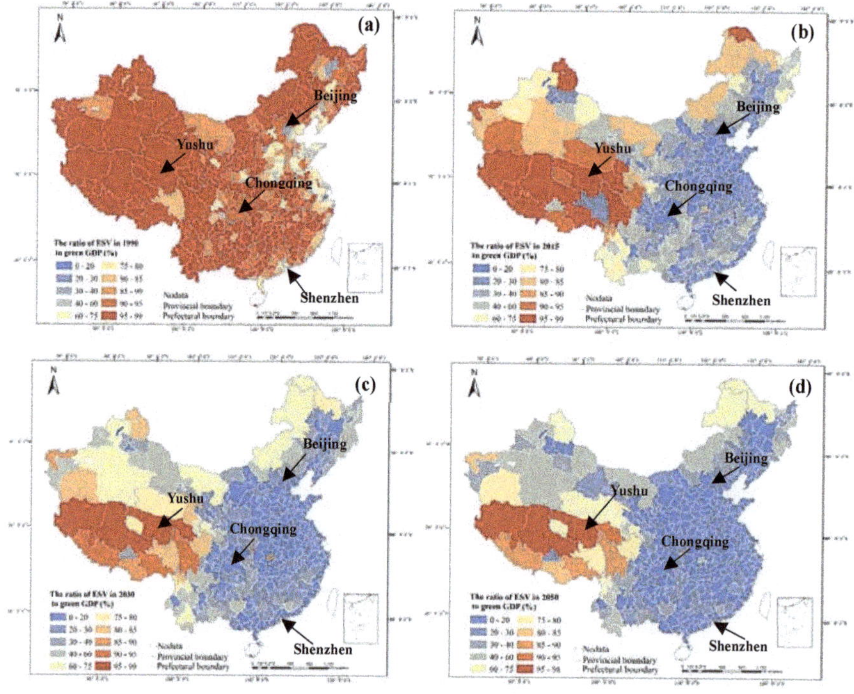

Figure 12. The ratio of ESV to green GDP, (**a**) the ratio of ESV to green GDP in 1990, (**b**) the ratio of ESV to green GDP in 2015, (**c**) the ratio of ESV to green GDP in 2030 and (**d**) the ratio of ESV to green GDP in 2050.

However, on the other hand, the green GDP in regions with low GDP was largely dependent on the variation of ESV and the drop in ESV would lead to a direct decrease in green GDP. For example, in the NWEZ, the ESV accounted for 99.06% of green GDP in 1990 and the contribution would drop to 47.10% in 2050. Consider, as an example, Yushu in Qinghai province, the contribution rate of ESV to green GDP of which was 95.20% in 1990 and 92.57% in 2050 and ESV has always occupied a high proportion.

5. Conclusions and Discussion

This study proposed a green GDP accounting and predicting system for prefecture-level cities for China. In the proposed method, we used the coefficient method to measure the ESV according to the output of local agricultural crops and land use patterns; on the other hand, the GDP value was obtained from statistical yearbooks in China. To measure the future green GDP value, we integrated the land use simulation model, CA-Markov, into the green GDP accounting system. The CA-Markov model was used to predict the future land use pattern, which has been taken as input in the coefficient method to measure the future ESV.

The results indicated that the green GDP value experienced a huge increase from 1990 to 2015 and the forecasting green GDP also displayed a dramatic growth from 2020 to 2050. A totally different increase rate has been found in regions of China. High increase rates of green GDP were concentrated in SCEZ, ECEZ and NCEZ, with growth rates of 87.96%, 93.33% and 93.90%, respectively, from 1990 to 2015 and 94.40%, 96.25% and 97.65%, respectively, from 2015 to 2050. However, in NWEZ and SWEZ, the lowest growth rates of 1.09% and 65.59%, respectively, from 1990 and 2015 and 33.56% and 80.92%, respectively, from 2020 to 2050 have been found. Consequently, the spatial differences in green GDP in regions became increasingly serious. The spatial differences of green GDP also have been detected in some other studies. Li and Fang [9] estimated the global ESV and GDP in 2009. According to the maps of global green GDP presented by Li and Fang [10], the coastal areas, such as Zhejiang and Jiangsu provinces, experienced the highest green GDPs—up to 10,000 to 17,299 US dollar per sq.km in 2009. However, in west China, the green GDP ranged from 0 to 200 US dollars per sq.km in 2009.

Meanwhile, the results indicated a dominant contribution of economic development to green GDP. In the ECEZ, NCEZ and SCEZ, the contribution of GDP to green GDP was up to 90%. A high contribution of GDP to green GDP would depress the importance of preserving ESV. On the contrary, in east China, such as Yunfu in Guangdong, Quzhou in Zhejiang and Huaian in Jiangsu, due to the slow economic development, the contributions of GDP to green GDP were merely 31.91%, 43.40% and 43.89% in 2050. In those regions, the green GDP is dependent on preservation of the ESV to a large extent. Being aware of the contribution of ESV to green GDP in different regions can help the government develop different policies for different regions. Li and Ding [13] evaluated the variations of ESV and GDP for Shananxi province from the 1980s to 2010 and found that disparities in GDP were much higher than those of ESV, which is consistent with our results.

The results proposed in this study can facilitate governmental decision making. Huge spatial differences in the future detected by our study would lead to abundant social problems. Being aware of the differences in green GDP, GDP and ESV in regions, a decision maker can determine the hotspots and make policies accordingly to preserve GDP or ESV.

Author Contributions: Conceptualization, Y.Y., M.Y. and L.L.; methodology, Y.Y., J.C. and D.L.; software, M.Y., J.C. and D.L.; validation, W.Z. and K.C.; formal analysis L.L.; resources, Y.Y.; writing-original draft, Y.Y.; writing-review & editing, W.Z. and K.C.; supervision, K.C.; funding acquisition, W.Z.

Funding: This research was funded by the National Natural Science Foundation of China (No. 41601415), the Open Fund of the Key Laboratory of Urban Land Resources Monitoring and Simulation, Ministry of Land and Resources (No. KF-2018-03-028) and the Student Research Fund of Huazhong Agricultural University (No. 2019099).

Acknowledgments: The authors would like to thank Data Central, Institute of Geographic Sciences and Natural Resources Research, the Chinese Academy of Sciences, for providing the land-use datasets.

Conflicts of Interest: The authors declare no conflicts of interest.

References

1. National Bureau of Statistics of China. *China Statistical Yearbook*; China Statistics Press: Beijing, China, 2016.
2. Sun, W.; Miao, Z.; Sun, W.; School, B. Change in the value of ecosystem services of Beijing-Tianjin-Hebei area and its relationship with economic growth. *Ecol. Econ.* **2015**, *31*, 59–62.
3. United Nations, E.C. *International Monetary Fund, Organisation for Economic Cooperation and Development*; World Bank, System of National Accounts: New York, NY, USA, 1993.
4. Bartelmus, P. SEEA-2003: Accounting for sustainable development? *Ecol. Econ.* **2007**, *61*, 613–616. [CrossRef]
5. Sutton, P.C.; Costanza, R. Global estimates of market and non-market values derived from nighttime satellite imagery, land cover, and ecosystem service valuation. *Ecol. Econ.* **2002**, *41*, 509–527. [CrossRef]
6. Kunanuntakij, K.; Varabuntoonvit, V.; Vorayos, N.; Panjapornpon, C.; Mungcharoen, T. Thailand green gdp assessment based on environmentally extended input-output model. *J. Clean. Prod.* **2017**, *167*, 970–977. [CrossRef]
7. Lin, W.B.; Yang, J.; Chen, B. Temporal and spatial analysis of integrated energy and environment efficiency in China based on a Green GDP index. *Energies* **2011**, *4*, 1376–1390. [CrossRef]
8. Talberth, J.; Bohara, A.K. Economic openness and green gdp. *Ecol. Econ.* **2006**, *58*, 743–758. [CrossRef]
9. Li, V.; Lang, G. China's "Green GDP" Experiment and the Struggle for Ecological Modernisation. *J. Contemp. Asia.* **2010**, *40*, 44–62. [CrossRef]
10. Li, G.; Fang, C. Global mapping and estimation of ecosystem services values and gross domestic product: A spatially explicit integration of national 'green GDP' accounting. *Ecol. Indic.* **2014**, *46*, 293–314. [CrossRef]
11. Yang, C.; Poon, J.P.H. Regional Analysis of China's Green GDP. *Eur. Geogr. Econ.* **2009**, *50*, 54–563. [CrossRef]
12. Xu, L.; Bing, Y.; Yue, W. A method of green GDP accounting based on eco-service and a case study of Wuyishan, China. *Procedia Environ. Sci.* **2010**, *2*, 1865–1872. [CrossRef]
13. Li, T.; Ding, Y. Spatial disparity dynamics of ecosystem service values and GDP in Shaanxi Province, China in the last 30 years. *PLoS ONE* **2017**, *12*, e0174562. [CrossRef] [PubMed]
14. Hou, Q.; Zhou, Q. Spatio-temporal relationships between urban growth and economic development: A case study of the Pearl River Delta of China. *Asian Geogr.* **2012**, *29*, 57–69. [CrossRef]
15. Arowolo, A.O.; Deng, X.; Olatunji, O.A.; Obayelu, A.E. Assessing changes in the value of ecosystem services in response to land-use/land-cover dynamics in nigeria. *Sci. Total Environ.* **2018**, *636*, 597–609. [CrossRef] [PubMed]
16. Xu, D.; Ding, X. Assessing the impact of desertification dynamics on regional ecosystem service value in north china from 1981 to 2010. *Ecosyst. Serv.* **2018**, *30*, 172–180. [CrossRef]
17. Anderson, S.J.; Ankor, B.L.; Sutton, P.C. Ecosystem service valuations of south africa using a variety of land cover data sources and resolutions. *Ecosyst. Serv.* **2017**, *27*, 173–178. [CrossRef]
18. Yang, X.; Zheng, X.Q.; Chen, R. A land use change model: Integrating landscape pattern indexes and markov-ca. *Ecological Model.* **2014**, *283*, 1–7. [CrossRef]
19. Halmy, M.W.A.W. Land use/land cover change detection and prediction in the north-western coastal desert of egypt using markov-ca. *Appl. Geogr.* **2015**, *63*, 101–112. [CrossRef]
20. Qing-Sheng, Y.; Xia, L. Integration of multi-agent systems with cellular automata for simulating urban land expansion. *Sci. Geogr. Sin.* **2017**, *27*, 542–548.
21. Kamusoko, C.; Aniya, M.; Adi, B.; Manjoro, M. Rural sustainability under threat in zimbabwe—Simulation of future land use/cover changes in the bindura district based on the markov-cellular automata model. *Appl. Geogr.* **2009**, *29*, 435–447. [CrossRef]
22. Burstedde, C.; Klauck, K.; Schadschneider, A.; Zittartz, J. Simulation of pedestrian dynamics using a two-dimensional cellular automaton. *Phys. A Stat. Mech. Appl.* **2001**, *295*, 507–525. [CrossRef]
23. Kirchner, A.; Schadschneider, A. Simulation of evacuation processes using a bionics-inspired cellular automaton model for pedestrian dynamics. *Phys. A Stat. Mech. Appl.* **2002**, *312*, 260–276. [CrossRef]
24. Barlovic, R.; Santen, L.; Schadschneider, A.; Schreckenberg, M. Metastable states incellular automata for traffic flow. *Eur. Phy. J. B.* **1998**, *5*, 793–800. [CrossRef]
25. Karafyllidis, I.; Thanailakis, A. A model for predicting forest fire spreading using cellular automata. *Ecol. Model.* **1997**, *99*, 87–97. [CrossRef]
26. Tobler, W.R. Cellular geography. In *Philosophy in Geography*; Gale, S., Olssen, G., Eds.; D. Reidel Publishing Co.: Dortrecht, Holland, 1979; pp. 379–386.

27. Clarke, K.C.; Gaydos, L.J. Loose-coupling a cellular automaton model and gis: Long-term urban growth prediction for san francisco and washington/baltimore. *Int. J. Geogr. Inf. Syst.* **1998**, *12*, 16. [CrossRef] [PubMed]

28. Li, X.; Yeh, A. Neural-network-based cellular automata for simulating multiple land use changes using gis. *Int. J. Geogr. Inf. Syst.* **2002**, *16*, 323–343. [CrossRef]

29. Etemadi, H.; Smoak, J.M.; Karami, J. Land use change assessment in coastal mangrove forests of iran utilizing satellite imagery and ca–markov algorithms to monitor and predict future change. *Environ. Earth Sci.* **2018**, *77*, 208. [CrossRef]

30. Shafizadeh Moghadam, H.; Helbich, M. Spatiotemporal urbanization processes in the megacity of mumbai, india: A markov chains-cellular automata urban growth model. *App. Geogr.* **2013**, *40*, 140–149. [CrossRef]

31. Xuan, F.H. The coupling development between tourism economy and ecological environment in harbin city. *Adv. Mater. Res.* **2013**, *4*, 726–731. [CrossRef]

32. Department of Rural Surveys, National Bureau of Statistics. *China Yearbook of Agricultural Price Survey*; China Statistics Press: Beijing, China, 2016.

33. Costanza, R.; d'Arge, R.; Groot, R.d.; Farberk, S.; Grasso, M.; Hannon, B.; Limburg, K.; Naeem, S.; Neill, R.V.O.; Paruelo, J.; et al. The value of the world's ecosystem services and natural capital. *Nature* **1997**, *387*, 253–260. [CrossRef]

34. Xie, G.D.; Zhang, C.X.; Zhang, C.S.; Xiao, Y.; Lu, C.X. The value of ecosystem services in China. *Resour. Sci.* **2015**, *37*, 1740–1746.

35. Hou, X.Y.; Chang, B.; Yu, X.F. Land use change in Hexi corridor based on CA-Markov methods. *Trans. Chin. Soc. Agric. Eng.* **2004**, *20*, 286–291.

36. Sang, L.; Zhang, C.; Yang, J.; Zhu, D.; Yun, W. Simulation of land use spatial pattern of towns and villages based on ca-markov model. *Math. Comput. Model.* **2011**, *54*, 938–943. [CrossRef]

37. Liu, X.P.; Li, X.; Yeh, G.O.; He, J.Q.; Tao, J. Discovery of transition rules for geographical cellular automata by using ant colony optimization. *Sci. China* **2007**, *50*, 1578–1588. [CrossRef]

38. Wang, S.Q.; Zheng, X.Q.; Zang, X.B. Accuracy assessments of land use change simulation based on markov-cellular automata model. *Procedia Environ. Sci.* **2012**, *13*, 1238–1245. [CrossRef]

39. Martinez, I. Rapd-typing of central and eastern north atlantic and western north pacific minke whales, balaenoptera acutorostrata. *ICES J. Mar. Sci.* **1999**, *56*, 640–651. [CrossRef]

40. Kim, J.; Son, S.K.; Son, J.W.; Kim, K.H.; Shim, W.J.; Kim, C.H.; Lee, K.Y. Venting sites along the fonualei and Northeast Lau spreading centers and evidence of hydrothermal activity at an off-axis caldera in the Northeastern Lau Basin. *Geochem. J.* **2009**, *43*, 1–13. [CrossRef]

41. Liu, J.Y.; Ning, J.; Wen, H.; Xu, X.L.; Zhang, S.Y.; Chang, Z.; Li, R.D.; Wu, S.X.; Hu, Y.F.; Du, G.M.; et al. Spatio-temporal patterns and characteristics of land-use change in china during 2010–2015. *Acta Geogr. Sin.* **2018**, *73*, 789–802.

42. Xu, L.Z.; Fang, C.; Feng, C.; Chen, W.; Hua, Y.U.; Huang, X.H.; Zeng, Y.F.; Li, X.; Hong, S.B.; Zhong, X.F. Spatial and temporal variation of near-ground CO concentration in the Eight Economic Regions in China in May and July, 2013. *Acta Sci. Circumst.* **2014**, *34*, 1934–1941.

43. Changchun, F.; Zanrong, Z.; Nana, C. The economic disparities and their spatio-temporal evolution in china since 2000. *Geogr. Res.* **2015**, *34*, 234–246.

44. Zhao, J.F.; Yan, X.D.; Jia, G.S. Simulation of carbon stocks of forest ecosystems in northeast china from 1981 to 2002. *Chin. J. Appl. Ecol.* **2009**, *20*, 241–249.

45. Gichuki, F.N. Makueni district profile: Soil management and conservation, 1989–1998. *Drylands Res. Work. Pap. Drylands Res.* **2000**, *37*, 1243–1257.

46. Landis, J.R.; Koch, G.G. The measurement of observer agreement for categorical data. *Biometrics* **1977**, *33*, 159–174. [CrossRef] [PubMed]

47. Muhammad, R. Detection of land use/land cover changes and urban sprawl in al-khobar, saudi arabia: An analysis of multi-temporal remote sensing data. *ISPRS Int. J. Geo-Inf.* **2016**, *5*, 15.

 *sustainability*

Article

Land Eco-Security Assessment Based on the Multi-Dimensional Connection Cloud Model

Qiuyan Liu, Mingwu Wang *, Xiao Wang, Fengqiang Shen and Juliang Jin

School of Civil and Hydraulic Engineering, Hefei University of Technology, Hefei 230009, China; qiuyanliu@yeah.net (Q.L.); wxiao_hfut@163.com (X.W.); shenfq@hfut.edu.cn (F.S.); jinjl66@126.com (J.J.)
* Correspondence: wanglab307@foxmail.com or mingwu_wang@hfut.edu.cn; Tel.: +86-551-62902066

Received: 18 May 2018; Accepted: 17 June 2018; Published: 20 June 2018

Abstract: The evaluation of land eco-security is challenging because it is involved with various uncertainty factors. Although the normal cloud model provides an idea for dealing with the randomness and fuzziness of indicators for the evaluation of land eco-security, it cannot simulate the distribution state of the evaluation indicators in a finite interval and their calculation process is complicated for multi-factor problems. Herein, a novel multi-dimensional connection cloud model is discussed to remedy these defects. In this model, combined with the range of evaluation factors in each grade, the identity-discrepancy-contrary principle of set pair theory is adopted to determine the digital characteristics of the multi-dimensional cloud model, which can uniformly describe the certainty and uncertainty relationships between the measured indices and the evaluation criteria and also improve the fuzzy-randomness of evaluation indicators closer to the actual distribution characteristics. The case study and the comparison of the proposed model with the normal cloud model and the matter element model were performed to confirm the validity and reliability of the proposed model. Results show that this model can overcome the subjectivity in determining the digital characteristics of the normal cloud model, providing a novel method for the comprehensive evaluation of land eco-security.

Keywords: land eco-security; multi-dimension; cloud model; set pair theory; evaluation

1. Introduction

Eco-security is defined as the "ecologically sustainable development that meets the environmental and ecological needs of the present generation without compromising the ability of future generations to meet their own environmental and ecological needs" [1,2], which was first introduced by the Insurance Accounting and Systems Association (IASA) in the global eco-security monitoring system of 1989 [3]. Land is a scarce resource at a global scale, and the space carrier of human activities, so it is fundamental in all human existence and development [4–6]. Ecological land offers important functions, such as soil and water conservation, wind prevention and sand-fixing, cleaning air, climate regulation, and biodiversity maintenance, and it has drawn more and more attention from many scholars [7–9]. Thus, land eco-security is an essential component of overall natural eco-security, a state in which it maintains a healthy and balanced structure and function within certain spatial and temporal ranges, and provides natural conditions in which human beings live [10,11]. However, with the large extension of urbanization and the great development of human societies in the past 50 years, some irrational activities of land use have posed a dangerous threat to the health and safety of the ecosystem, resulting in a series of environmental problems, such as biodiversity reduction, soil erosion, and land contamination [6,12–15]. Based on official statistics, there are 400,000 polluted sites in the European countries [16,17]. In America, approximately 600,000 hm^2 brown field sites have been polluted with heavy metals [17–19]. Soil erosion is frequently found in Northwestern

European catchments [20]. Natural disasters, such as the 2011 Tohuku earthquake and tsunami in Japan, and subsequent nuclear meltdowns at Fukushima, can also endanger land eco-security [21]. Extreme meteorological events and disasters, such as droughts, floods, and high temperatures, have frequently occurred worldwide [13,14,22–24]. According to the 2017 World Meteorological Organization Statement on the State of the Global Climate, 2017 has recorded the most serious economic loss in the world due to severe weather and climate events. The loss is estimated to be about US$320 billion. In China, the number of sandstorms in northern cities has continued to increase, and the cities in the south have suffered serious waterlogging. In 2017, high quality arable land only accounted for 27% of the national total arable land area. The total area of eroded soil was about 2.949 million km^2, accounting for 31.1% of the total area of the land census. As of 2014, the area of desertified land nationwide was 4,337,200 km^2. Such issues have drawn increasing attention of local governments, the public, and academic communities in China [2,25]. In 2018, the report of the 19th National Congress of the Communist Party of China explicitly set the goal of building a "rich, strong, democratic, civilized, harmonious, and beautiful modern socialist country", which also incorporated ecological civilization into the Constitution and established the Ministry of Ecology and Environment. It clearly shows that China has put emphasis on environmental protection. Thus, planners and policymakers of national development are in growing need of significant and scientific knowledge about the state of land eco-security [10].

Land eco-security evaluation is both the core and the foundation of the sustainable utilization of land resources [10]. Many scholars take full advantage of mathematics, computer science, and ecology in the study of land eco-security so that research results have been continuously enriched. In 1996, Canadian ecological economist William E. Rees proposed an ecological footprint calculation model to measure sustainable development. Based on the ecological footprint method, Li et al. [26] proposed an ecological pressure index method for the evaluation of land eco-security of Shandong Province. However, this method only measured the degree of ecological sustainability, emphasizing the impact of human development on the environmental system without considering economic, social, and technological sustainability [27–29]. Park et al. [30] suggested a linkage between the grey method and the artificial neural-network model in regional eco-environmental quality assessment. Additionally, the neural-network method needs historical data, which presents a problem in using existing domain knowledge in the learning process. To overcome these shortcomings, the fuzzy comprehensive evaluation method [31–33] and the matter element method [34–36] were used to analyse land eco-security, and the objective and reasonable results have been obtained. Although the fuzzy comprehensive evaluation method can quantify the fuzzy uncertainty, its result is dependent on the rational definition of membership function. The matter element method might miss some constraint conditions during the evaluation procedure so that it may lead to the deviation of the evaluation results from reality. Due to the effect of the subjective factor and multiple evaluation factors, the evaluation of land eco-security consists of uncertainties of both fuzziness and randomness. To depict fuzziness and randomness of evaluation indicators, Gao et al. [37] presented a one-dimensional normal cloud model for the land eco-security assessment. However, the traditional normal cloud model may have the disadvantages of being both cumbersome and slow in the calculation of the evaluation process with the increase in evaluation indices and samples.

This study is aimed at introducing a novel multi-dimensional connection cloud model for the assessment of land eco-security. The authors intend to develop a novel cloud model that can dialectically describe the fuzziness and randomness in evaluation indicators over a finite interval, and the certainty and uncertainty relationships between the measured evaluation indicators and each classification standard in a unified way on the basis of set pair theory. The proposed method can also express the conversion tendency of the result on the classification boundary. Moreover, the feasibility and validity of the proposed method are further discussed by case study taking the Wanjiang region as an example, followed by a comparison with the one-dimensional normal cloud model and the matter element model.

2. Methodology

The cloud model proposed by Li [38] is a useful tool for transforming a qualitative concept into quantitative data, and widely adopted to analyse problems with characteristics of fuzziness and randomness [39]. The traditional one-dimensional cloud model has the obvious deviation of the evaluation result from reality when there is a considerable difference in the interval spans of the evaluation levels. Therefore, the multi-dimensional normal cloud model is proposed to overcome the above problems by comprehensively considering all the evaluation factors. In addition, in the past, there was no substantial basis for the selection of the digital characteristics of the multi-dimensional normal cloud model, so that the simulation results may have considerable contingency and limitations. The set pair theory developed recently provides an idea for overcoming the deficiencies of the above normal cloud model since it has an advantage in terms of a unified description of the certainty and uncertainty relationships [40,41]. For this reason, this paper couples the normal cloud model with the set pair theory to present a multi-dimensional connection cloud model, which makes the selection of the digital characteristics of cloud model more objective, and makes the evaluation result more accurate and reliable. The definition of the calculation model is shown below.

Let $U = \{x_1, x_2, \ldots, x_m\}$ be an m dimension universe of discourse with precise values, and C be a qualitative concept in U. If $x \in U$ is a random instantiation of concept C, which satisfies $X(x_1, x_2, \ldots, x_m) \sim N(Ex(Ex_1, Ex_2, \ldots, Ex_m), (En'(En'_1, En'_2, \ldots, En'_m))^2)$, and, $En'(En'_1, En'_2, \ldots, En'_m) \sim N(En(En_1, En_2, \ldots, En_m), (He(He_1, He_2, \ldots, He_m))^2)$, and the certainty degree of x belonging to concept C satisfies:

$$\mu[x(x_1, x_2, \ldots, x_m)] = e^{-\sum_{j=1}^{m} \frac{(x_j - Ex_j)^2}{2(En'_j)^2}}, \quad (j = 1, 2, 3, \ldots, m), \tag{1}$$

then the distribution of x in the universe U is called the multi-dimensional normal cloud [42]. Here, set pair theory is introduced into the process of establishing the multi-dimensional cloud model, and this multi-dimensional cloud is called the multi-dimensional connection cloud. Namely, the entropy value En is determined according to the identity-discrepancy-contrary relationship between the simulation cloud drop and the evaluation interval, and the possibility that the simulated cloud drop is on the boundary of the discussed level and adjacent level is 0.5. Assuming that there are n ($i = 1, 2, \ldots, n$) grades of the evaluation of land eco-security, m ($j = 1, 2, \ldots, m$) evaluation indicators, the ith grade of the classification standard for m evaluation indicators can be represented through the m dimensional connection cloud of the digital characteristics $(Ex_1^i, En_1^i, He_1^i; Ex_2^i, En_2^i, He_2^i; \ldots; Ex_m^i, En_m^i, He_m^i)$ and the number of cloud drops N, and connection degree $\mu^i(x^i(x_1^i, x_2^i, \ldots, x_m^i))$ of the cloud drop $x^i\{x_1^i, x_2^i, \ldots, x_m^i\}$ satisfies:

$$\mu^i\left[x^i\left(x_1^i, x_2^i, \ldots, x_m^i\right)\right] = e^{-\sum_{j=1}^{m} \frac{(x_j^i - Ex_j^i)^2}{2(En_j^{i\prime})^2}}, \tag{2}$$

$$Ex_j^i = \frac{Cmin_j^i + Cmax_j^i}{2}, \tag{3}$$

$$He_j^i = \beta. \tag{4}$$

where Ex_j^i, En_j^i, He_j^i are the expected value, entropy, and hyper-entropy of grade i of evaluation indicator j, respectively. $Cmax_j^i$ and $Cmin_j^i$ are the upper limitation and lower limitation of the interval in the ith evaluation grade of evaluation indicator j, respectively. β is a parameter and amended by the fuzzy degree, $\beta = 0.01$.

3. Multi-Dimensional Connection Cloud Model-Based Evaluation Model

3.1. Basic Procedure

The basic evaluation principle of the proposed model was presented as follows: Classification standards and evaluation factors are firstly determined, respectively, and the range of evaluation factors for each grade $[\mathrm{Cmax}_j^i, \mathrm{Cmin}_j^i]$ is determined. Then, digital characteristics $\left(Ex_1^i, En_1^i, He_1^i; Ex_2^i, En_2^i, He_2^i; \ldots; Ex_m^i, En_m^i, He_m^i\right)$ are calculated. Based on *Ex*, *En*, and *He*, the *m* dimensional connection cloud in a finite interval is simulated. Next, the connection degrees of evaluation indicators are achieved. Finally, the sample evaluation level is expected. The corresponding evaluation process is shown in Figure 1.

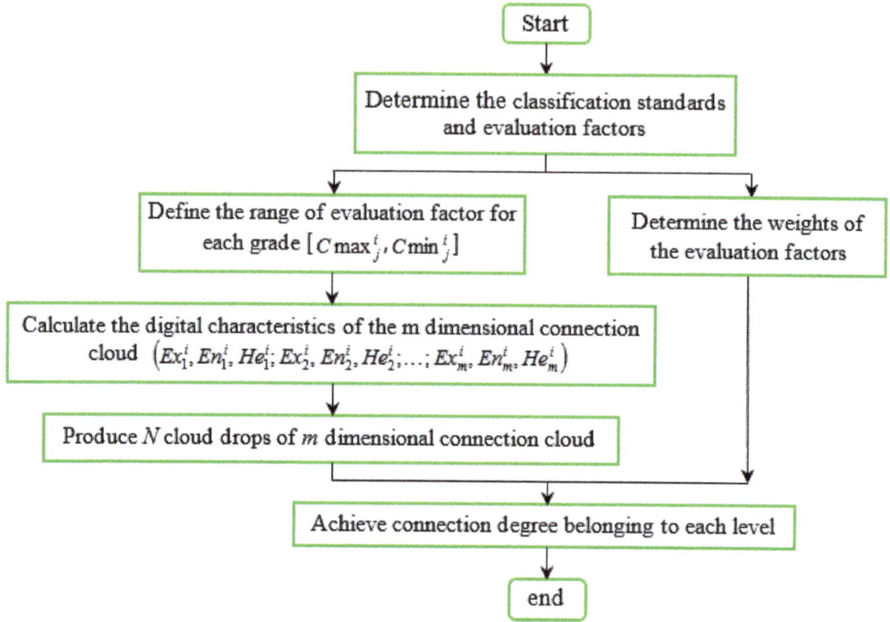

Figure 1. Flowchart of the proposed model.

3.2. Evaluation Process

The evaluation process consists of four steps, as follows:

Step 1: Select evaluation factors and classification standards. The selection of land eco-security assessment indicators is an important and complicated task. There is a lack of clear and unified standards so far, and the selection of indicators also has a great influence on the evaluation results.

The selection of indicators should not only take into account the state of the ecological environment, the safety of the land ecosystem itself, and sustainability for humanity, but also follow the principles of completeness and availability of data, etc. At least, the natural geography and socio-economic development of the Wanjiang region, especially the specific conditions of land resources and land usage should be considered. Concretely speaking, the following indicators should be emphasized: Firstly, these indicators which can reflect the natural environment conditions and confirm the land eco-security level with more objective and reasonableness, such as per capita cultivated area, per capita water resources, and forest coverage rate, etc.; Secondly, the indicators that reflect the negative impact on economic and social industry development, such as industrial wastewater discharge rate, industrial waste gas (SO_2) treatment rate, and the comprehensive utilization rate of industrial solid wastes, etc.; Finally, the indicators reflecting the sustainability for humanity and the ecosystem, such as unit area farmland fertilizer load and unit area farmland pesticide load. Here, 17 evaluation indicators, as shown in Table 1, were selected on the basis of national and local technical codes, and from the literature on land eco-security evaluation, environmental sustainable development evaluation, and ecological city evaluation [10,43–46].

Table 1. Evaluation index system and weight of land eco-security [36].

Evaluation Indicators	Weight
Per capita cultivated farmland (X_1, hm^2)	0.0371
Proportion of cultivated land on steep slopes with a slope of more than 25 (X_2, %)	0.2796
Forest coverage rate (X_3, %)	0.1450
Harmonious degree between water and soil (X_4, %)	0.0005
Forest pest control rate (X_5, %)	0.0272
Land natural disaster prevention rate (X_6, %)	0.0010
Proportion of natural disaster area (X_7, %)	0.0481
Population density (X_8, people per m^2)	0.0692
Land diversity index (X_9)	0.0176
Regional development index (X_{10})	0.0014
Per capita water resources (X_{11}, m^2)	0.1363
Sulphur dioxide emissions per unit of GDP (in ten thousand Yuan) (X_{12}, kg)	0.0884
Unit area farmland fertilizer load (X_{13})	0.0047
Unit area farmland pesticide load (X_{14})	0.0753
Industrial wastewater discharge rate (X_{15} %)	0.0004
Industrial waste gas (SO_2) treatment rate (X_{16}, %)	0.0582
Comprehensive utilization rate of industrial solid wastes (X_{17} %)	0.0100

Given the large number of evaluation indicators, it is difficult to directly determine the impact of each indicator on the ecological safety of the region (the weight of each indicator). In this study, the entropy weight method was utilized to calculate the weight of each index objectively [36]. At the same time, it is also convenient to compare with other methods.

As shown in Table 2, classification standards for evaluation indicators are divided into five levels, from very safe (I) through safe (II), marginally safe (III), marginally unsafe (IV), to unsafe (V). Based on the land eco-security classification standards, the evaluation factor range of each grade [$Cmax_j^i$, $Cmin_j^i$] is determined. For a factor that has only one side grade limit [Cmin, +∞], and its value monotonously increasing with the grade level, its upper limit value is specified as $Cmax = Ex^{n-1} + (Ex^{n-1} - Cmin^{n-2})$, while for a factor whose interval value monotonously decreases with the grade level, the upper limit value is $Cmax = Ex^{n+1} + (Ex^{n+1} - Cmax^{n+2})$.

Step 2: Calculate the digital characteristics. According to Equation (2), the connection degree at the boundary between grades i and $i - 1$ or $i + 1$ is 0.5. Namely, the sample at the classification boundary belonging to grade i or $i - 1$, and the possibility belonging to grade i or $i - 1$ is equal to each other. This particular feature is consistent with the identity-discrepancy-contrary principle of set pair theory. According to set pair theory, the identical-discrepancy-contrary relationship between the measured index and the ith level of the corresponding indicator j can be defined as follows: When the measured index is located in the level [$Cmin^i$, $Cmax^i$], it is defined as an identity relation. In addition, based on the criteria $3En$, the discrepancy and contrary relations are defined when the measured point lies on $\left[Ex^i - 3En^i, Cmin^i \right] \cup \left[Cmax^i, Ex^i + 3En^i \right]$ and the negative field, respectively. In order to quantitatively and uniformly describe the identity-discrepancy-contrary transformational potential corresponding to the connection degree of the connection cloud, En of evaluation indicator j for grade i can be calculated as follows:

$$En_j^i = \frac{Cmax_j^i - Ex_j^i}{\sqrt{\ln 4}}.$$

(5)

Taking as an example of X_1, the digital characteristics are as shown in Table 3.

Table 2. Classification standard for evaluation indicators [36,37].

Evaluation Indicators	I	II	III	IV	V
X_1	[0.10, +∞]	[0.07, 0.10]	[0.05, 0.07]	[0.02, 0.05]	[0.00, 0.02]
X_2	[0, 2]	[2, 7]	[7, 10]	[10, 24]	[24, +∞]
X_3	[47, +∞]	[34, 47]	[25, 34]	[10, 25]	[0, 10]
X_4	[95, +∞]	[90, 95]	[85.5, 90]	[80, 85.5]	[50, 80]
X_5	[97, +∞]	[85, 97]	[80, 85]	[70, 80]	[0, 70]
X_6	[97, +∞]	[85, 97]	[80, 85]	[70, 80]	[0, 70]
X_7	[0, 1]	[1, 8]	[8, 10]	[10, 50]	[50, +∞]
X_8	[100, 550]	[550, 1400]	[1400, 1600]	[1600, 2900]	[2900, +∞]
X_9	[0.90, +∞]	[0.55, 0.90]	[0.35, 0.55]	[0.10, 0.35]	[0.00, 0.10]
X_{10}	[0.95, +∞]	[0.89, 0.95]	[0.85, 0.89]	[0.74, 0.85]	[0.00, 0.74]
X_{11}	[1800, +∞]	[1450, 1800]	[1350, 1450]	[900, 1300]	[300, 900]
X_{12}	[0, 3]	[3, 8]	[8, 11]	[11, 15]	[15, +∞]
X_{13}	[150, 270]	[270, 380]	[380, 450]	[450, 650]	[650, +∞]
X_{14}	[0, 10]	[10, 20]	[20, 25]	[25, 40]	[40, +∞]
X_{15}	[97, +∞]	[85, 97]	[80, 85]	[70, 80]	[0, 70]
X_{16}	[97, +∞]	[85, 97]	[80, 85]	[70, 80]	[0, 70]
X_{17}	[97, +∞]	[85, 97]	[80, 85]	[70, 80]	[0, 70]

Table 3. Digital characteristics of the cloud model for X_1.

Cloud	*Ex*	*En*	*He*
I	0.110	0.008493	0.01
II	0.085	0.01274	0.01
III	0.060	0.008493	0.01
IV	0.035	0.01274	0.01
V	0.100	0.08493	0.01

Step 3: Produce N cloud drops of the m dimensional connection cloud. The multi-dimensional connection cloud in a finite interval is simulated on the basis of the corresponding digital characteristics (Ex, En, He). For the jth evaluation factor, there is a distinct difference in connection clouds between the internal grade ($i = 2, 3, \ldots, n - 1$) and both ends ($i = 1, n$). In the clouds at both ends, half of the clouds far from the intermediate grade actually have a uniform distribution of 1. When calculating, let $x_j^i = Ex_j^i$, then the contribution of the measured value to the overall connection degree is 1. The corresponding cloud drop generation algorithm is as follows:

(1) Generate a normally-distributed random number $En'(En_1', En_2', \ldots, En_m')$, with expectation $En(En_1, En_2, \ldots, En_m)$ and variance $He(He_1, He_2, \ldots, He_m)$.

(2) Generate a normally-distributed random number $x(x_1, x_2, \ldots, x_m)$, with expectation $Ex(Ex_1, Ex_2, \ldots, Ex_m)$ and variance $En'(En_1', En_2', \ldots, En_m')$.

(3) Calculate connection degree of the evaluation indicator according to Equation (1).

(4) Repeat steps (1)–(3) until N cloud drops are generated.

Step 4: Achieve connection degree belonging to each grade level. Based on the measured indicators of a sample, combined with the weight of the evaluation factors, the connection degree of a certain grade of land eco-security is calculated and the sample eco-security grade is determined. Since the normal cloud model satisfies the criteria $3En$ and the evaluation scope of the ith cloud model is $[Ex^i - 3En^i, Ex^i + 3En^i]$, when x_j^i is within the range of the contrary relationship, the contribution of the index x_j^i to the overall connection degree is equivalent to $x_j^i = Ex^i \pm 3En^i$, and $x_j^i = Ex^i \pm 3En^i$ and can be substituted into the calculation.

4. Case Study

4.1. Study Area

Located in Eastern China (29°41′–34°38′ N, 114°54′–119°37′ E), Anhui Province has a total area of approximately 139,600 km². At the end of 2017, the total permanent population of Anhui Province reached 70.592 million, and the total GDP of the province was about \$42.98 billion, which was about 43% higher than that of 2013. The Wanjiang region consists of nine cities, including Hefei, Wuhu, Ma'anshan, Tongling, Anqing, Chizhou, Chaohu, Chuzhou, and Xuancheng. It is an important hinterland and natural extension zone of the Yangtze River Delta economic circle, and also the most developed manufacturing base in Anhui Province. In the past 50 years, with the implementation of the extensive economic growth pattern, the introduction of chemical plants, cast iron and forging plants, and building materials plants has unbalanced the land eco-security in the Wanjiang region, caused a significant increase in population density and a clear decline in cultivated area and forest coverage, and there also exist significant differences within the nine cities. In 2010, the national government set up some industrial transfer demonstration districts in the Wanjiang region, focusing on new chemical industries, equipment manufacturing, and metallurgical industries. This provides a golden opportunity to the economic development of the Wanjiang region, but brings about a severe test to the ecological environment, which will determine if the sustainable economic development of the Wanjiang region can be realized.

4.2. Data

In order to verify the validity and feasibility of this proposed model, the data from the literature [36,37] were used to conduct the analysis. Measured values of evaluation indices are listed in Table 4, respectively.

Table 4. Measured values of indicators [36].

	Maanshan	Wuhu	Tongling	Chizhou	Anqing	Xuancheng	Chaohu	Chuzhou	Hefei
X_1	0.04	0.04	0.03	0.05	0.04	0.06	0.06	0.09	0.05
X_2	1.84	2.05	3.46	15.24	8.78	11.60	3.45	1.25	1.15
X_3	8.1	20.9	32.1	56.9	35.6	54.0	12.6	11.9	15.5
X_4	100	100	100	100	96.86	95.42	96.78	88.36	97.28
X_5	73.1	100	37.84	67.76	92.87	51.72	90.84	82.28	90.27
X_6	96.74	92.01	90.49	91.39	86.30	96.73	93.96	83.35	97.76
X_7	23.85	21.89	18.17	14.83	25.74	7.92	24.69	16.80	19.18
X_8	751	690	657	190	395	222	480	331	668
X_9	0.72	0.64	0.88	0.39	0.75	0.94	0.64	0.68	0.71
X_{10}	0.88	0.94	0.80	0.86	0.85	0.80	0.85	0.94	0.94
X_{11}	567.5	602.2	566.2	2851.1	856.1	2307.7	852.7	1238.3	928.84
X_{12}	12.19	5.08	14.14	13.36	15.29	12.89	16.70	5.32	2.54
X_{13}	537.72	783.90	765.58	711.27	804.50	690.20	774.11	779.36	800.47
X_{14}	52.98	26.44	40.61	61.86	57.08	30.78	21.42	15.69	21.22
X_{15}	97.48	97.84	99.58	99.74	94.16	94.11	99.27	88.98	96.14
X_{16}	10.14	10.08	11.02	5.73	15.12	8.81	17.10	6.18	4.89
X_{17}	91.53	100	71.54	98.77	96.88	57.46	95.39	100	99.61

4.3. Model Implementation

Based on the proposed model, 17 evaluation factors of the example are taken as 17 dimensions of the multidimensional connection cloud. Now, a comprehensive cloud model with a security grade of V is employed as an example to illustrate the process of establishing a 17-dimensional connection cloud. Through Equations (3)–(5), the digital characteristics of each evaluation factor of the V grade multi-dimensional connection cloud are calculated, as showed in Table 5. In order to facilitate the comparative analysis, the index weight value is the same as Yu et al. [36], ω = {0.0371, 0.2796, 0.1450,

0.0005, 0.0272, 0.0010, 0.0481, 0.0692, 0.0176, 0.0014, 0.1363, 0.0884, 0.0047, 0.0753, 0.0004, 0.0582, and 0.0100}. Taking the sample of Maanshan City as an example, the data in Table 5 and measured values were substituted into the Equation (2), and $\mu^V = 0.1209$ was obtained, indicating that the connection degree of V was 0.1209. The calculation process of other grades is the same, and the results of the evaluation are shown in Table 6.

Table 5. Digital characteristics and weight of each evaluation factor of the multi-dimensional connection cloud model of grade V.

Indicators	Ex	En	He	ω
X_1	825	148.63	0.01	0.0371
X_2	60	16.99	0.01	0.2796
X_3	20	4.25	0.01	0.1450
X_4	0.1	0.08	0.01	0.0005
X_5	600	254.80	0.01	0.0272
X_6	3450	467.13	0.01	0.0010
X_7	5	4.25	0.01	0.0481
X_8	65	12.74	0.01	0.0692
X_9	65	12.74	0.01	0.0176
X_{10}	0.05	0.04	0.01	0.0014
X_{11}	29.5	4.67	0.01	0.1363
X_{12}	35	29.73	0.01	0.0884
X_{13}	35	29.73	0.01	0.0047
X_{14}	35	29.73	0.01	0.0753
X_{15}	35	29.73	0.01	0.0004
X_{16}	35	29.73	0.01	0.0582
X_{17}	0.37	0.31	0.01	0.0100

Table 6. Evaluation results of the multi-dimensional connection cloud model and comparison with those of other methods.

Samples	μ					Proposed Model	One-Dimensional Cloud Model [37]	Matter Element Model [36]
	I	II	III	IV	V			
Maanshan	0.0777	0.0484	0.0146	0.0679	0.1209	V	V	V
Wuhu	0.0921	0.0838	0.0201	0.0641	0.0426	I	II	II
Tongling	0.0301	0.0880	0.0206	0.0748	0.0698	II	II	II
Chizhou	0.0605	0.0145	0.0139	0.0943	0.0692	IV	IV	IV
Anqing	0.0314	0.0642	0.0666	0.1037	0.0707	IV	III	IV
Xuancheng	0.0634	0.0230	0.0262	0.0972	0.0630	IV	I	I
Chaohu	0.0335	0.0877	0.0190	0.1157	0.0916	IV	II	II
Chuzhou	0.1303	0.1208	0.0128	0.0679	0.0625	I	I	I
Hefei	0.1457	0.0852	0.0177	0.0864	0.0509	I	I	I

5. Results and Discussion

It was discovered from Table 6 that the results from the proposed model were almost consistent with those from the one-dimensional cloud model and the matter element model. For Wuhu City, there are six indicators at I and five indicators at II. For $\mu^I = 0.0921$ and $\mu^{II} = 0.0838$, the data are relatively close to the proposed model, which also shows that the ecological environment of Wuhu City is relatively stable and the degree of safety is relatively high. Additionally, for Xuancheng City and Chaohu City, the results of the proposed method were IV, but I and II by the other methods. However, there are 10 and 8 measured indicators belonging to III, IV, and V grades for the two cities, respectively, so it is conservative that they were specified as IV. In addition, the economic growth of Chaohu and Xuancheng mainly shows extensive growth, and the traditional high-pollution, high-energy-consumption industries, such as cement, papermaking, chemicals, and steel are still dominant in the city's economy. While industrial gas and solid wastes have not been properly treated, causing pollutants, such as SO_2, to be excessive. At the same time, both cities are mountainous regions with less per capita cultivated farmland and excessive use of chemical fertilizers and pesticides. All of these threaten the land eco-security in the two cities [47–49]. The results indicated that the proposed method was feasible and more effective than other methods.

5.1. Comparison between the Multi-Dimensional Connection Cloud Model and One-Dimensional Cloud Model

The multi-dimensional connection cloud model is an improved model over the one-dimensional cloud model with respect to aspects of both the modelling process and selecting parameters.

(1) The multi-dimensional connection cloud model considers all evaluation indicators, but sets up only one comprehensive multidimensional connection cloud model, that is to say, five 17-dimensional connection clouds are set up to evaluate the eco-security grade based on the multi-dimensional connection cloud model. However, the one-dimensional cloud model needs to set up a one-dimensional normal cloud model at each evaluation grade of each evaluation factor. Namely, it needs to build 85 one-dimensional normal cloud models to assess the same land eco-security.

(2) The time of calculating connection degree of the multi-dimensional connection cloud model decreases. At the same grade, the multi-dimensional connection cloud model only needs to calculate one connection degree so that the last connection degree can be determined, while the one-dimensional cloud model needs to calculate 17 kinds of certainty, and then calculate these 17 kinds of certainty to determine the last certainty.

(3) The multi-dimensional connection cloud model improves the selection of digital features. *En* represents the evaluation scope corresponding to the eco-security level. According to the set pair theory and the criteria $3En$, the multi-dimensional connection cloud model chooses Equation (5), as *En* covers the scope of all the eco-security levels, which can quantitatively and uniformly describe the identity-discrepancy-contrary transformational potential corresponding to the connection degree of the connection cloud.

5.2. Comparison between the Multi-Dimensional Connection Cloud Model and Matter-Element Model

Compared with the matter-element model, the multi-dimensional connection cloud model comprehensively considers multiple evaluation factors, each evaluation factor is independent of the others, and then combines the weight, avoiding inaccurate evaluation results caused by the influence of a certain factor that is too large.

In addition, the challenge for the selection of land eco-security evaluation indicators was to achieve a balance between anthropogenic activities and land ecological carrying capacity in order to advance land ecological sustainability. In light of this, the "Driving forces–Pressures–State–Impacts–Responses" (DPSIR) framework, proposed by the Organisation for Economic Co-operation and Development and the United Nations Environment Programme, was chosen as the basis for defining indices to assess land eco-security. The driving forces are the load of human activities on the land ecological environment. Pressures are defined as the direct or indirect form of stresses arising from driving forces, which are anthropogenic in nature. The state is defined as the quality of the ecological environment, natural resources, and ecosystem. Impacts are defined as forms of vulnerability that humans and the natural environment faced due to the changes in the pressures existing in the states. Responses are actions and measures to tackle ecological and environmental issues. Land eco-security has a rich connotation, and the depth and breadth of researchers' understanding influence the construction of the land eco-security evaluation index system. From the analysis of the DPSIR framework and indicator selection principle in this paper the following can be found: Firstly, the selected indicators can reflect the health and sustainability of the land ecosystem from the perspective of the land ecosystem; secondly, the determined indicators have an ability to provide stable ecological services or guarantee functions for human beings by considering human development needs. According to the development status of different countries or regions, the selection of indicators may be different. The research of this paper has evident and significant reference for other regions.

Summing up, the case study has shown the superior accuracy and practical application of the model. We are confident that the results acquired in this study will contribute to the promotion of further future work for predicting the land eco-security.

6. Conclusions

Evaluation of land eco-security is a complex and uncertain problem since it is dependent on many factors. Based on the measured values of evaluation indicators, a novel multi-dimensional connection model coupled with set pair theory for the evaluation of land eco-security was addressed in this paper.

(1) The proposed model is a comprehensive connection cloud model and corresponds to each eco-security level on the consideration of all the evaluation factors. It is a clear modelling procedure with a concise algorithm and evaluation credibility. At the same time, the method for the selection of the cloud model parameters is improved. A case study of nine representative cities in Anhui, China indicates that the comprehensive cloud model proposed here is capable of assigning an eco-security level based on the consideration of randomness and fuzziness uncertainties of evaluation indicators distributed in a finite interval, and depict the certainty and uncertainty of indicators and the transformation tendency of the gained grade in a unified way. In addition, comparative analyses with the one-dimensional cloud model and the matter element model show that the proposed model is more comprehensive, objective and accurate to evaluate the land eco-security than other methods. Despite the above merits, the choice of β does not have a verdict, and a reasonable determination will be needed to be studied in the future.

(2) Evaluation results show that the land eco-security of the study area behaves in an unfavourable trend, and the marginally unsafe and unsafe levels are dominant. It should be noted that the negative impact of economic and social industry development on land eco-security accounts for a very large proportion. Excessive use of chemical fertilizers and pesticides, excessive industrial gas and solid wastes, and ineffective cleaning are the main reasons for the decrease of land eco-security. Under the supervision of the national government, some measures should be made for improving the land eco-security through environmental protection, reasonable employment of resources, proper adjustment of industrial structure, possible abatement of pollution, and so on.

(3) It is noted that, at present, there is no uniform standard for the selection of eco-security evaluation indicators. This study mainly considers the economic, environmental, and social factors, as well as the specific conditions of land resources and land use in the Wanjiang region. In this process, there may be a few factors that are ignored in establishing an indicator system and cause some deviations to the evaluation results. In fact, land eco-security involves multiple disciplines, so the authors also look forward to being able to combine sociology, economics, ecological environment, and cultural aspects to explore more indicators of land eco-security, and establish a universal and flexible indicator system.

In conclusion, this study provides a method for us to evaluate ecological environmental problems, and is a useful tool for decision-makers to judge the advantages and disadvantages of sustainable development, and to forecast land eco-security.

Author Contributions: For research articles with several authors, a short paragraph specifying their individual contributions must be provided. The following statements should be used "Conceptualization, M.W. and J.J.; Methodology, M.W.; Software, X.W.; Validation, M.W., Q.L. and F.S.; Formal Analysis, M.W.; Investigation, Q.L.; Resources, J.J.; Data Curation, Q.L.; Writing-Original Draft Preparation, Q.L.; Writing-Review & Editing, M.W. and F.S.; Visualization, M.W.; Supervision, M.W.; Project Administration, M.W.; Funding Acquisition, J.J.", please turn to the CRediT taxonomy for the term explanation. Authorship must be limited to those who have contributed substantially to the work reported.

Funding: The National Key Research and Development Program of China under grant nos. 2016YFC0401303 and 2017YFC1502405 and the National Natural Sciences Foundation, China under grant nos. 51579059 and 41172274.

Acknowledgments: Financial support provided by the National Key Research and Development Program of China under grant nos. 2016YFC0401303 and 2017YFC1502405 and the National Natural Sciences Foundation, China under grant nos. 51579059 and 41172274 is gratefully acknowledged. The authors would also like to express their sincere thanks to the reviewers for their thorough reviews and useful suggestions.

Conflicts of Interest: The authors declare that there is no conflict of interests regarding the publication of this paper.

References

1. Khramtsov, B. *A Primer on Ecological Security*; Delegates of the 34th National Student Commonwealth Forum: Ottawa, ON, Canada, 2006.
2. Feng, Y.; Liu, Y.; Liu, Y. Spatially explicit assessment of land ecological security with spatial variables and logistic regression modeling in Shanghai, China. *Stoch. Environ. Res. Risk Assess.* **2016**, *31*, 2235–2249. [CrossRef]
3. Ezeonu, I.C.; Ezeonu, F.C. The environment and global security. *Environmentalist* **2000**, *20*, 41–48. [CrossRef]
4. Lambin, E.F.; Gibbs, H.K.; Ferreira, L.; Grau, R.; Mayaux, P.; Meyfroidt, P. Estimating the world's potentially available cropland using a bottom-up approach. *Glob. Environ. Chang.* **2013**, *23*, 892–901. [CrossRef]
5. Scholz, M.; Hedmark, A.; Hartley, W. Recent advances in sustainable multifunctional land and urban management in Europe: A review. *J. Environ. Plan. Manag.* **2012**, *55*, 833–854. [CrossRef]
6. Xie, H.L.; Yao, G.R.; Liu, G.Y. Spatial evaluation of the ecological importance based on GIS for environmental management: A case study in Xingguo County of China. *Ecol. Indic.* **2015**, *51*, 3–12. [CrossRef]
7. Cramb, A. Fragile land: Scotland's environment. In *Fragile Land Scotland's Environment*; Cromwell Press: Trowbridge, UK, 1998.
8. Xie, H.; Liu, Z.; Peng, W.; Liu, G.; Lu, F. Exploring the mechanisms of ecological land change based on the spatial autoregressive model: A case study of the Poyang lake eco-economic zone, China. *Int. J. Environ. Res. Public Health* **2014**, *11*, 583–599. [CrossRef] [PubMed]
9. Yu, K.J.; Qiao, Q.; Li, D.H.; Yuan, H.; Wang, S.S. Ecological land use in three towns of eastern Beijing: A case study based on landscape security pattern analysis. *Chin. J. Appl. Ecol.* **2009**, *20*, 1932–1939.
10. Xu, L.; Hao, Y.; Li, Z.; Li, S. Land ecological security evaluation of Guangzhou, China. *Int. J. Environ. Res. Public Health* **2014**, *11*, 10537–10558. [CrossRef] [PubMed]
11. Zhao, Z.Z.Z.; Tao, X.T.X. An application of fuzzy matter-element theory in land ecological safety evaluation. In Proceedings of the 2010 3rd IEEE International Conference on Computer Science and Information Technology (ICCSIT), Chengdu, China, 9–11 July 2010.
12. Vitousek, P.M.; Mooney, H.A.; Lubchenco, J.; Melillo, J.M. Human domination of earth's ecosystems. *Science* **1997**, *277*, 494–499. [CrossRef]
13. Wessels, K.J.; Prince, S.D.; Frost, P.E.; Zyl, D.V. Assessing the effects of human-induced land degradation in the former homelands of northern south africa with a 1 km avhrr ndvi time series. *Remote Sens. Environ.* **2004**, *91*, 47–67. [CrossRef]
14. García, M.; Oyonarte, C.; Villagarcía, L.; Contreras, S.; Domingo, F.; Puigdefábregas, J. Monitoring land degradation risk using ASTER data: The non-evaporative fraction as an indicator of ecosystem function. *Remote Sens. Environ.* **2008**, *112*, 3720–3736. [CrossRef]
15. Salvati, L.; Bajocco, S. Land sensitivity to desertification across Italy: Past, present, and future. *Appl. Geogr.* **2011**, *31*, 223–231. [CrossRef]
16. Tóth, G.; Hermann, T.; Da, S.M.; Montanarella, L. Heavy metals in agricultural soils of the European Union with implications for food safety. *Environ. Int.* **2016**, *88*, 299–309. [CrossRef] [PubMed]
17. Mahar, A.; Wang, P.; Ali, A.; Awasthi, M.K.; Lahori, A.H.; Wang, Q.; Li, R.; Zhang, Z. Challenges and opportunities in the phytoremediation of heavy metals contaminated soils: A review. *Ecotoxicol. Environ. Saf.* **2016**, *126*, 111–121. [CrossRef] [PubMed]
18. Sousa, C.D.; Ghoshal, S. 6—Redevelopment of brownfield sites. *Metrop. Sustain.* **2012**, 99–117.
19. Orooj, S.; Sayeda, S.S.; Kinza, W.; Alvina, G.K. Phytoremediation of soils: Prospects and challenges. *Soil Remediat. Plants* **2015**. [CrossRef]
20. Boardman, J.; Poesen, J. *Soil Erosion in Europe*; John Wiley & Sons, Ltd.: Hoboken, NJ, USA, 2006.
21. Wada, T.; Nemoto, Y.; Shimamura, S.; Fujita, T.; Mizuno, T.; Sohtome, T.; Kamiyama, K.; Morita, T.; Igarashi, S. Effects of the nuclear disaster on marine products in Fukushima. *J. Environ. Radioact.* **2013**, *124*, 246–254. [CrossRef] [PubMed]
22. Li, F.; Lu, S.; Sun, Y.; Li, X.; Xi, B.; Liu, W. Integrated evaluation and scenario simulation for forest ecological security of Beijing based on system dynamics model. *Sustainability* **2015**, *7*, 13631–13659. [CrossRef]

23. Zhang, H.; Xu, E. An evaluation of the ecological and environmental security on China's terrestrial ecosystems. *Sci. Rep.* **2017**, *7*, 1–12. [CrossRef] [PubMed]

24. Zhang, S.H.; Wang, S.R.; Wang, Y.M. Impacts of climate disaster on Beijing's sustainable development and relevant strategies. *Acta Geol. Sin.* **2000**, *55*, 119–127.

25. He, D.; Wu, R.; Yan, F.; Li, Y.; Ding, C.; Wang, W. China's transboundary waters: New paradigms for water and ecological security through applied ecology. *J. Appl. Ecol.* **2014**, *51*, 1159–1168. [CrossRef] [PubMed]

26. Li, Z.C.; Zhang, X.L. Assessment of regional ecological security based on theory of ecological footprint—The case of Shandong Province. *Res. Agric. Mod.* **2008**, *29*, 465–467. (In Chinese)

27. Zhao, S.; Li, Z.; Li, W. A modified method of ecological footprint calculation and its application. *Ecol. Model.* **2005**, *185*, 65–75. [CrossRef]

28. Jin, S.Q.; Wang, J.X.; Song, G.J. A review of research on ecological footprint. *Environ. Sustain. Dev.* **2009**, *34*, 26–29. (In Chinese)

29. Wackernagel, M.; Rees, W. *Our Ecological Footprint: Reducing Human Impact on the Earth*; New Society Publishers: Gabriola, BC, Canada, 1996.

30. Park, Y.S.; Chon, T.S.; Kwak, I.S.; Lek, S. Hierarchical community classification and assessment of aquatic ecosystems using artificial neural networks. *Sci. Total Environ.* **2004**, *327*, 105–122. [CrossRef] [PubMed]

31. Meng, Z.X.; Li, C.Y.; Deng, Y.L. Ecological security early-warning and its ecological regulatory countermeasures in the Tuojiang river basin. *J. Ecol. Rural Environ.* **2009**, *25*, 1–8. (In Chinese)

32. Gong, J.Z.; Xia, B.C. Spatially fuzzy assessment of regional eco-security in Guangzhou, a fast-urbanizing area: A case study in Guangzhou city. *Acta Ecol. Sin.* **2008**, *28*, 4992–5001.

33. Zhang, X.H.; Gao, J.X.; Dong, W. The application of a nonlinear complex assessment model on ecological security assessment in longitudinal range-gorge region of Yunnan Province. *Res. Environ. Sci.* **2007**, *20*, 81–86. (In Chinese)

34. Zhang, H.B.; Liu, L.M.; Zhang, J.L.; Zhu, Z.Q. Matter-element model and its application to land resource eco-security assessment. *J. Zhejiang Univ.* **2007**, *33*, 222–229. (In Chinese)

35. Zhang, X.H.; Lei, G.P.; Yuan, L. Evaluation on ecological security of land based on entropy weight and matter-element model: A case study of Heilongjiang Province. *China Popul. Resour. Environ.* **2009**, *19*, 88–93. (In Chinese)

36. Yu, J.; Fang, L.; Cang, D.; Zhu, L.; Bian, Z. Evaluation of land eco-security in Wanjiang district base on entropy weight and matter element model. *Trans. Chin. Soc. Agric. Eng.* **2012**, *28*, 260–266. (In Chinese)

37. Gao, M.; Sun, T.; Zhao, T.; Dai, H.; Zhang, K. Application of normal cloud model on the comprehensive assessment of land eco-security in Wanjiang district. *J. Hunan Agric. Univ.* **2015**, *41*, 196–201. (In Chinese)

38. Li, D.Y. *Artificial Intelligence with Uncertainty*; Beijing National Defence Industry Press: Beijing, China, 2005. (In Chinese)

39. Wang, M.W.; Jin, J.L. *The Theory and Applications of Connection Numbers*; Beijing Science Press: Beijing, China, 2017. (In Chinese)

40. Wang, M.; Wei, D.; Li, J.; Jiang, H.; Jin, J. A novel clustering model based on set pair analysis for the energy consumption forecast in china. *Math. Probl. Eng.* **2014**, *2014*, 1–8. [CrossRef]

41. Wang, M.; Xu, X.; Li, J.; Jin, J.; Shen, F. A novel model of set pair analysis coupled with extenics for evaluation of surrounding rock stability. *Math. Probl. Eng.* **2015**, *2015*, 1–9. [CrossRef]

42. Wang, D.; Zeng, D.; Singh, V.P.; Xu, P.; Liu, D.; Wang, Y. A multidimension cloud model-based approach for water quality assessment. *Environ. Res.* **2016**, *148*, 24–35. [CrossRef] [PubMed]

43. Zhao, C.R.; Zhou, B.; Su, X. Evaluation of urban eco-security—A case study of Mianyang City, China. *Sustainability* **2014**, *6*, 2281–2299. [CrossRef]

44. Wang, Y.; Ding, Q.; Zhuang, D. An eco-city evaluation method based on spatial analysis technology: A case study of Jiangsu Province, China. *Ecol. Indic.* **2015**, *58*, 37–46. [CrossRef]

45. Liao, C.M.; Li, L.; Yan, Z.Q.; Hu, B.Q. Ecological security evaluation of sustainable agricultural development in karst mountainous area—A case study of Du'anyao autonomous county in Guangxi. *Chin. Geogr. Sci.* **2004**, *14*, 142–147. [CrossRef]

46. Loiseau, E.; Junqua, G.; Roux, P.; Bellon-Maurel, V. Environmental assessment of a territory: An overview of existing tools and methods. *J. Environ. Manag.* **2012**, *112*, 213–225. [CrossRef] [PubMed]

47. Yan, C.; Zhang, A.M.; Shi, R.R.; Huang, L.L. Research on land ecological security based on land use zoning taking Wanjiang urban belt as an example. *Resour. Dev. Mark.* **2015**, *31*, 820–824. (In Chinese)

48. Xu, L. Study on the necessity and feasibility of Xuancheng eco-city construction. *China Chem. Trade* **2014**, *6*, 227–228. (In Chinese)
49. Tian, J.; Wang, Z.X.; Wang, C.R. Study on ecological security evaluation of the Chaohu Lake Basin. *Ecol. Sci.* **2011**, *30*, 650–658. (In Chinese)

 *sustainability*

Article

Integrative Assessment of Land Use Conflicts

Zita Izakovičová [1], László Miklós [2] and Viktória Miklósová [1,*]

[1] Institute of Landscape Ecology of Slovak Academy of Sciences, 81499 Bratislava, Slovakia; zita.izakovicova@savba.sk

[2] Faculty of Ecology and Environmental Sciences, Technical University Zvolen, 96053 Zvolen, Slovakia; miklos@tuzvo.sk

* Correspondence: viktoria.miklosova@savba.sk; Tel.: +421-2-3229-3624

Received: 31 July 2018; Accepted: 5 September 2018; Published: 13 September 2018

Abstract: Changes in land use are reflected primary in changes of land cover, but subsequently cause conflict of interest of sectors and are the main initiation of many environmental problems. The basic tool for sustainable utilization of the landscape is integrated landscape management, which, in our understanding, is the environmentally biased harmonization of tools which regulate the spatial organization and functional utilization of the landscape to avoid the conflicts of interest of sectors. "Integrated" in this case means the systematic assessment of the interests of all relevant sectors from the environmental point of view. The scientific base of this approach is the understanding of the landscape as a geosystem, and, in particular, the proper interpretation of the mutual relations of primary, secondary and tertiary landscape structures and their role in the assessment of the conflicts of interest. This paper presents a theoretical and methodical base for the integrated approach to the assessment of the conflicts of interest of the sectors in the landscape. The theoretical-methodical base was applied to the model territory of the Trnava district (south-west Slovakia). Mutual conflicts of interest of endangering and endangered sectors cause diverse problems, which were ranked in three basic groups as: problems of endangering of the ecological stability of the landscape (including endangering of biodiversity and nature conservation areas); problems of endangering of natural resources (in particular forests, soils, waters); and, problems of endangering the immediate human environment (stress factors in residential and recreational areas). The result is the identification and analysis of the conflicts of interest in the territory and their projection to a map. This research should be followed by implementation of procedures of ecologically optimal spatial organization and utilization of the territory for regular spatial planning processes.

Keywords: land use conflicts; encounters of interests; landscape as geosystem; integrative landscape management; Trnava district

---

1. Introduction

European countries in the recent period have faced many substantial socio-economic changes, which are also reflected in the environment [1]. Changes in land use are often linked to the occurrence of environmental problems—qualitative and quantitative degradation of natural resources, decline of natural ecosystems, negative impact of abandoned agricultural fields on biodiversity, desertion of land, increase of synanthropic species etc. [2]. Changes in landscape structure are the major causes of climatic change, which, in addition to changes to ecosystems, are increasingly intensifying natural risks and hazards, such as floods, droughts, erosion, accumulation, landslides and others [3].

Environmental issues related to changes in the landscape were also highlighted by the European Environmental Agency which specified a total of 11 global megatrends (GMTs) in four clusters that need to be addressed urgently [4]. Within the environmental cluster, these are growing pressure on ecosystems and natural resources, climate change and growing environmental pollution. The impacts of the GMTs are not yet completely known, as many have begun to manifest themselves in recent years. The relationships between individual GMTs raise a large number of questions, and there are still many deficiencies in the knowledge of their impact on the landscape.

Several of these environmental global megatrends are initiated by inappropriate land use. According to Food and Agriculture Organization of the Unated Nations (FAO) information, up to 60% of the world's ecosystems are degraded and exploited unsustainably [5]. In the EU, only 17% of habitats and species and 11% of key ecosystems protected by European legislation show a favorable status [4–6]. This is despite the adoption of measures in 2001 to combat biodiversity loss. All scenarios of global and regional assessments show that biodiversity loss, degradation of ecosystems and threats to environmental conditions will continue or even accelerate [7]. Human activities, global population growth and changing consumption patterns are key factors responsible for this growing environmental burden, which primarily becomes evident in land use changes.

Climate change and its negative impacts, such as negative changes of the water cycle, are also often tied to inappropriate land use. Inappropriate use of river basins, such as deforestation and intensive agricultural production contributes to rapid water drainage and consequently to natural hazards such as floods and drought [8].

To find the cause and casualties of these changes is the base for development and implementation of methods of optimal land use. A key approach for the analysis of these problems is the integrated approach to land use assessment [9].

Land use conflicts are usually assessed by analyses and statistical models of changes of areas of land cover. There is a significant amount of such research. This step is needed, but the mere comparison of changes on a given area in certain time only shows the encounters of one pair of activities. The presented integrated approach, supported by a matrix model, allows the analysis for the same area at the same time all existing and planned encounters of all relevant activities by assessments of the harmony of primary landscape structure (PLS), secondary landscape structure (SLS) and tertiary landscape structure (TLS) [10–12] and [13] (see Chapter 2.1.d). This approach includes the mutual comparison of the abiotic conditions (PLS), land cover including biotic elements (SLS), as well as the legal conditions and limits for development of a territory (TLS). These methods have been developed at the Institute of Landscape Ecology of the Slovak Academy of Sciences since the early 1980s (including by the authors), and applied to a number of concrete territories [14–21]. The approach is based on geosystem theory [22–25] applied to the needs of landscape ecological planning [12,26]. The methodology and the individual methods have been continually improved, incorporating new techniques, such as GIS, remote sensing and other computer techniques [26–28].

The aim of this paper is to present the methodology for the assessment of environmental land use conflicts, based on this integrative approach. Since the problems of the conflict of land use in our understanding arise from the mutual relation of all structures and components of the geographical sphere, the object of our assessment is the integrated system of a spatial section of the geosphere—the landscape as a geosystem [29–31]. Obviously, this integrated system also includes the elements of the land-cover which are result of the land-use. Therefore, in this article we use the term "conflict of interests in the landscape", which naturally also includes conflicts of land use.

2. Materials and Methods

2.1. Theoretical Aspects of the Integrated Landscape Management

The term "integrated" in land-use management incorporates an extremely broad scope of topics with different understandings. The integrated approach is a slogan often used in environmental politics. Nevertheless, it is not a new issue at all. The base aspects were defined in AGENDA 21 from Rio Summit 1992, in Chapter 10 "Integrated approach to the management of land resources". The main statements presented in this chapter—still valid and obvious—are as follows:

- Several sectors claim land for their activities in the same territory.

- There is only one landscape space, which is to be accepted by each sector.
- Activities use the landscape, but are in conflict.
- Conflicts cause environmental, economic and other problems.
- To solve conflicts an integrated approach is needed for management of land resources [32].

Why use an integrated approach? The scientific basis is evident and quite simple: since the landscape is an integrated system of concrete spatial segments of the Earth, which integrate all material components of the landscape—such as the geological base, soil, water, relief, air, biota, and elements of man-influenced, man-altered and man-made land cover (i.e., the geosystem)—all changes to one single element of this system causes changes to all other elements. Both the space itself and the material elements are considered natural resources [30]. If one applies good management (e.g., for watershed protection), reducing biodiversity loss, soil erosion, and water pollution from agriculture and the microclimate can be achieved contemporaneously. Conversely, application of poor management may cause these problems.

These basic aspects presented by AGENDA 21 on an integrated approach were enhanced in a number of environmental-political documents [33–38].

To develop the integrative approach, keeping in mind the above-mentioned statements, we need to answer several basic questions, such as:

(a) Who manages the land and why are they in conflict?

The landscape is under the care/responsibility/management of different sectors, which realize their interests in the space, such as:

- forestry—manages the forests;
- agriculture—manages the agricultural land;
- water management—manages the waters and watersheds;
- nature conservation—manages the conservation areas and protected objects;
- recreation—manages the recreational activities and areas;
- urbanization and communal sphere—manages the built-up areas and communal areas;
- industry—manages the industrial areas.

The basic interest of these sectors is to fulfill their own goals, with less regard for other sectors; therefore, their activities can cause conflicts of interest, and, consequently, environmental and other problems.

Effective mitigation of these conflicts needs integrated landscape management [39–41].

(b) How are interests of a sector in the landscape realized and what is the best phase for integrative action in landscape management?

The process starts with a very basic question asked by humans: do we like the current structure of the landscape of a given territory? If so, we try to execute protective regulations to preserve its structure. More often, the case is that the development of society needs changes. In this case, the process (in civilized countries) follows the sequence: assessment of current situation, formulation of demands/interests/requests, planning/projecting, and realization/execution. All these steps are regulated by different institutional tools, such as economic and social conditions, legislation, methods, and normative and technical conditions [42].

Analyzing this process, it is obvious that the best chance to avoid conflicts of interests—that is, the best chance for harmonization of these interests (i.e., integration)—is in the early stages of the process. In practice, it is at the stage of planning.

Accordingly, land-use planning should act as one of the main tools of the integrated approach to the optimal spatial organization and functional utilization of landscape for all sectors. This is in line with the still-current provisions of AGENDA 21, which stated that integrated land-resource management needs integrated physical planning to act as a spatial frame and basis for each sectoral plan [32]. AGENDA 21 also defines the integration of environmental issues, such as:

> "Government on the appropriate level . . . should: Adopt planning and management systems that facilitate the integration of environmental components such as air, water, land and other natural resources, using landscape ecological planning (LANDEP) or other approaches that focus on, for example, ecosystem or a watershed."

Different countries adopt different approaches to the development of integrated planning tools [43–49]. These tools are marked with different terms, such as spatial planning, physical planning or land-use planning. In Slovakia, the most integrative tool which corresponds with the above-mentioned aspects of integration is so-called territorial planning ("územné plánovanie"), which, according to its legally defined provisions, should provide spatial organization and functional utilization

- of the whole territory without any missed areas;
- for all sectors requesting space in this territory.

Ecological aspects of territorial planning are present as landscape ecological planning and as the territorial system of ecological stability, both mentioned as obligatory regulations for each territorial plan [50].

(c)   What is the integration of landscape management?

In our understanding, the integration of landscape management means the harmonization of the tools which regulate the spatial organization and functional utilization of the landscape [51]. It is to be underlined, also, that the term "management" means the ruling device, the chain of planning, organizing/regulating, and controlling. Accordingly, integrated landscape management is the ruling device for harmonization of the planning, regulating and controlling tools of the spatial organization and functional utilization of the landscape. To clarify, we do not consider execution of physical actions to be integrated management (i.e., in forestry, in the agriculture, in water-management etc.) because there is no integrated grass-cutting, integrated tillage along the contour lines, integrated timber harvest, building flood preventive objects, etc. These are simply physical actions, which are, of course, needed and can lead to desired effects; however, integrated management is the ruling policy forcing the users to provide such actions [48,52–54].

(d)   What is the material basis for integrated landscape management?

Management is executed in the landscape as a geosystem, which is an integrated system of the space, geo-relief and all other natural, man-influenced and man-made components of the landscape—material resources in certain area—such as the geological base, water, soil geo-relief, biotic components, climate, land-use, man-made objects and non-material socio-economic phenomena in the landscape, and their relationships [29–31,55]. Of special importance for evaluating the conflicts of interest in the landscape is the definition of three structures of the landscape according to their genesis, physical matter, possibility to change and their role in the management:

- Primary landscape structure (PLS): is a set of material elements of the landscape and their relationships that constitute the original foundation and condition for the other two structures. These are the abiotic elements—the geological base and subsoils, soils, waters, geo-relief, and air. The principles of their functions are not changeable. This aspect is decisive in evaluating and solving the conflicts of interest.
- Secondary landscape structure (SLS): is constituted by man-influenced, reshaped and created elements of land-cover, which is the result of land-use. Here also belong the elements of real biota, man-made objects and constructions. A very frequent case of the conflict of interest is the disharmony between land-use and primary landscape structure.

- Tertiary landscape structure (TLS): is a set intangible (non-material) socio-economic factors/phenomena displayed to the landscape space as the interests, manifestations and consequences of the activities of individual sectors that are relevant to landscape. These are the protection and other functional zones of nature and natural resources protection, hygienic and safety zones of industrial and infrastructure objects, declared zones of specific environmental measures, administrative boundaries, etc. They are defined in acts, plans and other development documents. Since they are of non-material character they often overlap. This fact is one of the main causes of conflicts of interest.

Understanding the mutual relationships of the three different landscape structures is the theoretical basis for the evaluation and solution of the conflicts of interests in the landscape.

2.2. Methodical Approach

The method of the assessment of the interests of sectoral activities is based on the understanding of the landscape as a geosystem [29,56,57], in particular on the systematic interpretation of the mutual relationships of primary landscape structure (PLS), secondary landscape structure (SLS) and tertiary landscape structure (TLS). An elementary explanation of the relationships and roles of these structures in conflicts of interest is as follows.

The current use of the landscape—SLS—is conditioned by the features of the natural landscape structure—PLS—which represent a certain offer for the use of natural resources by humans. On the other hand, the use of the landscape is driven by the requirements, the demands, and the possibilities of using the natural resources by humans, as well as human's ability to protect and rationally exploit these sources, expressed by socio-economic factors and phenomena from the TLS. Needs, requirements, and options expressed by the TLS change over time (change of priorities, change of ownership relationships, change of technologies, change of socio-economic conditions, etc.). Changes in the claims and demands of society are consequently reflected in the change, forms and intensity of use of natural resources, expressed by the SLS. Changes can also be caused by the natural reaction of the PLS to the specific use of the landscape, if such use of the SLS is unsustainable. These are the negative impacts of inappropriate land use; for example, erosion processes, landslides, calamities, etc. If the use of the landscape SLS is not in accordance with the natural resources and potentials of the PLS, this leads to conflicts of interest and to environmental problems in the landscape.

The method comprises systematic evaluation of the spatial encounters of overlapping structures, where:

- The PLS represents the space and the material natural resources used by sectors.
- The SLS represents the current use of these resources.
- The TLS represents the regulations and legal provisions for use of these resources.

The method of assessment of the conflict of interests of the sectors in the landscape involves the following steps.

2.2.1. Spatial Projection of the Interests of the Sectors to the Landscape Space

In practice, this step means the spatial projection of relevant elements of all three landscape structures to maps. The PLS is represented by geological resources, water and soil resources; the SLS by land cover elements creating the characteristic landscape structure, including forests and other biotic elements; and the TLS by the displayed regulations of the use and protection of nature, natural resources and the environment. The map projection of these elements can be considered for display of the interests of the sectors to the landscape space.

The projected interests overlap in the space and thus creates an encounter of interests of very diverse characters.

2.2.2. Analyses, Syntheses and Evaluation of the Spatial Encounters of the Interests of the Sectors

This is the crucial step of the method. The core of the method is the objective delimitation, analysis and listing of combinations of the overlapping elements of all three structures. These can be combinations of several elements.

The activities of sectors for the purposes of the evaluation of conflicts of interest can be assessed from environmental/ecological point of view as [58]:

- Having a positive impact on the landscape and ecosystems. These are activities aimed at nature and landscape protection (conservation areas of different degrees) and natural resources protection (protected zones of geological, soil, water and forest resources). According to the division of ecosystem services [59] these zones usually provide protection for ecosystems which provide supportive, and in some cases even productive services; for example, the protection of soil resources is primarily focused on the use of production functions on high quality soils. When these activities are encountered with activities of other sectors (mainly with productive sectors) their execution is in most cases endangered.
- Having a negative impact on the landscape and ecosystems. These are activities of the stress factors that aggravate the quantity and quality of the landscape as a whole, as well as individual natural resources and, at the same time, they limit the use of their ecosystem services. These are activities from the industry, energy, mining, transport, and urbanization sectors. Negative impacts of human activities are manifested by spatial reduction of the areas of natural and semi-natural ecosystems and natural resources, reducing the overall ecological stability of the landscape, creating barriers to natural biota movement, producing contaminants, and degrading the environment; for example, damaged forest ecosystems and agri-ecosystems have reduced economic value and limited use due to stress factors. Contaminated soil is not hygienically suitable for growing crops for direct consumption. When these activities are encountered with activities of other sectors (mainly with protective sectors) they have the characteristic of endangering factors [60].
- Several sectoral activities are of a two-fold character. On one side they are important, or even crucial, for the provision of a high-quality environment or natural resources; on other side their execution endangers the optimum use of land, natural resources and ecosystem services.

These sectors are:

Agriculture: Agriculture is a very specific sector. On one hand, it is obviously responsible for the protection of its own production resource—soil resources—but, on other hand, the effort to increase production causes over-exhaustion of soils, often degradation by pollution, drought by ameliorations, erosion, compaction with heavy machines, etc. So, agriculture is a specific example of an internal conflict of interest within the same sector.

Forest management: Forest resources are undeniably crucial for ecological stability, biodiversity and environment. These aspects of forests are defined by law as their non-productive functions, and in practice by delineation of special-purpose and protective forests. However, at the same time, forest management that is oriented on timber production is sometimes considered a main enemy of nature conservation, causing diverse problems for biodiversity loss, erosion, accumulation, pollution, etc.

Water management: This is the key sector responsible for good quality and quantity of water resources. The activities and duties of water management are defined by law and, in practice, projected as protective zones of water resources of different character. On the other hand, the technical activities of water management, such as building dams, regulation of rivers and creating technical measures against floods, are often evaluated as a danger; for example, when water reservoirs overlap precious biotopes, the flow destroys the natural living conditions for animals when the measures cause considerable lowering or rise of underground water levels.

Urbanization: Residential areas—including all necessary accessories such as parks and other green surfaces—are considered resources for fulfilling human needs. On the other hand, communal activities, such as waste and sewage management, as well as the rise of residential areas and the population itself, cause significant problems by overlapping agricultural land, pollution, garbage, and non-controlled activities of the population with nearby nature.

Recreation: Recreational areas, spas, health resorts and provided recreational activities are considered a natural resource for human health. On the other hand, intensive recreational activities often mean an intensive impact on nature; for example, ski resorts in high mountains, dense tourist paths in national parks, overpopulated recreation centers in natural parks, etc., cause changes and destruction of natural habitats, lowering the ecological stability of these natural resources.

2.2.3. Assessment of the Character of Encounters of Interests

The evaluation of the encounters of interests is processed in a matrix (Figure 1).

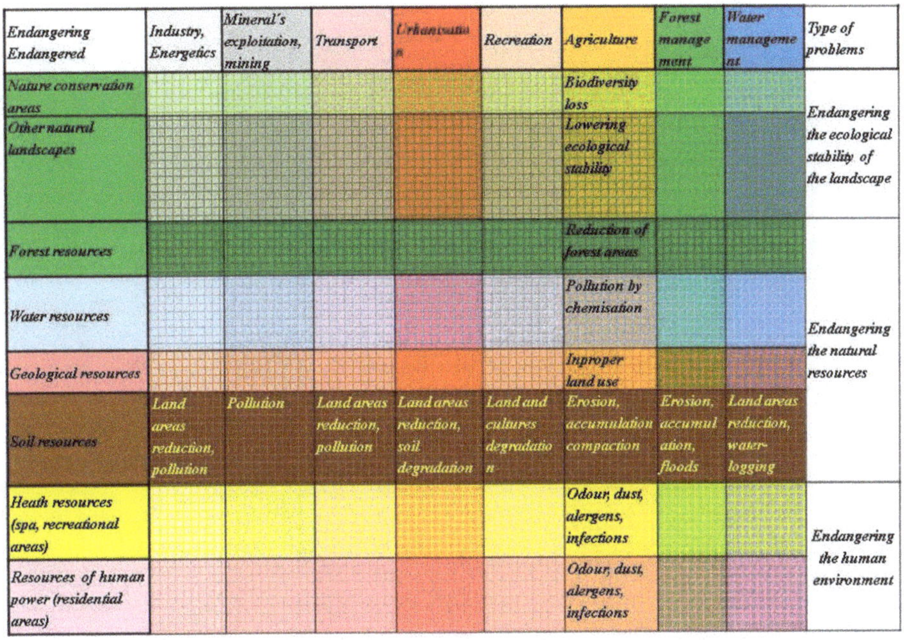

Endangering / Endangered	Industry, Energetics	Mineral's exploitation, mining	Transport	Urbanisation	Recreation	Agriculture	Forest management	Water management	Type of problems
Nature conservation areas						Biodiversity loss			Endangering the ecological stability of the landscape
Other natural landscapes						Lowering ecological stability			
Forest resources						Reduction of forest areas			
Water resources						Pollution by chemisation			Endangering the natural resources
Geological resources						Inproper land use			
Soil resources	Land areas reduction, pollution	Pollution	Land areas reduction, pollution	Land areas reduction, soil degradation	Land and cultures degradation	Erosion, accumulation compaction	Erosion, accumulation, floods	Land areas reduction, water-logging	
Heath resources (spa, recreational areas)						Odour, dust, alergens, infections			Endangering the human environment
Resources of human power (residential areas)						Odour, dust, alergens, infections			

Figure 1. Matrix of the encounters of interests in the landscape.

According to the above description of the sectoral activities, they are arranged in the matrix as follows:

- Rows of the matrix incorporate those sectoral activities which aim toward the protection and conservation of nature and landscape, protection of natural resources and human environment. From the ecological point of view these activities can be considered as endangered by other sectors.
- Columns of the matrix incorporate those sectoral activities which aim toward material production of goods and other technical services, such as industry, energy, mining, transport, urbanization, recreation, agriculture, forestry, and water management. From the ecological point of view these activities can be considered as endangering the sectors mentioned above.
- The squares of the matrix represent the encounters of the interests of the sectors.

There are two ways for evaluating these encounters:

- By evaluation down the columns we ask: how does an endangering activity threaten the endangered factors? Figure 1 describes this mode using the example of the evaluation of how agriculture threatens the particular endangered factors.
- By evaluation along the rows we ask: how is an endangered factor threatened by endangering activities? Figure 1 presents this mode using the example of the evaluation of how soil resources are threatened by particular endangering activities.

These encounters represent several types of encounters and conflicts, such as:

(a) The overlap of productive sectors—mainly industry, energy, mining, transport and urbanization—with all other sectors represents real conflicts where the basic problem arises from the efforts of these sectors to broaden their respective territories. Thus, these sectoral activities change the land use, destroy biotopes and build barriers. This is the type of conflict when the sectors "fight" for land of other sectors.

(b) Some overlap/encounters of sectoral interests have mutually supporting effects; e.g., the overlap of special-purpose forests with nature conservation, or of water protective zones with protective forests. These overlaps are actually just spatial encounters in the same area, rather than conflicts. On the other hand, the overlapping of productive forests with national parks, for example, represents real environmental conflicts for nature conservation. The designation and exclusive use of high quality soils exclusively for agricultural production can be clearly considered for protection of soil resources but, conversely, the overlapping of these soils with water protective zones causes real conflict because of the impacts on waterways of chemicals used in intensive agriculture. These types of conflicts can be considered as the encounters and conflicts when several sectors execute their activities on the same territory.

(c) Few sectoral activities cause conflict with their own productive resources. These include the conflict of agricultural activities with the soil resources, or timber production activities with forest resources. This is the type of internal conflict of interest within the same sector.

(d) The productive sectors—industry, energy, transport, agriculture—also produce pollution of the air and water, which does not respect any spatial limits. They cause real conflicts with the human environment and health resources, as well as conflicts with water, soil and forest resources [61]. This is the type of real conflict of polluting activities with environment and natural resources.

A few other important aspects of the process of evaluation in the matrix in Figure 1 should also be mentioned:

- The matrix may be considered as a model (even qualitative) of all possible mutual conflicts of all sectors. Although the presented graphical form of the matrix in paper is two-dimensional, in a real territory there are encounters not only of pairs of endangered and endangering activities, but there may also be diverse combinations of several activities. So, the evaluation of encounters and conflicts is much more complex.
- In the present work we applied Geographic Information System (GIS) technology, which allows the assessment of the conflicts of several sectoral activities on the same area in a certain time, if they occur. Therefore, the matrix is actually multidimensional. GIS technology allows the analysis of each conflict of interest quantitatively, according to its area size, including the conflict of the land use (SLS) with natural conditions (PLS).
- In a real territory the general designations of the sectors, elements and factors are concretized and replaced by real existing elements of PLS, SLS and TLS (as an example, see the matrix on the Figure 8, Section 3.4).

- Conflicts of land use were previously quantitatively documented by statistical data which mirror the changes of areas of land use. Methodologically, it is a simple procedure, and the literature includes many examples. However, these statistics document the changes, in a certain time period and a certain area, of pairs of land-use forms. The list of all binate changes can be considered as an integrated picture of the changes, but they remain simply a set of binate data. The present paper did not consider this method as a priority. The present paper describes the assessment of the mutual conflicts of all sectors, and, at the same time, the assessment of their relationships/conflicts with all three defined landscape structures. Thereafter, it also provides the characteristics of these conflict. These characteristics are qualitative. The evaluation and assessment of the characteristics of sectoral activities presented above is given from an environmental/ecological point of view. If the same encounter were assessed by an industrialist, an agriculturist or a forester, they would identify endangered and endangering activities differently. For example, ecologists consider timber production a significant threat to nature conservation; in contrast, foresters consider as an endangering factor something which threatens forestry, even nature conservation [62].

2.2.4. Classification of the Problems Caused by Conflicts of Interest in the Landscape

The evaluation of the conflicts of interest as described above allows a general classification of environmental problems caused by conflicts of interest in the landscape into three groups (see Figure 1) [58,63].

(a) Problems of endangering the ecological stability of the landscape

These problems arise as the consequence of the influence of stress factors to elements with high eco-stabilizing effects—nature conservation areas or other ecologically valuable areas of the landscape (forest, water areas, meadows, pastures, areas of public greenery etc.). These problems are represented in the first two rows of Figure 1.

(b) Problems of endangering the natural resources

These arise from spatial conflicts of stress factors and natural resources. The result of this conflict is deterioration of quality and quantity of natural resources. The most important problems in this category relate to endangering the territories with sources of drinking water, good quality soil, and productive forests. These problems are represented in the third, fourth, fifth and sixth rows of Figure 1.

(c) Problems of endangering the immediate human environment

These arise from the impact of stress factors on areas where people permanently reside (residential areas), or where they spend holidays or undergo medical treatment (recreational and health resorts). These problems are represented in the last two rows of Figure 1.

2.3. Brief Description of the Model Territory

The theoretical-methodical base described above was applied in the model territory of the Trnava district. This study area represents a typical agricultural landscape. The district is located in south-west Slovakia (Figure 2). Administratively, it consists of 45 rural settlements and the city of Trnava, which is also acting as the county's capital city. With an area of 741 km^2, it belongs to a medium-sized district of Slovakia. In the region, there live 131,644 inhabitants and the average population density is 178 inhabitants per kilometer [64].

Figure 2. Situation map of the model territory (the Trnava district).

The dominant position in the landscape structure comprises agricultural land. Up to 93.1% of agricultural land is intensively utilized as big-block structured arable land. Arable land covers the central and southern parts of the area of interest. Cereals dominate in the utilization of arable land. Forest vegetation is unevenly distributed in the given area. Acreage of the forests is 24,874 ha, and 16.8% of the territory is predominantly concentrated in sub-mountain villages in the northern part of the area (Figure 3). The territory ties to the Little Carpathians Protected Landscape Area (PLA).

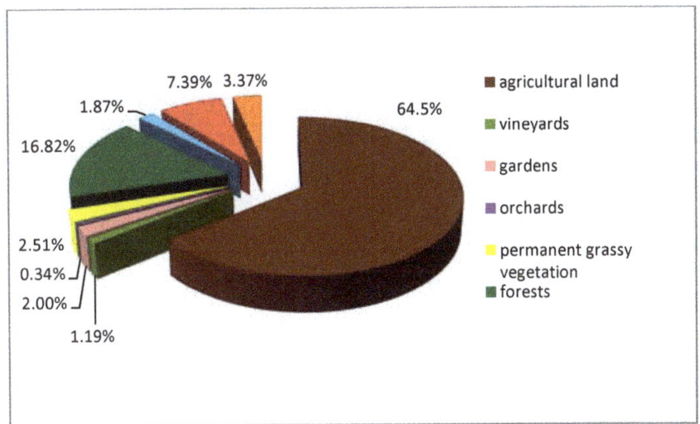

Figure 3. Land use in Trnava district in 2018 [64].

From a geomorphological point of view, the territory lies on Trnava loess plateau, the Danubian plain and the Little Carpathians mountains. The central and southern part belong to the Danubian lowland, i.e., the center of the territory. The Little Carpathians form the northwest edge of the study area. The geomorphology of this area also conditions the properties of other landscape components and the occurrence of natural resources, which then influence the functional use of the area. High quality soils dominate among the natural resources and, in conjunction with favorable climatic conditions, create significant potential for agricultural development.

In the northern part of the district are rich forest resources. Some of the forests are exploited for timber production, and parts of the forests provide soil protection, mainly on steep slopes of Little Carpathians. Other protective forests are linked to hygienic protection of water sources. Most of the forest areas belong to the Little Carpathians Protected Landscape Area (PLA). The main economic base is industry and agriculture. Industrial production is concentrated in the city of Trnava. Industrial production is dominated by mechanical engineering, followed by food and textile industries. Production of electricity in the Trnava district is noticeable at a national level, due to the presence of the Jaslovské Bohunice nuclear power plant.

3. Results

Each landscape unit has diverse potential for fulfilling ecosystem services and for realization of individual socio-economic activities. The same landscape unit can provide the potential for a number of ecosystem services: biomass production and drinking water supply, as well as various non-productive functions, such as soil protection, biodiversity protection, etc. The utilization of many ecosystem services is often associated with negative impacts on the landscape and its components; e.g., the production of agricultural biomass is associated with intensive chemical use and mechanization, which is the cause of contamination and degradation not only of soil but also of water resources.

Our study area serves as an example for the assessment of environmental problems that are the result of conflicts between land use and nature protection, protection of natural resources and protection of the environment. The potential of ecosystems to provide services is not always used effectively. Productive ecosystem services are used more intensively (i.e., biomass for food production, drinking water, technical purposes etc., and, more recently, cultivation of crops for energy production), because their use is connected with direct economic gains, easily expressed by money (i.e., costs of agricultural production, water, wood, energy, etc.). This is reflected in the threat to non-productive and cultural ecosystem services. When using ecosystem services, the main focus is on one ecosystem service and the needs of other ecosystem services in the area are not always taken into account. This often results in conflicts in the use of ecosystem services and consequently in environmental problems.

The use of some productive ecosystem services is linked to the production of undesirable polluting substances, i.e., stress factors that cause changes in ecological conditions of individual ecosystems (air pollution, soil contamination, vegetation damage, water contamination, noise and light pollution, etc.). Such altered environmental conditions limit the use of not only non-productive services, but also many productive services (soil contamination versus food production, loaded environment versus recreational activities, polluted water versus drinking water use, deterioration of environmental conditions versus biodiversity, etc.).

Transport of benefits from individual ecosystems to processors or end-users is also associated with negative impacts on surrounding ecosystems and thus makes it more difficult to use non-productive and productive ecosystem services. A typical negative example could be the threat to agri-ecosystems or cultural ecosystem services located near intensively loaded transport corridors (i.e., development of recreation due to negative transport impacts). The most busy transport corridors in the territory are: highway D61—Bratislava-Trnava-Piešťany; first-class road I/51—Trnava-Sereď, Trnava-Boleráz-Trstín; and, first-class road I/61—Cífer-Trnava-BučanyLeopoldov. The traffic intensity on these transport corridors reaches up to 50,000 vehicles per 24 h [65]. According to measurements, the equivalent noise level (LAeq) along the listed routes is 60–75 dB(A), and the maximum noise level (LAmax) is in the range of 80–100 dB(A), which is considered a significant noise load [66]. Such loaded areas are threats not only to biodiversity, but also restrict the development of socio-economic activities sensitive to hygienic environmental parameters, especially cultural and recreational services.

Although there have been positive trends in the production of emissions recently, there are still 184 large and medium sources of air pollution located in the territory of the Trnava district. In 2016, 92.23 metric tons of solid pollutants were released into the air; 144.37 tons of SO_2; 270.84 tons of No_x; and, 118.33 tons of CO. Most of the sources are concentrated in the industrial center of Trnava. A few sources are also located in rural settlements—Boleráz, Smolenice and others. Areas with air pollution form concentrated elliptical zones in the outskirts of Trnava city.

These pollutants subsequently penetrate other components of the environment—soil, water, river sediments, etc. Increased concentrations of some foreign substances in the soil have been reported in the study area (cadmium, chromium, arsenic, magnesium and copper). Increased concentrations of cadmium in the soil were recorded in the range of 0.3–0.6 mg·kg^{-1} in the southwestern part of the territory and in the Váh river alluvium. Across the region, higher chromium concentrations were recorded in the soil; these concentrations ranged from 85–101 mg·kg^{-1}, which is above allowed limits according to the Decree of the Ministry of Agriculture of the Slovak Republic no. 531/1994-540. Increased arsenic values occur both in the Little Carpathians, where they are of geogenic origin, but also in the lowlands where they are the result of intensive agricultural chemical use. Above-the-limit concentrations of manganese occur in the Little Carpathians, where they range from 0.14–0.41 mg·kg^{-1}. Increased content of copper is found in the study area in the wine-growing areas (i.e., the foothills of the Little Carpathians) [67].

Nitrates reach 62 mg/L in the Horné Orešany water resource. Elevated concentrations also occur in the central part of the region, which is the result of intensive agricultural production during the socialism period. The intensity of utilization of agricultural land significantly decreased after 1990 (after transition); a large number of hard accessible locations remained abandoned, there was a noticeable decrease in use of synthetic fertilizers, the intensity of mechanization decreased, etc. The consumption of synthetic fertilizers fell by almost half after transition, from 430 kg/ha to 243 kg/ha of agricultural land. Polluted water resources are a limiting factor for the development not only of non-productive services, but especially of productive ecosystem services. Polluted water is not suitable for drinking purposes. Water flows of the study area belong to the most polluted within Slovakia. Quantities exceeding limits of diatomic oxygen (O_2), biochemical oxygen demand, manganese, nitrogen and its forms, phosphates, total phosphorus, coliform bacteria and chlorophyll have been recorded in the main water flow of the Trnava district. Pollution is caused by discharging waste water from industry and urbanization. Agricultural production is also a major contributor to the pollution. A similarly unfavorable situation is also found on the Dolný Dudváh river, which shows quantities exceeding limits of phosphates, total phosphorus, coliform bacteria and chlorophyll.

The production of polluting substances also manifests in damage to the forest ecosystem. The most heavily loaded forest ecosystems are located in the northern part of the region, near Dobrá Voda municipality, where increased content of Mg, Ni, Se, Sr, Ca, Cu and F was recorded [68]. The loads of the area by stress factors is presented in Figure 4.

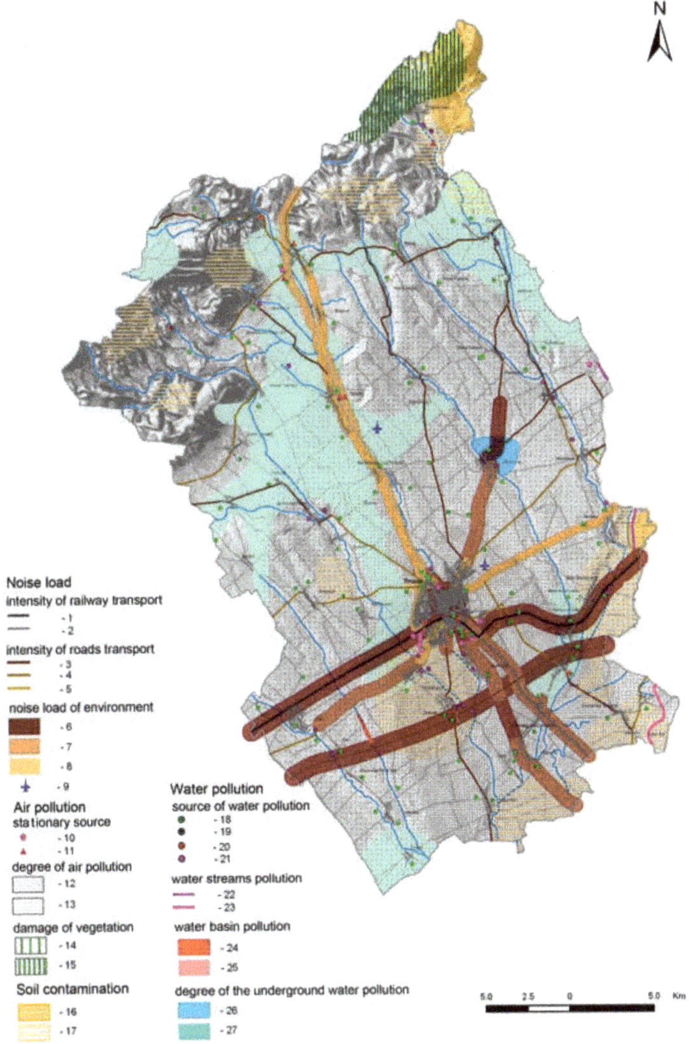

Figure 4. Anthropogenic stress factors (Noise load: 1—high, 2—medium, 3—very high, 4—high, 5—medium, 6—very high, 7—high, 8—medium, 9—airport; Air pollution: 10—large, 11—other important, 12—high, 13—medium, 14—high, 15—low; Soil contamination: 16—very high, 17—high; Water pollution: 18—from agriculture, 19—from industry and transport, 20—waste dump, 21—polluted water discharge site, 22—I. degree of pollution, 23—III. degree of pollution, 24—III. degree of pollution, 25—II. degree of pollution, 26—high, 27—medium).

The contaminated landscape components act as stressors towards other landscape constituents and also restrict and limit the development of individual socio-economic activities.

Issues associated with conflicts in the use of individual ecosystem services and natural resources in the study area can be divided into categories as follows.

3.1. Problems of Endangering of Ecological Stability of the Landscape

These problems arise from the effects of stress factors on areas with a high eco-stabilizing effect, such as protected areas, forests, wetlands, green areas and below. The following conflicts can be included in this group:

(a)   Endangering the spatial ecological stability of the landscape and endangering of the biodiversity by intensive agricultural production

Individual natural ecosystems have been gradually taken up, disposed of and replaced by agri-ecosystems due to the development of intensive agriculture [69]. This has led to the gradual transformation of the Trnava region area into a monofunctional intensely utilized agricultural landscape with a low degree of ecological stability. The current landscape structure is dominated by arable land, which occupies 61% of the total area of the region. Intensive agricultural production (intensive chemical use and mechanization) is a significant stress factor in terms of biodiversity. The assessment of the ecological stability of the study area was carried out on the basis of the change of natural representative geo-ecosystems (REPGES). REPGES are landscape units based on abiotic complexes and potential vegetation. Altogether 96 REPGES were allocated in the area of the region [70]. The assessment of the preservation/anthropogenic change of REPGES was performed on the basis of the coefficient of naturalness [71], which expresses the actual percentage of natural vegetation within the different types of REPGES. Up to 17.7% of the REPGES types do not have any natural vegetation and up to 27% of REPGES have no natural vegetation percentage greater than 1%. 30% of REPGES types have a 50% representation of natural vegetation; these are REPGES located in the northern part of the region within the Little Carpathian PLA. The structure of the REPGES types based on the coefficient of naturalness is expressed in Figure 5.

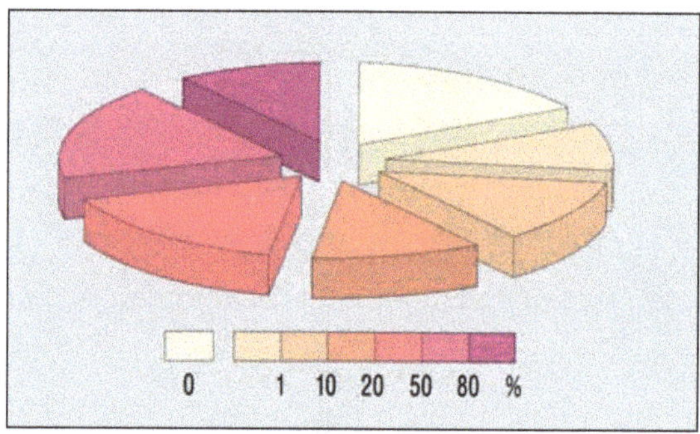

Figure 5.   Percentage representation of natural ecosystems within individual representative geo-ecosystems (REPGES)/coefficient of naturalness.

(b)   Endangering the spatial ecological stability of the landscape by the elimination of landscape greenery

Numerous harvests of natural linear habitats have been recorded in the area, and these habitats were then replaced by robinia and poplar artificial monocultures. As a result of the intensification of agricultural land, many boundaries, windbreaks and last remnants of linear vegetation, which prevented the use of heavy machinery, have been eliminated. Many of these have been replaced by artificial, non-original ecosystems. After the transformation (after 1989), agricultural production has declined and many agricultural parcels, mainly those that are isolated, have been abandoned. Many areas are currently transforming into built-up areas in the lowlands, and into forests in the foothills. The development of the share of agricultural land, forests and built-up areas is shown in Figure 6. The proportion of vegetation within the boundaries of cities and towns has decreased [72]. The share of public vegetation per capita in the region is 29.9 m^2, which is below the Slovak average (34 m^2 per inhabitant).

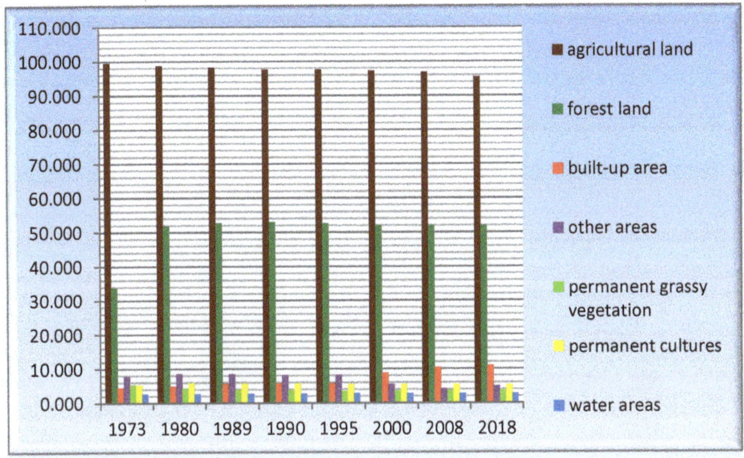

Figure 6. Development of land use in Trnava district (ha).

(c)   Endangering of human environment by the intensive agriculture and spreading of ruderal vegetation

Agrocenoses brought and spread non-original plant species, mainly weeds from the Secalietea class. Species such as Consolida regalis, Tithymalus exiguus, Tithymalus falcatus, Mercurialis annua, Lathyrus tuberosus, Stachys annua, Mercurialis annua, Convolvulus arvensis and others are common in this area. This vegetation occurs in man-affected sites or artificial habitats. The largest expansion of this vegetation has been recorded in intravilan, but it is also often found in extravilan, especially next to field trails, agricultural objects and dumps. Many species also penetrate into relatively natural habitats. Among the species most commonly found in ruderal habitats are: Urtica dioica, Ballota nigra, Agropyron repens, Cirsium arvense, Lactuca serriola, Convolvulus arvensis, Conium maculatum, Falcaria vulgaris, Artemisia vulgaris, Amaranthus retroflexus, Chenopodium spec. div., Atriplex spec. div., Arctium lappa, Panicum miliaceum, Cannabis ruderalis, Aster novi-belgii agg., Conyza canadensis and others. These species also reduce the aesthetic value of the environment. Among the invasive species can be found Robinia pseudoacacia, Negundo aceroides, Lycium barbarum, Ailanthus altissima, Conyza canadensis, and Fallopia japonica [73].

(d)   Endangering of the biodiversity by timber production

This is a threat to protected areas and elements of the territorial system of ecological stability (TSES) due to intensive wood harvesting, including in Vlčkovský háj protected area, Buková nature reserve, Kamenec biocentre, Šarkan, Farský Mlyn and Holý vrch biocentres, and Podmalokarpatský and Parná bio-corridors. The problem is also wood harvesting in the Little Carpathians PLA, which has a main function of biodiversity protection. The area of the economic forests in the region is 70% of the total forest size, which determines their exploitation for wood production. The problem is that up to 90% of forests are located in the Little Carpathians PLA and its protection zone, where there are conflicts of interest between nature conservation and forest owners and users. Wood harvesting is about 50,000 m^3 per year; in 2015 it was 58,376 m^3 and in 2017 it was 53,563 m^3.

(e)   Endangering of the biodiversity by water resources exploitation

Excessive water consumption and unnecessary waste of water leads to a reduction in surface and ground waters, resulting in a change in the ecological conditions of many ecosystems, especially wetlands. Climate change also contributes to decreasing water flow rates. Water depth in the Trnávka stream in the mountains reaches an average value of 105 cm; in the lowlands its depth is only 17 cm. In particular, floodplain forests in the vicinity of the region's water flows are threatened as a result of an uncontrollable water supply for irrigation of gardens, especially during dry seasons in summer. Although water consumption after the transformation has decreased, it can still be considered high. In 2015, specific water consumption was 165 liters per inhabitant per day [74]. Unmanageable use directly from watercourses or from private wells, which contribute to a drop in surface or groundwater, are particularly dangerous.

(f)   Conflicts between nature conservation and recreational resources

Relationships between non-productive and cultural-recreational services are reciprocal. On one hand, the intensive and inefficient use of cultural-recreational services is a threat to nature and biodiversity protection; on the other hand, the protection zones and protected areas declared to protect non-production ecosystem services are limitations and constraints for the use of cultural ecosystem services. The intensive and uncontrollable development of recreation is a negative factor in terms of biodiversity protection and protection of the nature and landscape of the Trnava region. Plant ecosystems are endangered by trampling, plucking rare plants, storing waste, etc. For fauna, the main stress factors are noise, disturbance, destruction of nesting sites and poaching [75]. IN particular, areas of the Little Carpathians Protected Landscape Area (PLA) and the Trnavské rybníky Protected Area (PA) are threatened by recreation. Trnava ponds PA is negatively impacted by intensive fish farming and sport fishing, as well as by increasing tourism and expanding construction in the surroundings. Recently, individual housing construction has occurred significantly closer to the boundaries of this protected area. The Little Carpathians PLA is threatened due to uncontrolled cottage construction. An additional problem is also the storage of waste in protected areas and the creation of unorganized waste dumps.

3.2. Problems of Endangering the Natural Resources

These arise due to the effect of stress factors on individual natural resources. The result is a qualitative and quantitative threat to individual natural resources.

(a)   Endangering the soil resources by industrialization and urbanization

Socio-economic development in the region is accompanied by pressures on agro-ecosystems, often with the highest quality soils. Up to 70% of agriculture soil fund belong to the fourth-best categories within Slovakia [76]. After the changes doe to transformation in 1990, the region has undergone intensive socio-economic development characterized by significant spatial impacts. Construction of industrial and logistic facilities, transport networks, shopping centers and residential areas takes up the largest areas of soil. Construction works often utilize best quality soil. An example of such a negative impact on agri-ecosystems is the construction of PSA Peugeot Citroën Slovakia and several electrotechnical plants, of which Samsung Voderady could be considered as the most dominant. While these units have contributed to improving socio-economic conditions of the region, their development is also associated with negative impacts, especially in the environmental field. New industrial plants have been built, often on "green meadows", and are causing significant pressures on natural and semi-natural ecosystems and on natural resources. PSA Peugeot Citroën Slovakia was built on 192 ha of quality soils (the best quality in Slovakia).

The decline of arable land is an environmentally negative phenomenon, especially in relation to removing arable land from the agricultural land fund and re-allocating it to the built-up area [77].

At the same time, the closure and liquidation of agricultural cooperatives and industrial plants has caused an increase in abandoned buildings and ruins. After Bratislava, Trnava is one of the regions with the greatest increase in built-up areas in Slovakia [78], as new industrial and commercial activities are located on regional and trans-regional axes, as well as in the city center. New technical buildings have also been constructed in rural settlements, where atypical garden suburbs are created. A similar situation exists in other European countries. Agricultural land is reduced. In the EU during the past 15 years, about 6.5% (12.5 million ha) was occupied. Daily surface sealing in Germany is approx. 71 ha, and in the Czech Republic approx. 25 ha [79]. This construction is linked not only with a whole range of environmental issues, but also with social problems. Immigrant populations are often difficult to assimilate into the settlement community, and new elements are emerging in settlement communities—isolation, closeness, separation, constraint of communication, egoism, preference of own local interests, loss of interest in public affairs, etc., which is reflected in a change of lifestyle and consequently in the quality of community life. New psychosocial phenomena such as divorce, unrestrained lifestyles, crime, drugs, etc., are beginning to permeate [80]. "Consumer egoism" can also be seen in the increase in the built-up area size per capita (Figure 7) [1].

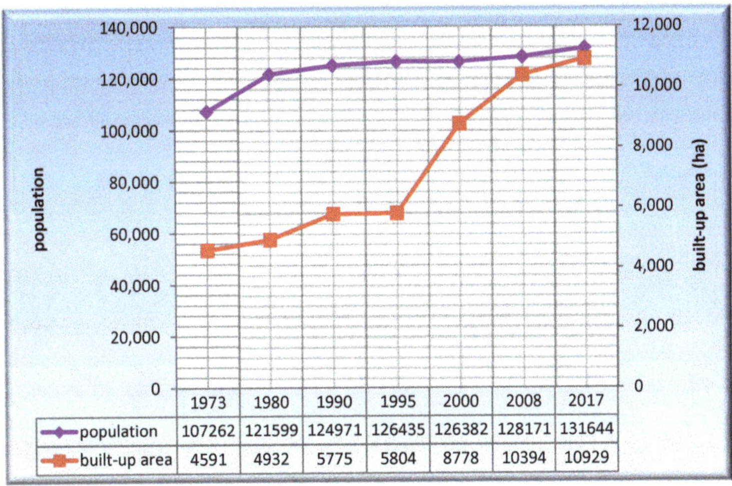

	1973	1980	1990	1995	2000	2008	2017
population	107262	121599	124971	126435	126382	128171	131644
built-up area	4591	4932	5775	5804	8778	10394	10929

Figure 7. Built-up area vs. population in Trnava district.

(b)   Endangering the soil resources by intensive agriculture

Inappropriate management of the soil by its owners and users of the land in biomass production causes physical degradation of the soil, especially by the effects of hard packed soil and erosion processes. Such disturbed soil weakens non-productive functions, such as soil biodiversity protection, soil quality protection, etc. Degraded soil also reduces biomass production. In the Trnava district, up to 80% of ASF are at risk due to hard packed soil, of which up to 65% are at risk due to inappropriate management of the soil [76]. Approximately 30% of ASF are threatened by wind erosion, of which 7.9% are in the category of severe erosion and 1% in the category of extremely endangered soils by wind erosion. Wind erosion causes secondary dustiness in settlements lying in the lowlands of the region. About 30% of ASF are threatened by water erosion, of which heavy erosion occurs on 18% of ASF and 2.5% of ASF are in the category of extreme erosion with soil transport (40 t/ha/year) [76]. Flat windy parts of the Trnava plain are at risk in terms of spatial wind erosion, while water erosion dominates in the mountainous areas of the Little Carpathians.

(c)   Conflicts between nature conservation, water protection and use of soil resources

Protected zones and areas associated with the protection and use of many non-productive functions restrict and limit intensive development of agricultural production and wood harvesting and thus make it more difficult to use agri-ecosystems and forest ecosystems for biomass and wood production (e.g., protected areas and their protected zones, zones of hygienic protection of water sources, etc., are limitations for the wood harvesting development). Regarding the protection of waters by the Nitrates Directive (91/676/EC), 44.3% of the territory was classified as land with a middle-class constraint for land management and 0.05% is in the category with the highest level of farming constraint on the land [76]. Protected and special purpose forests constitute 13% and 17%, respectively, which acts as a limit for forestry development. Protected areas and various protected zones, mostly kept for the protection of non-productive functions, are limiting the development of cultural and recreational services and the development of recreational activities, similar to the case for productive services.

3.3. Problems of the Endangering of Human Environment

These problems arise from the effects of stress factors on the human environment—the effect of stress factors on residential and recreation areas. The following issues can be included in this category.

(a)   Conflicts between development of socio-economic activities and the environment

As we have already mentioned above, the realization of many socio-economic activities is associated with the production of polluting substances which contaminate the individual components of the environment, and which subsequently results in deterioration of the quality of the environment. Impaired environmental quality includes that of the Trnava city center, where several sources of air and noise pollution are located. There are also threatened settlements lying around Trnava (see Figure 4). Inappropriate localization of production facilities, as well as abandonment of rubbish, is reflected in the reduced aesthetic value of individual settlements.

(b)   Conflicts between agriculture and recreational resources

Areas with a loaded environment are not suitable for the development of recreation and tourism in terms of hygienic conditions. Similarly, an agricultural landscape with a low degree of ecological stability that is used intensively and one-sidedly is not attractive for recreation and tourism development. There is no developed tourism in the entire central part of the Trnava region, which is extensively farmed. Tourism is concentrated only in parts of the Little Carpathians PLA, where it is in conflict with nature conservation. In agricultural areas, tourism is concentrated in water reservoirs—Suchá nad Parnou, Boleráz, Buková. Boleráz and Suchá nad Parnou water reservoirs were originally built for agricultural irrigation, but currently are used mainly for water regulation in the country and for sport purposes. In particular, sport fishing is popular there. Swimming is limited due to poor water quality, because the content of coliform bacteria, faecal streptococci, phenols, nonpolar extracts and cyanobacteria is excessive. These increased indicators point to an increased degree of eutrophication caused by agricultural activity and, in particular, by municipal pollution [81].

(c)   Conflicts between forestry and recreational resources

Wood harvesting is also associated with negative environmental impacts, such as degradation of forest ecosystems, increased noise and dustiness, etc. These factors represent the main stressors and limits of the development of cultural and recreational activities. This conflict is typical for the Little Carpathians PLA. The degree of limitation depends on the nature of the protected zone and the degree of protection. With an increasing degree of protection, the degree of limitation for the development of individual recreational activities is also increasing [82].

3.4. Integrated Assessment of the Conflicts of Interests in Trnava District

The map in Figure 8 presents the integration of all spatial encounters of the interests of all sectors described above.

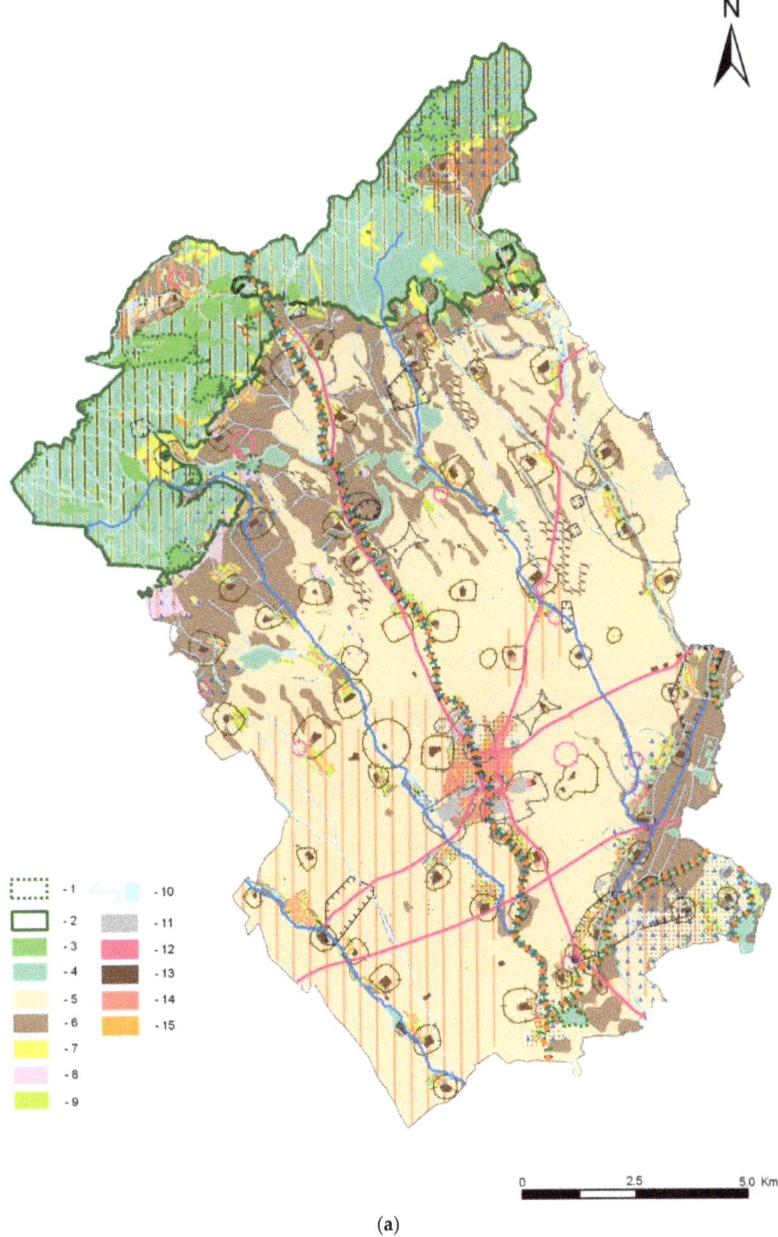

(a)

Figure 8. *Cont.*

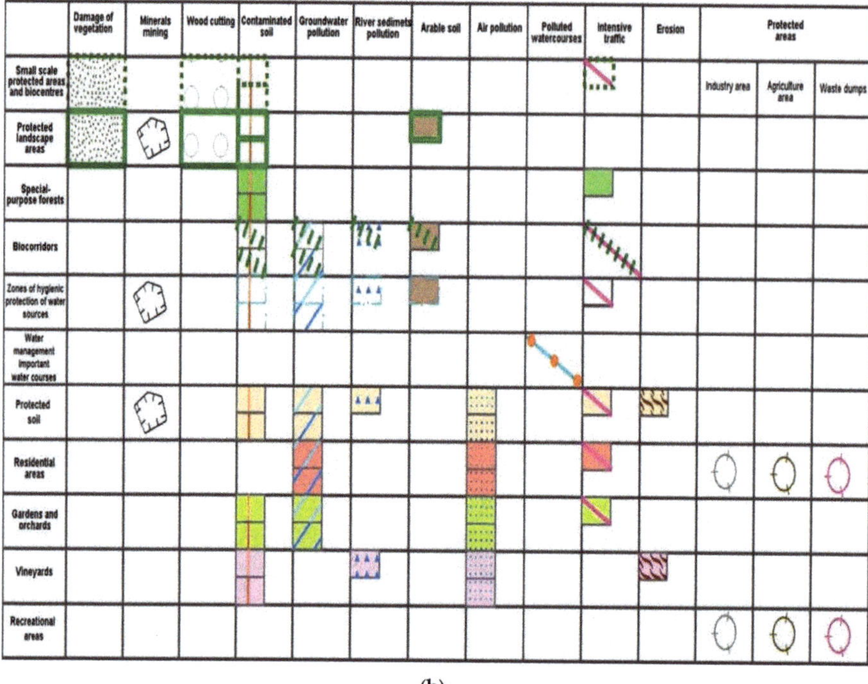

(b)

Figure 8. Conflicts of interests in the landscape—environmental problems: (**a**) The input characteristics. 1—Small scale protected area, 2—Protected landscape area Malé Karpaty, 3—Special-purpose forests, 4—Productive and other forests and non-forest woody vegetation, 5—Protected soils, 6—Other types of soils, 7—Permanent grassy vegetation, 8—Vineyards, 9—Gardens and orchards, 10—Watercourses and all surface water, 11—Industrial areas, 12—Waste dump, 13—Agricultural areas, 14—Residential areas, 15—Recreational areas; (**b**) Matrix of conflicts of interests in the landscape—environmental problems. Rows represent environmentally positive factors; columns represent environmentally negative factors; squares represent real existing encounters of positive and negative factors in the model territory.

The input data are the relevant elements of all three structures of geosystems—PLS, SLS and TLS—which occur in the Trnava district.

The decisive elements of the primary landscape structure (PLS) in the district are soils of high quality ranked to 2 classes (keys 5 and 6 in the legend to Figure 8), and the water courses and other surface waters (key 10).

The decisive elements of the secondary landscape structure (SLS) are forests and other woody vegetation in the landscape (keys 3 and 4), arable land (keys 5 and 6), grasslands (key 7), vineyards (key 8), gardens and orchards (key 9), industrial areas (key 11), waste dumps (key 12), agricultural objects and areas (key 13), and residential areas (key 14).

The decisive elements of the tertiary landscape structure (TLS) in the district are bounded to nature conservation (key 1 and 2), to protection of forest resources (key 3) and soil resources (key 5).

It is obvious that these elements are mutually interrelated and they partially overlap each other. For example:

- The high quality and other productive soils are used as arable land (key 5 and 6). Moreover, the highest quality soils were declared for protected soils.
- The nature conservation areas are declared for forests; thus, these forests were declared as special-purpose forests (key 1, 2, 3 and 4).

- The industrial areas, agricultural objects, waste dumps are bound with their safety and hygienic zones, etc.

In addition to these permanent elements, the map also shows the spatial distribution of the contamination by industry.

The spatial projection of the factors described above allows the assessment of their spatial encounters from the ecological/environmental point of view as characterized in Section 3:

- rows represent environmentally positive factors that are endangered by negative factors;
- columns represent environmentally negative factors that endanger the positive factors;
- squares represent the real existing encounters of positive and negative factors in the model territory.

The analysis and characteristics of the encounters and conflicts is the base for finding solutions for their mitigation.

4. Discussion

As shown with the example of the model area, the environmental problems issuing from non-harmonized approaches to land use cause real environmental problems. Environmental issues related to landscape changes and the need for solutions have been highlighted by the European Environmental Agency (EEA) and many other authors who point to the continuing threatening of natural ecosystems, either direct or by changing their ecological conditions [83–93].

The encounter of interests of sectors in the model territory appear basically in 4 modes:

(a) The sectors "fight" for land to broaden their own territory. This is expressed in the building of new objects for industry, transport and residential areas. An important specific problem of the district is that most of its soils are of the highest quality and productivity, so are protected. Nevertheless, developers are able to find ways to change existing territorial plans and gain land, even if they must pay compensation for the occupied land. The problem of occupation of the best soils for industry, shopping centers and residential complexes has been the topic of much research [94–97]. Another specific problem is the motorway, which creates a massive barrier against migration of the biota. This issue is to be addressed according to the result of the project of the Regional Territorial System of Ecological Stability of the district [98].

(b) The sectors of forestry, agriculture, water management, nature conservation and recreation utilize large areas where their activities overlap. Characteristic of this group is the spatial overlap of the Protected Landscape Area Small Carpathians with special purpose and protective forest and water protective zones, which are mutually supportive. The opposite case is the spatial overlap of high quality soils in the water protective zones, which has a characteristic of conflict since the use of agrichemicals in order to exhaust the productivity potential endangers and pollutes water resources. This conflict should be mitigated by respecting the legal limitations for agriculture. Nevertheless, as the monitored data show, the pollution of water resources still occurs.

(c) "Internal" conflicts within the sectors, i.e., conflict of the activity with its own land resources. This is a typical problem of Trnava district. Intensive agriculture on arable land on steep slopes and on shallow soils causes erosion. This problem is to be solved urgently by acceptance of projects for land consolidation, which also include a proposal for a local territorial system of ecological stability; this is, in fact, a detailed proposal for landscape greenery, which provides complex eco-stabilizing functions, including protection against erosion [99–101].

(d) Production of pollution: most typical are air pollution, which influences the residential and recreational areas, and water pollution, which endangers the water quality both of surface and underground water. A very specific feature of the district is the presence of the nuclear power plant, Jaslovské Bohunice. According to the Environmental Regionalization of the Slovak Republic, the territory of the district of Trnava belongs to the most environmentally burdened regions of Slovakia [77].

Some Characteristic Issues Concerning the Model Territory

The most important driving forces in the area of interest were the period of socialization and collectivization, and the process of transformation. The collectivization period (1950s–1960s) was characterized by the establishment of cooperatives. Thus, the process of concentration and consolidation of estates began, and the creation of a monofunctional agricultural landscape started. This led to the disappearance of many valuable representative ecosystem types. A similar process was also recorded in forestry. Intensive wood harvesting took place and inappropriate forms of forest management began, which subsequently threatened the natural character and species composition of forest ecosystems. Typical for this period was also the thoughtless removal of orchards, gardens and small-blocked vineyards, thus eliminating traditional gardening and fruit growing. Similarly, the last remaining ecosystems were removed, and the whole area of lowlands was transformed into monotone, intensively used arable land with a low degree of ecological stability [102].

The period after 1989 is characterized by the reduction of central planning and the transition to a market economy. During this period, strong pressure from investors to build new industrial facilities and residential areas is typical. This is exemplified by automotive and technology parks, logistics centers and transport infrastructure. The region has become one of the most industrialized areas in Slovakia. This investment is quantitatively reflected in sequestration of natural resources and semi-natural ecosystems, and qualitatively by threatening both environmental quality and natural resources. The growth in built-up areas has altered natural and semi-natural ecosystems and changed agricultural and forest-based land use. The accompanying loss of arable land, planting of forests with increased unused agricultural areas, and the elimination of vineyard and fruit orchard crops have changed the landscape of the model territory [78,103,104].

Similar development was also observed in regions in other post-communist countries (Czech Republic, Poland, Hungary and Bulgaria), where original activities were replaced by new ones. These took the form of commercialization [105,106], construction of new apartments [107], brownfield regeneration [108,109] and the establishment of new commercial centers [110,111]. The emergence of brownfields and decline of some housing estates constructed during socialism highlighted major problems remaining in post-socialist cities [112].

The basic tool for eliminating these problems in the landscape is the application of integrated landscape management aimed at the ecologically optimum spatial organization, utilization and protection of the landscape, which results in a proposal for the most suitable localization of demanded human activities within the given territory (what and where?) and successively to the proposal of necessary measurements ensuring the ecologically proper functioning of those activities on the given locality (how?). The European Union has shown a strong effort towards the application of integrative tools towards the optimum utilization of natural resources. In particular the Integrated Pollution Prevention and Control (IPPC) Directive, the Network of nature protection areas (NATURA 2000), and the Water Framework Directive should be mentioned in this respect. These are intended to be supported by the Infrastructure for Spatial Information in the European Community (INSPIRE). Directive, which enforces the integrated spatial information system as the inevitable basis of an integrated approach. The Water Framework Directive defines integrated river basin management *as the process of coordinating conservation, management and development of water, land and related resources across sectors within a given river basin.* Other concepts of integrative character include the Europäische Raumordnungs Charta, the Alpine Convention, the Carpathian Convention, the Danube River Protection Convention and the European Landscape Convention of the EEC, and, naturally, also the EUSDR and other regional strategies.

5. Conclusions

As characterized in the paper, the encounters, conflicts and environmental problems show significant diversity, which underlines the need for integrated solutions. The presented assessment procedure resulted in the identification and analysis of the conflicts of interests in the territory and their projection to a map. However, this is only the first step in the solution of problems, and should be followed by implementation of procedures for ecologically optimum spatial organization and utilization of the territory in regular spatial planning processes.

Land use changes are also the main root of many environmental problems [113–115]. This is the consequence of the sectoral approach to land use. Therefore, the assessment of land use changes should be followed by the evaluation of their environmental impacts. This evaluation requires new methods that consider not only the complexity of socio-economic dynamics, but also interdependencies between drivers, impacts, and responses to these dynamics [116].

In line with the basic provisions of AGENDA 21 and newer environmental-political documents enhancing the integrative approaches, we consider land-use planning the main tool for influencing sectoral activities in an early stage of planning and, thus, as the tool for harmonization of sectoral approaches. Local government should respond to all current challenges of spatial development, trends and environmental issues [73,117–120].

Slovakia has well-developed landscape-ecological planning [1]. The application of landscape documents (landscape plans and ecological networks) to spatial planning processes is legally anchored. However, in practice, landscape-ecological principles are difficult to integrate into spatial planning documentation for a range of different reasons. In conclusion, these include the significant pressures of investors and landowners to invest on "green meadows", which leads to the occupation of natural areas. The owners are no longer interested in the management of small lots of land or have no technical, financial or human options for management, and thus their aim is to evaluate properties by realizing investment plans. Similar situations occur in other post-communist countries [105,121–124]. In Slovakia, Poland, the Czech Republic, and Hungary the process of shaping the structure of land ownership started anew after 1990, with dynamic changes taking place in land use [123–126]. This was especially visible in metropolitan areas, where competition for land was the highest, and the development of private sector enterprises and growing urbanization resulted from conversion of agricultural land [125–127].

One other problem is the low environmental awareness of local authorities who approve development projects. Investment objectives and growing urbanization were priorities in most municipalities, while ecological interests and priorities are at the edge of residential development goals [128–131].

Therefore, in spite of professional and legal support, the actual implementation of comprehensive physical planning (in different countries referred to as land-use planning, spatial planning, landscape planning, etc.)—which should provide a complex framework for harmonization of the activities of all sectors—is in practice inadequate and needs significant strengthening. The sectoral approach is still prevailing, and the developer's plans are aimed firstly at localization and realization of industrial, transport, business and housing activities without adequate consideration of other sectors, such as agriculture and nature protection. These problems occur all around the word as the consequence of global megatrends.

The theoretical, methodical and legal conditions to approximate the integrated approach and harmonization of land use in order to avoid conflicts of interests of sectors causing environmental problems are given here. The problem remaining to be solved is the strengthening of their application into planning practice and realization.

Author Contributions: Z.I. contributed to the creation of the theoretical context of work and also wrote the majority of the discussion and conclusion. She processed the assessment of environmental problems on study area. L.M. compiled the methodology and theoretical context of work and also contributed to create a discussion and conclusion. V.M. participated in the evaluation of interest encounters in the study area and processed graphical attachments and maps in the GIS.

Funding: This research was funded by grant agency VEGA—Ministry of Education SR and Slovak Academy of Sciences (No. 2/0066/15 Green Infrastructure of Slovakia and No. 1/0096/16: Ecosystem services of the landscape ecological complexes in the area of the UNESCO World Cultural and Natural Heritage site Banská Štiavnica and surrounding technical monuments).

Acknowledgments: This research was funded by grant agency VEGA—Ministry of Education SR and Slovak Academy of Sciences (No. 2/0066/15 Green Infrastructure of Slovakia and No. 1/0096/16: Ecosystem services of the landscape ecological complexes in the area of the UNESCO World Cultural and Natural Heritage site Banská Štiavnica and surrounding technical monuments).

Conflicts of Interest: The authors declare no conflict of interest.

References

1. Izakovičová, Z.; Mederly, P.; Petrovič, F. Long-term land use changes driven by urbanisation and their environmental effects (example of Trnava City, Slovakia). *Sustainability* **2017**, *9*, 553. [CrossRef]

2. Munteanu, C.; Kuemmerle, T.; Boltiziar, M.; Butsic, V.; Gimmi, U.; Kaim, D.; Király, G.; Konkoly-Gyuró, É.; Kozak, J.; Lieskovský, J.; et al. Forest and agricultural land change in the Carpathian region—Ameta-analysis of long-term patterns and drivers of changes. *Land Use Policy* **2014**, *38*, 685–697. [CrossRef]

3. Lieskovský, J.; Burgi, M. Persistence in cultural landscapes: A pan-European analysis. *Reg. Environ. Chang.* **2018**, *18*, 175–187. [CrossRef]

4. EEA. *The European Environment State and Outlook 2015. Assessment of Global Megatrends*; European Environmental Agency Copenhagen: København, Denmark, 2015; p. 134.

5. FAO. *The State of the World's Forests 2018—Forest Pathways to Sustainable Development*; Licence: Rome, Italy, 2018.

6. Mitchley, J.; Tzanopoulos, J.; Cooper, T. Reconciling conservation of biodiversity with declining agricultural use in the mountains of Europe. In *Interdisciplinary Research and Management in Mountain Areas*; Taylor, L., Ryall, A., Eds.; Banff Centre Canada: Banff, AB, Canada, 2005; pp. 61–65.

7. The EU Biodiversity Strategy to 2020. Available online: http://ec.europa.eu/environment/nature/biodiversity/strategy/index_en.htm (accessed on 10 September 2018).

8. Grunewald, K.; Bastian, O. (Eds.) *Ecosystem Services. Concept, Methods and Case Studies*; Springer: Berlin, Germany, 2015; p. 312.

9. Fürst, C.; Helming, K.; Lorz, C.; Müller, F.; Verburg, P. Integrated land use and regional resource management—A cross-disciplinary dialogue on future perspectives for a sustainable development of regional resources. *J. Environ. Manag.* **2013**, *127*, 1–5. [CrossRef] [PubMed]

10. Krcho, J. Structure and spatial differentiation of the physical-geographic sphere as a cybernetic system. Bratislava. *Geogr. Čas.* **1974**, *26*, 132–162.

11. Haase, G. Izučenie topičeskich I choričeskich struktur, ich dinamiki I razvitija v landšaftnych sistemach. In *Structura, Dinamika I Razvitije Landšaftov*; Institut Geografii. AN SSSR: Moskva, Russia, 1980; pp. 57–81.

12. Miklós, L.; Izakovičová, Z. *Landscape as Geosystem*; VEDA, SAV: Bratislava, Slovakia, 1997; p. 152.

13. Miklós, L.; Izakovičová, Z.; Boltižiar, M.; Diviakova, A.; Grotkovská, L.; Hrnčiarová, T.; Imrichová, Z.; Kočická, E.; Kočický, D.; Kenderessy, O.; et al. *Atlas of Representative Geosystem of Slovakia*; UKE SAV: Bratislava, Slovakia, 2016; p. 123, ISBN 80-969272-4-8.

14. Ružička, M.; Miklós, L. Landscape-ecological Planning (LANDEP) in the Process of Territorial Planning. *Ecol. CSSR Cas. Ekologicke Probl.* **1982**, *1*, 297–312.

15. Miklós, L. *Ecological Generel, ČSSR Part: SSR. I. Stage. Spatial Differentiation of the Area from the Ecological Point of View*; ČSŽP Bratislava, ÚEBE CBEV SAV Bratislava, Stavoprojekt: Banská Bystrica, Slovakia, 1985; p. 125.

16. Miklós, L.; Kozová, M.; Ružička, M. *Ecological Plan for the Use of the East Slovak Lowland at 1:25,000. Ecological Optimization of Use of VSN*; ÚEBE SAV: Bratislava, Slovakia, 1986; pp. 5–312.

17. Izakovičová, Z. Ecological interpretations and evaluation of encounters of interests in landscape. *Ekológia* **1995**, *14*, 261–275.

18. Izakovičová, Z. Evaluation of the stress factors in the landscape. *Ekológia* **2000**, *19*, 92–103.

19. Izakovičová, Z. Integrated approach to the assessment of the agricultural landscape. *Geogr. Rev. Geogr. Geoekologické Štúd.* **2006**, *2*, 333–339.

20. Izakovičová, Z. *Integrated Landscape Management*; Institute of Landscape Ecology, SAS: Bratislava, Slovakia, 2006; p. 232.

21. Miklós, L.; Diviaková, A.; Izakovičová, Z. *Ecological Networks and Territorial System of Ecological Stability*; Springer International Publishing: Berlin, Germany, 2018; p. 159.

22. Krcho, J. The natural part of the geosphere as a cybernetic system and its expression in the map. *Geogr. Čas.* **1968**, *20*, 115–130.

23. Chorley, R.; Kennedy, B. *Physical Geography—A System Approach*; Prentice Hall Interantional: London, UK, 1971; p. 243.

24. Demek, J. The landscape as a geosystem. *Geoforum* **1978**, *9*, 29–34. [CrossRef]

25. Sochava, V.B. *Vvedenie v Učenije o Geosistemach (An Introduction to the Science of Geosystems)*; Nauka: Novosibirsk, Russia, 1978; p. 319.

26. Miklós, L.; Kočická, E.; Izakovičová, Z.; Kočický, D.; Špinerová, A.; Diviaková, A.; Miklósová, V. *Landscape as a Geosystem*; Springer International Publishing: Berlin, Germany, 2018; p. 245.

27. Izakovičová, Z.; Miklós, L. New concept of nature protection in the Slovak Republic. In *Nature Conservation Management: From Idea to Practical Results*; Chmielewski, T.J., Ed.; PWZN Print 6; Lublin-Lódz-Helsinki: Aarhus, Denmark, 2007; pp. 86–98, ISBN 83-87414-98-0.

28. Miklós, L.; Špinerová, A. *Landscape-Ecological Planning LANDE*; Springer: Berlin, Germany, 2018; p. 254.

29. Izakovičová, Z.; Miklós, L.; Drdoš, J. *Landscape-Ecological Conditions of Sustainable Development*; Veda: Bratislava, Slovakia, 1997; p. 183, ISBN 80-224-0485-3.

30. Kočická, E. Príklad Ohraničovania Komplexných Priestorových Jednotiek pre Krajinno-Ekologické Hodnotenia. *Acta Facultatis Ecologiae* **2011**, *24–25*, 55–65. Available online: https://fee.tuzvo.sk/sites/default/files/acta-fee-24-25-2011.pdf (accessed on 12 September 2018).

31. Miklós, L.; Izakovičová, Z.; Oferrtálerová, M.; Miklósová, V. The institutional tools of integrated landscape management in Slovakia for mitigation of climate change and other natural disasters. *Eur. Countrys.* **2017**, *9*, 647–657. [CrossRef]

32. AGENDA 21. United Nations Conference on Environment and Development. Rio de Janeiro (United Nations), A/Conf. 151/4; 1992. Available online: https://sustainabledevelopment.un.org/outcomedocuments/agenda21 (accessed on 10 September 2018).

33. Office for Official Publications of the European Communities. *European Spatial Development Perspective ESDP. Towards Balanced and Sustainable Development of the Territory of the EU*; Office for Official Publications of the European Communities: Luxembourg, 1999; ISBN 92-828-7658-6.

34. Directive of the 2000/60/EC of the European Parliament and of the Council—Water Framework Directive. 2000. Available online: http://data.europa.eu/eli/dir/2000/60/oj (accessed on 10 September 2018).

35. Council of Europe. *European Landscape Convention*; Council of Europe: Florence, Italy, 2000.

36. United Nations. *United Nations Millennium Declaration. Resolution Adopted by the General Assembly*; A/55/L.2; United Nations: New York, NY, USA, 2000.

37. Economic Comission for Europe. *Spatial Planning: Key Instrument for Development and Effective Governance with Special Reference to Countries in Transition*; ECE/HBP/146; UN Publication, Economic Comission for Europe: Geneva, Switzerland, 2008.

38. United Nations. *Transforming Our World: The 2030 Agenda for Sustainable Development*; United Nations: New York, NY, USA, 2015; Available online: http://www.un.org/ga/search/view_doc.asp?symbol=A/RES/70/1&Lang=E (accessed on 10 September 2018).

39. Bezák, P. Integrated approach to the evaluation landscape on the example of research in National Park Poloniny. In *Integrated Landscape Management—Basic Tool of the Implementation of the Sustainable Development*; Izakovičová, Z., Ed.; Slovak Academy of Sciences, Ministry of the Environment: Bratislava, Slovakia, 2006; pp. 125–130.

40. Cairns, J., Jr.; Crawford, T.V.; Salwasser, H. (Eds.) *Implementing Integrated Environmental Management*; Virginia Polytechnic Institute and State University: Blacksburg, VA, USA, 1994; p. 137.

41. Ružička, M.; Miklós, L. Basic premises and methods in landscape-ecological planning and optimisation. In *Changing Landscapes: An Ecological Perspectives*; Zonnenveld, I.S., Forman, R.T.T., Eds.; Springer: New York, NY, USA, 1990; pp. 233–260.

42. Bezák, P.; Mederly, P.; Izakovičová, Z.; Špulerová, J.; Schleyer, C. Divergence and conflicts in landscape planning across spatial scales in Slovakia: An opportunity for an ecosystem services-based approach? *Int. J. Biodivers. Sci. Ecosyst. Serv. Manag.* **2017**, *13*, 119–135. [CrossRef]

43. Miklós, L.; Kočická, E.; Diviaková, A.; Belaňová, E. *Integrated Landscape Management. Institutional Tools*; VKÚ. a.s.: Harmanec, Slovakia, 2011; p. 196.

44. Estrada-Carmona, N.; Hart, A.K.; DeClerck, F.A.J.; Harvey, C.A.; Milder, J.C. Integrated landscape management for agriculture, rural livelihoods, and ecosystem conservation: An assessment of experience from Latin America and the Caribbean. *Landsc. Urban Plan.* **2014**, *129*, 1–11. [CrossRef]

45. García-Martín, M.; Bieling, C.; Hart, A.; Plieninger, T. Integrated landscape initiatives in Europe: Multi-sector collaboration in multi-functional landscapes. *Land Use Policy* **2016**, *58*, 43–53. [CrossRef]

46. Milder, J.C.; Hart, A.K.; Dobie, P.; Minai, J.; Zaleski, C. Integrated landscape initiatives for African agriculture, development, and conservation: A region-wide assessment. *World Dev.* **2014**, *54*, 68–80. [CrossRef]

47. Scherr, S.J.; Heiner, K. *Towards an Approach to Integrated Landscape Modeling for the SDGs*; EcoAgriculture Partners: Washington, DC, USA, 2016.

48. Scherr, S.J.; Shames, S.; Friedman, R. Defining Integrated Landscape Management for Policy Makers. Ecoagriculture Policy Focus No. 10. 2013. Available online: http://www.un.org/esa/ffd/wp-content/uploads/sites/2/2015/10/IntegratedLandscapeManagementforPolicymakers_Brief_Final_Oct24_2013_smallfile.pdf (accessed on 11 September 2018).

49. Zanzanaini, C.; Tran, B.T.; Singh, C.; Hart, A.; Milder, J.; DeClerck, F. Integrated landscape initiatives for agriculture, livelihoods and ecosystem conservation: An assessment of experiences from South and Southeast Asia. *Landsc. Urban Plan.* **2017**, *165*, 11–21. [CrossRef]

50. Act of the National Council of Slovak Republic No. 237/2000 Coll., Amending and Supplementing the Act No. 50/1976. On Territorial Planning and Building Code (Building Act). Available online: https://www.ujd.gov.sk/files/legislativa/145_2010_EN.pdf (accessed on 10 September 2018).

51. UNCCD/Global Land Outlook Working Paper/Integrated Landscape Management. 2017. Available online: https://www.researchgate.net/publication/320211462 (accessed on 10 September 2018).

52. Kočická, E.; Kočický, D. Database of abiotic complexes as a landscape and ecological basis for integrated landscape management in the Slovak Republic. *Acta Fac. Ecol.* **2014**, *31*, 35–50.

53. Belaňová, E.; Kočická, E.; Diviaková, A. *Implementation of Integrated Landscape Management Tools at Regional and Local Level*; Vydavateľstvo Technickej Univerzity vo Zvolene: Zvolen, Slovakia, 2014; p. 117.

54. Grunewald, K.; Bastian, O. Special issue: "Maintaining Ecosystem Services to Support Urban Needs". *Sustainability* **2017**, *9*, 1647. [CrossRef]

55. Antrop, M.; Van Eetvelde, V. *Landscape Perspectives the Holistic Nature of Landscape (Landscape Series 23)*; Springer: Dordrecht, The Netherlands, 2017; p. 436, ISBN 9789402411812.

56. Naveh, Z.; Lieberman, A.S. *Landscape Ecology Theory and Applications*, 2nd ed.; Springer: New York, NY, USA, 1993; p. 360.

57. Mezősi, G.; Blanka, V.; Bata, T.; Ladanyi, Z.; Kemeny, G.; Meyer, B.C. Assessment of future scenarios for wind erosion sensitivity changes based on ALADIN and REMO regional climate model simulation data. *Open Geosci.* **2016**, *8*, 465–477. [CrossRef]

58. Izakovičová, Z. Evaluation of the anthropogenic change of the landscape structure. *Ekológia* **1997**, *16*, 73–80.

59. Haines-Young, R.; Potschin, M. Common International Classification of Ecosystem Services (CICES) Consultation on Version 4, August–December 2012. Available online: https://unstats.un.org/unsd/envaccounting/seearev/GCComments/CICES_Report.pdf (accessed on 10 September 2018).

60. Izakovičová, Z.; Oszlányi, J. The impact of stress factors, landscape loads and human activities: Implications for sustainable development. *Int. J. Environ. Waste Manag.* **2013**, *11*, 111–128. [CrossRef]

61. Chavez-Tafur, J.; Zagt, R.J. (Eds.) *Towards Productive Landscapes*; Tropenbos International: Wageningen, The Netherlands, 2014; Volume 56, p. 246.

62. Sayer, J.A.; Campbell, B. Research to integrate productivity enhancement, environmental protection, and human development. *Conserv. Ecol.* **2001**, *5*, 32. [CrossRef]

63. Miklós, L. The concept of the territorial system of ecological stability in Slovakia. In *Ecological and Landscape Consequences of Land Use Change in Europe*; Jongman, R.H.G., Ed.; ECNC: Tilburg, The Netherlands, 1996; pp. 385–406.

64. *Statistic Yearbook of Trnava District*, 1st ed.; Úrad Geodézie, Kartografie a Katastra Slovenskej Republiky: Bratislava, Slovakia, 2018; ISBN 978-80-89831-06-7.

65. Slovak Road Administration, 2015, Slovenská Správa Ciest, 2015. Available online: http://www.ssc.sk/sk/technicke-predpisy-rezortu/databaza-hydroizolacnych-systemov-na-mostoch/3-2015.ssc (accessed on 10 September 2018).

66. Moyzeová, M.; Miklós, L.; Šatalová, B.; Izakovičová, Z.; Oszlányi, J.; Kenderessy, P.; Štefunková, D.; Krnáčová, Z. *Quality of Environment Assessment for Rural Settlements (Example of Trnava District)*; ÚKE SAV: Bratislava, Slovakia, 2015; ISBN 978-80-89325-26-9.

67. Čurlík, J.; Šefčík, P.; Jambor, P. *Geochemical Atlas of the Slovak Republic. Part V, Soils*; Ministry of the Environment of the Slovak Republic: Bratislava, Slovakia, 1999; ISBN 80-88833-14-0.

68. Maňkovská, B.; Izakovičová, Z.; Oszlányi, J.; Frontasyeva, M. Temporal and spatial trends (1990–2010) of heavy metal accumulation in mosses in Slovakia. *Boil. Divers. Conserv.* **2017**, *10*, 28–32.

69. Bezák, P.; Izakovičová, Z.; Miklós, L. *Representative Types of Slovak Landscape*; ÚKE SAV: Bratislava, Slovakia, 2010; p. 165.

70. Izakovičová, Z.; Miklós, L.; Moyzeová, M.; Špilárová, I.; Kočický, D.; Halada, L.; Gajdoš, P.; Špulerová, J.; Baránková, Z.; Štefunková, D. *Model of Representative Geo-Ecosystems on Regional Level*; ÚKE SAV: Bratislava, Slovakia, 2011; p. 88.

71. Miklós, L.; Izakovičová, Z.; Kanka, R.; Ivanič, B.; Kočický, D.; Špinerová, A.; David, S.; Piscová, V.; Štefunková, D.; Oszlányi, J. *Geographic Information System of Ipel Basin*; Slovak Academy of Sciences: Zvolen, Slovakia, 2011; p. 43.

72. Dobrucká, A.; Mederly, P. *Territorial System of Ecological Stability of Trnava Town—Actualisation*; Atelier Dobrucka Ltd. Regioplan: Nitra, Slovakia, 2009; p. 366.

73. Mederly, P.; Bezák, P.; Izakovičová, Z. Ecosystem services assessment methods—Examples and perspectives for planning and decision making. In *Flows, Spaces and Societies in Central Europe*; Comenius University: Bratislava, Slovakia, 2017; p. 62, ISBN 978-80-223-4350-3.

74. Water Management in the Slovak Republic in 2012. Available online: http://www.vuvh.sk/Documents/vodne_hospodarstvo/MS2014EN.pdf (accessed on 10 September 2018).

75. Hrnčiarová, T.; Kenderessy, P.; Špulerová, J.; Vlachovičová, M.; Piscová, V.; Dobrovodská, M. Status and outlook of hiking trails in the central part of the Low Tatra Mountains in Slovakia between 1980–1981 and 2013–2014. *J. Mt. Sci.* **2018**, *15*, 1615–1632. [CrossRef]

76. National Agriculture and Food Center. Available online: http://www.vupop.sk/eng/index.php (accessed on 5 June 2018).

77. Enviroportal of the Slovak Republic. Available online: https://www.enviroportal.sk/ (accessed on 30 July 2017).

78. Šveda, M.; Vigašová, D. Land use changes in the hinterland of major Slovak cities. *Geografie* **2010**, *115*, 413–439. (In Slovak)

79. Agency for Nature Conservation and Landscape Protection of the Czech Republic. Available online: http://www.ochranaprirody.cz/en/ (accessed on 10 September 2018).

80. Moyzeová, M. Territorial development and social relationships. *Ekologické Štúd.* **2017**, *8*, 69–76.

81. Environmental Status Report of Trnava District. 2002. Available online: https://www.enviroportal.sk/uploads/report/ktt02s.pdf (accessed on 10 September 2018).

82. Izakovičová, Z.; Špulerová, J.; Petrovič, F. Integrated approach to sustainable land use management. *Environ. Open Access J. Environ. Conserv. Technol.* **2018**, *5*, 37. [CrossRef]

83. Leadley, P.; Pereira, H.M.; Alkemade, R.; Fernandez-Manjarrés, J.F.; Proença, V.; Scharlemann, J.P.W.; Walpole, M.J. Biodiversity Scenarios: Projections of 21st Century Change. In *Biodiversity and Associated Ecosystem Services*; Technical Series No. 50; Secretariat of the Convention on Biological Diversity: Montreal, QC, Canada, 2010; p. 132.

84. IEEP; Alterra; Ecologic; PBL; UNEP-WCMC. *Scenarios and Models for Exploring Future Trends of Biodiversity and Ecosystem Services Changes*; Institute for European Environmental Policy, Alterra Wageningen UR, Ecologic, Netherlands Environmental Assessment Agency, United Nations Environment Programme World Conservation Monitoring Centre: Wageningen, The Netherlands, 2009.

85. Sachs, J.D.; Baillie, J.E.M.; Sutherland, W.J.; Armsworth, P.R.; Ash, N.; Beddington, J.; Blackburn, T.M.; Collen, B.; Gardiner, B.; Gaston, K.J.; et al. Biodiversity Conservation and the Millennium Development Goals. *Science* **2009**, *325*, 1502–1503. [CrossRef] [PubMed]

86. UNDP. *Human Development Report 2011, Sustainability and Equity: A Better Future for All*; United Nations Development Programme: New York, NY, USA, 2011.

87. CBD, Global Biodiversity Outlook 3. Available online: http://gbo3.cbd.int/ (accessed on 20 December 2010).

88. EEA. Asessing Biodiversity in Europe—The 2010 Report. 2010. Available online: http://www.eea.europa. eu/publications/assessing-biodiversity-in-europe-84 (accessed on 12 December 2010).

89. FAO—GFRA Global Forest Resource Assessment 2005. FAO Forest Report. Food and Agriculture Organization of the United States, 2006. Available online: http://www.fao.org/DOCREP/008/a0400e/ a0400e00.htm (accessed on 10 September 2008).

90. Mooney, H.; Larigauderie, A.; Cesario, M.; ELmquist, T.; Hoegh-Guldberg, O.; Lavorel, S.; Mace, G.T.; Palmer, M.; Scholes, R.; Yahara, T. Biodiversity, climate change and ecosystem services. Current Opinion in Environmental. *Curr. Opin. Environ. Sustain.* **2016**, *1*, 46–54. [CrossRef]

91. Reid, W.V.; Mooney, H.A.; Cropper, A.; Capistrano, D.; Carpenter, S.R.; Chopra, K.; Dasgupta, P.; Dietz, T.; Duraiappah, A.K.; Hassan, R.; et al. *Millenium Ecosystem Assesment—Ecosystems and Human Well-Being. Syntheses*; Univerzita Karlova v Prahe: Praha, Czech Republic, 2005; p. 138.

92. Sabo, P.; Urban, P.; Turisová, I.; Považan, R.; Herian, K. *Threat and Protection of Biodiversity. Selected Chapters of Global Environmental Problems*; Centrum Vedy a Výskumu., Inštitút Výskumu Krajiny a Regiónov, Univerzita Mateja Bela v Banskej Bystrici, Katedra Biológie a Ekológie, Fakulta Prírodných vied, Univerzita Mateja Bela v Banskej Bystrici & Občianske Združenie Živica: Banská Bystrica, Slovakia, 2011; p. 328, ISBN 978-80-968989-6-5.

93. Yafei, L.; Gaohuan, L. Characterizing Spatiotemporal Pattern of Land Use Change and Its Driving Force Based on GIS and Landscape Analysis Techniques in Tianjin during 2000–2015. *Sustainability* **2017**, *9*, 894. [CrossRef]

94. Kopecká, M.; Rosina, K. Identification of changes in urbanized landscape based on VHR satellite data: Study area of Trnava (in Slovak). *Geogr. Čas.* **2014**, *66*, 247–267.

95. Izakovičová, Z.; Bezák, P.; Mederly, P. Implementing ecosystem services in planning and decision-making. In *Gospodarka Przestrzenna, Stan Obecny i Wyzwania Przyszłości—Ujęcie Interdyscyplinarne*; Uniwersytet Przyrodniczy we Wrocławiu: Wrocław, Poland, 2017; pp. 18–19.

96. Izakovičová, Z.; Považan, R. Increasing pressure on ecosystems. In *Global Megatrends: Evaluation and Challenges from the Perspective of the Slovak Republic*; Centrum Spoločenských a Psychologických Vied, SAV: Staré Mesto-Bratislava, Slovakia, 2016; p. 327.

97. Izakovičová, Z.; Moyzeová, M.; Oszlányi, J. Problems in Agricultural Landscape Management Arising from Conflicts of Interest—A Study in the Trnava Region, Slovak Republic. In *Innovations in European Rural Landscapes*; Wiggering, H., Ende, H.P., Knierim, A., Pintar, M., Eds.; Springer: Berlin/Heidelberg, Germany, 2010; pp. 77–95.

98. Aurex. *Territorial Plan of the Region Trnava. Self-Governing Region. Directional Part*; TTSK: Trnava, Slovakia, 2014; p. 387.

99. Muchová, Z.; Jusková, K. Stakeholders' perception of defragmentation of new plots in a land consolidation project: Given the surprisingly different Slovak and Czech approaches. *Land Use Policy* **2017**, *6*, 356–363. [CrossRef]

100. Pauditšová, E.; Slabeciusova, B. Modelling as a Platform for Landscape Planning. In *Geoconference on Informatics. Geoinformatics and Remote Sensing*; Stef92 Technology Ltd.: Sofia, Bulgaria, 2014; Volume III, pp. 753–760.

101. Muchova, Z.; Leitmanova, M.; Petrovic, F. Possibilities of optimal land use as a consequence of lessons learned from land consolidation projects (Slovakia). *Ecol. Eng.* **2016**, *90*, 294–306. [CrossRef]

102. *Atlas of Slovak Landscape*; MŽP SR, Bratislava, SAŽP: Banská Bystrica, Slovakia, 2002; p. 344, ISBN 80-88833-27-2.

103. Poudevigne, I.; Alard, D. Agricultural landscape dynamic: A case study in the Odessa region, the Ukraine and comparative analysis with the Brionne basin case study France. *Ekológia* **1997**, *16*, 295–308.

104. Matlovič, R. Transformation processes and their effects in intra-urban structures of post-communist cities (in Slovak). *Acta Fac. Rerum Nat. Univ. Matthiae Belii* **2001**, *8*, 73–81.

105. Sýkora, L. Changes in the internal spatial structure of post-communist Prague. *GeoJournal* **1999**, *49*, 79–89. [CrossRef]

106. Hirt, S. Landscapes of post-modernity: Changes in the built fabric of Belgrade and Sofia since the end of socialism. *Urban Geogr.* **2008**, *29*, 785–809. [CrossRef]

107. Medvedkov, Y.; Medvedkov, O. Upscale housing in post-Soviet Moscow and its environs. In *The Post-Socialist City: Urban Form and Space Transformations in Central and Eastern Europe after Socialism*; Stanilov, K., Ed.; Springer: Dordrecht, The Netherlands, 2007; pp. 245–265.

108. Feldman, M. Urban waterfront regeneration and local governance in Tallinn. *Eur. Asia Stud.* **2000**, *52*, 829–850. [CrossRef]

109. Kiss, E. Spatial impacts of post-socialist industrial transformation in the major Hungarian cities. *Eur. Urban Reg. Stud.* **2004**, *11*, 81–87. [CrossRef]

110. Sýkora, L.; Bouzarowski, S. Multiple Transformations: Conceptualising the Post-communist Urban Transition. *Urban Stud.* **2011**, *49*, 1–18. [CrossRef]

111. Temelová, J.; Novák, J. From industrial neighbourhood to modern urban centre: Transformation in physical and functional environment of central Smíchov. *Geogr-Sb. CGS* **2007**, *112*, 315–333.

112. Sýkora, L. Gentrification in post-communist cities. In *The New Urban Colonialism: Gentrification in a Global Context*; Atkinson, R., Bridge, G., Eds.; Routledge: London, UK, 2005; pp. 90–105.

113. Forman, R.T.T.; Godron, M. *Landscape Ecology*; John Wiley and Sons: New York, NY, USA, 1986; p. 619.

114. Mander, Ü.; Murka, M. Landscape coherence: A new criterion for evaluating impacts of land use changes. In *Multifunctional Landscapes. Vol. III—Continuity and Change*; Mander, Ü., Antrop, M., Eds.; WIT Press: Southampton, UK, 2003; pp. 15–32.

115. Pauleit, S.; Ennos, R.; Golding, Y. Modelling the environmental impacts of urban land use and land cover change—A study in Merseyside, UK. *Landsc. Urban Plan.* **2005**, *71*, 295–310. [CrossRef]

116. Haase, D.; Frantzeskaki, N.; Elmqvist, T. Ecosystem services in urban landscape. Practical applications and governance implications. *Ambio* **2014**, *43*, 407–412. [CrossRef] [PubMed]

117. Haase, D.; Kabisch, N.; Haase, A. Endless Urban Growth? On the Mismatch of Population, Household and Urban Land Area Growth and Its Effects on the Urban Debate. *PLoS ONE* **2013**, *8*, e66531. [CrossRef] [PubMed]

118. Albert, C.; Aronson, J.; Fürst, C.; Opdam, P. Integrating ecosystem services in landscape planning: Requirements, approaches, and impacts. *Landsc. Ecol.* **2014**, *29*, 1277–1285. [CrossRef]

119. Haase, D.; Haase, A.; Kabisch, N.; Kabisch, S.; Rink, D. Actors and factors in land-use simulation: The challenge of urban shrinkage. *Environ. Model. Softw.* **2012**, *35*, 92–103. [CrossRef]

120. Fei, Z.H.; Johnson, C.V. Assessment of Land-Cover/Land-Use Change and Landscape Patterns in the Two National Nature Reserves of Ebinur Lake Watershed, Xinjiang, China. *Sustainability* **2017**, *9*, 1553. [CrossRef]

121. Bičík, I.; Jeleček, L.; Štěpánek, V. Land-Use Changes and their Social Driving Forces in Czech in the 19th and 20th Centuries. *Land Use Policy* **2001**, *18*, 65–73. [CrossRef]

122. Mendel, M. Residential developers and investors in Central Europe: Boom and bust. *Monu Mag. Urban.* **2010**, *12*, 23–30.

123. Václavík, T.; Rogan, J. Identifying Trends in Land Use/Land Cover Changes in the Context of Post-Socialist Transformation in Central Europe: A Case Study of the Greater Olomouc Region. *GISci. Remote. Sens.* **2006**, *46*, 54–76. [CrossRef]

124. Nuissl, H.; Haase, D.; Wittmer, H.; Lanzendorf, M. Environmental impact assessment of urban land use transitions—A context-sensitive approach. *Land Use Policy* **2009**, *26*, 414–424. [CrossRef]

125. Bariski, J. The consequences of changes of ownership for agricultural land use in Central European countries following the collapse of the Eastern Bloc. *Land Use Policy* **2017**, *66*, 120–130.

126. Sroka, W.; Mikolajczyk, J.; Wojewodzic, T.; Kwoczynska, B. Agricultural Land vs. Urbanisation in Chosen Polish Metropolitan Areas: A Spatial Analysis Based on Regression Trees. *Sustainability* **2018**, *10*, 837. [CrossRef]

127. Muchová, Z.; Hrnčiarová, T.; Petrovič, F. *Local Territorial System of Ecological Stability for Land Consolidation*; Slovenská Polnohospodárska Univerzita v Nitre: Nitra, Slovakia, 2013; p. 87, ISBN 978-80-552-1127-5.

128. Izakovičová, Z.; Bezák, P.; Mederly, P.; Špulerová, J. Implementation of the ecosystem services concept in planning and management practice in the Slovak Republic—Results of the Open NESS project in Trnava case study. In *Životné Prostredie*; Institute of Landscape Ecology of Slovak Academy of Sciences: Bratislava, Slovakia, 2017; pp. 198–204. ISSN 0044-4863.

129. Huang, D.; Jin, H.; Zhao, X.; Liu, S. Factors influencing the conversion of arable land to urban use and policy implications in Beijing, China. *Sustainability* **2015**, *7*, 180–194. [CrossRef]

Sustainability **2018**, *10*, 3270

130. Mazzocchi, C.; Sali, G.; Corsi, S. Land use conversion in metropolitan areas and the permanence of agriculture: Sensitivity Index of Agricultural Land (SIAL), a tool for territorial analysis. *Land Use Policy* **2013**, *35*, 155–162. [CrossRef]
131. Deng, X.; Huang, J.; Rozelle, S.; Zhang, J.; Li, Z. Impact of urbanization on cultivated land changes in China. *Land Use Policy* **2015**, *45*, 1–7. [CrossRef]

MDPI
St. Alban-Anlage 66
4052 Basel
Switzerland
Tel. +41 61 683 77 34
Fax +41 61 302 89 18
www.mdpi.com

Sustainability Editorial Office
E-mail: sustainability@mdpi.com
www.mdpi.com/journal/sustainability